AF544975

GAMMA-RAY LINE ASTROPHYSICS

AIP CONFERENCE PROCEEDINGS 232

GAMMA-RAY LINE ASTROPHYSICS

PARIS-SACLAY, FRANCE 1990

EDITORS:
PHILIPPE DUROUCHOUX
NIKOS PRANTZOS
CENTRE NATIONAL DE LA
RECHERCHE SCIENTIFIQUE, FRANCE

American Institute of Physics New York

L.C. Catalog Card No. 91-55492
ISBN 0-88318-875-9
DOE CONF-901275

Printed in the United States of America.

CONTENTS

DIFFUSE 511 keV AND 1.8 MeV EMISSION IN THE GALACTIC PLANE

NOVAE

SN1987A AND SUPERNOVAE

COMPACT OBJECTS

SOLAR GAMMA-RAY LINES

POT-POURRI

INSTRUMENTAL

* **Invited Paper**

Preface

Gamma-Ray Line Astrophysics fully came of age in the 80s, due to several remarkable discoveries, like: the variability of the 511 keV line (due to electron-positron annihilation) from the direction of the Galactic Center, suggesting the existence of compact sources of relativistic electron-positron plasma; the diffuse galactic emission at 511 keV and 1.8 MeV (due to the decay of Al-26), showing that large amounts of radioactive nuclei were recently synthesized in the Galaxy; the ^{56}Co decay lines at 0.847 MeV and 1.238 MeV from SN1987A in the Large Magellanic Cloud, confirming early speculations on explosive nucleosynthesis and the important role of radioactivity in powering supernovae; line features at tens of keV from gamma-ray bursts (probably due to cyclotron emission in terragauss magnetic fields) suggesting that the bursts are produced in the magnetospheres of neutron stars, and the systematic detection of a variety of gamma-ray lines from solar flares, showing that ions and relativistic electrons are routinely accelerated there. The unique signatures of the detected gamma-ray line photons have, in the last ten years, enriched in a spectacular way our knowledge of those astrophysical sites and of the corresponding emission processes.

The International Symposium on Gamma Ray Line Astrophysics, held in the Centre d'Etudes de Saclay (France) from December 10–13, 1990 convincingly demonstrated the "health" of this young and rapidly growing discipline. The Scientific Program, established with the help of the Scientific Organizing Committee (C. Cesarsky–*Service d'Astrophysique, CE Saclay*, K. Nomoto–*Tokyo University*, N. Prantzos–*Institut d'Astrophysique, Paris*, R. Ramaty–*NASA, Goddard Space Flight Center*, R. Sunyaev–*IKI Moskow*, G. Vedrenne– *CESR Toulouse*, and S. Woosley–*University of California Santa Cruz*), covered all topics relevant to the spectroscopy of astrophysical gamma-ray line sources, whether detected or potentially interesting: the galactic center and plane, solar flares, compact objects (e.g., neutron stars and gamma-ray bursts), stellar explosions (e.g., novae, supernovae, and SN1987A in particular), quiescently burning stars (like red giants and Wolf–Rayet stars) and extragalactic sources. In each of those topics, new and exciting results were presented at the Symposium, along with recent progress in instrumentation, that will eventually bring our discipline into the next century. Thirty-eight oral presentations (of which seventeen were invited talks) and twenty-eight poster papers fueled lively discussions among the 120 participants.

This volume contains all the contributions to the Symposium and essentially summarizes the current status of, and the prospects for, Gamma-Ray Line Astrophysics. We hope that the reader will appreciate the quality of the various contributions, authored by almost all active research groups in the discipline. Several review articles, written by leading experts in the field, offer a synthetic view of those recent developments, accessible to astrophysics students as well as researchers in other fields.

Among the many impressive results reported at the Symposium, the localization of the 511 keV source in the Galactic Center region by the SIGMA experiment should undoubtedly stand as a milestone in modern High Energy Astrophysics, as exclaimed an enthusiastic Marvin Leventhal, the most active "hunter" of the "beast" in the past 15 years. This result of the French–Soviet scientific collaboration in space clearly shows that an important potential for research in Gamma-Ray Line Astrophysics exists outside the original "home" of that discipline, namely the United States. It is the hope of the organizers that the success of the Symposium, the first of such size ever held outside the USA, will contribute to the further development of that potential.

It is well known, of course, that the success of a meeting depends not only on the quality of its scientific content, but also on the various individuals and institutions that contribute to its organizations. We gratefully acknowledge important financial support from the CEA (CE Saclay), CNRS, CNES, and NASA. It is with great pleasure that we acknowledge the help of Reuven

Ramaty, constantly providing advice, inspiration, and enthusiasm. Finally, we wish to express, once more, our gratitude to the numerous colleagues and students of the Service d'Astrophysique due CE Saclay for their help, and especially to the Secretary of the Symposium, Genevieve Thierry, for her devoted and quite efficient work during many months.

Philippe Durouchoux
Nikos Prantzos
Local Organizing Committee
Paris, March 10, 1991

PARTICIPANTS

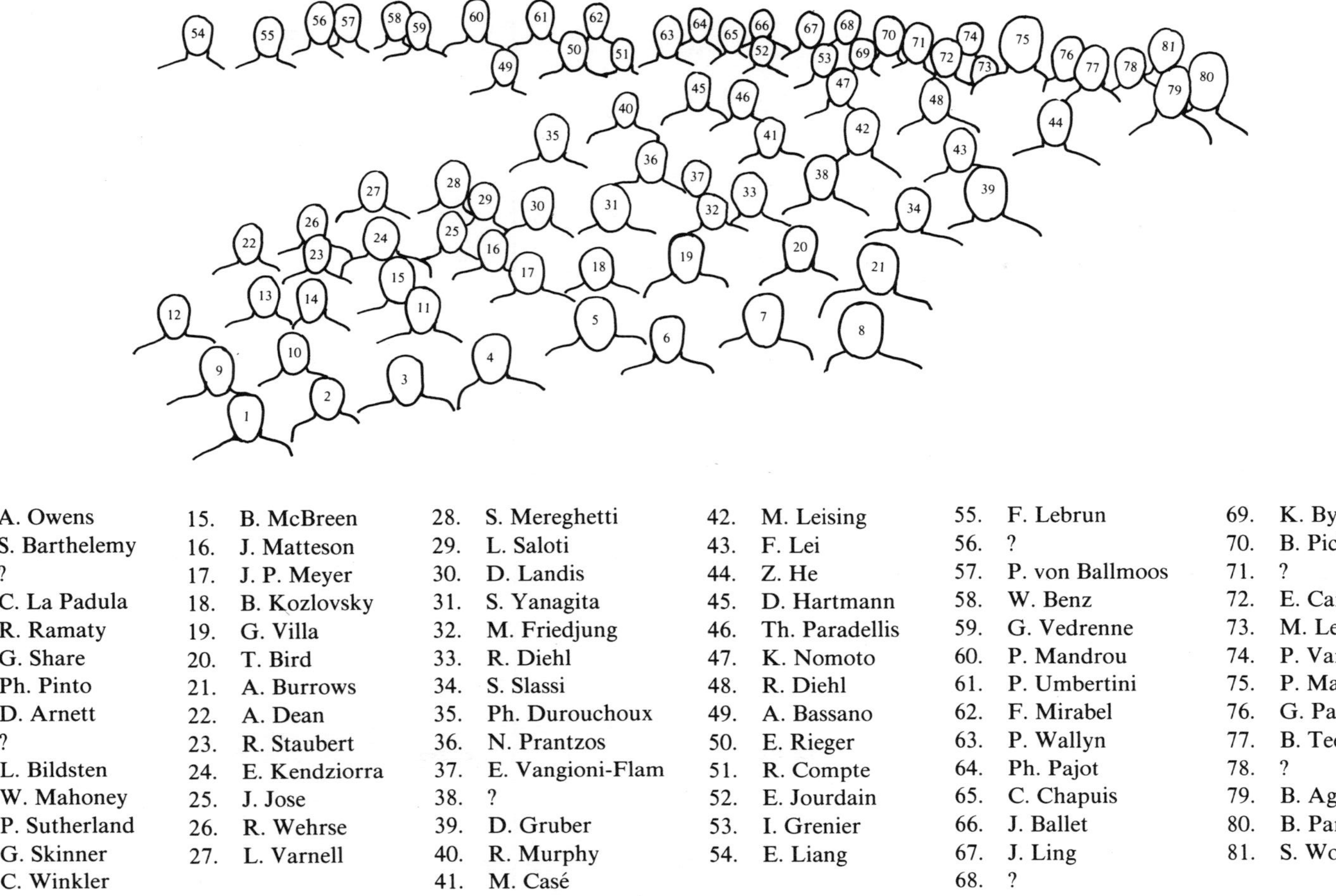

1. A. Owens
2. S. Barthelemy
3. ?
4. C. La Padula
5. R. Ramaty
6. G. Share
7. Ph. Pinto
8. D. Arnett
9. ?
10. L. Bildsten
11. W. Mahoney
12. P. Sutherland
13. G. Skinner
14. C. Winkler
15. B. McBreen
16. J. Matteson
17. J. P. Meyer
18. B. Kozlovsky
19. G. Villa
20. T. Bird
21. A. Burrows
22. A. Dean
23. R. Staubert
24. E. Kendziorra
25. J. Jose
26. R. Wehrse
27. L. Varnell
28. S. Mereghetti
29. L. Saloti
30. D. Landis
31. S. Yanagita
32. M. Friedjung
33. R. Diehl
34. S. Slassi
35. Ph. Durouchoux
36. N. Prantzos
37. E. Vangioni-Flam
38. ?
39. D. Gruber
40. R. Murphy
41. M. Casé
42. M. Leising
43. F. Lei
44. Z. He
45. D. Hartmann
46. Th. Paradellis
47. K. Nomoto
48. R. Diehl
49. A. Bassano
50. E. Rieger
51. R. Compte
52. E. Jourdain
53. I. Grenier
54. E. Liang
55. F. Lebrun
56. ?
57. P. von Ballmoos
58. W. Benz
59. G. Vedrenne
60. P. Mandrou
61. P. Umbertini
62. F. Mirabel
63. P. Wallyn
64. Ph. Pajot
65. C. Chapuis
66. J. Ballet
67. J. Ling
68. ?
69. K. Byard
70. B. Pichon
71. ?
72. E. Caroli
73. M. Leventhal
74. P. Varendorff
75. P. Malaguti
76. G. Paulus
77. B. Teegarden
78. ?
79. B. Agrinier
80. B. Parlier
81. S. Woosley

The Organizers apologize for the few persons that have not been identified in the photo.

List of Participants

S. Grebenev
IKI
Moscow
(USSR)

Maurice Gros
SAp
CEN Saclay
(France)

Duane E. Gruber
UCSD/CASS
LaJolla
(USA)

Dieter Hartmann
Phys. Dept.
Lawrence Livermore Lab.
Livermore
(USA)

Zong He
SAp
CEN Saclay
(France)

Willem Hermsen
Lab. Space Research
Leiden
(Netherlands)

Jordi Jose
Dept. Astron.
University of Barcelona
(Spain)

Elisabeth Jourdain
CESR
Toulouse
(France)

Gottfried Kanbach
Max-Plank-Inst. Extrater. Phys.
Garching
(Germany)

Eckhard Kendziorra
Astronomisches Institut
University Tübingen
(Germany)

Hans-Volker Klapdor
Max-Planck-Inst. Kernphysik
Heidelberg
(Germany)

Lydie Koch-Miramond
SAp
CEN Saclay
(France)

Ben-Zion Kozlovsky
School Phys. Astronomy
Tel-Aviv University
(Israel)

D. Landis
University of California
(USA)

F. L. Lang
NASA/GSFC
Greenbelt
(USA)

Philippe Laurent
SAp
CEN Saclay
(France)

Jean-Michel Lavigne
CESR
Toulouse
(France)

François Lebrun
SAp
CEN Saclay
(France)

Lefebvre
Université Paris Sud
Orsay
(France)

Mark Leising
Naval Research Lab.
Washington
(USA)

Marvin Leventhal
AT&T Bell Laboratories
Murray Hill
(USA)

Edison Liang
Lawrence Livermore Lab.
Livermore
(USA)

Giselher Lichti
Max-Planck-Inst. Extrater. Phys.
Garching
(Germany)

James Ling
NASA Headquarters
Washington
(USA)

William Mahoney
JET Propulsion Lab.
Pasadena
(USA)

Giuseppe Malaguti
Phys. Dept.
Southampton University
(UK)

Pierre Mandrou
CESR
Toulouse
(France)

Jean-Louis Masnou
DARC
Observatoire de Meudon
(France)

James Matteson
CASS/UCSD
La Jolla
(USA)

Brian McBreen
University of Dublin
(Ireland)

Sandro Mereghetti
CESR
Toulouse
(France)

Jean Paul Meyer
SAp
CEN Saclay
(France)

Félix Mirabel
SAp
CEN Saclay
(France)

Ronald Murphy
Naval Research Laboratory
Washington
(USA)

Lorenzo Natalucci
CESR
Toulouse
(France)

Ken'ichi Nomoto
Dept. Astronomy
Univ. Tokyo
(Japan)

Jean-François Olive
CESR
Toulouse
(France)

Alan Owens
NASA/GSFC
Greenbelt
(USA)

Themis Paradellis
Democritus Centre for Nuclear Research
Athens
(Greece)

Bruno Parlier
SAp
CEN Saclay
(France)

Jacques Paul
SAp/GERES
CEN Saclay
(France)

Guy Paulus
Inst. Astron. Astrophysics
Univ. Libre
Bruxelles
(Belgium)

Geoff Pendleton
NASA/Marshall Space Flight Center
Huntsville
(USA)

Nicolas Petrou
SAp
CEN Saclay
(France)

Bernard Pichon
DARC-LAM
Obs. Meudon
(France)

Philip Pinto
Center for Astrophysics
Cambridge
(USA)

Graziella Pizzichini
TESRE/CNR
Bologna
(Italy)

Nikos Prantzos
Institut d'Astrophysique
Paris
(France)

Reuven Ramaty
NASA/GSFC
Greenbelt
(USA)

Erich Rieger
Max-Planck-Inst. Extrater. Phys.
Garching
(Germany)

Jacques Roland
Institut d'Astrophysique
Paris
(France)

Emanuel Rossetti
SAp
Saclay
(France)

Robert Rothenflug
SAp
CEN Saclay
(France)

Pilar Ruiz-Lapuente
University of Barcelona
(Spain)

Lola Sabau
Laboratorio de Fisica Espacial
Madrid
(Spain)

Bruno Sacco
IFCAI/CNR
Palermo
(Italy)

Pierre Salatti
LAPP
Annecy-le-Vieux
(France)

Lucas Salotti
SAp
CEN Saclay
(France)

Livio Scarsi
Istituto Fisica Cosmica Informatica
Palermo
(Italy)

Volker Schonfelder
Max-Planck-Inst. Extrater. Phys.
Garching
(Germany)

Saïd Slassi Sennou
CESR
Toulouse
(France)

Gerald Share
Naval Research Lab.
Washington
(USA)

Gerry Skinner
University of Birmingham
(UK)

Aimé Soutoul
SAp
CEN Saclay
(France)

Rüdiger Staubert
Astronomisches Institut Univ.
Tübingen
(Germany)

Peter Sutherland
Physics Dept.
McMaster Univ.
Hamilton
(Canada)

Michel Tagger
SAp
CEN Saclay
(France)

Bonnard Teegarden
NASA/GSFC
Greenbelt
(USA)

Lhi-Sin The
Dept. Physics Astronomy
Clemson Univ.

Jack Tueller
NASA/GSFC
Greenbelt
(USA)

Pietro Ubertini
IAS/CNR
Frascati
(Italy)

G. Vacanti
SAp
CEN Saclay
(France)

Elisabeth Vangioni-Flam
Institut d'Astrophysique
Paris
(France)

Martin Varendorff
Max-Planck-Inst. Extrater. Phys.
Garching
(Germany)

Larry Varnell
Jet Propulsion Lab.
Pasadena
(USA)

Gilbert Vedrenne
CESR
Toulouse
(France)

Gabriele Villa
IFCTR/CNR
Milano
(Italy)

Nicole Vilmer
DASOP
Obs. de Paris
Meudon
(France)

Pierre Wallyn
SAp
CEN Saclay
(France)

Rainer Wehrse
Institut Theoreticische Astrophysik
Heidelberg
(Germany)

Carl Werntz
Catholic Univ. of America
Washington
(USA)

Christoph Winkler
ESA/ESTEC
Space Science Dept.
Noordwijk
(Netherlands)

Stan Woosley
Lick Obs. and UCSC
Santa Cruz
(USA)

Raphael Yahel
School of Technology
The Open Univ.
Tel Aviv
(Israel)

Shohei Yanagita
Dept. of Earth Sciences
Ibaraki University
(Japan)

THE GALACTIC CENTER

Observations of the Galactic Center 511 keV Line

Neil Gehrels
Laboratory for High Energy Astrophysics
NASA / Goddard Space Flight Center
Greenbelt, MD 20771 USA

ABSTRACT

A review of observations of the positron annihilation 511 keV line from the galactic center and plane is presented. Emphasis is given to the new data from balloon observations in 1988 and 1989. The positron annihilation source in the direction of the center was found to be turned on again in 1988 October after being observed in a quiescent state since the early 1980's. There is evidence for variability between May and October in 1988. The new observations give the first data in which the line is spectrally resolved and in which the galactic center and plane components of the annihilation radiation are independently measured. These data support the two-component model of the emission with a variable point source near the center and a steady-state source distributed along the galactic plane. Comparing measurements of narrow and wide field instruments in the 1980's gives a flux of ~8×10^{-4} photons cm^{-2} s^{-1} for the point source in its high state and a flux of ~1.2×10^{-3} photons cm^{-2} s^{-1} rad^{-1} for the galactic plane component with a relatively flat distribution over the central radian. The line width for the point source component is ~3 keV FWHM.

INTRODUCTION

It has been known since the 1970's that the galactic center region harbors a powerful positron source (> 10^{42} positrons produced per sec), but its location and nature are still not well understood. The positrons annihilate with electrons in the source region to produce a spectral line at 511 keV and a continuum below the line due to three-photon annihilation of a positron atom. The line was first detected in 1970[1], but was not uniquely identified as annihilation radiation until the first high resolution (germanium detector) observation in 1977[2]. To date, more than 20 observations have been made (see reference list in Lingenfelter and Ramaty[3]). The 511 keV line flux for all measurements is shown in Figure 1, where square points are used for wide (> 40°) field-of-view measurements. Table 1 lists the high resolution observations. During its high state in the 1970's the positron source was emitting a flux of more than 10^{-3} photons cm^{-2} s^{-1} or a luminosity of more than 7×10^{36} ergs s^{-1} in the narrow 511 keV line. The narrow field-of-view observations show the source decreasing in intensity in 1980[6] and being in a low state during observations in 1981[7,8] and 1984[9].

The wide field-of-view instruments have systematically measured higher fluxes for the 511 keV line as shown by the square points in Figure 1. Most notable are the observations throughout the 1980's by the spectrometer on the Solar Maximum Mission (SMM)[12,13] with its 130° FWHM field-of-view. The strong, relatively constant flux measured by SMM during periods when the narrow-field instruments detected no line is strong evidence for diffuse emission of annihilation radiation from the galactic plane. This has lead to a two-component model for the source[3]: a variable point source located at or near the center and a steady-state source distributed along the plane. Possible production mechanisms for the positrons are γ–γ interactions near a black hole for the point source component[14] and β^+ decays of radionuclides generated in supernovae over the past ~10^5 years for the distributed source[15,16].

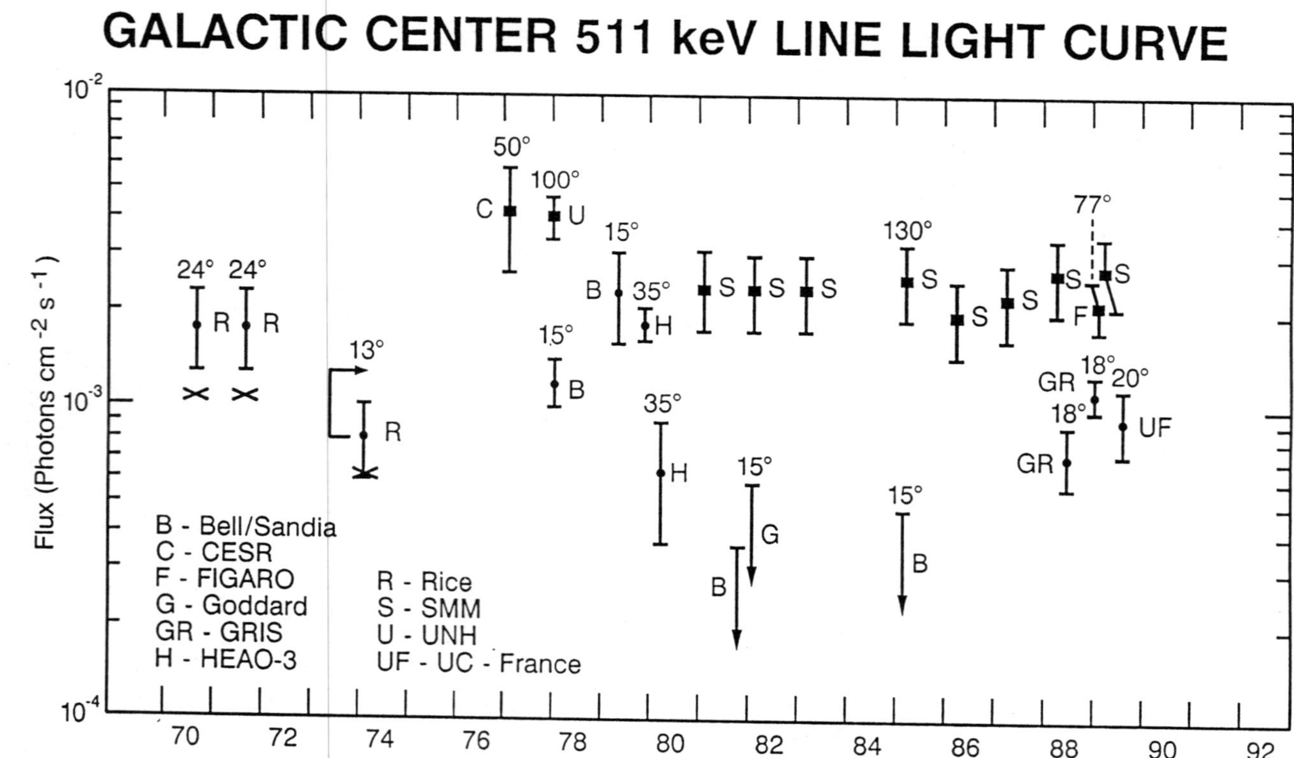

Figure 1 - Flux in the 511 keV line for all observations. Square data points are for instruments with fields-of-view greater than 40° FWHM. The crosses show fluxes corrected for contamination from positronium continuum and the arrow for the 1974 measurement is to correct for instrument pointing ~5° away from the galactic center. The SMM points are from Figure 2b of Share *et al.*[12] using an effective area of 150 cm^2 and including a 20% systematic error.

Table 1

HIGH-RESOLUTION OBSERVATIONS OF THE GALACTIC CENTER 511 keV Line

Date	Line Flux x 10^{-3} (ph/cm^2-s)	Centroid (keV)	Width (keV)	Reference
1977.13	4.18±1.56	-	-	Albernhe *et al.*[4]
1977.86	1.22±0.22	510.7±0.5	< 3.2	Leventhal *et al.*[2]
1979.29	2.35±0.71 [a]	-	-	Leventhal *et al.*[5]
	1.24±0.43 [a]	-	-	
1979.8	1.85±0.21	510.90±0.25	1.6(+0.9/-1.6)	Riegler *et al.*[6]
1980.2	0.65±0.27	-	-	"
1981.89	0.0±0.6	-	-	Paciesas *et al.*[7]
1981.89	0.0±0.38	-	-	Leventhal *et al.*[8]
1984.89	0.0±0.5	-	-	Leventhal *et al.*[9]
1988.33	0.75±0.17	511.46±0.38	-	Gehrels *et al.*[10]
1988.83	1.18±0.16	510.97±0.25	2.9±0.6	"
1989.39	0.89±0.21	511.54±0.38	1.1(+1.4/-1.1)	Chapuis *et al.*[11]

[a] Numbers for first 2 hrs and all 6 hrs of flight.

As first reported at this Symposium[17], the SIGMA gamma-ray imaging instrument on the GRANAT spacecraft has observed a spectral feature near 511 keV emanating from 1E1740.7-2942, a source 0.7° from the galactic center. The feature was present during a single 16 hour observation on 1990 October and was not detected in observations 3 days earlier or 1 day later. This may indicate that 1E1740.7-2942 is the point source responsible for the variable 511 keV line from the galactic center.

The case for the two-component model is considerably strengthened by the new observations reviewed in this paper made by three balloon instruments in 1988 and 1989. These instruments are the Gamma Ray Imaging Spectrometer (GRIS) which observed the galactic center and plane in 1988 May and October, the FIGARO II experiment which observed the center in 1988 November, and HEXAGONE which observed the center in 1989 May .

OBSERVATIONS

The GRIS instrument[18] is one of two new generation, high-resolution balloon spectrometers. The collaborating institutions are NASA/Goddard Space Flight Center, AT&T Bell Laboratories and the University of New Mexico. It has a detector array consisting of seven large high-purity, n-type coaxial germanium detectors. The total active detector volume is ~1530 cm^3 and the total effective area at 511 keV is ~85 cm^2. The energy resolution was the nominal 1.8 keV FWHM at 511 keV for the October flight, but was degraded by an electronics problem in the maiden May flight to ~5 keV. The anticoincidence shield surrounding the detectors is 15 cm thick sodium iodide with aperture holes defining an 18° FWHM (at 511 keV) field-of-view. Observations of the galactic center were made on 1988 May 1 and on 1988 October 28-29 and of a point 25° west of the center (l^{II}=335°, b^{II}=0°; chosen to exclude the center from the field-of-view) on 1988 October 29-30. The flights were from Alice Springs, Australia at mean atmospheric depths of 4.7 and 6.0 g cm^{-2} for the May and October galactic center observations and 4.5 g cm^{-2} for the galactic plane observation. Observations were performed using the standard on-source, off-source method. The observing time was divided into 20 minute intervals alternating pointing direction of the instrument between the target and a blank region of the sky at the same zenith angle as the target.

The HEXAGONE instrument[19] is the other new generation high resolution balloon spectrometer. The collaborating institutions are the University of California at San Diego, the University of California at Berkeley, Lawrence Berkeley Laboratories, CESR in Toulouse and CENS in Saclay. It has a detector array consisting of twelve high-purity coaxial germanium detectors. In the 1989 flight, five of the detectors had segmented contacts and pulse-shaped-discrimination electronics for background reduction. The total active detector volume is ~1560 cm^3 and total effective area at 511 keV is 84.5 cm^2. The energy resolution is 2.2 keV FWHM at 511 keV. The anticoincidence shield is 5 cm thick bismuth germanate on the bottom and sides and 10 cm thick cesium iodide on the top. Aperture holes in the top piece define a 20° FWHM field-of-view. An observation of the galactic center was made on 1989 May 22. The flight was from Alice Springs, Australia at an average atmospheric depth of 4.2 g cm^{-2}. The observation was performed using the on-source, off-source method described above.

The FIGARO II instrument[20], a collaboration among eight institutions in Italy and France, has an array of nine sodium iodide scintillation detectors surrounded by an active shield made of sodium iodide and plastic scintillators. The total detector volume is 18,000 cm^3 and the effective area at 511 keV is ~2000 cm^2. The energy resolution at 511 keV is ~12% or 60 keV FWHM and the field-of-view is ~77° FWHM. An observation was made of the galactic center on 1988 November 25-26 during a balloon flight from Charleville, Australia. The observation was performed using the drift-scan method in which the instrument points at the zenith for the entire observation. The target rises into the field-of-view, transits near the center of the field and then sets out of the field.

RESULTS

GRIS

The GRIS analysis methods and results are summarized here. Details on the analysis are given by Barthelmy *et al.*[21] and Tueller *et al.*[22] GRIS results on the galactic center are given by Gehrels *et al.*[10] The analysis is performed by subtracting background data from target data to obtain a counts spectrum. Model fits to the data are then derived by multiplying a model photon spectrum by the instrument response matrix and adjusting parameters in the model to obtain a minimum chi-squared. The response matrix contains the instrument effective area, energy resolution, off-diagonal scattering terms and the atmospheric attenuation. A Gaussian was used for the instrument resolution function for the October flight. For the May flight, the electronic problem caused a double-peak line shape which was modelled as two Gaussians and a rectangular fill function.

The model photon spectrum used to fit the data has three terms: 1) a power-law continuum of the form A $(E/100\ \mathrm{keV})^{-\gamma}$, 2) an orthopositronium three-photon continuum[23] with an amplitude parameter $F_{3\gamma}$ (integral flux) and 3) a Gaussian line near 511 keV with parameters for the flux, F_{511}, the line width, W_{511}, and the line centroid, E_{511}. The parameters for the power-law continuum are determined primarily by the spectrum below ~300 keV, which is attributed to the combined flux from hard X-ray sources in the galactic center region[10,24]. In other measurements[25,26] the data above 511 keV were sufficiently different from the extrapolated low-energy power-law to warrant inclusion of added terms in the model. The GRIS data in this energy range, however, are of low statistical significance and are consistent with the low-energy power-law, so we have not included additional terms.

The data and model fits for the three GRIS observations are shown in Figure 2, with the energy bands near 511 keV expanded in Figure 3. The best-fit model parameters and statistical errors are listed in Table 2. Also listed are derived values for the positronium fraction[27], which is defined as the fraction of positrons annihilating via the formation of a positronium atom, and is equal to $4\ F_{3\gamma}/(4.5\ F_{511}+3\ F_{3\gamma})$.

HEXAGONE

The HEXAGONE analysis methods and results for the 1989 May 22 galactic center observations are given in this Proceedings in the papers by Chapuis *et al.*[11] and Matteson *et al.*[28] The results are summarized here. A strong 511 keV narrow line was detected from the galactic center region at a flux of $(8.9 \pm 2.7)\times10^{-4}$ photons $\mathrm{cm}^{-2}\ \mathrm{s}^{-1}$ and a width of 1.1 (+1.4/-1.1) keV FWHM. The spectrum in the vicinity of the line is shown in Figure 4. There is emission below the line that can be interpreted as a positronium three-photon continuum or as a continuum due to Compton scattering of 511 keV photon in material near the source. A strong feature seen in the spectrum at ~170 keV may be a backscatter peak from Compton scattering.

FIGARO

The FIGARO II data analysis and results for the drift scan observations of the galactic center are given by Niel *et al.*[29] and summarized here. The background and source signals can be separated for a drift scan by plotting the counting rate as a function of time (or collimator transmission for the source of interest). The change in

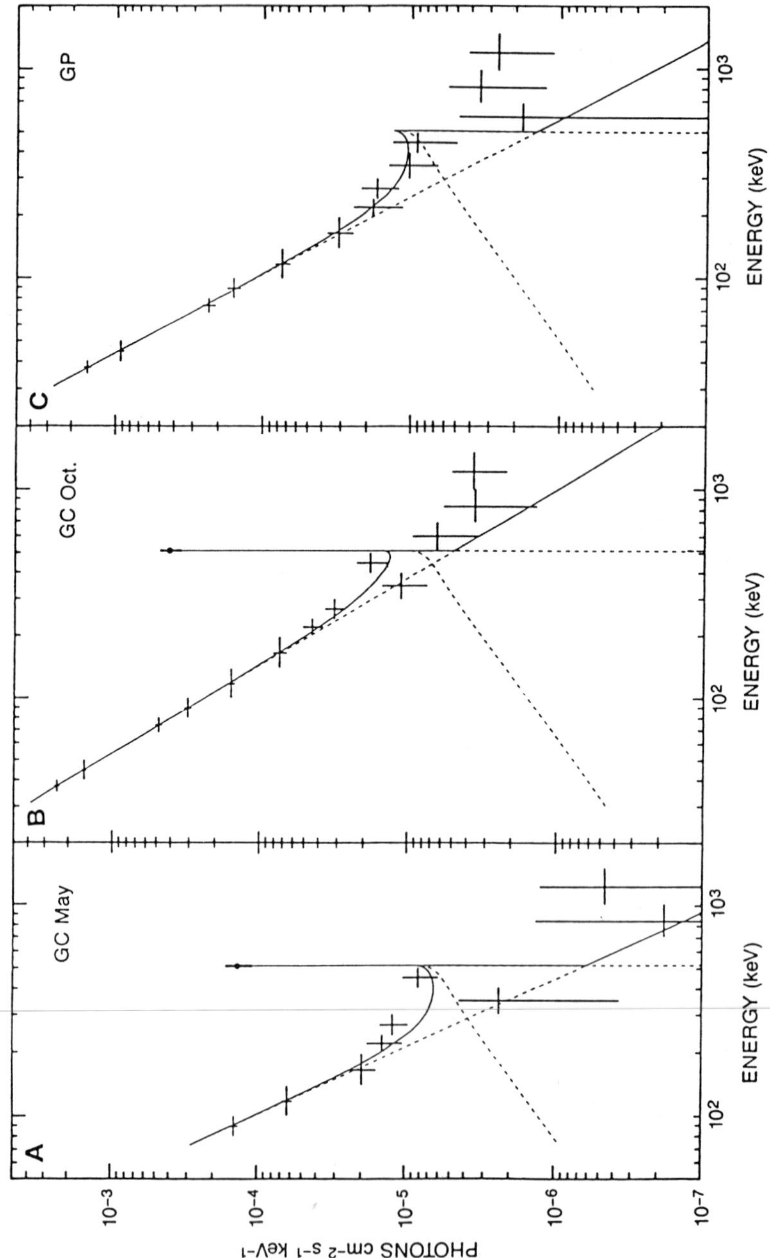

Figure 2 - Spectra and model fits for the 1988 GRIS observations of the galactic center (GC) and galactic plane (GP). Data between 52 and 68 keV have large statistical and systematic uncertainties due to the presence of strong background lines in this range and are not plotted. From Gehrels, *et al.*[10]

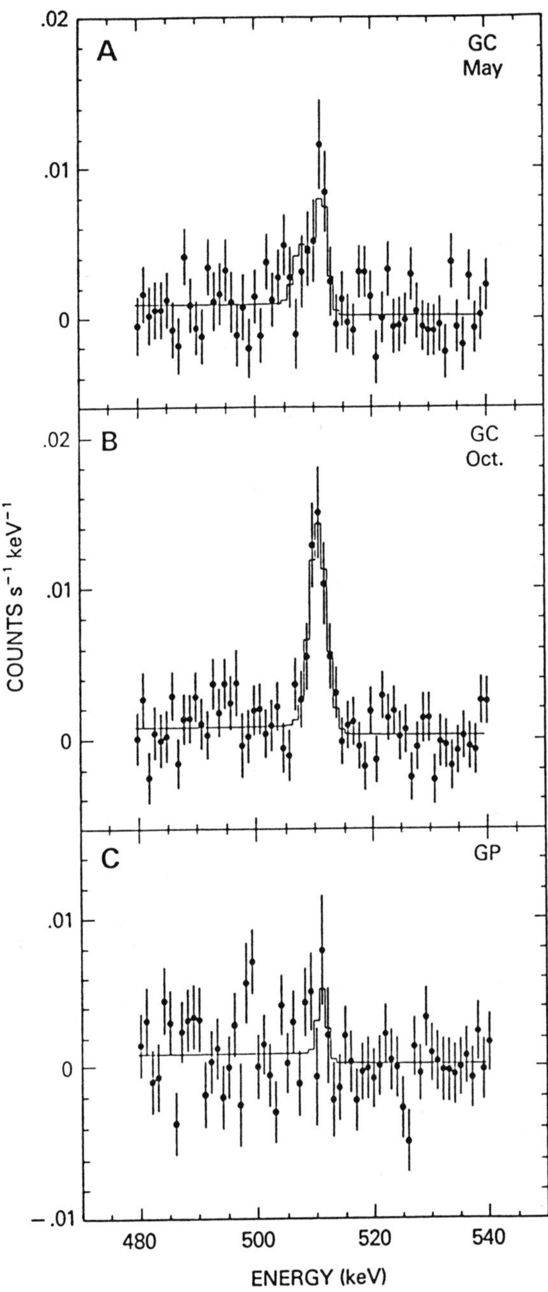

Figure 3 - Spectra and model fits in the vicinity of the 511 keV line for the 1988 GRIS observations. From Gehrels *et al.*[10]

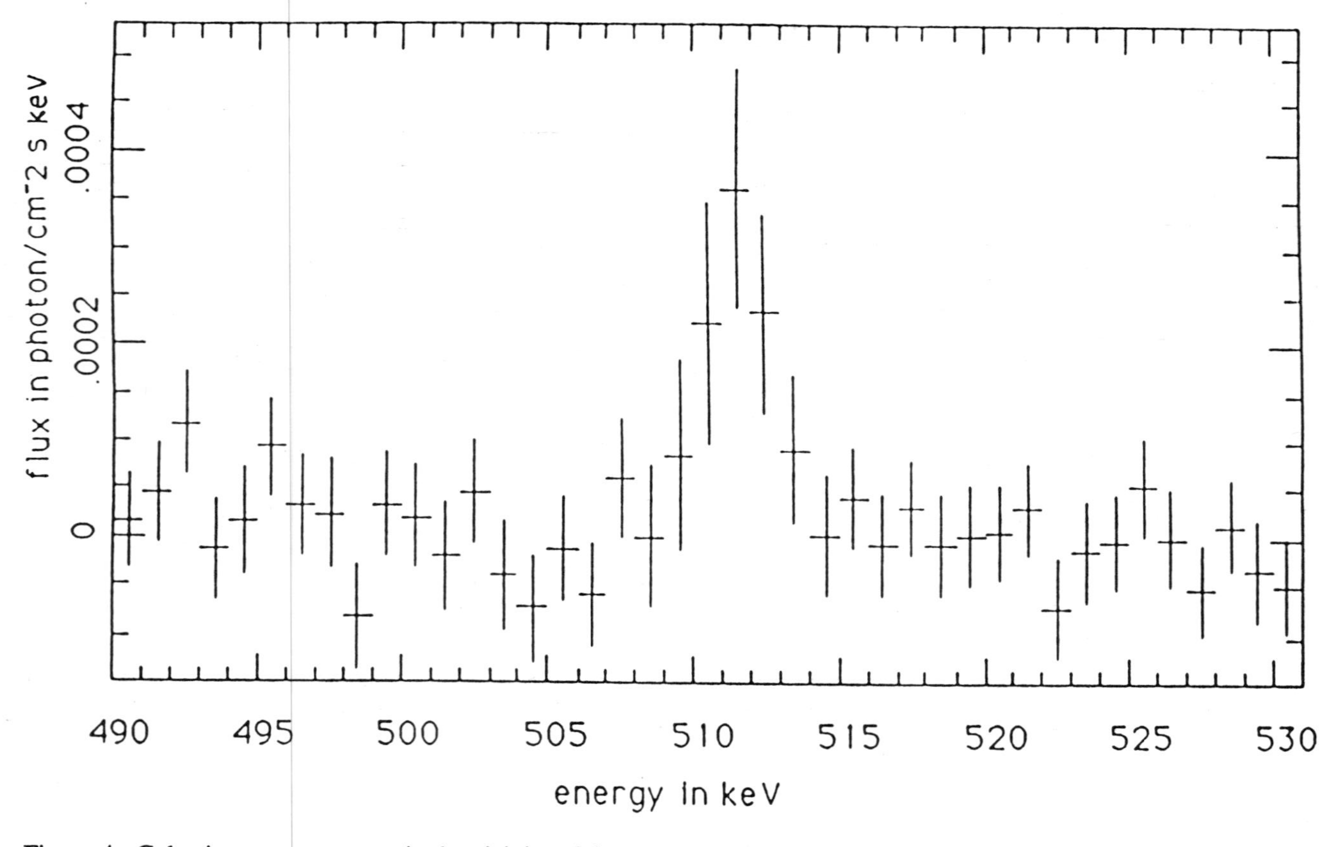

Figure 4 - Galactic center spectrum in the vicinity of the 511 keV line for the 1989 May HEXAGONE observation. From Chapius *et al.*[11]

Table 2
Model Fit Parameters for the GRIS Data

Parameter	1988 May Galactic Center	1988 October Galactic Center	1988 October Galactic Plane
F_{511} (ph cm^{-2} s^{-1})	$(7.5 \pm 1.7) \times 10^{-4}$	$(11.8 \pm 1.6) \times 10^{-4}$	$(1.8 \pm 1.1) \times 10^{-4}$ $<4.0 \times 10^{-4}$ (2σ)
Centroid, E_{511} (keV)	511.46 (+0.33/-0.43)	510.97 (+0.22/-0.29)	511.0 fixed
Width, W_{511} (keV)	0.1 + 2.4	2.9 ± 0.6	0 + 3.6
Powerlaw A[a]	$(1.90 \pm 0.09) \times 10^{-4}$	$(2.28 \pm 0.06) \times 10^{-4}$	$(1.06 \pm 0.06) \times 10^{-4}$
Powerlaw γ[a]	3.15 ± 0.25	2.36 ± 0.04	2.62 ± 0.08
$F_{3\gamma}$ (ph cm^{-2} s^{-1})	$(3.3 \pm 0.8) \times 10^{-3}$	$(2.1 \pm 0.8) \times 10^{-3}$	$(2.6 \pm 0.7) \times 10^{-3}$
Positronium Fraction (%)	99 ± 8	72 ± 13	121 ± 8
χ^2_{min} / deg. of freedom	31 / 24	22 / 28	27 / 29

[a] Powerlaw = A $(E/100\ keV)^{-\gamma}$

the counting rate can be fitted to a calculated change for a given point source or source distribution to determine the source flux. Such a plot for the FIGARO galactic center 511 keV line is shown in Figure 5. The fit shown by the solid line is for a point source at the galactic center and the dashed line is for the source distributed along the galactic center with the same longitude dependence as the CO column profile of Burton *et al.*[30] as analytically represented by Leising and Clayton[31].

Results of the FIGARO analysis show a strong 511 keV line detection on 1988 November 25-26. The flux for an assumed point source at the galactic center is $(2.2 \pm 0.4) \times 10^{-3}$ photons cm^{-2} s^{-1}, and for an extended galactic source is $(1.9 \pm 0.3) \times 10^{-3}$ photons cm^{-2} s^{-1} rad^{-1} at $l^{II}=0^{o}$. This high flux and wide field-of-view (77° FWHM) are consistent with the two-component model. The FIGARO team has analyzed the data assuming a two-component model and using the near-simultaneous GRIS 1988 October measurements to obtain a value of $(1.2 \pm 0.2) \times 10^{-3}$ photons cm^{-2} s^{-1} rad^{-1} for the intensity of the distributed component and $(8.6 \pm 2.8) \times 10^{-4}$ photons cm^{-2} s^{-1} for the point source.

DISCUSSION

The 511 keV line was detected from the galactic center with good ($>3\sigma$) significance in 1988 May and October by GRIS[10], 1988 November by FIGARO[29] and 1989 May by HEXAGONE[11,28]. Since each instrument has a different field-of-view it is difficult to compare the measurements to look for variations in the point-source

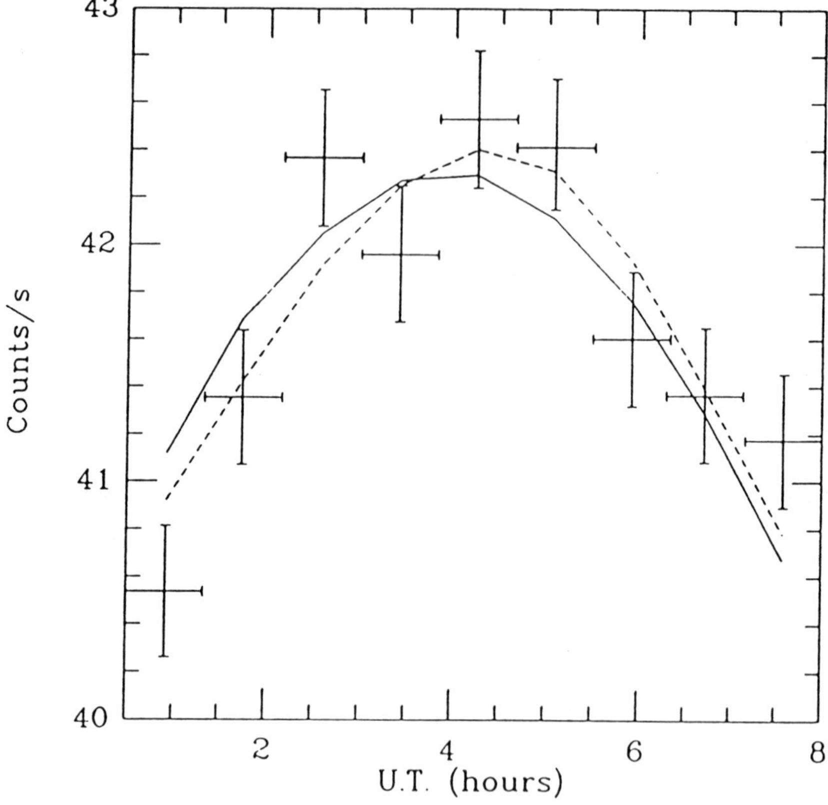

Figure 5 - FIGARO counting rate in the 511 keV line as a function of time during galactic center drift scan observations on 1988 November 25-26. Contamination by positronium three-photon continuum is estimated to be less than 15%[29]. Best fit expected counting rates are shown for a point-source at the galactic center (solid lines) and for a source distributed along the galactic plane with the CO longitude profile (dashed line). From Niel *et al.*[29]

component. This can only be done for the two GRIS observations, with the result that the 1988 May and October line fluxes are inconsistent with a steady-state mean value at the 95% confidence level. This is direct evidence for 511 keV line variability on time scales of < 6 months.

In order to compare all measurements we have derived values for the point-source component by subtracting the contribution of an assumed diffuse component in the apertures. We have included in this comparison the 1981 and 1984 balloon upper limits[7-9] which give a combined 511 keV line flux for the galactic center of $(0.0 \pm 2.9)\times10^{-4}$ photons cm^{-2} s^{-1}. The numbers in Table 3 are for three different assumed diffuse flux distributions. The first two are CO distributions from the SMM analysis of Harris *et al.*[26] The third distribution was chosen to be flat at a flux of 1.2×10^{-3} photons cm^{-2} s^{-1} rad^{-1}. The results for one of the CO distributions and for the flat distribution are shown in Figure 6. The statistically significant large negative numbers in the table for GRIS-GP and 1981-84 show that the narrow-field balloon data are inconsistent with the CO models at the best-fit SMM flux levels. We have found that a better fit to the narrow-field data is a flat distribution at a relatively low flux level of $\sim1.2\times10^{-3}$ photons cm^{-2} s^{-1} rad^{-1}. For this model, the significance of both negative values is reduced to $\leq2\sigma$. The only strong point source flux is for the GRIS 1988 October observation at a value of 8×10^{-4} photons cm^{-2} s^{-1}. The 1988 May and 1989 May point source fluxes are both positive, but each has less than $\sim2\sigma$ significance. There are three pieces of supporting evidence for these two-component flux levels:

1) The FIGARO 511 keV line observation, when analyzed jointly with the GRIS October observation (see Results section) gives a similar point source flux of 8.6×10^{-4} photons cm^{-2} s^{-1} and a similar distributed flux of 1.2×10^{-3} photons cm^{-2} s^{-1} rad^{-1} [29].

2) A reanalysis of the 1980 HEAO-3 galactic center observation[34] assuming a dominant distributed emission gives the same flux of $(1.2 \pm 0.3)\times10^{-3}$ photons cm^{-2} s^{-1} rad^{-1}.

3) A flat source distribution at $\sim1.2\times10^{-3}$ photons cm^{-2} s^{-1} rad^{-1} gives a total flux in the 130° SMM aperture of 2.7×10^{-3} photons cm^{-2} s^{-1}, which is equal to the equivalent point source flux observed [26]. This was calculated using a triangular aperture response function which ignores shield leakage. Also, the SMM data itself is not consistent with a flat distribution over the entire field. But the point here is to illustrate that a flatter, less intense distribution than that used by SMM can give the integral flux in their field and be in better agreement with balloon observations.

The GRIS October observation is the first time the galactic center 511 keV line has been spectroscopically resolved. In addition to the overall line width listed in Table 1, we have derived a width for the point source component by fitting the data with a composite line with three Gaussian components. Two of the lines are for the assumed diffuse component and had fixed parameters derived from models of Guessoum, Ramaty and Lingenfelter[35], while the line for the point source had free parameters determined by the best fit. There are two cases for the diffuse line[35] depending on the regions of the interstellar medium (ISM) in which the annihilation takes place (width, W, in keV; flux, F, in 10^{-4} photons cm^{-2} s^{-1}): 1) uniform ISM - line 1 W=6.4 F=2.4, line 2 W=1.5 F=1.6; 2)cold cores of clouds excluded - line 1 W=6.4 F=0.4, line 2 W=1.5 F=3.6. The results for the point source component width are 2.7 ± 0.8 keV for case 1 and 3.6 ± 0.9 keV for case 2. These widths of ~3 keV imply a temperature for the point source annihilation region of $\sim7\times10^{4}$ K if the width is due to thermal broadening or imply an orbit of $\sim10^{5}$ Schwarzschild radii if the width is due to annihilation in a Keplerian ring around a black hole.[36]

The values in Table 2 for the positronium fractions for the May and October observations are also consistent with the two-component model and with previous data[3]. It is postulated that the lower value for the October galactic center observation is due to suppression or destruction of positronium in the vicinity of a compact object.

Table 3
Calculated Point Source 511 keV Flux in Narrow-Field Spectrometers
(units = 10^{-4} photons cm^{-2} s^{-1})

	GRIS GC May '88	GRIS GC Oct. '88	GRIS GP Oct. '88	Balloon GC '81-'84	HEXAGONE GC May '89
FOV	18°	18°	18°	~15°	20°
Total Measured	7.5±1.7	11.8±1.6	1.8±1.1	0.0±2.9 [a]	8.9±2.7 [b]
Model					
SMM CO1 [c]	-2.6±1.7	1.7±1.6	-4.1±1.1	-8.4±2.9	-2.7±2.7
SMM CO3 [d]	0.9±1.7	5.2±1.6	-5.2±1.1	-5.1±2.9	0.6±2.7
Flat $1.2x10^{-3}$	3.5±1.7	7.8±1.6	-2.2±1.1	-3.1±2.9	3.9±2.7

[a] Combined upper limit for three observations in 1981 and 1984 (Leventhal *et al.*[8,9]; Paciesas *et al.*[7])
[b] Chapuis *et al.*[11]
[c] Dame *et al.*[32] distribution, flux normalization= $2.34x10^{-3}$ photons cm^{-2} s^{-1} averaged over central radian (Harris[33])
[d] Leising and Clayton[31] distribution, flux normalization= $2.07x10^{-3}$ photons cm^{-2} s^{-1} averaged over central radian (Harris[33])

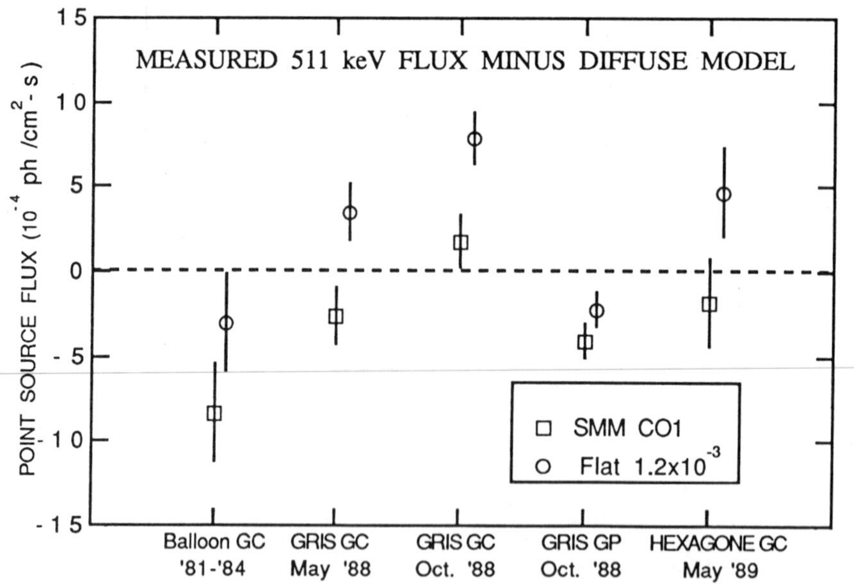

Figure 6 - Calculated point source fluxes for assumed SMM CO1 and flat galactic plane distributions. From Table 3.

Possible mechanisms include annihilation in a region of high density, high magnetic field strength, a dusty medium or in the presence of photoionizing UV radiation[16].

ACKNOWLEDGEMENTS

The author acknowledges the tremendous effort of the other members of the GRIS team, S. Barthelmy, M. Leventhal, C. MacCallum, B. Teegarden and J. Tueller, in obtaining the GRIS data presented in this paper. He appreciates receiving information on the SMM observations from M. Harris and on the HEXAGONE observations from J. Matteson and C. Chapuis. He is grateful for useful discussions with A. Owens, R. Ramaty, G. Share, and the GRIS team. He thanks T. Waclo for assisting in the preparation of the manuscript.

REFERENCES

1. W.N. Johnson, F.R. Harnden, and R.C. Haymes, Ap. J. (Letters), 172, L1 (1972).
2. M. Leventhal, C.J. MacCallum, and P.D. Stang, Ap. J. (Letters), 225, L11 (1978).
3. R.E. Lingenfelter and R. Ramaty, Ap. J., 343, 686 (1989).
4. F. Albernhe, *et al.*, Astr. Ap., 94, 214 (1981).
5. M. Leventhal, *et al.*, Ap. J., 240, 338 (1980).
6. G.R. Riegler, *et al.*, Ap. J. (Letters), 248, L13 (1981).
7. W.S. Paciesas, *et al.*, Ap. J. (Letters), 260, L7 (1982).
8. M. Leventhal, *et al.*, Ap. J. (Letters), 262, L1 (1982).
9. M. Leventhal, *et al.*, Ap. J., 302, 459 (1986).
10. N. Gehrels, *et al.*, Ap. J. (Letters), submitted (1991).
11. C. Chapuis, *et al.*, this Proceedings (1991).
12. G.H. Share, *et al.*, Ap. J. (Letters), 358, L45 (1990).
13. M. Harris, *et al.*, Ap. J., 362, 135 (1990).
14. R.E. Lingenfelter and R. Ramaty, in Positron-Electron Pairs in Astrophysics, ed. M.L. Burns, A.K. Harding, and R. Ramaty (New York: AIP), p. 267 (1983).
15. M. Sigmore and G. Vedrenne, Astronomy and Astophysics, 201, 379 (1988).
16. R.E. Lingenfelter and R. Ramaty, Nuclear Phys. B (Proc. Suppl.), 10B, 67 (1989).
17. J. Paul, *et al.*, this Proceedings (1991).
18. J. Tueller, *et al.*, in Nuclear Spectroscopy of Astrophysical Sources, ed. N. Gehrels and G.H. Share, (New York: AIP), p. 43 (1988).
19. J.L. Matteson, *et al.*, Proc. 19th. Int. Cosmic Ray Conf.3, 326 (1985).
20. G. Agnetta, *et al.*, Nuclear Inst. Meth., A281, 197 (1989).
21. S.D. Barthelemy, *et al.*, this Proceedings (1991).
22. J. Tueller, *et al.*, Ap. J. (Letters), 351, L41 (1990).
23. M. Leventhal, Ap. J. (Letters), 183, L147 (1973).
24. J.L. Matteson, in The Galactic Center, ed. G.R. Riegler and R.D. Blandford, (New York: AIP), p. 109 (1982).
25. G.R. Riegler, *et al.*, Ap. J. (Letters), 294, L13 (1985).
26. M. Harris, *et al.*, Ap. J., 362, 135 (1990).
27. B.L. Brown and M. Leventhal, Ap. J., 391, 637 (1987).
28. J.L. Matteson, *et al.* this Proceedings (1991).
29. M. Niel, *et al.*, Ap. J. (Letters), 356, L21 (1990).
30. W.B. Burton , *et al.*, Ap. J., 202, 30 (1975).
31. M.D. Leising and D.D. Clayton, Ap. J., 294, 591 (1985).
32. T.M. Dame, *et al.*, Ap. J., 322, 706 (1987).
33. M. Harris, private communication (1990).
34. W.A. Mahoney, in Nuclear Spectroscopy of Astrophysical Sources, ed. N. Gehrels and G.H. Share (New York: AIP), p. 149 (1989).
35. N. Guessoum, R. Ramaty, and R.E. Lingenfelter, Ap. J., submitted (1990).
36. D. Bhattacharya and N. Gehrels, Advances in Space Research, in press (1991).

THE SOFT GAMMA-RAY SOURCES IDENTIFIED BY SIGMA IN THE GALACTIC CENTER REGION

J. Paul, B. Cordier, A. Goldwurm and F. Lebrun
Service d'Astrophysique
Centre d'Etudes Nucléaires de Saclay
91191 Gif-sur-Yvette Cedex, France

P. Mandrou, J.P. Roques, G. Vedrenne and L. Bouchet
Centre d'Etude Spatiale des Rayonnements
9, Avenue du Colonel Roche
BP 4346, 31029 Toulouse Cedex, France

E. Churazov, M. Gilfanov, A. Kuznetzov, R. Sunyaev,
I. Chulkov, A. Dyachkov, N. Khavenson and B. Novikov
Institute for Cosmic Research (IKI)
USSR Academy of Sciences
Profsoyouznaya, 84/32, Moscow 117296, USSR

ABSTRACT

Series of observations of the Galactic Center region in the soft gamma-ray regime (up to 1.3 MeV) have been performed by the SIGMA telescope aboard the GRANAT space observatory on March, April, September and October 1990. Analysis of the March and April data has revealed the presence of two prominent point sources: 1E 1740.7-2942, a bright soft gamma-ray source 0.9° away from the galactic nucleus, and a previously unknown source, GRS 1758-258, 0.7° apart from GX 5-1. This paper reports on all the observational material so far analyzed, including spectral evidence for transient electron-positron annihilation radiation from the soft gamma-ray source 1E 1740.7-2942.

INTRODUCTION

In contrast with the X-ray band (between 1 and 30 keV), where many sources have been identified in the Galactic Center (GC) region [1, 2, 3, 4], earlier observations performed in the soft gamma-ray domain have given a somewhat confused view of this region. Thanks to the precise images obtained by the French imaging telescope SIGMA, the first coded-aperture space telescope operating in the soft gamma-ray regime, a clarified picture of the GC region is now emerging. In this paper, we report on the discovery of GRS 1758-258, a bright soft gamma-ray source 0.7° apart from GX 5-1, and on the identification of 1E 1740.7-2942 with a bright soft gamma-ray source located 0.9° away from the galactic nucleus. The continuum spectrum of 1E 1740.7-2942, as derived from a series of observations performed by SIGMA in March and April 1990, is very similar to that of Cygnus X-1, suggesting that 1E 1740.7-2942 is an accreting black hole of stellar mass. We report also on the transient soft gamma-ray feature detected by SIGMA from 1E 1740.7-2942 on october 13, i.e. an intense emission of electron-positron annihilation radiation, a phenomenon which is naturally predicted for an accreting black hole.

OBSERVATIONS AND DATA ANALYSIS

The French coded-aperture telescope SIGMA, one of the main payloads of the Soviet high-energy space-observatory GRANAT, launched on December 1st, 1989, have been designed to produce high-resolution images of the soft gamma-ray sky. The experiment, detailed in Paul *et al.* [5], was developed in the framework of the French-Soviet Space Co-operation Program. Featuring a coded mask placed 2.50 m in front of a NaI(Tl) position-sensitive detector based on the Anger gamma-camera principle, the instrument provides sky images in the energy range 35 keV to 1.3 MeV, with an angular resolution of approximately 15 arcmin over a field-of-view consisting of a central area ($4.7^\circ \times 4.3^\circ$), the totally coded field-of-view, in which the sensitivity is uniform and maximum, surrounded by a wide field of decreasing sensitivity, the partially coded field-of-view (the half-sensitivity boundary is a $11.5^\circ \times 10.9^\circ$ rectangle).

The first SIGMA survey of the GC region was carried out on March and April 1990. It consists of a series of four observation sessions (mean duration: 24 h), during which the satellite was pointed towards the galactic nucleus (two sessions), the X-ray source GX 1+4 (one session), and the X-ray source GX 5-1 (one session). A second series of seven observation sessions with the soft gamma-ray source 1E 1740.7-2942 as main target was performed on September and October 1990.

During all the GC sessions, the SIGMA telescope was operating in the "spectral-imaging" mode. Two sets of images are recorded simultaneously by the instrument in this operating mode: the "fine images", consisting of maps with pixel size of 1.6 arcmin in four contiguous wide energy bands, and the "spectral images", a series of maps with pixel size of 3.2 arcmin in 95 energy channels between 35 keV and 1.3 MeV. The recorded images have been corrected for the spatial non-uniformities intrinsic to the detector or due to the background, and then deconvolved using standard techniques [6].

In the particular case of the images recorded in sessions 70 and 178, since one of the 61 photomultiplier (PM) tubes of the gamma-camera did not operate during both entire sessions, a sensitivity decrease by about 25% has been experienced.

THE DISCOVERY OF GRS 1758-258

During the four one-day observation sessions devoted to the GC region in March and April 1990, a significant excess has been detected near the position of the X-ray source GX 5-1, in the sky maps derived from the SIGMA data in the 40-160 keV energy band [7] (see Table 1). This new source, GRS 1758-258 (GRANAT Source), was also detected by ART-P, the soviet imaging hard X-ray telescope on board GRANAT, in the two sessions during which the X-ray source GX 5-1 was within the ART-P field-of-view [8].

Table 1 Main characteristics of the March and April 1990 SIGMA observations of GRS 1758-258.

session	date	exposure (sec)	sensitivity factor	S/N	source rate (count s^{-1})
063	24 03 90	65004	0.72	7.4	0.52 ± 0.07
066	4 04 90	66725	0.25	5.4	0.58 ± 0.10
068	8 04 90	72348	0.66	5.5	0.47 ± 0.08
070	12 04 90	60303	1.0	9.3	0.52 ± 0.06

Notes to Table 1: the sensitivity factor takes into account the partial modulation of the source flux when the source is outside the totally coded field-of-view. The signal-to-noise ratio (S/N) is derived in the 40-160 keV energy band; it is the ratio between source counts and the statistical error. The source count rates, corrected for the sensitivity factor, have been also derived in the 40-160 keV band; an additional correcting factor was used in the case of session 70 to account for the reduced sensitivity resulting from the PM tube failure.

In the 40-160 keV energy band, the best fit position of GRS 1758-258 is at right ascension $\alpha = 269.52°$ and declination $\delta = -25.70°$ in 1950.0 coordinates (90 % error circle radius: 2.8 arcmin), i.e. 2.2 arcmin away from the best fit position derived in the 3-30 keV ART-P energy band at $\alpha = 269.54°$ and $\delta = -25.73°$ in 1950.0 coordinates (90% error circle radius: 2 arcmin).

It is worth noting that a reanalysis of the 3-30 keV data collected in July and August 1985 by the Spacelab 2 coded-mask telescope revealed a significant excess clearly to be identified with GRS 1758-258 [9]. The Spacelab 2 source is at $\alpha = 269.525°$ and $\delta = -25.747°$ in 1950.0 coordinates (90 % error circle radius: 0.75 arcmin), i.e. 2.7 arcmin away from the position of the soft gamma-ray source detected by SIGMA and 1.4 arcmin away from the position of the hard X-ray source detected by ART-P.

Note also that Levine et al. [10], on the basis of the HEAO-1 A4 survey, suspected that the soft gamma-ray emission occasionally assigned to GX 5-1 may originate from another nearby harder source. Without any doubt, all these measurements refer to the same source (i.e. GRS 1758-258) which is then not simply a transient but a persistent source in both hard X-ray and soft gamma-ray regimes.

Characterized by a hard spectrum, GRS 1758-258 is a rather unusual object among the numerous X-ray sources in the GC region. The photon statistics is too poor to derive firm conclusions from the spectral analysis of the SIGMA data : the comptonized disk model [11] does not provide a better fit to the spectrum of GRS 1758-258 as measured by SIGMA, than a simple power law (see Table 2).

Table 2 Spectral analysis of GRS 1758-258

spectral model	best fit results
Comptonized disk	
reduced chi square	1.2 (3 d.o.f.)
kT_e (keV)	52 ± 22
optical depth	2.8 ± 0.8
power law	
reduced chi square	1.3 (4 d.o.f.)
photon spectrum index	2.0 ± 0.3

Note to Table 2: the spectral analysis of GRS 1758-258 has been derived from the data collected during session 70.

However, it should be stressed that if placed at the GC distance of 8.5 kpc, the 35-200 keV luminosity of GRS 1758-258 is comparable to that of the black hole candidate Cygnus X-1. This latter remark plus the hard spectrum of GRS 1758-258 both argue to consider this source as a genuine black hole candidate.

THE BRIGHT SOFT GAMMA-RAY SOURCE NEAR THE GALACTIC CENTER

The first soft gamma-ray images with arcmin accuracy of the GC region, obtained by SIGMA during the March and April 1990 observation sessions devoted to the GC region (see Table 3), have revealed a strong source, 0.9° away from the galactic nucleus [12]. In the 40-160 keV energy band, the best fit position of the bright soft gamma-ray source is at $\alpha = 265.18°$ and $\delta = -29.73°$ in 1950.0 coordinates (90% error circle radius: 2 arcmin), i.e. ≈ 1 arcmin away from the weak X-ray source 1E 1740.7-2942 detected during the *Einstein* galactic plane survey [2], at $\alpha = 265.178^\circ$ and $\delta = -29.713^\circ$ in 1950.0 coordinates (90% error circle radius: 0.9 arcmin).

Table 3 Main characteristics of the March and April 1990 *SIGMA* observations of 1E 1740.7-2942.

session	date	exposure (sec)	sensitivity factor	S/N	source rate (count s^{-1})
063	24 03 90	65004	1.00	16.4	0.84 ± 0.05
066	4 04 90	66725	0.78	11.3	0.73 ± 0.07
068	8 04 90	72348	1.00	12.6	0.71 ± 0.06
070	12 04 90	60303	0.68	10.3	0.78 ± 0.07

Notes to Table 3: the sensitivity factor takes into account the partial modulation of the source flux when the source is outside the totally coded field-of-view. The signal-to-noise ratio (S/N) is derived in the 40-160 keV energy band; it is the ratio between source counts and the statistical error. The source count rates, corrected for the sensitivity factor, have been also derived in the 40-160 keV band; an additional correcting factor was used in the case of session 70 to account for the reduced sensitivity resulting from the PM tube failure.

The hard aspect of the emission of 1E 1740.7-2942 was already reported by several observations of the GC region performed beyond the *Einstein* energy window, including the observations performed in June 1985 by *SPARTAN 1* in the energy band 1-15 keV [3], and in July/August 1985 by *Spacelab 2* in the 3-30 keV band [4]. In the soft gamma-ray regime, the emission from 1E 1740.7-2942 was certainly detected in the 13-180 keV band in 1977 and 1978, as the "near galactic center" source reported at $\alpha = 265.7^\circ$ and $\delta = -29.5^\circ$ in the *HEAO-1 A4* catalog [10]. More recently, the first coded-aperture image (with 1.1° resolution) of the GC region obtained in April 1988 by the balloon-borne telescope *GRIP* in the 35-200 keV band [13], revealed a single strong source of soft gamma rays, tentatively identified with 1E 1740.7-2942.

As all these measurements refer to the same source, it is beyond doubt that 1E 1740.7-2942 is a persistent source, in both X-ray and soft gamma-ray regimes. The source flux, as measured by *SIGMA* in the 35-200 keV band during both 1990 spring and 1990 fall GC surveys, shows no tangible time variability. Moreover, the 35-200 keV flux as measured by *GRIP* in April 1988 is also comparable to that measured by *SIGMA* two years later, implying again no evidence for time variability at energy $E < 200$ keV in 1E 1740.7-2942.

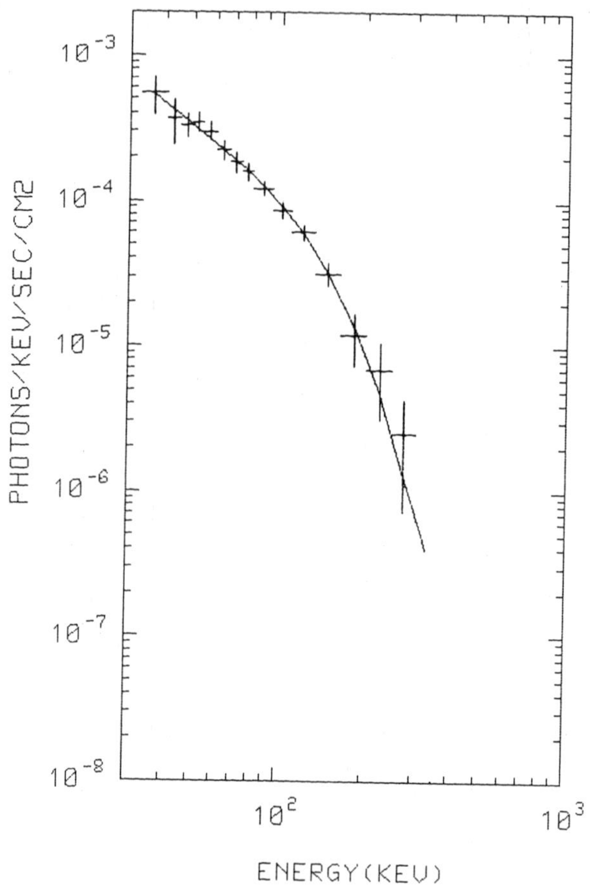

Figure 1 Photon spectrum of 1E 1740.7-2942 derived from the SIGMA data collected during session 63 (March 24, 1990), combined with session 68 (April 8, 1990). The solid line corresponds to the best fit comptonized disk model.

The photon energy spectrum of 1E 1740.7-2942 as derived from the SIGMA data from session 63, March 24 1990, combined with session 68, April 8, 1990, is shown in Figure 1. Firstly, it should be noted that the absence of time variability in 1E 1740.7-2942 allows to combine data from non contemporaneous observations to built a single photon spectrum from X-rays to soft gamma rays. It appears from such a composite spectrum that the hard Spacelab 2 spectrum [14] is seen to continue to 150 keV in the SIGMA data, as it was the case for the GRIP data [15]. On the contrary, in the 3-6 keV band, the Spacelab 2 spectrum deviates significantly from the overall trend (i.e. a single power law) which prevails from 10 keV up to 150 keV, suggesting that the X-rays from 1E 1740.7-2942 are severely absorbed either in the source itself, or more probably by interstellar material.

If indeed the genuine source spectrum extends down to few keV as a single power law, H-atom column densities in excess of 2-3 10^{23} cm^{-2} are required to account for the observed absorption, implying that 1E 1740.7-2942 is at approximately the same distance as the GC. If placed at the GC distance of 8.5 kpc, both the luminosity and the spectral shape of 1E 1740.7-2942 in the 35-300 keV energy domain are all but similar to that of Cygnus X-1, which is widely believed to be a stellar-mass black hole.

If one excepts the gamma-ray burst sources, only two classes of galactic objects are known to radiate hard photons: Crab-like radio pulsars, which produces high-energy photons via inverse Compton and curvature radiation mechanisms, and black holes powered by matter accretion in X-ray binaries like Cygnus X-1 [16, 17]. The extremely hard spectrum of 1E 1740.7-2942 and its similarity in shape and intensity to that of Cygnus X-1 both argue to include this source into the list of the galactic black hole candidate.

Table 2 Spectral analysis of 1E 1740.7-2942

spectral model	best fit results
Comptonized disk	
reduced chi square	1.26 (53 d.o.f.)
kT_e (keV)	30.5 ± 6.0
optical depth	1.80 ± 0.35
power law	
reduced chi square	1.50 (54 d.o.f.)
photon spectrum index	2.20 ± 0.2

Note to Table 2: the spectral analysis of 1E 1740.7-2942 has been derived from the SIGMA data collected during session 63 combined with session 68.

The analysis performed on the spectral material collected in March and April 1990 (see Table 2) confirms that as for Cygnus X-1, the spectrum of 1E 1740.7-2942 as measured by SIGMA is adequately explained in terms of disk accretion onto a stellar-mass black hole, in the framework of the comptonized disk model [11] (the solid line in Figure 1 is the best fit comptonized disk model spectrum with electron temperature kT_e = 30 keV and optical depth τ = 1.8).

THE HARD STATE OF 1E 1740.7-2942

The source 1E 1740.7-2942 was observed by the _SIGMA_ telescope in the course of the second GC survey performed by the _GRANAT_ space observatory from the end of August 1990 to the middle of October 1990. At the time of writing, the data collected during the second GC survey are far from being fully analyzed. Our preliminary data reduction was focussed on the source 1E 1740.7-2942, which was within the fully coded field-of-view of the instrument over seven observing sessions for a total exposure duration of $\approx 4 \times 10^5$ sec.

During six of these observation sessions devoted to 1E 1740.7-2942, the source spectrum, as derived from the _SIGMA_ data, was roughly similar to that measured in the course of the first _SIGMA_ GC survey, with no significant emission detected beyond 300 keV. On the contrary, the source emission as measured by _SIGMA_ during the October 13 session extends well beyond 300 keV (see Table 5). In the sky map derived from the _SIGMA_ data in the 330-570 keV energy band, a significant excess (5.8 σ) is present at the position of the source 1E 1740.7-2942. As shown in Figure 1, the spectral shape of the 330-570 keV radiation responsible for the excess detected in the sky map, appears as a conspicuous bump in the source spectrum derived from the data collected during the October 13 session.

Table 5 _SIGMA_ observation of the 330-570 keV emission from 1E 1740.7-2942 in September/October 1990

session	date	exposure (sec)	detection in the 330-570 keV band
162	19 09 90	97650	NO
169	29 09 90	55800	NO
173	9 10 90	37800	??
174	10 10 90	40950	NO
177	13 10 90	45360	YES (5.8 σ)
178	14 10 90	81000	NO
179	18 10 90	49500	NO

Notes to Table 5: at the time of writing, the analysis of the data collected during session 173 is not completed enough to derive firm conclusions. In the particular case of session 178, the overall sensitivity of the telescope was reduced by ≈ 25% since one of the 61 PM tubes of the gamma-camera did not operate during the entire session.

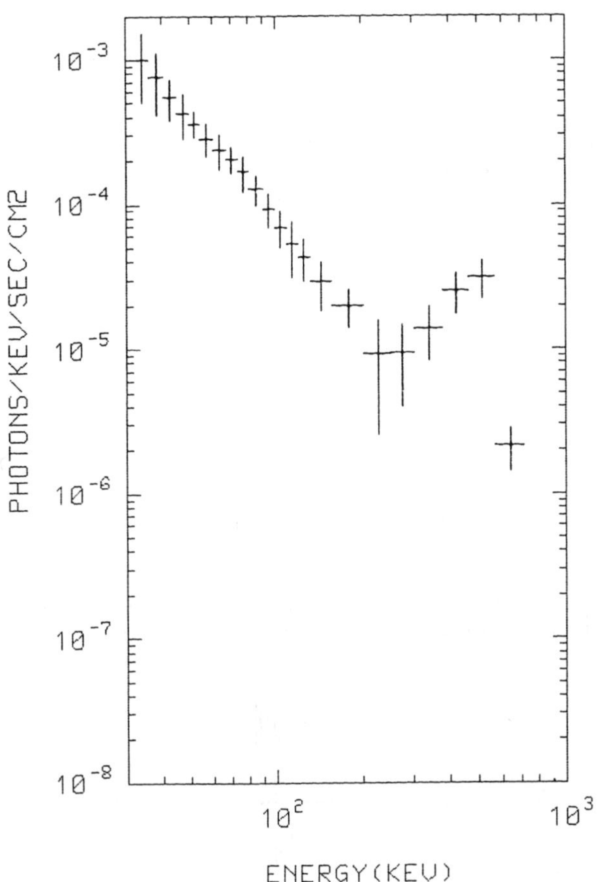

Figure 2 Photon spectrum of the source 1E 1740.7-2942, derived from the SIGMA data collected during session 177 (October 13, 1990).

The presence of a spectral feature around 500 keV suggests strongly that 1E 1740.7-2942 is a source of electron-positron annihilation radiation. However, since the spectral resolution of the SIGMA telescope is of the order of 8% around 500 keV [18], the unusual shape of the high-energy spectral feature, characterized by a monotonous increase from 300 keV to 500 keV followed by a rather sharp decrease beyond 500 keV, cannot result simply from a narrow 511 keV electron-positron annihilation line, convolved with the detector spectral resolution.

Other hypothesis, including positronium formation and subsequent annihilation [19] may account for the asymmetric shape of the high-energy feature, which can also be interpreted in the framework of the pair-dominated plasma model developed by Liang and Dermer [20] to interpret the Cygnus X-1 gamma-ray bump (see also Ramaty and Mészàros [21]). In any case, the observation of a strong gamma-ray feature, the luminosity of which exceeds that observed in the hard X-rays (as it is the case for Cygnus X-1 in its Γ_1 state [22]), is considered as a unique signature of black holes [23]. This further strengthens the hypothesis that the soft gamma-ray source 1E 1740.7-2942 is an authentic black hole candidate.

These results leave little doubt about the location and the nature of the compact source whose presence near the GC is requested to account for the time-dependent positron-annihilation radiation observed since 1970 from the GC region [24, 25, 26, 27]. The case for a modest (< 10^3 solar mass) black hole as the compact source of annihilating positrons has been presented by Lingenfelter and Ramaty [28].

More recently, they have analyzed the observations of the time-dependent positron-annihilation radiation and gamma-ray continuum emission from the GC region, and conclude on the absence of positronium in the compact source emission [29]. At face value, this prediction seems in contradiction with our findings if 1E 1740.7-2942 is indeed the compact source of annihilating positrons and if the positrons annihilate via positronium. On the contrary, if it is correct, another mechanism has to be found to account for the asymmetric shape of the high-energy feature observed in the source spectrum.

CONCLUSIONS

Thanks to the first arcmin resolution images obtained by the SIGMA telescope in the soft gamma-ray regime, it become now possible to derive a disentangled picture of the high-energy phenomenons at work in the GC region. Even if the analysis of the data recorded during the second GRANAT survey of the GC region should be considered as preliminary, the observational material so far processed prompts three chief conclusions.

Firstly, in contrast with the 0.5-3 keV energy band, where dozens of point sources were detected in the GC region by Einstein [1, 2], only two bright persistent point sources have been detected by SIGMA in the 40-160 keV energy range.

The precise estimate of the sensitivity limit of the overall SIGMA GC region survey, including 1990 spring plus 1990 fall data, is not yet available. However, it could be stressed that in the 40-160 keV range, no persistent source weaker than 1E 1740.7-2942 and GRS 1758-258 has been detected within the GC region down to a source flux limit equal to 10% of that of the brightest one (1E 1740.7-2942). In particular, the X-ray pulsar GX 1+4, recently suggested as a possible source of 511 keV annihilation line emission [30], was never detected in the imaging data, both in the single observations and in their total sum [31].

Second, the energy spectrum of 1E 1740.7-2942, the most intense source beyond 20 keV of the GC region, located only 0.9° away from the galactic nucleus, shows remarkable agreement with that of Cygnus X-1, which is widely believed to be a stellar-mass black hole. It is then not surprising that the spectrum of 1E 1740.7-2942 as measured by SIGMA is well explained in terms of disk accretion onto a black hole, in the framework of the comptonized disk model.

Third, the electron-positron annihilation spectral feature observed on October 13, 1990, strengthens the hypothesis that the soft gamma-ray source 1E 1740.7-2942 is an authentic black hole. Another direct consequence of this observation is that 1E 1740.7-2942 is beyond any doubt the compact source of annihilating positrons whose presence near the GC is requested to account for the time-dependent positron annihilation radiation observed since 1970 from the GC region.

ACKNOWLEDGEMENTS

We acknowledge the paramount contribution of the SIGMA Project Group of the CNES Toulouse Space Center to the overall success of the mission. We thank the staff of the Lavotchine Space Company, of the Babakin Space Center, of the Baikonour Space Center and of the Evpatoria Ground Station for their unfailing support.

REFERENCES

1. Watson, M.G. et al. Astrophys. J. **250**, 142 (1981).
2. Hertz, P. & Grindlay, J.E. Astrophys. J. **278**, 137 (1984).
3. Kawai, N. et al. Astrophys. J. **330**, 130 (1988).
4. Skinner, G.K. et al. Nature **330**, 544 (1987).
5. Paul, J. et al. in Advances in Space Research, proc. of XVIII COSPAR Symp. ME.4 Imaging and Spectroscopy for Gamma-Ray Space Missions (ed. G.G. Palumbo, Pergamon Press, Oxford, in the press).
6. Laudet, P. & Roques, J.P. Nucl. Instr. and Meth. **A267**, 212 (1988).
7. Mandrou, P. IAU Circular **5032** (1990).
8. Sunyaev, R.A. et al., in preparation.
9. Skinner, G.K. in these proceedings.
10. Levine, A.M. et al. Astrophys. J. Suppl. Ser. **54**, 581 (1984).

11. Sunyaev, R.A. & Titarchuk, L.G. Astron. Astrophys. **86**, 121 (1980).
12. Cordier, B. et al. in Advances in Space Research, proc. of XVIII COSPAR Symp. ME.8, Highlights in X-ray and Gamma-ray Astronomy (ed. G. Vedrenne, Pergamon Press, Oxford, in the press).
13. Cooks, W.R. et al. in The Center of the Galaxy, proc. of IAU Symp. 136, (ed. M. Morris, Kluwer Academic Publishers, Dordrecht, 1988), p. 581.
14. Skinner, G.K. et al. in Proc. Gamma Ray Obs. Sci. Work. (ed. N. Johnson, 1989), p. 191.
15. Cooks, W.R. et al. in Advances in Space Research, proc. of XVIII COSPAR Symp. ME.4 Imaging and Spectroscopy for Gamma-Ray Space Missions (ed. G.G. Palumbo, Pergamon Press, Oxford, in the press).
16. Sunyaev, R.A. & Trümper, J. Nature **279**, 506 (1979)
17. Ling, J.C. et al. Astrophys. J. **275**, 307 (1983).
18. Mandrou, P. et al. in these proceedings.
19. Leventhal, M. Astrophys. J. Lett. **183**, L147 (1973).
20. Liang, E.P. & Dermer, C.D. Astrophys. J. Lett. **325**, L39 (1988).
21. Ramaty, R. & Mészàros, P. Astrophys. J. **250**, 384 (1981).
22. Ling, J.C. et al. Astrophys. J. Lett. **321**, L117, (1987).
23. Ling, J.C. in these proceedings.
24. Riegler, G.R. et al. Astrophys. J. Lett. **248**, L13, (1981).
25. Leventhal, M. et al. Astrophys. J. Lett. **260**, L1, (1982).
26. Paciesas, W.S. et al. Astrophys. J. Lett. **260**, L7, (1982).
27. Leventhal, M. et al. Nature **339**, 36 (1989).
28. Lingenfelter, R.E. & Ramaty, R. in The Galactic Center (ed. G.R. Riegler and R.D. Blandford, Am. Inst. Phys., New-York, 1982), p. 148.
29. Lingenfelter, R.E. & Ramaty, R. Astrophys. J. **343**, 686 (1989).
30. McClintock, J.E. & Leventhal, M. Astrophys. J. **346**, 143 (1989).
31. Barret, D. et al. in these proceedings.

GRANAT IMAGES OF THE GALACTIC CENTER REGION IN 4-1300 KEV BAND: LOCALIZATION OF THE POSSIBLE CANDIDATE FOR 511 KEV SOURCE.

R.Sunyaev, E.Churazov, M.Gilfanov, M.Pavlinsky,
S.Grebenev, I.Dekhanov, A.Kuznetsov, N.Yamburenko
Space Research Institute, Academy of Sciences, Moscow, USSR
J.Ballet, P.Laurent, J.Paul, L.Salotti
Service d'Astrophysique, CEA Saclay, France
L.Natalucci, M.Niel, J.P.Roques, P.Mandrou
Centre d'Etude Spatiel des Rayonnements, Toulouse, France.

ABSTRACT.

The most important results of the Galactic Center observations by GRANAT during 1990 are: discovery of new very hard X-ray source GRS1758-258 in 40 arcmin from GX5-1, strong variability of Sgr A^* in 4-30 keV energy range, unambiguous detection of 1E1740.7-2942 as the brightest hard X-ray source in the vicinity of Galactic Center and discovery of the variable hard component in its spectrum. The shape of the latter allows one to suppose the relation of this component with electron-positron annihilation. The hardness of the spectra of both 1E1740.7-2942 and GRS1758-258 gives strong arguments to include them to the list of black hole candidates.

INTRODUCTION.

The localization of the source of narrow 511 keV annihilation line in the Center of our Galaxy (G.C.) was considered as the main objective of the GRANAT international observatory and, particularly, SIGMA hard X-ray and low gamma-ray telescope. After a year of operation of the observatory we can not assert, that this goal is achieved completely. Nevertheless, SIGMA telescope localized the source with strong variable spectral feature in the vicinity of 511 keV (Fig.1,2). Interpretation of this feature in terms of annihilation in cold electron-positron plasma via positronium formation[1] leads to the flux in narrow 511 keV line at the level of $\sim 1.5 \cdot 10^{-3}$ phot/sec/cm^2. The variability of this feature implies the compactness of this source (size < 10^{15}-10^{16} cm). The position of the source coincides within errors with the position of well known source 1E1740.7-2942. Unusual hardness of this source was noted earlier.[2-5]

Another important result of intensive observations of G.C. region with GRANAT in 1990 is the fact, that only two X-ray sources (well known 1E1740.7-2942, mentioned above, and newly discovered GRS1758-258, situated in 40 arcmin away from GX5-1) were undoubtedly detected by SIGMA telescope for ~700,000 sec of total observation time (Fig.3).[4-6] This confirms, that sources with anomalously hard spectra are rather rare in our Galaxy. In the central 60 square degrees of the Galaxy several dozens of sources are known to be bright enough in standard X-rays with luminosities $\sim 10^{35}$-10^{38} erg/sec (here and below the G.C. distance is assumed 8.5 kpc). At the energies above 35 keV, however, SIGMA detected only two sources with

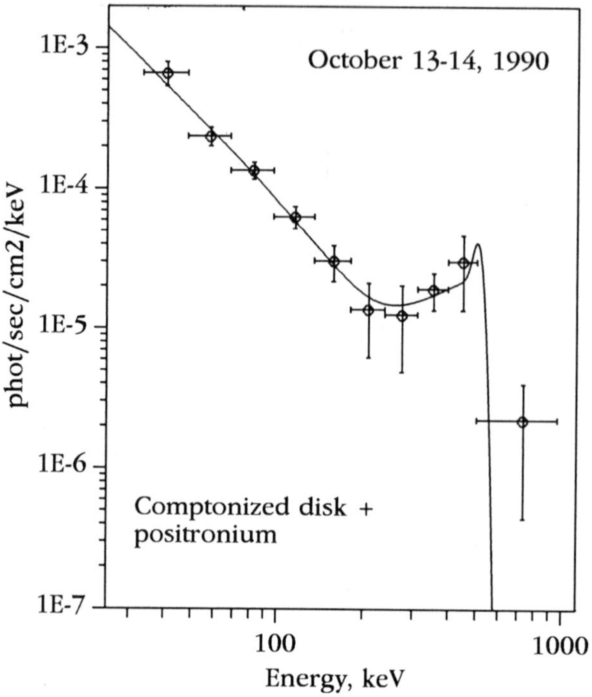

Fig.1 The spectrum of 1E1740.7-2942 during hard state, convolved with instrument spectral resolution. The spectrum was collected over ~15 hours observation of SIGMA telescope on October, 13-14. The solid line represents the best fit to the observed spectrum by the three-component model including comptonized disk spectrum, orthopositronium continuum and narrow annihilation line.

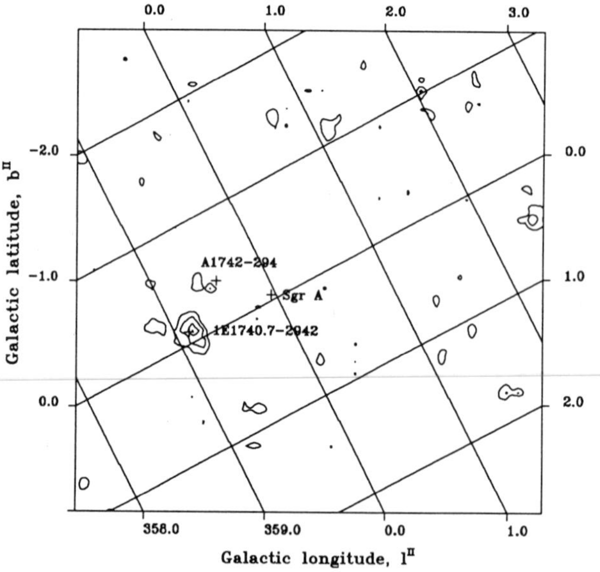

Fig.2 Localization of low gamma-ray source, responsible for strong variable feature near 511 keV. The image of the G.C. region in 360-670 keV band, obtained by SIGMA telescope during observation on October, 13-14. Levels are 2,3,4,5,6 standard deviations. Solid line grid corresponds to galactic coordinates.

35-300 keV luminosities at the level ~$(2\text{-}3)\cdot10^{37}$ erg/sec. All other sources, if any, are 5-7 times less luminous. This supports the suggestion, that these two sources gave the main contribution to the hard X-ray flux observed from the region of G.C. in numerous nonimaging experiments in 70-80-th. Of course, one can not exclude

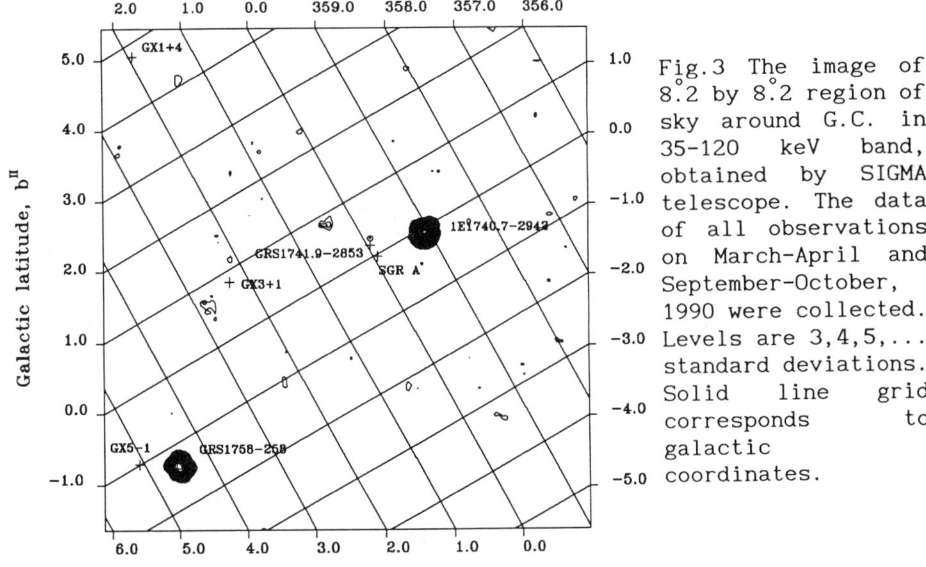

Fig.3 The image of 8°.2 by 8°.2 region of sky around G.C. in 35-120 keV band, obtained by SIGMA telescope. The data of all observations on March-April and September-October, 1990 were collected. Levels are 3,4,5,... standard deviations. Solid line grid corresponds to galactic coordinates.

the possibility of bright transients like GS2023+338, GS2000+25, Oph X-ray Nova, Musca Nova, etc. to contribute to flux observed for a short periods of time.

Thus, the SIGMA observations provided upper limits on hard X-ray and low gamma-ray flux in any part of 35-1300 keV energy band from any source in the 4-5 degrees vicinity of G.C. It is important in the context of recent models, suggesting hard X-ray and low gamma-ray emission from LMXB's.[7]

ART-P telescope has detected relatively weak source, position of which coincides with 1 arcmin accuracy with the Center of our Galaxy - Sgr A*.[4,8] ART-P observes X-ray emission from this source up to 25 keV. Notable is the extremely low luminosity of this source ($\sim 1.5\cdot 10^{36}$ erg/sec in 4-20 keV band). This luminosity is many orders of magnitude less than observed from a number of AGN's and much less than the luminosity of nucleus of M31. Moreover, the SIGMA telescope did not detect any hard X-ray emission from several central arcmin's of our Galaxy with upper limit for 35-100 keV luminosity $\sim 10^{36}$ erg/sec. It is worth to mention, that the luminosity of the source detected by ART-P is ~4 times higher, than measured by XRT telescope on board Spacelab-2 mission 5 years before.[2] This fact argues strongly, that observed emission originates from a few (most probably one) sources. The variability of this source (or sources) and it's spectrum can be naturally explained in terms of accretion on neutron star in low mass binary system. Therefore X-ray measurements until now do not give any evidence for existence of supermassive black hole in the Galactic Center.

In the course of G.C. observations in 1990 ART-P telescope discovered 5 new sources (one of them is GRS1758-258, observed simultaneously by SIGMA) and detected a number of X-ray bursts from GX3+1, A1742-294, SLX1744-300.[4] It is first undoubted localization of

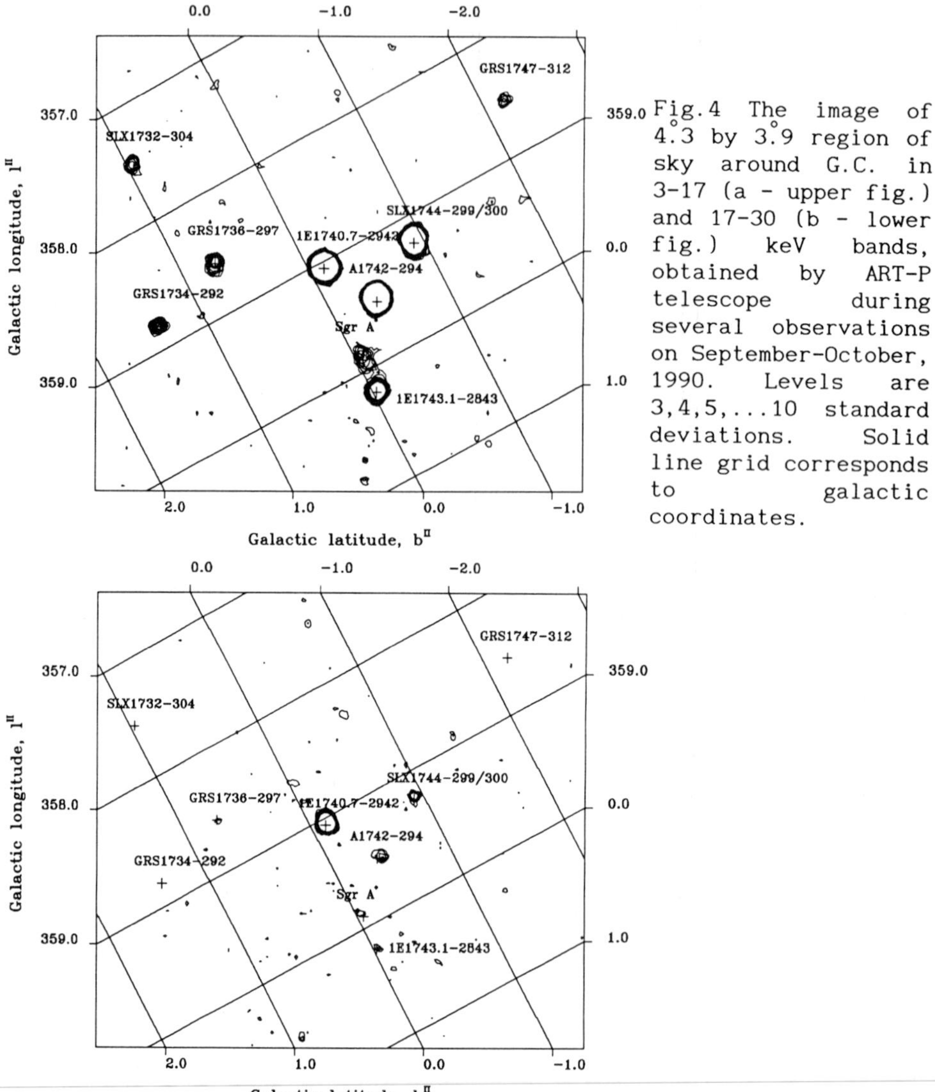

Fig.4 The image of $4^\circ.3$ by $3^\circ.9$ region of sky around G.C. in 3-17 (a - upper fig.) and 17-30 (b - lower fig.) keV bands, obtained by ART-P telescope during several observations on September-October, 1990. Levels are 3,4,5,...10 standard deviations. Solid line grid corresponds to galactic coordinates.

X-ray bursts from GX3+1 and A1742-294 with accuracy of ~1 arcmin, although both were suggested earlier to be X-ray bursters.

INSTRUMENTS AND OBSERVATIONS

The ART-P instrument is an array of four identical coded-mask telescopes (field of view (FOV) $3^\circ.6$ x $3^\circ.6$, angular resolution: 5 arcmin), sensitive in the 4-30 keV energy range, each featuring a 625 cm^2 position sensitive proportional counter.[9] The SIGMA coded mask telescope, detailed in Paul et al.,[10,11] provides sky images (~ 15° x 15° FOV) in the energy range 35-1300 keV with an angular resolution

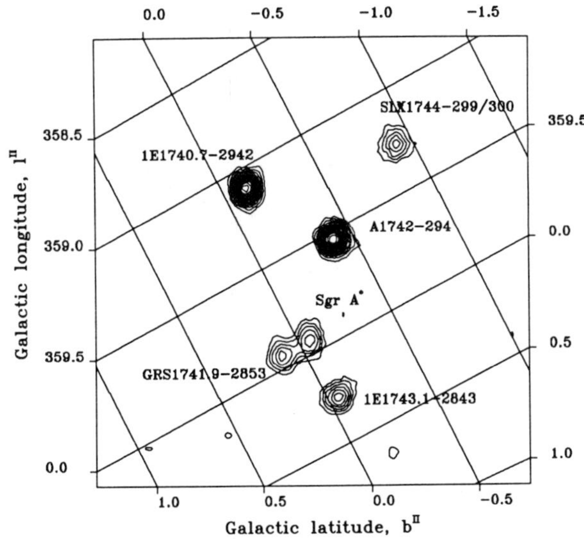

Fig. 4 (c) The image of 2°.3 by 2°.3 region around G.C. in 3-15 keV energy band, obtained by ART-P in the course of March, 23-24 and April, 7-8 observations.

of 13 arcmin approximately. The point source localization accuracy of both instruments is better then their nominal angular resolution (corresponding to the mask pixel size) and can achieve ~1-3 arcmin for bright sources.

Two sets of intensive G.C. observations were performed by GRANAT in 1990: in March-April and September-October. Totally, ~ 20 sessions of observations of G.C. region were performed with averaged duration 10-20 hours. The coverage of the Galactic Center and Plane by telescopes FOVs is given in paper of Sunyaev et al.[4]

IMAGES OF GALACTIC CENTER IN STANDARD, INTERMEDIATE AND HARD X-RAY BANDS.

X-ray maps of the 4°.3 by 3°.9 rectangular region around G.C. in 3-17 and 17-30 keV bands are shown in Fig.4 (a,b). This maps were obtained by ART-P telescope during five days of observations in September-October, 1990. Nine sources have been significantly detected below 17 keV (Fig. 4a), three of them - newly discovered sources (see also table 1). At higher energies (Fig. 4b) the most of the sources go down. It is worth to mention, that X-ray sky is extremely variable in ART-P band. The image of this region, obtained by ART-P during March-April observations, contains additional bright source in the vicinity of Sgr A* (Fig. 4c). During September-October observations ART-P did not detect X-ray flux from this source.

Hard X-ray image in 35-120 keV band, shown in Fig. 3, was obtained by SIGMA telescope (the sum of almost all observations in March-April and September-October, 1990). Note, that the total sky area of SIGMA images presented (8°.2 by 8°.2) is four times larger, than area of ART-P images. Only two sources were detected in hard X-rays in ~60 square degrees around G.C. above 6 standard deviations threshold: well known 1E1740.7-2942 (~50 arcmin from Sgr A*), which is also present on ART-P images (Fig. 4 a,b,c) and GRS1758-258 (~40

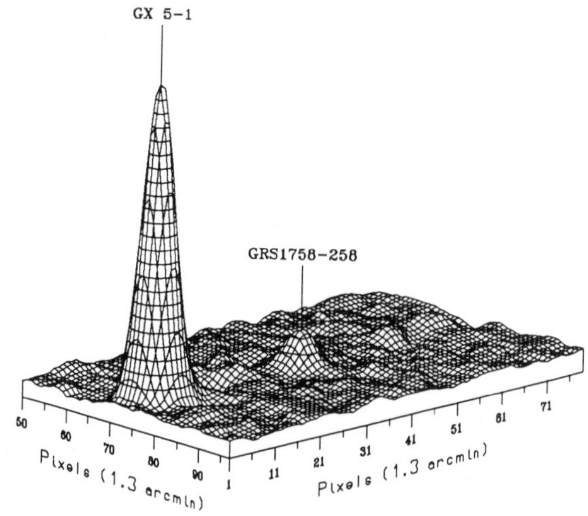

Fig.5 The slice (1°.2 × 1°.2) of ART-P image in 4-20 keV energy band, containing two sources: GX 5-1 and GRS1758-258. The peak height is proportional to the significance of source detection.

arcmin from GX 5-1), which is beyond of the sky region, shown in Fig.4. The latter is the new X-ray source (GRanat Source) with extremely hard spectrum (see below) discovered by both telescopes in the course of March-April observations (Fig. 3, 5).[4-6] It was localized at: R.A. = $17^h58^m07^s$, Dec.= $-25^\circ44.4$ (1950 equinox, error circle radius ~ 1 arcmin). Note that the position of the source has been recalculated more accurately with respect to previously reported,[6] see also.[12]

The list of sources newly discovered by ART-P and SIGMA in the vicinity of the G.C. is presented in Table.1. More details on these sources can be found elsewhere.[4,8] One of these sources - GRS1758-258 has been simultaneously detected by both telescopes. There are also few positions on integrated hard X-ray image (Fig.3), at which number of counts exceeds 4σ threshold. If they correspond to real sources, these sources are at least 5-6 times less luminous, than 1E1740.7-2942 and GRS1758-258. This work is now in progress.

Table 1 The List of the Sources Discovered by GRANAT. in the region of GC.

Source Name	Position (1950) R.A.	Dec.	Error box
GRS1734-292	$17^h34^m14^s$	$-29^\circ09'02''$	1'.0
GRS1736-297	$17^h36^m21^s$	$-29^\circ41'50''$	1'.0
GRS1741.5-2851	$17^h41^m29^s$	$-28^\circ51'$	3'.0
GRS1741.9-2853	$17^h41^m52^s$	$-28^\circ52'55''$	1'.6
GRS1747-312	$17^h47^m38^s$	$-31^\circ15'25''$	1'.0
GRS1758-258	$17^h58^m07^s$	$-25^\circ44'22''$	1'.0

THE SPECTRA OF TWO HARD X-RAY SOURCES.

GRS1758-258. The spectrum of GRS1758-258 is shown on Fig.6. The data above 30 keV (SIGMA) were collected over the whole set of March-April and September-October, 1990 observations. Two spectra below 30 keV correspond to ART-P observations on April, 11-12 and 16. The variability of the spectrum is clearly visible - the 4-30 keV luminosity was $(8.2\pm0.9)\cdot10^{36}$ erg/s and $(4.6\pm0.7)\cdot10^{36}$ erg/s during April 11-12 and 16 observations respectively (the source distance was assumed 8.5 kpc). The 4-300 keV luminosity, averaged over March/April observations is $(2.2\pm0.2)\cdot10^{37}$ erg/s. Solid curve in Fig.6 corresponds to the best power law fit to data in 30-400 keV band with photon index $\alpha=2.0\pm0.2$ and spectral flux at 100 keV $F_{100}=(4.8\pm0.3)\cdot10^{-5}$ phot/s/cm^2/keV.

1E1740.7-2942 - normal state. The averaged spectrum of 1E1740.7-2942 is shown in Fig. 7. The solid curve represents the best fit approximation of the experimental data in 4-400 keV by comptonized disk model[13] with electron temperature of 39±2 keV, half thickness of the disk $\tau=1.3\pm0.1$ and hydrogen column density $N_HL=(1.8\pm0.9)\cdot10^{23}$ cm^{-2}. The averaged luminosity of the source in 4-300 keV energy range is $L=(3.1\pm0.2)\cdot10^{37}$ erg/s. In contrast to GRS1758-258, the source spectrum (Fig.7) shows clearly steepening above ~150 keV. It is much better described by comptonized disk model (reduced χ^2 0.7 for 11 d.o.f.) than by simple power law (reduced χ^2 3.6 for 12 d.o.f.).

ON THE NATURE OF THE HARD X-RAY EMISSION FROM 1E1740.7-2942 AND GRS1758-258.

In contrast to standard X-ray band where dozens of X-ray sources are observed in ~100 deg^2 area around GC, we see that only two of them are bright enough in hard X-rays.

There are only three types of celestial objects known to produce hard X-rays up to at least 200 - 300 keV as 1E1740.7-2942 or GRS1758-258 :

1. Black hole candidates, powered by matter accretion in X-Ray binaries like Cyg X-1[14,15], A0620+00[16], X-ray nova in Oph[17], GS2000+25[18], GS2023+338[19], Musca Nova (GRS1124-684) and active galactic nuclei and quasars like 3C273[20], Cen A[21], NGC4151[22] etc.

2. Crab - like X-ray pulsars, which produce hard X-rays due to inverse Compton and curvature radiation.

3. Gamma-ray burst sources, which are bright only in the short time span.[23]

Accreting neutron stars do not usually produce hard X-ray tails by thermal mechanisms. A part of hard X-ray photons originated from hot layer (like boundary layer between accretion disk and the neutron star) illuminates neutron star surface and heats it due to the recoil effect and photoabsorption. The neutron star surface reradiates this energy through soft photons which effectively cool the hot layer via comptonization. This process does not allow temperature in hot layer to rise. Similar mechanism works in the hot spots on the bottom of accretion column of X-ray pulsars with strong magnetic field.[24] This

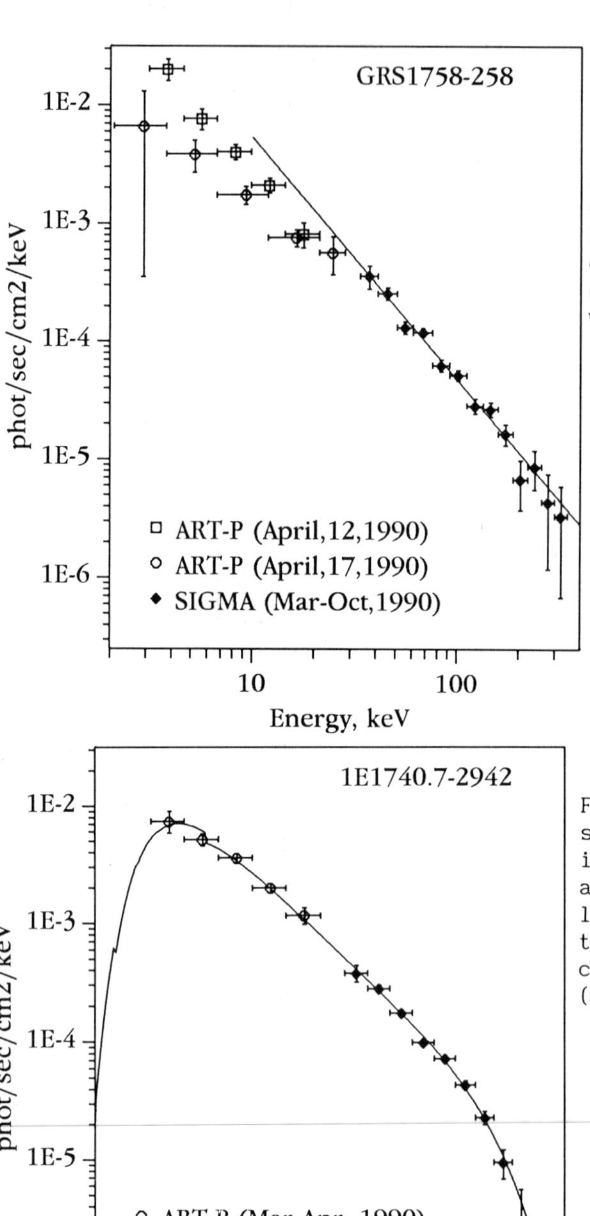

Fig.6 The spectrum of GRS1758-258 in 4-300 keV band (ART-P and SIGMA data). Solid line corresponds to power law with photon index ~2.

Fig.7 The normal state spectrum of 1E1740.7-2942 in 4-300 keV band (ART-P and SIGMA data). Solid line represents fit to the data by the comptonized disk model (see text).

picture is confirmed by detailed numerical calculations in[25]. Only non accreting Crab-like pulsars can produce extremely hard beamed X-rays rather far from the neutron star surface via generation of nonthermal particles.

On the other hand accretion on the black hole naturally produces hard X-rays, since due to the lack of soft photons for comptonization the temperature of plasma might be very high.[26-28] This fact does not depend upon the mass of the central object and could be applied to both stellar mass black holes and extragalactic objects like Cen A or 3C273. One can note that the luminosities of both GRS1758-258 and 1E1740.7-258 in 30 - 300 keV (if this objects are really at the distance of 8.5 kpc) are close to one of the well known black-hole candidate Cyg X-1.[14,15] Actually X-ray spectrum of 1E1740.7-2942 in 4 - 400 kev energy range (combined ART-P and SIGMA data) is much better described by comptonized disk model[13] than simple power law. This might be a reason to consider this source as a new black hole candidate. The GRS1758-258 spectrum has constant power law slope up to energies ~300 keV without any evidence for exponential cut-off. Hard spectrum of this source also argues to include it to the list of black hole candidates.

The possible origin of low energy γ-rays from low-mass X-ray binaries has been discussed recently.[7] SIGMA observations of the GC region, containing a number of LMXB, have given the following upper limits (3σ, for Crab-like spectrum) on the flux from any source in the field: $6\cdot10^{-6}$ phot/s/cm^2/keV (100-300 keV), $5\cdot10^{-6}$ phot/s/cm^2/keV (300-600 keV), $2\cdot10^{-6}$ phot/s/cm^2/keV (600-1300 keV). These upper limits could be applied to any source in the central 50 square degrees around GC. These values provide observational constraints for any theory predicting the origin of low energy γ-rays from LMXB.

Finally, we would like to note, that both GRS1758-258 and 1E1740.7-2942 do not show any evidence of strong soft component similar to observed in typical soft X-ray novae like A0620-00, X-ray Nova in Oph, GS2000+25, Musca Nova (GRS1124-684) and high state of Cyg X-1. From this point of view observed spectra are much more similar to spectra of GS2023+338 and Cyg X-1 in low (or hard) state.

DISCOVERY OF VARIABLE HARD X-RAY AND LOW GAMMA-RAY COMPONENT IN THE SPECTRUM OF 1E1740.7-2942.

During observations of October, 13-14 SIGMA telescope has detected drastic hardening of 1E1740.7-2942 spectrum. On the map of this region in 360-670 keV band, obtained for ~6 hours of observations on October 13-14, the peak at six standard deviations level at the position of 1E1740.7-2942 is clearly seen (Fig. 2). The peak height corresponds to the source intensity of $(6.4\pm1.1)\cdot10^{-4}$ counts/sec/cm^2. Important is that the previous and following observations of this region, held on October 10-11 and 14-15, did not give any evidence for flux from this source above 300 keV with upper limit (90% confidence) of $2.2\cdot10^{-4}$ counts/sec/cm^2 (360-670 keV band). Moreover, the analysis of the whole set of observations on March-April and September-October, 1990 gives upper limit on 360-670 keV flux from the source at the level of $7\cdot10^{-5}$ counts/sec/cm^2 (90% confidence).

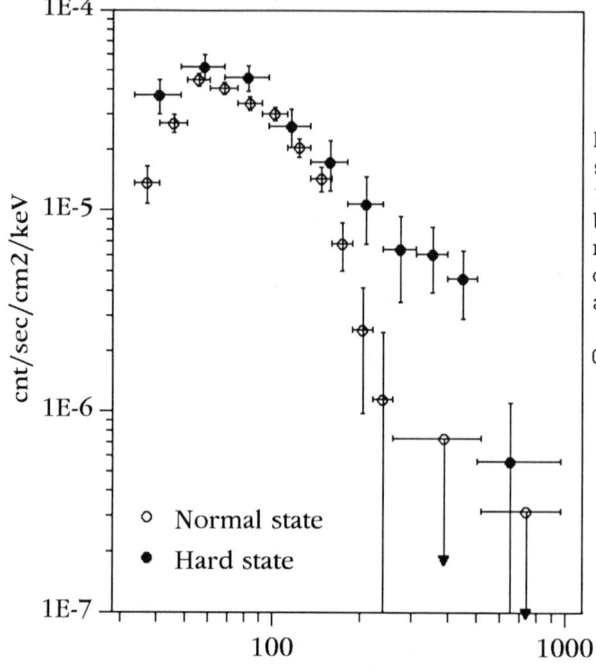

Fig.8 The counts spectrum of 1E1740.7-2942, obtained by SIGMA telescope in normal (all available data were collected) and hard states (observation of October, 13-14).

The counts spectra of 1E1740.7-2942 in two states are shown in Fig. 8. The spectrum, termed "Hard state" was obtained by SIGMA telescope during October, 13-14 observations, the "Normal state" corresponds to all other SIGMA observations available.

DISCUSSION.

The characteristic feature of the spectrum, observed by SIGMA on October, 13-14, is changing slope above ~200 keV and has a maximum at 400-500 keV (Fig.1). Also, a cut-off at energies higher than ~500 keV is clearly seen. The width of the cut-off, according to the existing data, seems to be narrower than ~100 keV.

Three obvious possible explanations can be proposed for this spectral feature:

1. The simplest one: There is the high temperature region in accretion disk with sufficient optical depth, producing Wien-like spectrum with the maximum occasionally at 500 keV.

2. Hard component is related with the emission of hot electron-positron plasma in the accretion disk, producing spectral feature due to the inflight annihilation.

3. The annihilation of comparatively cold positrons take place in cold matter, far from the region of positrons production.

One can note also that Crab-like pulsar-driven models can be ruled out immediately, since the very low ratio of the positron rest mass to the initial energy of accelerated particle. The most of the energy in this case should be released in another energy band. We will show below, that it is difficult to match explanations 1 and 2 with actual observations. The third explanation is of special interest, since the narrow 511 keV line, predicted for this model, has been observed in numerous balloon experiments from the G.C. region.[29,30] We have no reliable arguments that observed component is related with the source of narrow 511 keV line, but it is worth mentioning that among more than 10 sessions of G.C. observations, only the October 13-14 gave undoubtedly the detection of hard source in 300-600 keV band with the flux, comparable with observed in balloon flights.

The width of the cut-off in the spectrum is more abrupt, than the exponential cut-off in the Wien type spectra with the maximum at 400-500 keV, requiring the temperature kT~150 keV[31]. The spectrum of annihilation of hot e^+e^- plasma has a maximum, shifted upwards due to maxwellian distribution of annihilating particles at $m_e c^2+(2\div3)kT_e$ and is broaden due to Doppler effect. Even at kT~10 keV the width of the peak is defined by $\Delta E \sim m_e c^2 \cdot v/c \sim 100$ keV. It is obvious, that practically for any mechanism of e^+e^- production, e^+e^- plasma born will have the temperature much higher, than 10 keV. From another hand, the cut-off of the spectrum at energies higher 511 keV as well as the slope change at ~200 keV are considered as a specific features of e^+e^- annihilation via positronium formation. Indeed, the spectrum, obtained by SIGMA, is consistent with the model, including 3 spectral components:

1)comptonized disk spectrum with kT=37 keV and Thomson half thickness of the disk $\tau \sim 1.0$

2)three-photon orthopositronium continuum

3)narrow 511 keV annihilation line of parapositronium.

In the frame of this assumption, the best fit value for 511 keV line flux is $\sim 1.5 \cdot 10^{-3}$ phot/sec/cm^2 under the assumption that all e^+e^- annihilations occur via positronium formation. This value of flux is consistent with observed by Ge instruments in numerous balloon flights in 70-80-th. Note, that the total luminosity of the source in 200-600 keV band is $3 \cdot 10^{37}$ erg/s.

The source, responsible for 200-500 keV feature in the spectrum of 1E1740.7-2942, was "on" during one observational session, i.e. about ~20 hours on October, 13-14. Undoubtedly, the hard source was "off" during observations of October, 10-11 and 14-15, i.e. 2 days before and 1 day after the detection of the "on" state. This gives obvious constrains on the size of the emitting region: $l < c \cdot \Delta t \sim 10^{15} - 10^{16}$cm. If the source, observed by GRANAT, and the 511 keV source, observed many times before in balloon flights with Ge detectors are the same or of the same nature, this region should satisfy the following constrains:

1) this region must be cold and dense enough, to decelerate and cool down e^+e^- plasma to produce narrow and unshifted 511 keV line, observed in balloon flights.

2) this region must be transparent for photon energies of several keV, as observed by ART-P, i.e. $N_H L$ is less than $\sim 10^{23}$ cm^{-2} This gives immediately the density estimate for any given size of the region.

A simple qualitative picture of the origin of narrow unshifted

511 keV line is proposed below.

Due to change of the accretion rate or due to deficiency of low frequency photons in the maximum energy release regions of the disk, the temperature of inner parts of the disk increases. In principle, it could become high enough, that the significant part of the photons outcoming from the disk will have energies 500-1000 keV. For the disk luminosity at the level of 1/10 of the Eddington luminosity these photons will produce e^+e^- pairs due to $\gamma+\gamma=e^++e^-$ processes. If the disk has low optical depth $\tau_T\leq1$ and is geometrically thin, majority of the pairs are born above the disk, where the density of protons is negligibly small. Note, that the Eddington luminosity for e^+e^- plasma is ~1000 times less, than for p-e plasma. So, e^+e^- plasma born above the disk will be accelerated by the radiation, outcoming from the disk, until the averaged momentum of radiation field in the comoving frame becomes zero. For uniform infinite plane this corresponds to the velocity of 2/3 *c*. Thus, weakly relativistic, with Lorentz factor γ~1.5, e^+e^- wind will be formed. If the Thomson optical depth of this wind becomes ~1, e^+e^- production will be more efficient due to isotropization of the hard photons field. Then, e^+e^- wind, born above the disk, should be decelerated to very low energies (much less, than kT~10^5-10^6 K) in order to give narrow unshifted 511 keV line. The size of the deceleration region should be less than 10^{15}-10^{16} cm, as it follows from the variability. Very high density of low temperature gas in this region is required. Probably, it is cold stellar wind from red supergiant. During the deceleration, kinetic energy of pairs of ~ several 10^{37} erg/sec will be released and reradiated by gas and dust in other spectral bands, most probably, in IR and submillimeter bands. Thus, independently on the luminosity of the optical star, strong and variable IR or submillimeter source with luminosity ~ half or more of the luminosity observed in 511 keV line and associated three-photon continuum, i.e. ~ several 10^{37} erg/sec must appear at this position.

Finally, we would like to note, that since the source 1E1740.7-2942 has galactic coordinates $l^{II}=-0.^{\circ}9$ and $b^{II}=-0.^{\circ}1$ it is located more than 130 pc from the dynamic center of our Galaxy. Therefore, discussing this source as the possible origin site for 511 keV line, one can expect stellar mass object as a main engine of it.

REFERENCES

1. Leventhal M., 1973, Ap.J., 183, L147.
2. Skinner G.K. et al., 1987, Nature, 330, 544.
3. Cook W.R. et al., Proc. of COSPAR XXVIII, 1990, Report ME4.1.3.
4. Sunyaev R.A. et al., Proc. of COSPAR XXVIII, 1990, Report ME8.5.5.
5. Sunyaev R.A. et al., 1991, Soviet Astron. Let., v.17, No.2, p.116.
6. Mandrou et al., 1990, IAU Circ. No.5032.
7. Kluzniak et al., 1988, Nature, 336, 558.
8. Sunyaev R.A. et al., 1991, Soviet Astron. Lett., v.17, No.2, p.99.
9. Sunyaev, R. et al. 1990, in Adv. Sp. Res. v.10. No.2. p.(2) 233.
10. Paul J. et al., Proc. of COSPAR XXVIII, 1990, Report ME4.2.2.
11. Paul J. et al., 1990, in this Proceedings.

12. Skinner G.K. et al., 1990, in this Proceedings.
13. Sunyaev, R. & Titarchuk L., 1980, Astron.Astrophys., 86, 121.
14. Sunyaev, R.A. & Truemper, J., 1979, Nature, 279, 506.
15. Ling, J.C. et al., 1987, Ap. J. Lett. 241, L117.
16. Coe, M.J. et al., 1976, Nature, 259, 544.
17. Wilson, C.K. & Rothschild, R.E., 1983, Ap.J., 274, 717.
18. Sunyaev R.A. et al., 1988, Soviet Astron. Lett. v.14,No.9,p.771.
19. Aref'ev, V. et al., 1989, Proc. of the 23rd ESLAB Symposium (ed. J.Hunt & B.Battrick, ESA Publications Division), vol.1, 255.
20. Primini, F. et al., 1979, Nature 278, 234.
21. Baity, W.A. et al., 1981, Ap. J. 244, 429.
22. Maisack, M. et al., 1989, Proc. of the 23rd ESLAB Symposium (ed. J.Hunt & B.Battrick, ESA Publications Division), vol.2, 975.
23. Mazets, E.P. & Golenetsky, S.V., 1981, in Astrophysics and Space Physics Rev. (ed.R.Sunyaev., Harwood Acad. Publ.) V.1, 205.
24. Basko, M. & Sunyaev, R., 1973, Ap. Space Sci., 23, 117.
25. Sunyaev, R.A. & Titarchuk, L.G., 1989, Proc. of the 23rd ESLAB Symp. (ed. J.Hunt & B.Battrick, ESA Publications Division), v.1, 627.
26. Shakura, N. & Sunyaev, R., 1973, Astron.Astrophys. 24, 337.
27. Shakura, N. & Sunyaev, R., 1976, M.N.R.A.S. 175, 613.
28. Shapiro, S.L. et al., 1976, Ap. J. 204, 187.
29. Leventhal, M. & MacCallum, C.J., 1982, in: The Galactic Center, eds. G.Riegler and R.Blandford, New York, p.132.
30. Lingenfelter, R.E. & Ramaty, R., 1989, Ap.J. 343, 686.
31. Pozdnyakov, L.A. et al., 1983, in: Astrophysics and Space Science Physics. Space Science Reviews. ed. by R.Sunyaev, Harwood Academic Publishers, V.2, P.189.

HIGH ENERGY OBSERVATIONS OF GX1+4 WITH THE SIGMA IMAGING TELESCOPE

D.Barret, S.Mereghetti*, L.Natalucci, J.P.Roques
Centre d'étude spatiale des rayonnements, Toulouse, France

P.Laurent, J.Ballet, B.Cordier, A.Lambert
Service d'astrophysique, CEA Saclay, France

E.Churazov, M.Gilfanov, A.Kusnetzov, R.Sunyaev,
I.Chulkov, A.Dyachkov, N.Khavenson, B.Novikov
IKI, Moscow, USSR

ABSTRACT

The x-ray pulsar GX1+4, located near the galactic center, is a candidate for the 511 keV line source observed from this region of sky. Recent results suggest that GX1+4 is returning to a state similar to that observed during the seventies, when also the 511 keV point source in the galactic center was active. Here we present the results of SIGMA observations of GX1+4 in the energy range 35 keV-1.3 MeV, indicating a decrease in the spin-down rate. The SIGMA unprecedented angular resolution in this energy range allows us to place an upper limit on the 511 keV flux from this particular source.

INTRODUCTION

The x-ray pulsar GX1+4 was discovered as one of the brightest x-ray sources in the galactic center region by Lewin et al[1]. During the years 1970-1980 its X-ray luminosity remained near to the Eddington limit (10^{38} erg sec^{-1}). In 1980, GX1+4 entered a low intensity state, and it was not visible in the soft x-ray energy band until observations made by GINGA[2] in 1987. McClintock and Leventhal[3] pointed out that this long term variation trend resembled that reported for the 511 keV point source in the galactic center region[4]. On the basis of this correlation, and of the unique nature of this binary system (a neutron star with an M6III companion), these authors proposed GX1+4 as a possible candidate for the annihilation line from the galactic center.

On several occasions between March and October 1990, GX1+4 was in the field of view of the hard x-ray / gamma-ray SIGMA imaging telescope. Here we report the preliminary results obtained from these observations.

OBSERVATIONS AND RESULTS

The soft gamma-ray telescope SIGMA[5], using a coded mask associated with a NaI position sensitive detector, provides sky images with an angular resolution of 15 arcmin over a wide field of view (4°.7x4°.3 at full sensitivity). In "spectral-imaging" mode, the telescope records simultaneously two sets of images: a) 4 high resolution images of 248 x 232 pixels in 4 adjacent energy bands covering the range 35 keV-1.3 MeV. b) 95 images of 124 x 116 pixels in 95 consecutive energy channels.

So far, two SIGMA observations were pointed on GX1+4 (April 4, and August 24, 1990). Additional useful data were provided by several other observations in which this source was in the partially coded field of view of the instrument. GX1+4 was never detected in the imaging data, both in the single observations and in their total sum. Table I shows the 3σ upper limits on its flux. The upper limits for the total March-October period were computed by using all the available observations. They can be considered as limits on the time-averaged flux from GX1+4, by making the hypothesis that the source luminosity did not increase in the periods between successive observations.

* *on leave from Istituto di Fisica Cosmica del CNR, Milano, Italy*

Using the sum of all the available data on GX1+4, we have also computed an upper limit on its flux at the energy of the electron/positron annihilation line. Assuming a line centered at 511 keV and with a width given by the instrumental energy resolution (40 keV HWHM), we obtain a 3σ upper limit of 8×10^{-4} photons cm^{-2} s^{-1}.

Table I : 3σ upper limits on the flux from GX1+4 (photons cm^{-2} s^{-1}).

	35-52 keV	*52-77 keV*	*77-120 keV*
April, 4	$4.1\ 10^{-3}$	$1.9\ 10^{-3}$	$1.4\ 10^{-3}$
August, 24	$5.2\ 10^{-3}$	$2.4\ 10^{-3}$	$1.7\ 10^{-3}$
March to October	$2.2\ 10^{-3}$	$1.0\ 10^{-3}$	$0.7\ 10^{-3}$

In parallel to the images acquisition, the telescope records also the total number of counts detected every 4 seconds in the same energy bands defined for the high resolution images. These "slow variability" data can be used to perform a periodicity analysis. Note that, in the case of sources whose flux is modulated in time, they can provide a sensitivity greater than that given by the imaging data. In fact, while a fraction of the source counts in the images is lost due to the non perfect spatial resolution of the detector, the whole pulsed flux can be recovered in the slow variability data.

In order to search for the pulse period, we performed a standard chi-squared analysis on the observations in which GX1+4 was in the totally coded field of view. We detected a significant signal only in the case of the observation of August 24. The period corresponding to the maximum chi-squared value was 114.58 +/- 0.2 s (reduced chi-squared=3.8, for 9 d.o.f.). The periodicity was visible only in the first energy band (40 - 80 keV). The light curve is shown in figure 1. This period value was confirmed by a simultaneous observation at lower energy performed with the ART-P experiment, which yielded the value 114.60 +/- 0.05 s at the solar system barycenter[6]. The pulsed flux estimated from the light curve of figure 1 is compatible with the upper limit on the total flux measured from the simultaneous imaging data. By combining these two measurements, we can place a lower limit of about 0.5 on the pulsed fraction. The negative result for the longer observation of April, 4, indicates a variation in the pulsed flux or in the shape of the light curve.

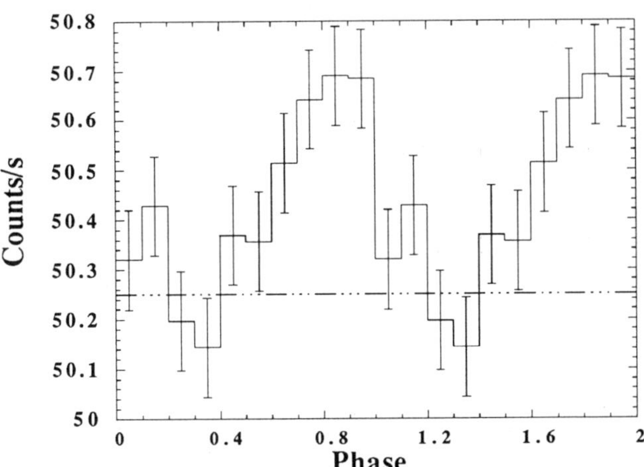

Figure 1 : Light curve of GX1+4 on August 24 (P=114.58 s, 40-80keV) . 1σ error bars are shown. The dashed line indicates the estimated level of the background + unpulsed source flux

DISCUSSION

Observations of GX1+4 made with the GINGA satellite on March 1987, March 1988 and August 1989, indicate a monotonic increase in the 2-20 keV luminosity of this source[7]. Had this trend continued until the time of our observations, GX1+4 would have reached the same luminosity it had in the seventies, before the transition to the low state. On the contrary, the upper limits we have derived, as well as the ART-P flux measurement, (2.3 +/- 0.2x10^{36} erg s^{-1}, d=8.5 kpc, 3-20 keV)[6] indicate a luminosity lower than the March 1988 value of GINGA (8.3x10^{36} erg s^{-1}, d=8.5 kpc, 2-20 keV)[7].

Before the discovery of spin-down in 1987, GX1+4 was considered as one of the best examples of disk-fed accretion, due to its fast and regular spin-up on a very short time scale of about 50 years. On the other hand, its recent period evolution can hardly be reconciled with the "standard" models of disk accretion torques[8]. First, a strong magnetic field, of the order of 10^{14} G, would be required. Second, it would be difficult to understand its regular, luminosity-independent spin-down. The alternative scenario of accretion via stellar wind cannot therefore be excluded. Note, however, that our period measurement, compared to the last GINGA value of Aug 27, 1989, indicates a spin-down rate of 3.1 10^{-8} s s^{-1}, smaller than the value of 4.5 10^{-8} s s^{-1} measured since 1987 with GINGA and KVANT[7,9]. This variation in the acretion torque could possibly be related to the luminosity decrease after 1989.

Exploiting the imaging capabilities of the SIGMA telescope, it has been possible to compute an upper limit on the 511 keV flux from GX1+4. Contrary to the point source flux estimates obtained with wide field of view, non imaging instruments, our value does not depend on the assumed spatial distribution of the underlying diffuse 511 keV emission. Although a safe comparison is hampered by the large statistical and systematic uncertainties of the flux values reported so far, we note that our upper limit is inconsistent with some of the previous observations, and in particular with a continuation of the trend suggested by the most recent results[4,10,11].

REFERENCES

1. Lewin et al, 1971, Ap.J. Letters, **169**, L17
2. Makishima, K. et al, 1988, Nature, **333**, 746
3. McClintock, J.E and Leventhal, M. 1989, Ap. J.,**346**,143
4. Leventhal et al, 1989, Nature, **339**, 36
5. Paul, J. et al, 1990, Advances in Space Research, in press
6. Sunyaev & GRANAT Team, 1990, IAU Circ n° 5104
7. Sakao, T. et al 1990, M.N.R.A.S. **246**, 11P
8. Ghosh, P. and Lamb, F.K, 1979, Ap. J., **234**, 296
9. Mony, B. et al, 1989, Proc. 23rd ESLAB Symp., 541
10. Niel, M., et al, 1990, Ap. J., **356**, L21
11. Matteson, J. et al 1990, these proceedings

AN OBSERVATION OF THE GALACTIC CENTER REGION WITH THE HEXAGONE HIGH RESOLUTION GAMMA-RAY SPECTROMETER

J. Matteson, M. Pelling, B. Bowman, M. Briggs, D. Gruber, R. Lingenfelter, L. Peterson
Center for Astrophysics and Space Sciences
University of California, San Diego
La Jolla, CA 92093–0111 USA

R. Lin, D. Smith, P. Feffer, K. Hurley
Space Sciences Laboratory
University of California, Berkeley
Berkeley, CA 94720 USA

C. Cork, D. Landis, P. Luke, N. Madden, D. Malone, R. Pehl, M. Pollard
Lawrence Berkeley Laboratory
University of California, Berkeley
Berkeley, CA 94720 USA

P. von Ballmoos, M. Niel, S. Slassi, G. Vedrenne
Centre d'Etude Spatiale Des Rayonnements
31029 Toulouse Cedex, France

P. Durouchoux, and C. Chapuis
Service d'Astrophysique
Département de Physique Générale
Centre d'Etudes Nucléaires de Saclay
91191 Gif-sur-Yvette Cedex, France

ABSTRACT

The galactic center region was observed for 6 hours on 22 May 1989 from a high altitude balloon with the HEXAGONE high resolution gamma-ray spectrometer. The instrument had a 285 cm^2 array of cooled germanium detectors with an energy resolution of 2.2 keV at 511 keV and an 18° FWHM field of view. 511 keV gamma-rays from electron-positron annihilation and 1809 keV gamma-rays from the radioactive decay of ^{26}Al were observed to have fluxes of 8.9 x 10^{-4} and 1.9 x 10^{-4} ph/cm^2-s, respectively. Continuum emission was detected from 20 to 800 keV and preliminary results have been obtained for the spectrum. Below 120 keV this is well described by power law with a slope of -2.6. In the 120–250 keV band the spectrum contains a broad line-like feature with a flux of (2 to 6) $\times$ 10^{-3} ph/cm^2-s, depending on the assumed underlying continuum. This is interpreted as the result of Compton backscattering of $\sim$ 511 keV photons from a compact source of electron-positron annihilation radiation.

INTRODUCTION

Observations from balloons and satellites over the past two decades have shown that the galactic center (GC) region contains sources of high energy X-rays and gamma-rays. These have been studied with ever-improving sensitivity, angular resolution and energy resolution in order to identify and understand the nature of their sources.

In the late 1970's the UCSD/MIT instrument on the HEAO–1 spacecraft mapped the 13 – 180 keV radiation from the sky with 1.5° resolution by using scanned fan-beam collimators [1]. This work isolated ~ 15 sources within ~ 15° of the GC. A source within ~ 1° of the GC had the hardest spectrum, was variable on a six month time scale and at times accounted for most of the > 100 keV flux within 15° of the GC [2]. From 1985 to 1988 coded mask imaging of the GC region was done in the 3 – 30 keV range with 3' and 12' resolution [3,4] and in the 30 – 200 keV range with 1.1° resolution [5]. These observations showed that the dominant source above 20 keV in the GC region was 1E1740.7-2942, which is located ~ 0.8° from the GC and thus is likely to be the source seen by the HEAO–1.

Above ~ 200 keV the spectral continuum flux from the GC region is highly variable [2,6], implying that it is dominated by one, or at most just a few, sources. Because of their high luminosity and location in the Galaxy these may form a new class of astrophysical objects. Over the past year the SIGMA instrument on the GRANAT spacecraft has performed ~ 10 imaging observations of the GC region which spanned 35 keV to several Mev and had ~ 10' resolution. It found [7,8,9] both 1E1740.7-2942 and a new source, GRS1758-258 which is ~ 5° from the GC, to be important above 100 keV, with GRS 1758-258's spectrum extending up to 1 MeV and being greater than 1E1740.7-2943's quiescent state above 200 keV. During one of the observations in October 1990 a new phenomenon was discovered [8,9,10]. 1E1740.7-2942 was seen to undergo a flare that apparently lasted at least several hours and produced > 250 keV radiation with a spectrum that peaked at about 500 keV and continued up to 1 MeV . This is in contrast to its quiescent spectrum which cuts off above ~ 300 keV. When it was observed several days later the quiescent spectrum had returned.

With the advent of high resolution gamma-ray spectroscopy in astrophysics, Leventhal et al. [11] showed in 1977 that the GC region's spectral line feature at ~ 500 keV discovered [12] by Haymes and his co-workers in 1970, was a narrow line at precisely 511 keV with a flux of $(12.2 \pm 2.2) \times 10^{-4}$ ph/cm^2-s in a 15° FWHM field of view (FOV). This could only be due to electron-positron annihilation. Then multiple observations by the same instruments showed [13,14,15] that the line flux is variable, requiring that a large fraction of the flux from within ~ 15° of the galactic center comes from a compact source(s). However, results from different instruments showed a trend for larger aperture instruments to observe more flux, which required an extended distribution of sources. This was clearly established by the 1980–1986 SMM observations [16], with a 130° FWHM FOV, of a flux of $(21 \pm 4) \times 10^{-4}$ ph/cm^2-s from the central third or so of the Galaxy. This was much greater than the upper limits on the flux within 15° of the GC obtained [14,15,17] from balloons during this period. The relative roles of the compact and extended sources in the observations were studied by Lingenfelter and Ramaty [18] who showed that all the observations through 1988 could be accounted for by a variable source, with a maximum flux of ~ 10×10^{-4} ph/cm^2-s in the late 1970's, and an extended or diffuse source, with a flux of ~ 15×10^{-4} ph/cm^2-s-rad of galactic longitude in the direction of the GC. An analysis of the SMM data for evidence of a variable source [19] placed a limit of 8×10^{-4} ph/cm^2-s on its yearly average flux in the 1987–1988 and 1988–1989 GC transits.

Although the SIGMA results suggest that 1E1740.7-2942 may be the long sought source of variable electron-positron annihilation radiation, there are some difficulties with this hypothesis. The SIGMA results suggest that the duration of its flaring phase is short, less than a few days, while the balloon and satellite results imply periods of weeks to years of high activity for the compact source. Since this object is not at the galactic center, it is probable that it is not unique and thus more than one source may have produced the variability observed over the past 20 years. In addition, the role of the two GC region sources seen by the SIGMA in producing the variable continuum emission seen in other observations is unclear.

Thus many questions remain to be answered, such as the location of other compact source(s) and the spectra and variability of all of them, and the distribution of the diffuse emission, which is essentially unknown. High spectral resolution observations that map the Galaxy are needed to distinguish between the compact and extended sources and their narrow 511 keV line, positronium and Compton scattered spectral components as well as to determine the width and profile of the line itself.

To study these and other problems in astrophysics, the American–French collaboration in high resolution gamma-ray spectroscopy has developed a new, balloon-carried instrument. This is called HEXAGONE, for High Energy X-Ray and Gamma-Ray Observatory for Nuclear Emissions. When fully developed the HEXAGONE will have a sensitivity to narrow gamma-ray lines of $< 10^{-4}$ ph/cm^2-s in a balloon flight observation. In its first balloon flight in May 1989 it observed the GC region in order to study the 511 and 1809 keV lines and the continuum spectrum. The timing of the balloon flight was fortuituous since after being undetectably weak for 8 years, the GC region's 511 keV line had been observed [20,21,22] by a high resolution gamma-ray spectrometer to once again be strong in May and October 1988.

INSTRUMENT AND OBSERVATION

The HEXAGONE balloon-borne, high resolution gamma-ray spectrometer [23] was developed by a collaboration of American and French scientists in order to observe astrophysical sources of gamma-ray lines, with a sensitivity of $< 10^{-4}$ ph/cm^2-s, and to prove new instrumental techniques that may be applied to future space missions. Gamma rays are detected in an array of 12 high purity germanium detectors which measure $\sim$ 55 mm in length and $\sim$ 55 mm in diameter. They are cooled to 85 K by liquid nitrogen and have an energy resolution in the laboratory of 1 keV below 100 keV and 2 keV at 1 MeV. In the balloon gondola their resolution was slightly degraded, to 2.2 keV at 511 keV. The detectors' total volume is 1568 cm^3 and total area is 285 cm^2. Five of the detectors are equipped with segmentation and pulse shape discrimination [24,25,26] for the rejection of β-decay background due to induced radioactivity in the detectors. These capabilities were not used in the results reported here. Energy losses are processed by dual-range 4096 channel pulse height analyzers which have 0.67 keV/channel up to 2.74 MeV and 3.35 keV/channel up to 13.7 MeV. Data for accepted energy losses are telemetered in an event-by-event format at 40 kbps.

The detectors are surrounded by an anticoincidence shield made of 240 kg of 5 cm thick BGO, at the sides and rear, and 55 kg of 10 cm thick CsI(Na), at the front. Apertures in the latter define an 18° FWHM FOV. 61 photomultipliers view the shield and provide an anticoincidence threshold of 30 keV. The entire instrument mass is 540 kg. Alt-azimuth pointing control is provided by the balloon gondola, whose total mass, including the instrument, is 1134 kg.

The balloon flight was conducted from Alice Springs, Australia on 22 May 1989. Four of the detectors had fully operational segmentation and pulse shape discrimination. Carried on a Raven 28 million cubic foot balloon, the gondola floated at 3.5–4.7 g/cm^2 for 17.4 hours and observed SN1987A for 9.9 hours, the galactic center for 6.3 hours, and the Crab Nebula and the transient x-ray source 0535+26 for 1.3 hours. Observations were performed by a series of target and background pointings, each for 20 minutes. Pointing aspect was verified by a star camera. Three of the detectors had abnormal behavior and did not produce useful data.

DATA ANALYSIS AND RESULTS

In order to combine the data from the various detectors, it was necessary to map their pulse heights into standard, 1 keV energy channels. This was performed by using the stronger background lines to determine gain and zero offset, and their drift, as well as nonlinearity. For energies below 800 keV, the 198.3 and 511.0 keV lines were used for a linear correction giving a final calibration accurate to < 0.1 keV. At higher energies 10 lines were used with energies up to 2614.5 keV to obtain a nonlinear correction which is accurate to ~ 0.2 keV.

A gain-corrected spectrum that represents the sum of the 9 operating detectors was formed for each 20 minute period. The difference spectrum, source minus background, for each source/background pair was then determined and corrected for atmospheric transmission. Finally, the set of difference spectra were combined and corrected for detector efficiency and the transmission of the gondola materials.

An extensive set of detector efficiency measurements were performed at energies from 43 to 835 keV, with the absolute efficiency at 511 keV being determined to < 1 percent accuracy. These results were used to normalize a Monte Carlo model of the instrument, which gives the efficiency at any energy. In this work it was determined that the Monte Carlo calculations typically overestimated the true efficiency by 8 percent.

At the present stage of analysis the instrument response matrix is assumed to be diagonal, with the efficiency determined above, plus the energy resolution broadening. While this will not affect the results on lines, it will result in a slight overestimate of the continuum flux. Monte Carlo studies show that the overestimate is < 5 percent for typical power law spectra with a slope of ~ -2.

The measured spectrum of the GC region is shown in Figure 1. It contains a prominent narrow line at 511 keV, due to electron-positron annihilation, with a flux of $(8.9 \pm 2.7) \times 10^{-4}$ ph/cm^2-s. This is discussed in detail in these proceedings by Chapuis et al. [27] There is also a narrow line at 1809 keV from the radioactive decay of ^{26}Al which was observed to have a flux of $(1.9 \pm 0.9) \times 10^{-4}$ ph/cm^2-s. This is discussed in detail in these proceedings by Malet et al. [28]

Continuum emission was measured from 20 keV to ~ 800 keV and limits were obtained up to 10 MeV. A single power law plus the 2 lines cannot provide a satisfactory fit to the data. From 20 to 120 keV the spectrum is well described by the power law, $n(E)=(2.50 \pm 0.14) \times 10^{-4} \times (E/70\ \text{keV})^{(-2.61 \pm 0.08)}$ ph/cm^2-s-keV. This is indicated by the solid line in Figure 1. But this lies well below the measurements above 120 keV. Between 120 and 200 keV the spectrum becomes constant or perhaps even rises with energy, having a peak at ~ 170 keV. Above 200 keV the spectrum falls approximately as E^{-2}.

The spectrum requires an additional component above the extrapolation of the 20 to 130 keV power law. This has the appearance of a broad line with a peak at ~ 170 keV. The line's flux and width depend on what is assumed to be the underlying continuum and the long dashed lines in Figure 1 indicate what we take as extreme examples for this. For the high and low assumed continuum, respectively, the line has a 120–260 keV flux of $(1.9 \pm 0.7) \times 10^{-3}$ and $(5.9 \pm 0.7) \times 10^{-3}$ ph/cm^2-s and a width of ~ 40 and ~ 140 keV.

Backscattered 511 keV radiation in the instrument could, in principle, produce a line at ~ 170 keV in the spectrum of the GC region. We studied this possibility with Monte Carlo models of the instrument and concluded that this effect is negligible. It could produce no more than a few percent of the broad line that we observed.

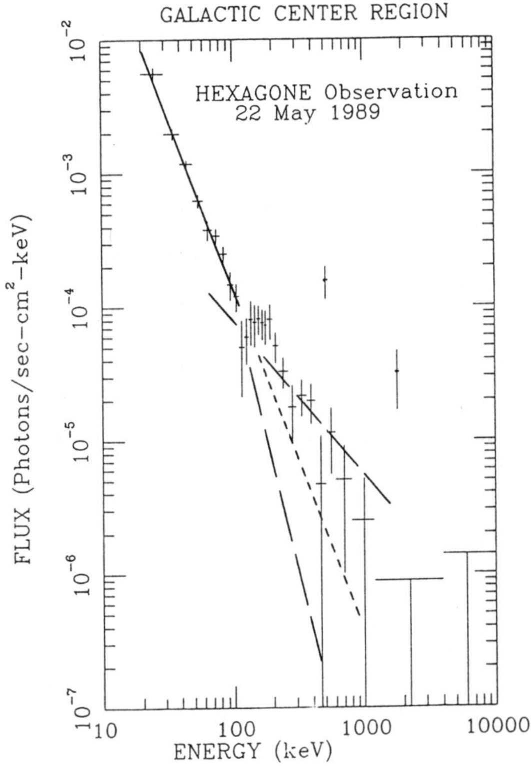

Figure 1. The spectrum of the galactic center region measured by the HEXAGONE high resolution gamma-ray spectrometer on 22 May 1989. The solid line indicates the power law that gives the best fit to the 20 – 120 keV data. This is extended to higher energies as the short dashed line. The spectrum above 120 keV clearly lies well above this. The two long dashed lines indicate the underlying continuum cases that were assumed in order to estimate the flux and width of the broad line feature centered at ~ 170 keV. The points at 511 and 1809 keV are averages over 6 keV. 2 σ upper limits, which exclude the 6 keV band at 1809 keV, are plotted above 1200 keV.

DISCUSSION

The 511 and 1809 keV line results are discussed elsewhere in these proceedings [27,28]. Here we discuss the spectral continuum and 170 keV line results.

We studied previous observations of the GC region's 80–500 keV spectrum for evidence of variability and spectral structure. In order to have a meaningful comparison, we selected results from high resolution gamma-ray spectrometers with fields of view similar to HEXAGONE's. These are all balloon-borne instruments, namely Bell/Sandia in November 1977 [11,29] and November 1984 [15] (15 ° FOV), LEGS in November 1981 [17] (15 °) and GRIS in May and October 1988 [22] (17 °). These results show that the continuum flux varies by a factor of 2 and above 200 keV our result agrees with the maximum previously observed. However, below 130 keV our spectrum is below all of the previous observations, which may indicate that the ensemble of high energy x-ray sources in the GC region were particularly inactive. In 4 of the cases the spectrum's shape is only slightly changed from one observation to the next and lines are not reported nor are they apparent in the spectrum. But the Bell/Sandia 1977 observation [29] contains a 12 keV wide line centered at 170 keV with a flux of $(7.4 \pm 1.8) \times 10^{-4}$ ph/cm^2-s. The line we observed had the same energy, but $\sim$ 3 to 10 times the width and 3 to 8 times the flux. Our results confirm the spectral line observed in 1977 and show that it can be very broad, at least 40 keV and perhaps as great as 140 keV, and have a large flux, $(2 \text{ to } 6) \times 10^{-3}$ ph/cm^2-s. It is interesting that there are hints of structure in the vicinity of 170 keV in the spectra measured by the HEAO–1 in 1977 [2], the HEAO–3 in 1980 [6] and the GRIP in 1988 [5]. These are all of low statistical significance and were not addresses by the authors.

Recently Lingenfelter and Hua [30] modeled the spectrum that would be observed when a source of 511 keV photons is surrounded by a torus of material in which Compton scattering occurs. Photons scattered at large angles produced a broad line feature which was peaked at 170 to 200 keV with a width of $\sim$ 15 to 50 keV, depending on the specific geometry. Indeed, in a 180° scattering the recoil photon's energy is 170.3 keV. The line is strongest when the torus is optically thick and the observer is near the plane of the torus. The line's width is a consequence of Compton scattering's angle-dependent cross section and energy-scattering angle relation, and which scattering angles the geometry makes accessible to the observer. Additional broadening may be necessary to account for the line width reported here. There are many possibilities, such as a broad distribution of photons from the central source, perhaps due to annihilation in a hot or rapidly moving medium. Inclusion of an associated positronium continuum would also cause broadening as would a differential gravitational redshift or Compton scattering in a hot or rapidly moving medium.

The Compton scattering model appears to provide an appealing explanation for the 170 keV line feature that we observed. It also provides a context within which future observations may be interpreted to obtain new insights into the nature and environment of compact sources of electron-positron annihilation radiation.

ACKNOWLEDGEMENTS

The successful development and balloon flight of the HEXAGONE was the result of the work of many people. The authors express their appreciation for the efforts of F.

Duttweiler, C. James, G. Huszar, G. Allen and G. Beriones at UCSD, H. Primbsch, S. McBride, J. Penegor and B. Campbell at UCB, F. Cotin, J. Couteret and J. Coutelier at CESR and J. Poulalion at CEN-Saclay. The balloon flight and supporting logistics were ably handled by R. Kubara, D. Gage, D. Ball and the rest of the NSBF crew in Alice Springs and R. Sood of the UNSW. This work was supported, in part, by NASA Grant NAGW-449.

REFERENCES

1. A.M. Levine et al., Ap.J. Suppl. **54**, 581 (1984).
2. J.L. Matteson, in The Galactic Center, ed. G.R. Riegler and R.D. Blandford, AIP Conf. Prof. No. 83, (AIP: New York) 109 (1982).
3. G.K. Skinner et al., Nature **330**, 544 (1987).
4. G.K. Skinner et al., in Proc. of the Gamma Ray Observatory Science Workshop, ed. W.N. Johnson (NASA, Greenbelt, 1989) p. 4–191.
5. W.R. Cook et al., Proc. 21st Int. Cosmic Ray Conf. (Adelaide, 1990) **1**, p. 216.
6. G.R. Riegler et al., Ap.J. (Letters) **294**, L13 (1985).
7. P. Mandrou, IAU Circular No. 5032 (1990).
8. J. Paul et al., in these proceedings.
9. E. Churazov et al., in these proceedings.
10. P. Mandrou et al., IAU Circular No. 5140 (1990).
11. M. Leventhal et al., Ap.J. (Letters) **225**, L11 (1978).
12. W.N. Johnson et al., Ap.J. (Letters) **172**, L1 (1972).
13. G.R. Riegler et al., Ap.J. (Letters) **248**, L13 (1981).
14. M. Leventhal et al., Ap.J. (Letters) **260**, L1 (1982).
15. M. Leventhal et al., Ap.J. **302**, 459 (1986).
16. G.H. Share et al., Ap.J. **326**, 717 (1988).
17. W.S. Paciesas et al., Ap.J. (Letters) **260**, L7 (1982).
18. R.E. Lingenfelter and R. Ramaty, Ap.J. **343**, 686 (1989).
19. G.H. Share et al., Ap.J. (Letters) **358**, L45 (1990).
20. M. Leventhal et al., Nature **339**, 36 (1989).
21. N. Gehrels, in these proceedings.
22. N. Gehrels et al., submitted to Ap.J. (1991).
23. J.L. Matteson et al., Proc. 19th Int. Cosmic Ray Conf. (San Diego, 1985) **3**, p. 326.
24. J. Roth et al., IEEE Trans. Nucl. Science **NS–31**, 367 (1984).
25. D.M. Smith et al., in Nuclear Spectroscopy of Astrophysical Sciences, ed. N. Gehrels and G.H. Share (New York, AIP, 1988) p. 484.
26. P.T. Feffer et al., SPIE **1159**, 287 (1989).
27. C. Chapuis et al., in these proceedings.
28. I. Malet et al., in these proceedings.
29. M. Leventhal and C.J. MacCallum, Ann. N.Y. Acad. Sci. **336**, 248 (1980).
30. R.E. Lingenfelter and X.-M. Hua, submitted to Ap.J. (1991).

OBSERVATION OF THE 511 keV ANNIHILATION LINE IN THE DIRECTION OF THE GALACTIC CENTER WITH HEXAGONE

C.G.L. Chapuis, P. Wallyn, Ph. Durouchoux
Service d'Astrophysique
Centre d'Etudes de Saclay
91191 Gif sur Yvette Cedex FRANCE

J. Matteson, M. Pelling, B. Bowman, M. Briggs, D. Gruber, L. Peterson
Center for Astrophysics and Space Sciences
University of California, San Diego
La Jolla, CA 92093 USA

C. Cork, D. Landis, P. Luke, N. Madden, D. Malone, R. Pehl, M. Pollard
Lawrence Berkeley Laboratory
University of California, Berkeley
Berkeley, CA 94720 USA

R. Lin, D. Smith, P. Feffer, K. Hurley
Space Science Laboratory
University of California, Berkeley
Berkeley, CA 94720 USA

G. Vedrenne, M. Niel, P.v. Ballmoos, I. Malet
Centre d'Etude Spatiale des Rayonnements
31029 Toulouse Cedex, France

ABSTRACT

The HEXAGONE balloon-borne spectrometer has flown on 22 May 1989. HEXAGONE is a high resolution gamma-ray spectrometer and consists of an array of twelve cooled germanium detectors. One of the observed targets was the Galactic Center and its vicinity (field of view 19° at 511keV) and it was seen during 6.3 hours. The 511 keV annihilation line was observed with a flux of $(8.88\pm2.67)\times10^{-4}\ \gamma cm^{-2}s^{-1}$, a width 1.09+1.38,-1.09 keV and its centroid at 511.54±0.38 keV. The results are consistent with an upper limit of 8.3×10^{4}K for the temperature of the annihilation medium of the positrons.

INTRODUCTION

The 511keV line comes from annihilation of positrons. The positrons can annihilate with electrons or neutral matter like hydrogen. Two photons of 511 keV can be produced directly after the annihilation. A bound state (positronium) can be formed by radiative recombination with free electrons or by charge exchange with a gas atom or molecule. Two kinds of positronium can be produced: orthopositronium (triplet state) and parapositronium (singlet state). The annihilation of this bound state gives different spectra. The decay of parapositronium produces two photons of 511 keV energy, whereas the orthopositronium decay produces a three-photon continuum below 511keV.

The 511keV line was first detected in the early 70's by the Rice University experiment[1]. This result was confirmed by numerous experiments (see [2] for a review) and the measurements show a dependence of the observed flux with time and with the field of view. This could be explained by the fact that the total detected flux should be the sum of a steady diffuse flux depending on the Galactic longitude and a variable compact source flux. The compact object could be a black hole[2]. A recent work shows the influence of the physical conditions on the width of the line, like the temperature or the ionization degree[3] and emphasizes that the positron annihilation does not take place exclusively in either the cold or the hot phase.

INSTRUMENT AND OBSERVATIONS

The balloon-borne experiment contains twelve cooled germanium crystals. Four unsegmented and five segmented performed properly during the flight. Each detector has a volume of approximately 120 cm^3 and has an area of 23.76 cm^2. The detectors are surrounded by a bismuth germanate (BGO) anticoincidence 5 cm thick. The field of view (F.O.V.) is 19 degrees full width half maximum (F.W.H.M.) at 511 keV because of a 10 cm thick cesium iodide active collimator[5]. The energy resolution is 2.26 keV at 511 keV.

The balloon was launched in Alice Springs (Australia) on 22 May 1989. The targets were the Galactic Center (6.3 hours), SN1987a (9.9 hours), Crab Nebula and A0535+26 (1.3 hours). Source and background pointings were alternated at 20 minutes intervals at the same elevation angle but usually about 90 degrees away in azimuth. The elevation angles varied between 18 and 83 degrees while the residual atmospheric depth varied between 3.7 to 4.7 g cm^{-2} during the observation of the Galactic Center.

DATA ANALYSIS AND RESULTS

The data were gain corrected using a calibration method taking into account two peaks of energy 198 and 511 keV. The counts are then summed for a given energy over the active detectors, taking into account the duty cycle. Difference spectra are generated using each ON spectrum and the associated OFF spectrum.

The difference spectra are corrected from atmospheric transmission, instrument windows, area and efficiency. We use the table 3.-27 of NSRDS[6] for the cross sections of air, the local g=9.6925mxs^{-2} and the laboratory measurements of HEXAGONE transmissions, efficiencies and effective area for each of the twelve detectors . The difference spectra were summed to provide the final spectrum of the Galactic Center (see figure 1.). The final spectrum includes the resolution of the instrument. The data were fitted with a gaussian and the resulting flux is:

$$F_{511} = (8.88 \pm 2.67) \times 10^{-4} \text{ photons cm}^{-2} \text{ s}^{-1}$$

The fitted FWHM of the line is 2.51±0.84 keV and the centroid is located at energy 511.54±0.38 keV. The blueshift is not statistically significant. The resolution of the detector in flight was 2.26±0.07 keV. Substracting this in quadrature from a gaussian fit, results in a width for Galactic Center 511 keV photons of 1.09 +1.38,-1.09 keV. The upper limits to the real FWHM are 3.16 keV at a 95% confidence level and 3.58 keV at a 98% confidence level.

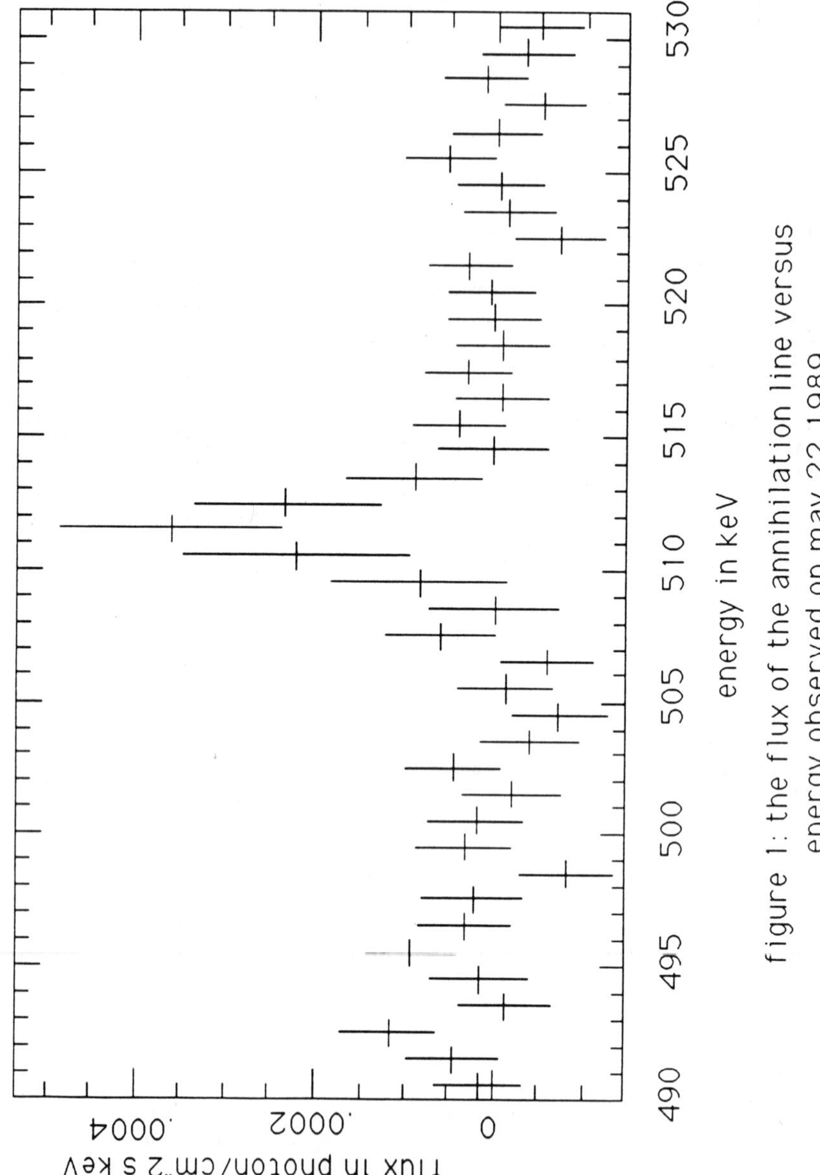

figure 1: the flux of the annihilation line versus energy observed on may 22 1989

DISCUSSION

The observed 3.16 keV upper limit of the FWHM at a 95% confidence level allows to reject the cold phase and the uniform ISM[3]. The upper limit to the temperature is 8.3×10^4K for radiative combination and direct annihilation on free electrons[4], 1.3×10^5K for direct annihilation and 2.3×10^5K for radiative recombination[3].

The flux results from different experiments, taking into account their field of views, show that this line must come from two different origins: a steady diffuse component and a variable point source.

During the long period when the point source was off, one experiment regularly measured the flux of the 511keV line: SMM. This experiment has a large FOV (130°FWHM) and detected no significant variation of the count rate during the period 1980-1988 and also when the point source was on[13,14,15]. At the same time, other experiments have flown and found different fluxes when the source was off (1980-1988) with a high signifiance level and with different FOV. These results are consistent with a steady diffuse source of 511keV. This put in light two points:

A) A large FOV experiment has difficulties to detect a contribution of a point source if it exists because of the large contribution of the diffuse component.

B) Such an experiment is of great interest to confirm the existence of a steady diffuse source of 511 keV line with other experiments of different FOV.

The best criterion to prove the existence of a point source is to analyze the results of similar experiments during a period from 1977 to 1989. The Bell/Sandia group[7,8] (15° FOV FWHM) measured different fluxes between 1977 to 1979 with a high confidence level: $(1.22 \pm 0.22) \times 10^{-3}$ photons cm^{-2} s^{-1} in November 1977 and $(2.35 \pm 0.71) \times 10^{-3}$ photons cm^{-2} s^{-1} in April 1979. The HEAO 3 experiment (35° FOV FWHM) detected this line between fall 1979 to spring 1980. The flux decreased suddenly: $(1.85 \pm 0.21) \times 10^{-3}$ photons cm^{-2} s^{-1} in 1979 to $(0.65 \pm 0.27) \times 10^{-3}$ photons cm^{-2} s^{-1} in March 1980[9]. This indicates that the point source dropped out between this two periods. This point source reappeared in october 1988: the Bell/Sandia experiment and GRIS show a big difference between 1981 to October 1988[10,11,12]: the flux was $(0.0 \pm 0.38) \times 10^{-3}$ photons cm^{-2} s^{-1} in November 1981 and increased up to $(0.75 \pm 0.17) \times 10^{-3}$ photons cm^{-2} s^{-1} in May 1988, and $(1.21 \pm 0.16) \times 10^{-3}$ photons cm^{-2} s^{-1} in October 1988. We plotted the measured fluxes for same type experiments (Ge detectors) and fields of view (15 to 20°)[7,8,10,11,12,20] versus time (see figure 2). This clearly shows that the flux increases between 1977 to 1979, and then stays in a steady low state between 1980 to april 1988, to rise to a high state since october 1988 to may 1989. This is consistent with a point source.

HEXAGONE has a field of view close to GRIS and confirms that the ON state continued for a year (see figure 2). This shows a time variability for the point source and a recent observation of the telescope SIGMA aboard the GRANAT satellite suggests that the variable point source of 511 keV[16] could be 1E1740.7-2942. The apparent disagreement between SMM and balloon experiments could be explained by a compact source variability of time scale less than 1 day or less. We have looked into this possibility during the 6 hours observation but we see no significant variability. The SIGMA instrument has sufficient time and mapping capabilities to detect such an effect and it could indicate the presence of a black hole in the direction of the Galactic Center[2,17,18,19].

CONCLUSION

Our experiment shows that the Galactic Center high state continued for a year. The 3.16 keV upper limit at a 95% confidence level FWHM of the line obtained by HEXAGONE implies that a large fraction of positrons should annihilate in a medium of

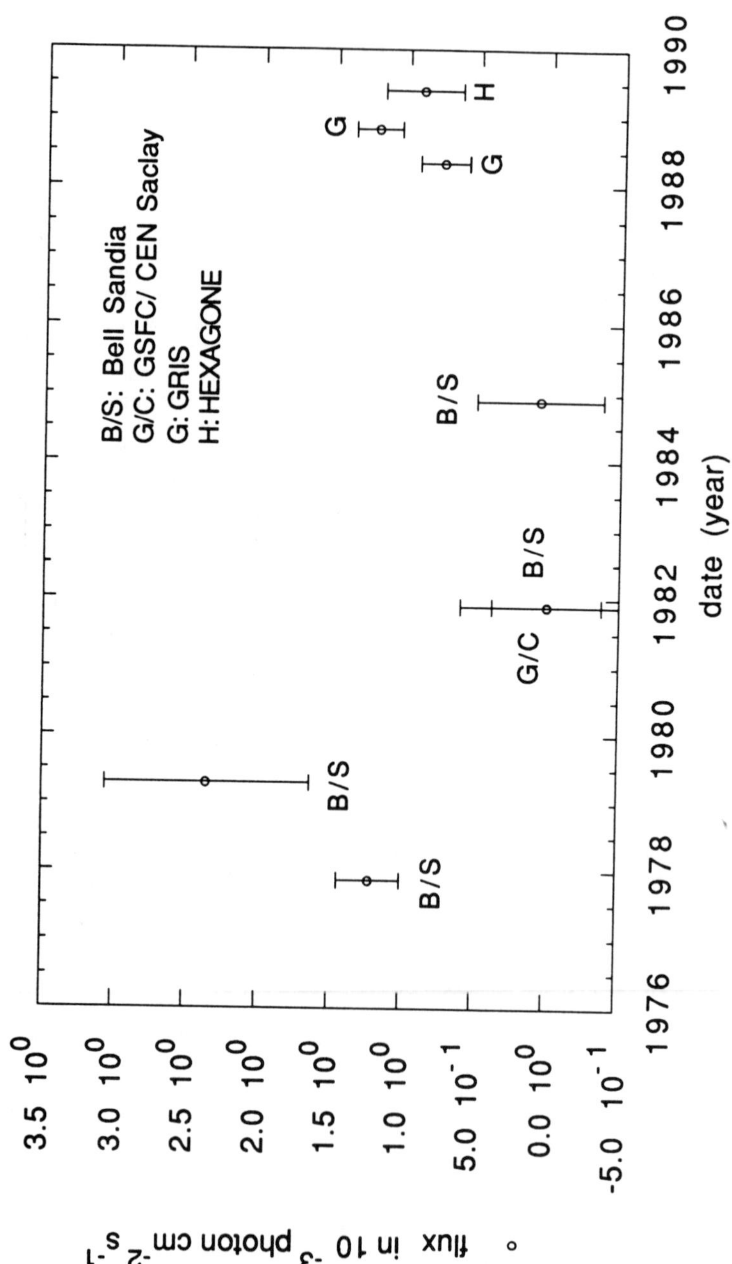

figure 2: fluxes for the 511 keV line for 15-20° FOV FWHM germanium experiment

temperature less than $8.3x10^4$K for radiative combination and direct annihilation with electrons, $1.3x10^5$K for direct annihilation and $2.3x10^5$K for radiative combination. The blueshift is not significant and seems to prove that positrons annihilate in a quite static medium.

ACKNOWLEDGEMENTS

We thank George Huszar (UCSD) and the UCSD/CASS team, Pierre Masse, Nicolas Pétrou and Bertrand Morin (CEN Saclay) for their helpful and valuable discussions about the computer forest.

REFERENCES

1.Johnson III, W.N., Harmden, F.R., Haymes, R.C., *Ap.J.*, **172**, L1-L7 1972 February 15
2.Lingelfelter, R.E., Ramaty, R., *Ap.J.*, **343**, 686, 1989 August 15
3.Guessoum,N., Ramaty, R., Lingelfelter, R.E., accepted to *Ap.J.*
4.Cranell, C.J. *et al.*, *Ap.J.*, **210**, 582, 1976
5.Matteson, J. *et al.* this symposium, 1990
6.National Standard Ref. Data Ser.
7.Leventhal, M., MacCallum, C.J., Stang, P.D., *Ap.J.*, **225**, L11-L14, 1978 October 1
8.Leventhal, M., MacCallum, C.J., Huters, A.F., Stang, P.D., *Ap.J.*, **240**, 338-343, 1980 August 15
9.Riegler, G.R., Ling, J.C., Mahoney, W.A., Wheaton, W.A., Willett, J.B., Jacobson, A.S., Prince, T.A., *Ap.J.*, **248**, L13-L16, 1981 August 15
10.Leventhal, M., MacCallum, C.J., Huters, A.F., Stang, P.D., *Ap.J.*, **260**, L1-L5, 1982 September 1
11.Leventhal, M., MacCallum, C.J., Huters, A.F., Stang, P.D., *Ap.J.*, **302**, 459-461, 1986 March 1
12.Leventhal, M., MacCallum, C.J., Barthelmy, S.D., Gehrels, N., Teegarden, B.J., Tueller, J., *Nature*, **339**, 36, 1989 May 4
13.Harris, M.J., Share, G.H., Leising, M.D., Kinzer, R.L., Messina, D.C., *Ap.J.*, **362**, 1990 October 10
14.Share, G.H., Kinzer, R.L., Kurfess, J.D., Messina, D.C., Purcell, W.R., Chupp, E.L., Forrest, D.J., Reppin, C., *Ap.J.*, **326**, 717, 1988 March 15
15.Share, G.H., Leising, M.D., Messina, D.C., Purcell, W.R., *Ap.J.*, **358**, L45-L48, 1990 August 1
16.Mandrou, P. *et al.* IAU circular n°5140, 1990
17.Ramaty, R., Lingelfelter, R.E., Annals of the New York Academy of sciences 1989
18.Lingelfelter et al, high resolution gamma ray cosmology eds Cline and Fenyves Nuclear Physics B, Proc. Suppl.,1989 in press-- preprint
19.Lingelfelter, R.E., Ramaty, R., The Center of the Galaxy, 587-605, 1989
20.Paciesas, W.S., Cline, T.L., Teegarden, B.J., Tueller, J., Durouchoux, P., Hameury, J.M., *Ap.J.*, **260**, L7-L10, 1982 September 1

OBSERVATION OF ANNIHILATION RADIATION FROM THE GALACTIC CENTER: THE POINT AND RIDGE COMPONENTS

D. E. Gruber and M. S. Briggs
CASS, UCSD, La Jolla, Ca. 92075

ABSTRACT

Using the all-sky map obtained with the Medium Energy Detectors (80 kev - 2 MeV) of the UCSD/MIT Hard X-Ray and Low-Energy Gamma Ray Experiment on HEAO-1, a joint determination of the point source and the galactic ridge components of 511 keV annihilation radiation has been made. The ridge determination assumes the longitudinal distribution measured from 90-270 keV, which has a strong central concentration of width about 120°. Spectral fits assumed an unresolved annihilation line centered at 511 keV and orthopositronium continuum with positronium fraction f=0.9. Detector resolution was 9% at 511 keV, and field of view was 16° FWHM. The point source flux is significant at 95% confidence, and is in reasonable agreement with that measured with HEAO-3 one to two years later in 1980. The ridge component is not significantly measured, but is consistent with a measurement by GRIS at 25° longitude, and the derived two sigma limit lies well below the $\sim$ 2.5 × 10^{-3} photon/(cm^2-s-rad), averaged over the central radian, obtained with SMM observations.

INTRODUCTION

Measurable 511 keV positron annihilation radiation and an associated 3-gamma continuum from orthopositronium decays in the galaxy may arise in the vicinity of one or more collapsed stars. The distribution for collapsed stars, as for normal stars, should peak at the galactic center. Diffuse emission arising from supernova ejecta and cosmic ray interactions in the interstellar medium, should also peak at the center. Numerous observations (see Lingenfelter and Ramaty[1], Harris et al.[2]) over 20 years have been made of annihilation radiation near the galactic center, but controversy remains over whether point-source or diffuse emission dominates. The satellite-borne instruments capable of surveying the plane have very wide fields of view (SMM 130°, HEAO-3 35° FWHM) with little mapping ability. Most other observations, from balloon-borne instruments, have tighter fields of view, 15-20° FWHM, but with one exception (Leventhal et al.[3]) have viewed only the galactic center and not the plane. Rapid time variability on scales of a year or less have been reported (e. g. Riegler et al.[4]) and the historical record has been used to solve for point and diffuse contributions by Lingenfelter and Ramaty[1], although not without challenge (Share et al.[5]). Recently an imaging observation[6] with the SIGMA experiment has strongly established both the presence of point source emission near the center and rapid variability. Even so, the level of diffuse emission is still not well determined. We report here a different approach to separation of diffuse and point contributions based on a sky map. Due to statistical limitations, the conclusions we reach must depend on an assumed distribution function for the diffuse emission. Our result gives a very low flux for the diffuse annihilation radiation, consistent with the GRIS[3] observation but not with the diffuse flux reported from SMM[2].

OBSERVATIONS

Between 1977 and 1979 HEAO-1 made 2-1/2 scans over the complete sky. The four Medium Energy Detectors of the UCSD/MIT experiment were sensitive from 85 keV to 2 MeV, with an area of 43 cm^2 each and a conical field of view of 16 degrees FWHM. Greatest sensitivity was achieved at the low end of the range — nominally 90 to 165 keV. A map was constructed from three of these detectors, using pixels of 25 square degrees. Coverage was fairly uniform, although naturally peaked at the ecliptic poles.

The 16° FWHM of the Medium Energy Detectors of the UCSD/MIT detectors on HEAO-1 permit a quite reasonable spatial separation of point source emission at the Galactic Center from the more extended diffuse galactic emission. The longitude distribution of the diffuse emission has been directly measured[7] to good precision between 90 and 280 keV. It has a broad central maximum with width 120° FWHM and a center to anticenter ratio of about 10. We have used this longitude distribution also for the diffuse annihilation radiation; even though the true distribution may well differ, the present assortment in use (see Harris et al.[2]) have the same general character, with the addition of a central spike for some. We cannot separate a central spike from a point source in our map, thus we restrict ourselves to a phenomenological separation of a broad component from a sharp central component, which may include one or more point sources and the spike in the central ten degrees or so. Standard fitting procedures on the pixels of the skymap at each of 15 energy bands between 80 and 1100 keV have been used to determine jointly (but nearly independently) the distributed flux and the condensed Center flux. The resulting spectra are analyzed for the signature of annihilation radiation: the 511 keV line and the 3-γ continuum, with positronium fraction fixed at 0.9. Standard errors include only Poisson statistics; other noise processes have been extensively examined[8,9] and found to be negligible for this problem.

RESULTS

The 80-1100 keV average spectra (Fig. 1) for the three epochs of observation were each fit (Table I) to a model with power law plus annihilation radiation with positronium fraction 0.9. Chi-squares were acceptable. The power law continuum is probably the sum emission from several sources, and for the ridge is dominated by cosmic ray electron bremsstrahlung. The point source spectrum show a weak but reasonably convincing annihilation line and 3-γ continuum, while the ridge spectrum shows only the faintest trace of the annihilation features.

Table I. Spectral Fits

Parameter	Value	Units
	Point source	
Power Law Flux (100 keV)	2.195 ± 0.055	10^{-4} ph/(cm^2-s-keV)
Power Law Index	2.75 ± 0.12	
511 line flux	5.2 ± 2.2	10^{-4} ph/(cm^2-s)
	Ridge source	
Power Law Flux (100 keV)†	1.581 ± 0.061	10^{-4} ph/(cm^2-s-keV-rad)
Power Law Index	2.03 ± 0.13	
511 line flux†	1.0 ± 2.7	10^{-4} ph/(cm^2-s-rad)

† averaged over central radian

FIGURE 1a

POINT SOURCE AT

GALACTIC CENTER

ϕ_{511} 5.2±2.2×10⁻⁴ ph/(cm²−s)

photons/(cm²−s−keV)

keV

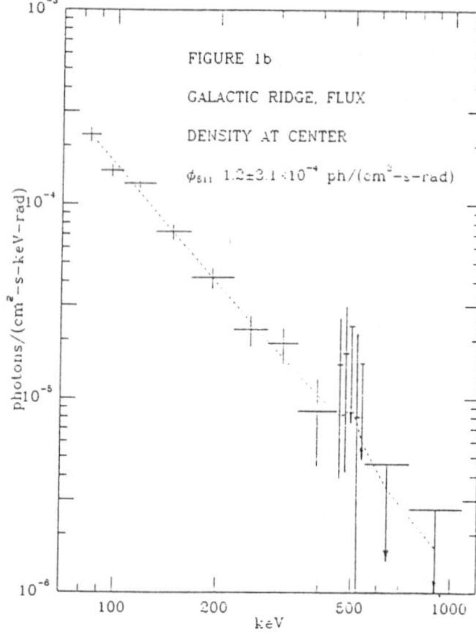

DISCUSSION

The sum of point flux and the central radian of the diffuse is $6.3 \pm 3.5 \times 10^{-4}$ photons/(cm^2-s-keV), which is significantly below the $27 \pm 6 \times 10^{-4}$ reported from SMM, assuming a point source only. Our very weak ridge emission is, however, consistent with a measurement with GRIS at 25°. One resolution of this conflict is the possibility of another source of annihilation radiation at higher galactic latitude, say 10° or more, which would not be seen well with the HEAO-1 or GRIS field of view, but could contribute fully to the SMM counting rate.

This work was supported by NASA Grant NAG 8-499.

REFERENCES

1. R. E. Lingenfelter and R. Ramaty, Ap J. 343, 686 (1989).
2. M. J. Harris, G. H. Share, M. D. Leising, R. L. Kinzer, and D. C. Messina, Ap. J. 362, 135 (1990).
3. M. Leventhal, C. J. MacCallum, S. D. Barthelmy, N. Gehrels, B. J. Teegarden and J. Tueller, Nature 339, 36 (1989).
4. G. R. Riegler, J. C. Ling, W. A. Mahoney, W. A. Wheaton and A. S. Jacobson, Ap. J. (Letters) 294, L13 (1985).
5. G. H. Share, M. D. Leising, D. C. Messina and W. R. Purcell, Ap. J. (Letters) 358, L45 (1990).
6. E. Churazov et al., this symposium.
7. L. E. Peterson, D. E. Gruber, G. V. Jung and J. L. Matteson, 21st International Cosmic Ray Conference, OG 3.1-12 (1989).
8. G. V. Jung, unpublished PhD dissertation, UCSD (1988).
9. D. E. Gruber, G. V. Jung and J. L. Matteson, Proceedings of the Conference on High Energy Radiation Background in Space, eds. C. Rester and J. Trombka (New York:AIP, 1989), p 232.

HARD X-RAY OBSERVATION OF GALACTIC CENTER REGION

A. Bazzano, C. La Padula, P. Ubertini and R.K. Sood*

Istituto di Astrofisica Spaziale (CNR), Frascati, Italy

*University of N.S.W., ADFA, Canberra, Australia

ABSTRACT

The Galactic Center region, including the high energy emitter 1E1740.7-2942, has been observed on may 17, 1989 in the energy range 15 - 150 keV. The measurement was performed with the balloon borne experiment "POKER" based on an array of three high pressure Multiwire Proportional Counters (MWPC) with energy resolution of 10% at 120 keV. The total geometric area was 8,000 cm^2 and the experiment was passively collimated with copper hexagonal collimator with 1.9 degree FWHM field of view.

Data analysis is still in progress and preliminary results are presented.

I. INTRODUCTION

The 1985 SpaceLab 2 observation of the Galactic Center region[1] pointed out the origin of the high energy (>20 KeV) was from the source 1E1740.7-29.42[2] which was the strongest revealed source in the 6^0x 6° field in the 20-32 KeV range. The first coded aperture images of the Galactic Center region for energies above 35 KeV were obtained by the Caltech Gamma-Ray Imaging Payload (GRIP) during 1988[3]. This instrument revealed two strong gamma ray sources: one located 0.7 ± 0.1^0 far from the nucleus and identified with the Einstein source and the other with the x-ray source GX 354+0 while no hard x-ray emission was detected from Sgr A*[4]. In the same energy range non imaging instrument revealed 0.511 MeV positron annihilation line and gamma-ray continuum emission from the direction of the galactic Center[5,6,7,8,9,10,11] Because of the time variability on time scale as short as six months of both line and continuum components and of the similarity with the Cyg X-1 spectrum in the same energy range, a massive black hole as been suggested as compact nucleus of the source[12,13]

II. THE INSTRUMENT

The "POKER" experiment, described in detail elsewhere[14,15], was designed to perform high sensitivity observations of cosmic sources in the hard X-Ray range (10÷180 keV). The telescope consists of an array of three high pressure Xenon Multiwire Proportional Counters (hereinafter MWPC) with a sensitive area of 2,500 cm^2 each. The detectors named SPC1, SPC3 and SPC4 were filled at a pressure of 2.6, 3.6 and 3.8 bar respectively with a mixture of Xe/CO_2(SPC1) and Xe/Isobutane (SPC3 and SPC4) to provide an efficiency greater than 20% respectively for 13 keV and 120 keV photons. The MWPCs spectral resolution was 14% at 60 keV. The fields of view of the three MWPCs were coaligned and limited by means of hexagonal copper collimators with an aperture of 1.9 degree FWHM. The X-ray telescope is mounted on the central cage of the steering platform which includes the elevation drive system and azimuthal control. The azimuthal reference is obtained with the use of a three axes flux gate magnetometer measuring the earths magnetic field. The elevation angle is controlled by a shaft encoder relying on the local vertical, measured with a set of three independent inclinometers, on each detector, to measure the elevation and cross-elevation angle. Changes of the payload azimuth beneath the balloon are achieved by the combined use of a reaction wheel mounted on the payload and a torsion relief drive in the payload bearing. The signals from the magnetometers are on-board processed to compute the azimuth angle of the payload. This angle is compared with the desired target azimuth either telecommanded or calculated by the on-board microprocessor. The actual and target azimuth data are used to generate an error signal to drive the motors. The reaction wheel is directly driven by the signal error while the torsion relief system is driven with the use of velocity feedback from a tacho generator integrally fitted to the reaction wheel drive. Two CCD cameras, coaligned with the detector axis, have been used to check the absolute telescope aspect in real time. The two cameras have been equipped with different objectives and filters to be used differently, the first one as a Sun Sensor, over a very narrow f.o.v. of 1.9 deg, and the second one as a Star Sensor with a 4 deg f.o.v. The video data are transmitted via a 1 Mhz bandwidth high power transmitter. During the first part of the flight aspect check were performed on the moon. Drift scan and tracking observations of SCO X-1, performed just before the Galactic Center observation, were used to check the pointing and stability accuracy of the stabilization system. The in-flight positioning accuracy and stability was better than 0.1 deg in elevation and azimuth. The orientation of the telescope axis was known with an accuracy of ± 0.1 degree.

III. OBSERVATION AND DATA ANALYSIS

The telescope was launched from the Balloon Launching Station of Alice Springs (Australia) at 09.30 U.T. on May 17, 1989. The Galactic Center region was pointed between 14:20 and 16:05 U.T. using tracking observations and during this period the energy response of the instrumentation was found to be stable within 2%.

The Galactic Center region was observed during a period of 5,400 s while rocking the detectors array off-source every 15 minutes toward a sky area free of known hard X-ray sources to collect the background data. Correlation between the background counting rate and potential sources of systematic variation such as altitude, longitude, latitude and steering coordinates were carefully investigated to allow the appropriate correction to the background noise subtraction. During the flight the only significant source of systematic background variation was found to be related to the float altitude changes. The Galactic Center and the associated background data were collected at a float altitude corresponding to an atmospheric pressure ranging between 3.50 and 3.60 mbar. This variation resulted in a negligible effect if compared with the errors due to the statistics.

The detector matrices were determined by taking into account the detector parameters (gas mixture, geometry of the absorption cell, escape effects etc.), the electronics thresholds, the collimator effect, and the atmospheric absorption. The precision of this deconvolution method has been confirmed by observation of the Crab Nebula during other flights[16,17,18]

In this paper we present the preliminary data analysis of about two third of the whole data.

The data from each MWPC are separately analyzed and the results statistically combined. The raw detected excess, corresponding to a statistical significance of 19.5 σ, was interpreted as originating from the source 1E1740.7-29.42 and folded through the matrix describing the detector with a standard reiterative χ^2 procedure. The best fit was obtained with a power law as follows:

$$dN/dE = (2.85\pm0.24) \times 10^{-4}\ (E/30keV)^{-1.95\pm0.27} \quad (ph\ cm^{-2}s^{-1}keV^{-1})$$

with a reduced χ^2 of 2.0 for 25 degrees of freedom. The errors are computed at the 68% confidence level for joint parameters estimation[19]. The final aspect reconstruction for this observation is not yet completed, then we estimate an additional uncertainty of ±10% of the source flux. Furthermore, we cannot rule out a contribution at low energy due to the presence in the field of view of the source A1742-294. This object has been already observed at energy ≤ 20keV at flux level similar to the 1E source. During our observation, the average efficiency to this source was ≤ 55%.

Figure 1 shows the deconvolved spectra obtained from detectors SPC1 and SPC3 together with the power laws best fit; both spectra show an excess in the energy bins around 58-59 keV. The contribution to the χ^2 of the two data points in the bin 56-61 keV for SPC1 and 57-63 kev for SPC2 is 40% (0.8 per degree of freedom).

The fit to a thermal bremsstrahlung emission give a poorer χ^2 of 2.6 for 25 degrees of freedom and the corresponding temperature obtained is $KT = 57^{+46}_{-21}$.

Figure 2 shows the combined spectra (SPC1 + SPC3) compared with the Spacelab 2 and the CALTEC data. Our measurement shows the same intensity of the Spacelab 2 data in the overlapping region (15-30 keV) while is inconsistent with the CALTEC 1988 because of the lower detected flux during 1989 corresponding to 60 mCrab at 100 keV.

Recently the source has been detected at a similar flux level of 1988 with the GRANAT satellite[20]

The low emission detected in May 1989 from the galactic center region imply a flux variability (at 2σ level) at least of a factor of 2 at 100 keV over a time scale of one year. On the contrary there is no strong indication of spectral slope change.

The complex spectrum measured by our experiment also suggest the presence of an emission feature around 58 keV (4σ level) with an intensity of $\simeq 9 \times 10^{-4}$ph $cm^{-2}s^{-1}keV^{-1}$, or an absorption feature at $\simeq$47 keV. The data analysis of detector SPC4 is underway.

Continuum monitoring of the region and spatially resolved high sensitivity observations, performed with high spectral resolution are necessary to clarify the intrigued scenario of the Galactic Center region and to confirm the behaviour of 1E1740.7-29.42 that at the moment seems to be the main high energy emitter in the central part of the Galaxy.

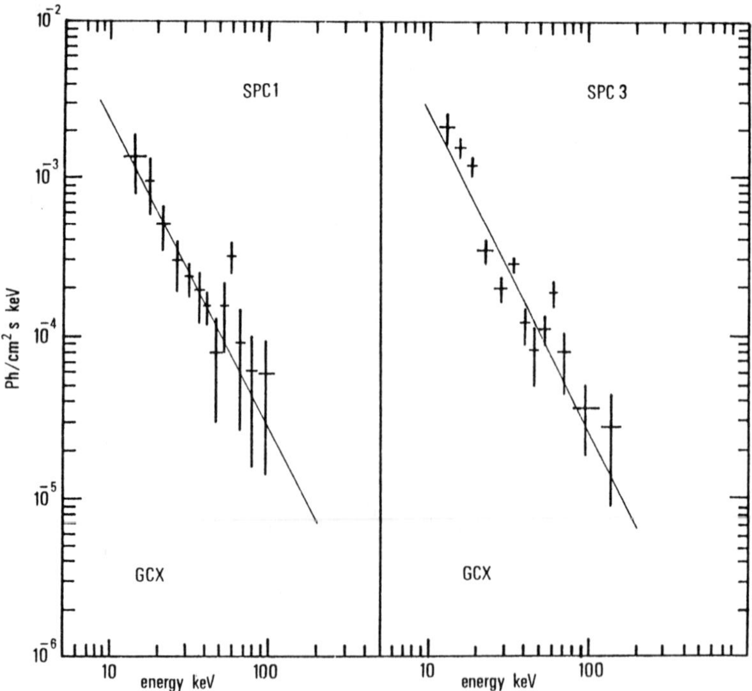

Figure 1. The GCX Spectrum reduced at the top of the atmosphere

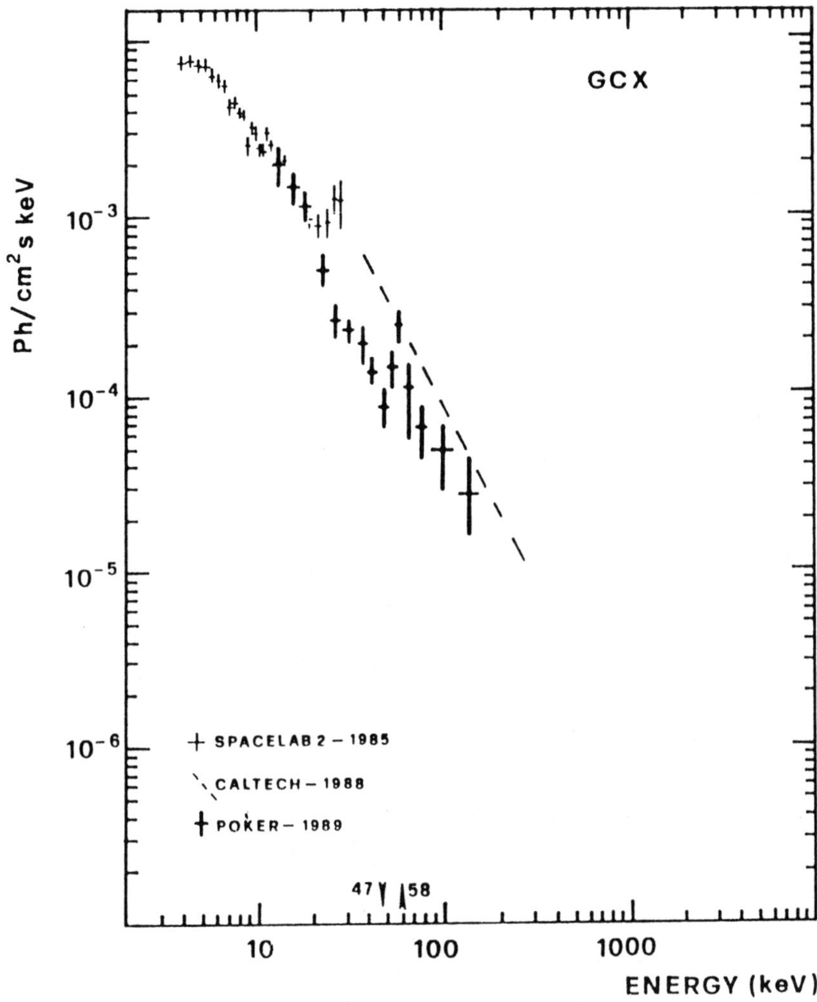

Figure 2: "POKER" data compared with the Spacelab 2 data.

Acknowledgements

We thank the technical staff of I.A.S. for the decennal support to this project and in particular during the last stressing campaign.

We also are grateful to the NSBF/NASA crew, led by R. Kubara, for the flight operations conducted with high professionalism during a long and difficult campaign.

REFERENCES

1. Skinner, G.K., *et al.*, 1987, Nature, **330**, 544
2. Hertz P. and Grindlay J.E., 1987, Ap. J., **278**, 137
3. Cook, W.R., *et al.*, 1988, Proc. of the IAU Symp. N. 136
4. Cook, W.R., *et al.*, 1990, Proc. of 21st I.C.R.C., **1**, 217
5. Matteson, J.L., 1982, The Galactic Center, Eds.Riegler and Blandford, AIP, New York, **109**, 122
6. Riegler, G.L., *et al.*, 1981, Ap. J., **248**, L13
7. Leventhal, M., *et al.*, 1982,Ap.J., **260**, L1
8. Riegler, G.L., *et al.*, 1985, Ap.J., **294**, L13
9. Leventhal, M., *et al.*,1989, Nature, **339**, 36
10. Matteson, J.L., *et al.*, Proc. of the 21st ICRC, **2**, 174
11. Niel, M., *et al.*, 1990, Ap. J.Letters, **356**, L21
12. Lingefelter, R.E. and Ramaty R., Ap.J., **343**, 689
13. Ling, J.C. and Wheaton, W.A., Proc.of GRO Science Workshop Eds. Johnson W.N., 1989, **4**, 282
14. Bazzano, A. *et al.*, 1983, Nucl. Instr. & Meth., **214**, 481.
15. Ubertini, P. *et al.*, 1985, SPIE, **579**, 74.
16. Ubertini, P., Bazzano., A., Sood, R., Staubert, R., Sumner, T.J. and Fry, G., 1989, Ap.J. (Letters), **337**, L19
17. Bazzano, A., *et al.*, 1989, SPIE, **1159**, 126
18. Ubertini, P., *et al.*, 1990, Submitted to Ap.J.
19. Lampton, M.,Margon,B.,and Bowyer,S., 1976, Ap. J.(Letters), **208**, 177
20. Paul, J., *et al.*, 1991, these proceedings

POSITRON ANNIHILATION RADIATION FROM THE GALACTIC CENTER REGION

R. Ramaty
NASA/Goddard Space Flight Center, Greenbelt, MD 20771, USA

R. E. Lingenfelter
University of California, San Diego, La Jolla, CA 92093, USA

ABSTRACT

Observations show that there are two components of positron annihilation radiation from the region of the Galactic Center: a variable component resulting from one or just a few compact sources at or near the Galactic Center and a steady, diffuse component resulting from positron annihilation in the galactic disk. We model the diffuse component using the observed longitude distributions of 70–150 MeV gamma rays, CO, and hot plasma revealed by Fe line emission.

We review recent results on positron annihilation in the interstellar medium and discuss the implications of the annihilation processes on the fraction of positrons annihilating via positronium and on the shape of the 511 keV annihilation line. We also review the sources of diffuse galactic positrons and discuss the nature of the compact source of annihilation radiation near the Galactic Center.

INTRODUCTION

Positron annihilation radiation from the direction of the Galactic Center has been observed since 1970. It was first measured[1,2,3] by Haymes and coworkers in a series of balloon flights in Brazil, but it was not until 1977 that the line energy was accurately determined[4] by Leventhal *et al.* with a high resolution Ge instrument flown on a balloon from Australia. The observed line-center energy, 510.7±0.5 keV, clearly established that the radiation was due to the annihilation of positrons. We list in Table 1 all of the annihilation radiation observations from the direction of the Galactic Center to-date, except the SIGMA observation, which was first reported at this meeting and is presented elsewhere in this volume. Positron annihilation radiation has also been observed from solar flares[20], gamma ray bursts[21], and possibly from the black hole candidate Cygnus X-1[22], but in the present paper we deal only with the Galactic Center observations.

Evidence that the 511 keV line flux varies with time was first provided by the HEAO-3 observations[8], which showed that the line flux decreased by about a factor of 3 in less than 0.5 years. The variable nature of the emission was confirmed with balloon-borne detectors, which showed that the 511 keV line flux decreased[11,12,13] after 1979 and increased[16] between 1984 and 1988. These time variations require[23,24] one, or at most a few compact, variable sources in the central region of the Galaxy. The balloon observations constrain the position of these sources to within 15° of the Galactic Center, while the HEAO-3 data further constrain their position to within 4° of the center. The recent SIGMA results (this volume) have shown that the dominant source at the present epoch is the hard X-ray source 1E1740.7-2942 located about 120 pc from the Galactic Center[25,26].

TABLE 1
Annihilation Radiation Observations from the Galactic Center Region

Date	511 keV Line Flux† (10^{-3} ph/cm^{2}s)	Name	FOV (FWHM)	References
Direction of the Galactic Center, $l = 0°$				
1970.90	0.67 ±0.19	RICE	24°	Johnson, Harnden and Haymes[1]
1971.89	0.67 ±0.19	RICE	24°	Johnson and Haymes[2]
1974.25	0.8 ±0.23	RICE	13°	Haymes *et al.*[3]
1977.13	4.18 ±1.56	CESR	50°	Albernhe *et al.*[5]
1977.86	1.22 ±0.22	BELL-SANDIA	15°	Leventhal, MacCallum and Stang[4]
1977.89	4.0 ±0.6	UNH	100°	Gardner *et al.*[6]
1979.29	2.35 ±0.71	BELL-SANDIA	15°	Leventhal *et al.*[7]
1979.8	1.85 ±0.21	HEAO-3	35°	Riegler *et al.*[8]; Mahoney[9]
1980.2	0.65 ±0.27	HEAO-3	35°	Riegler *et al.*[8]; Mahoney[9]
1981–1987	$2.7^{+0.6}_{-0.7}$	SMM	130°	Share *et al.*[10,14], Harris *et al.*[15]
1981.89	0.0 ±0.6	GSFC	15°	Paciesas *et al.*[11]
1981.89	0.0 ±0.38	BELL-SANDIA	15°	Leventhal *et al.*[12]
1984.89	0.0 ±0.5	BELL-SANDIA	15°	Leventhal *et al.*[13]
1988.33	0.75 ±0.17	GRIS	20°	Leventhal *et al.*[16]; Gehrels[17]
1988.83	1.18 ±0.16	GRIS	20°	Leventhal *et al.*[16]; Gehrels[17]
1988.90	2.24 ±0.41	FIGARO	77°	Niel *et al.*[18]
1989.39	0.89 ±0.27	HEXAGONE	20°	Matteson *et al.*[19]
Direction $l = -25°$ in the Galactic Plane				
1988.83	0.18 ±0.11	GRIS	20°	Leventhal *et al.* [16]; Gehrels[17]

†The listed fluxes were derived assuming emission from a point source

A correlation between the observed 511 keV fluxes from the direction of the Galactic Center and the fields of view of the detectors was first pointed out[27] by Dunphy *et al.* and subsequently demonstrated[10] by Share *et al.* using their SMM observations. As suggested by Dunphy *et al.*, this correlation requires the existence of a spatially distributed galactic source of 511 keV emission. We have proposed[28,29] a two component model, in which the data can be interpreted in terms of the superposition of emission from such a diffuse source and a variable, compact source at or near the Galactic Center. Annihilation radiation, probably due to the diffuse component, was detected[16] with the GRIS instrument (Table 1) pointing in the galactic plane at $l = -25°$.

We first review the two component model in the light of the most recent observations. We then present the results of a recent study of positron annihilation in the interstellar medium and its implications on the fraction of positrons annihilating via positronium and on the shape of the annihilation line. Next, we review the positron sources that are thought to be responsible for the diffuse galactic emission and we discuss the nature of the compact source of annihilation radiation near the Galactic Center.

A TWO COMPONENT MODEL

We have plotted in Figure 1 the various observed 511 keV fluxes from the direction of the Galactic Center versus the fields of view of the detectors (Table 1). We have discussed these data in our previous summary[29], except for the BS/G (BELL-SANDIA/GSFC) upper limit, which is a 2σ bound based on the two 1981.89 and the 1984.89 balloon observations, and the more recent SMM[15], GRIS[17], FIGARO[18], and HEXAGONE[19] results. For display purposes, we have plotted the HEXAGONE flux at 22°. All of the fluxes shown in Table 1 and plotted in Figure 1 have been obtained by assuming a single point source in the fields of view of the instruments. The correlation of the observed fluxes with detector field of view is quite evident, particularly when the high fluxes represented by the broad field of view SMM, UNH and FIGARO observations are compared with the narrow field of view BS/G upper limit. The time variations are also evident, as can be seen by comparing observations with similar instruments at different times, in particular the high BELL-SANDIA 1977.86 and 1979.29 fluxes versus the low BS/G limit at 15°, and the two HEAO-3 observations (1979.8 and 1980.2) at 35°. The HEAO-3, SMM and FIGARO observations have also been analyzed[9,15,18] under the assumption that the observed emission originates from a distributed source, or from the superposition[18] of a point source and a distributed component. We present these results in Table 2. As can be seen, for the HEAO-3 results, the fluxes derived from this analysis, which is not restricted to a 35° field of view around the Galactic Center, yield a smaller variation than that shown in Figure 1.

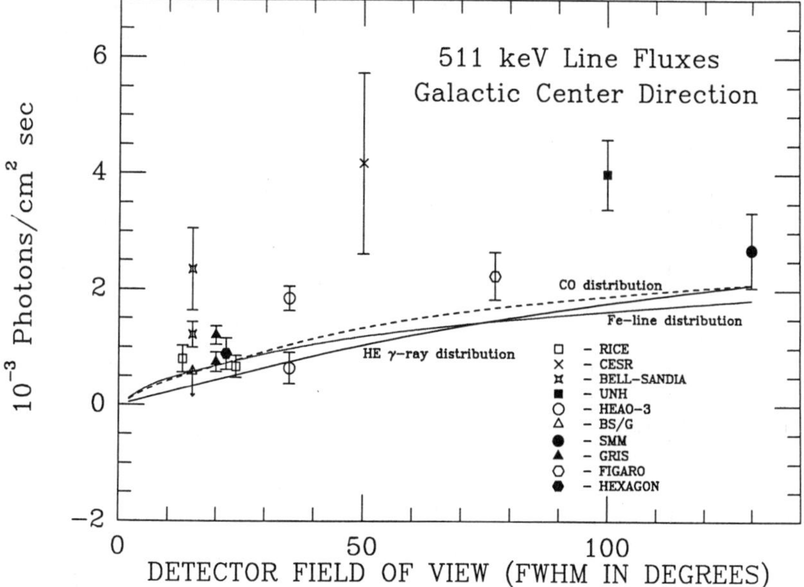

Fig. 1. Observed and calculated line fluxes versus detector fields of view

TABLE 2
Diffuse 511 keV Line Fluxes from the Central Radian of the Galaxy in 10^{-3} Photons cm^{-2} sec^{-1} rad^{-1}

HEAO-3 fall[9] (1979.8) (COS B distribution)	1.50 ± 0.33
HEAO-3 spring[9] (1980.2) (COS B distribution)	1.23 ± 0.29
SMM[15] (CO distribution)	$2.3^{+0.5}_{-0.8}$
FIGARO[18] (CO distribution)	1.91 ± 0.32
FIGARO[18] (point source + CO distribution)	1.23 ± 0.24
COS B Distribution	1.2
CO Distribution	1.6
Fe-Line Distribution	1.4

The other values shown in Table 2 are the result of our present calculations which follow a procedure similar to that used in our previous analysis[29]. For the dependence of the diffuse 511 keV line emission on galactic longitude we adopted: (1) the COS B (70–150 MeV) gamma ray distribution[30]; (2) the CO distribution[31]; and (3) the recently discovered[32] distribution of hot plasma revealed by Fe-line emission. The calculated diffuse 511 keV fluxes are given by

$$\Phi_{511}(\theta, l) = \int_{-\pi}^{+\pi} dl' \phi(l') r(\mid l' - l \mid, \theta), \tag{1}$$

where l is the longitude of the direction of observation, $\phi(l')$ is the assumed distribution of the 511 keV emission per radian of galactic longitude, and r is the detector response function assumed to have a simple triangular shape, $r(\mid l' - l \mid, \theta) = 1 - \mid l' - l \mid / \theta$ with θ the full width at half maximum. In Figure 1 we show Φ_{511} as function of θ for $l = 0°$, while in Figure 2 we show Φ_{511} as a function of l for two values of θ. In both figures the $\phi(l')$ functions have the same normalization for each assumed longitude dependence. We have normalized $\phi(l')$ such that the predicted diffuse emissions equal the lowest observational 511 keV flux upper limits (2σ). For a diffuse emission which is independent of time, such a normalization yields the highest allowed diffuse flux. Thus, the COS B and CO distributions are normalized to the 1988.83 upper limit of 0.4×10^{-3} photons cm^{-2} sec^{-1} at $l = -25°$ (Figure 2, upper panel). This normalization ensures consistency with the BS/G upper limit, as can be seen in the lower panel. Because of its sharper peak at the Galactic Center, the Fe-line distribution has to be normalized to the BS/G upper limit at $l = 0°$ (lower panel). As seen in the upper panel, this normalization ensures consistency with the 1988.83 2σ limit at $l = -25°$.

As is evident from Figure 1, all of the observed 511 keV fluxes are, within errors, either equal to or greater than the calculated diffuse fluxes. This result is consistent with the basic premise of the two component model, i.e. that the observed fluxes are the superposition of contributions from a steady diffuse component and a variable source. The need for two components is more clearly illustrated in Figure 2, where a subset of the data (corresponding to narrow field of view Ge detector observations) are compared with the calculated diffuse emission. The (1977.86;1979.29) flux (lower panel) is the combined mean flux for these two observations. It is quite evident that this flux, and the 1988.83 GRIS flux (upper panel) must have contained very significant contributions from the variable source. On the other hand, the 1988.33 GRIS flux and the 1989.39 HEXAGONE flux are consistent with the diffuse component, at least for two of our assumed models.

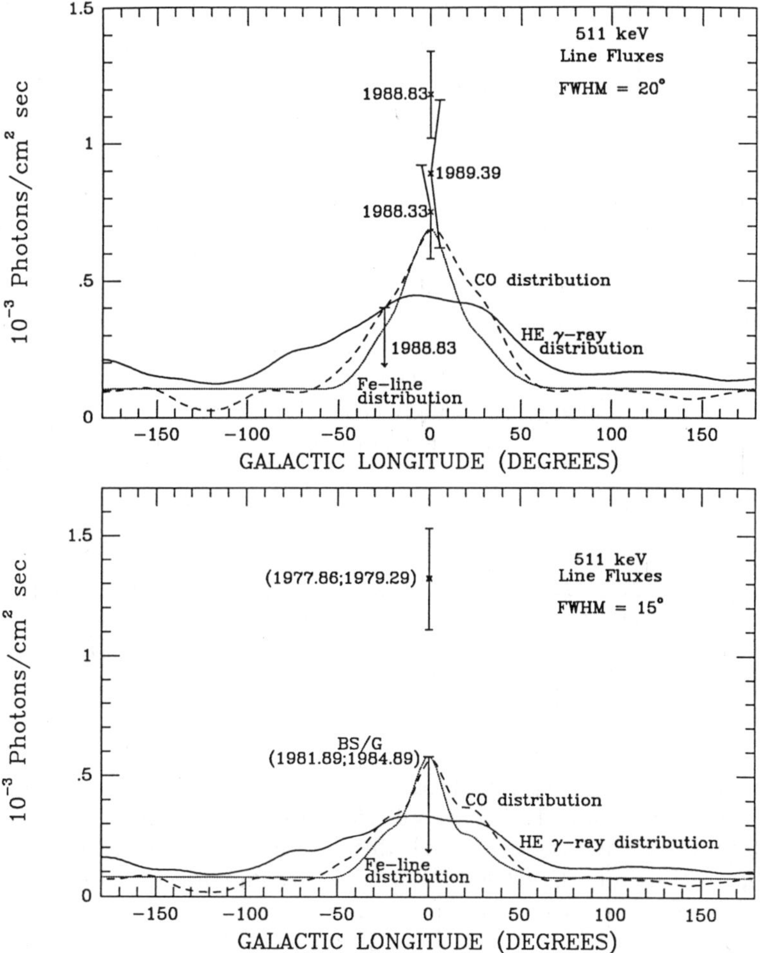

Fig. 2. Observed and calculated line fluxes versus galactic longitude

The 511 keV fluxes from the central radian of the Galaxy implied by the normalizations of the three assumed longitude distributions are given in Table 2. We see that these results are consistent with the other measurements of the diffuse flux for similar assumed distributions. Although the SMM flux for a CO distribution is nearly twice as large as the spring 1980 HEAO-3 flux for a COS B distribution, the HEAO-3 value is nearly identical to our value for a COS B distribution, and the SMM value is consistent within 1σ of both our value for a CO distribution and the FIGARO diffuse flux for a CO distribution alone. Moreover, if both a point source and a distributed source is assumed, the FIGARO diffuse flux is greatly reduced, becoming identical to the spring 1980 HEAO-3 flux. We suggest that because of its very broad field of view and very long integration time (several months), the SMM flux may also contain significant transient contributions from the compact, variable source or sources near the Galactic Center.

For our deduced normalizations, a galactic diffuse source scale height of 60 to 200 pc, a distance[33] to the Galactic Center of 7.7 ±0.7 kpc, and a galactic disk radius of 15 kpc, the total 511 keV luminosity of the Galaxy is (1.4 ±0.2) $\times\ 10^{43}$ photons sec^{-1} (from a calculation by J. Skibo, private communication 1991). For a positronium fraction of 0.9 ±0.1 (see below), this implies a total diffuse galactic positron annihilation rate of (2.1 ±0.6) $\times 10^{43}$ sec^{-1}. But because we have normalized the assumed longitude distributions to upper limits on the 511 keV line flux (the 1988.83 observation in the galactic plane at $l = -25°$ and the BS/G limit), this annihilation rate could be lower. We discuss below the potential sources of the diffuse annihilation radiation, and we estimate lower limits on the total diffuse galactic annihilation rate.

TABLE 3
Galactic Center 511 keV Line Fluxes
(10^{-3} Photons cm^{-2} sec^{-1})

Date	Observed	Diffuse Component			Compact Source		
	Flux	HE	CO	Fe	HE	CO	Fe
1977.86	1.22 ±0.22	0.33	0.56	0.58	0.89 ±0.22	0.66 ±0.22	0.64 ±0.22
1979.29	2.35 ±0.71	0.33	0.56	0.58	1.59 ±0.71	1.30 ±0.71	1.48 ±0.71
1979.8	1.85 ±0.21	0.76	1.05	0.98	1.09 ±0.21	0.80 ±0.21	0.87 ±0.21
1988.33	0.75 ±0.17	0.44	0.68	0.69	0.31 ±0.17	0.07 ±0.17	0.06 ±0.17
1988.83	1.18 ±0.16	0.44	0.68	0.69	0.74 ±0.16	0.50 ±0.16	0.49 ±0.16
1989.39	0.89 ±0.27	0.44	0.68	0.69	0.45 ±0.27	0.21 ±0.27	0.20 ±0.27

Having established three models for the diffuse 511 keV emission, we evaluate fluxes from the compact source by subtracting the contribution of the diffuse emission from the total observed fluxes. Results for the narrow field of view Ge balloon observations and the HEAO-3 fall data are summarized in Table 3, where the total observed fluxes (obtained by assuming emission from a point source) are shown together with the expected contributions of the diffuse component for the three models. We see that for the 1977.86 and 1988.83 observations, there is a statistically significant 511 keV line flux from the compact object for all three assumed diffuse component models. The 1979.8 HEAO-3 observation also yields a significant compact source flux, although this result depends[9] on whether a point source or a diffuse source was assumed. The 1988.33 and 1989.39 observations do not indicate significant emission from the compact source, unless the diffuse emission is less than 1/2 of the upper bounds assumed here. Because of its limited statistics, the 1979.29 observation is inconclusive. Not shown in Table 3 is the 1988.90 broad field of view FIGARO observation, which yielded[18] a significant compact source flux of (0.86 ±0.28) $\times 10^{-3}$ photons cm^{-2} sec^{-1}. Two additional broad field of view observations, those of 1977.13 and 1977.89, yielded large fluxes (Table 1) which probably included contributions from the compact source. Thus, this source appears to have been active between 1977 and 1979, it transited into a low state in early 1980, was inactive on 1981.89 and 1984.89, and was found to be active again on 1988.83. There may have also been transient activity at intermediate times, including periods of SMM observations as suggested above. Since the HEXAGONE flux on 1989.39, for at least two of our three models, appears to be consistent with the diffuse component, the compact source may have become again inactive on this occasion, although this conclusion depends on actual flux in the diffuse component whose value is still uncertain.

A line flux of $\sim 10^{-3}$ photons cm^{-2} sec^{-1}, characteristic of the high state of the compact object, implies the production of $\sim 7 \times 10^{42}$ line photons sec^{-1} at the source, assumed to be at a distance[33] of 7.7 kpc. This requires the annihilation of $\sim 7 \times 10^{42}/(2 - 1.5f)$ positrons sec^{-1}, where $0 < f < 1$ is the positronium fraction (see below).

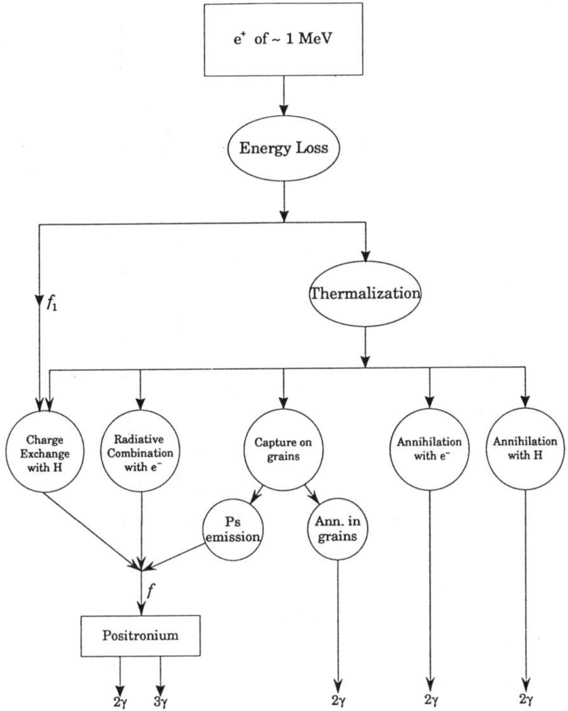

Fig. 3. Positron annihilation paths

POSITRONIUM FRACTION

The hallmark of positron annihilation via positronium is the characteristic 3-photon continuum at energies < 511 keV due to orthopositronium annihilation. This 3-photon continuum, derived by fitting the observed continuum below and above 511 keV, yields the fraction f of positrons annihilating via positronium, which, in turn, can provide information on the properties of the annihilation site. Before analyzing the observational results, we review the annihilation processes[34,35,36].

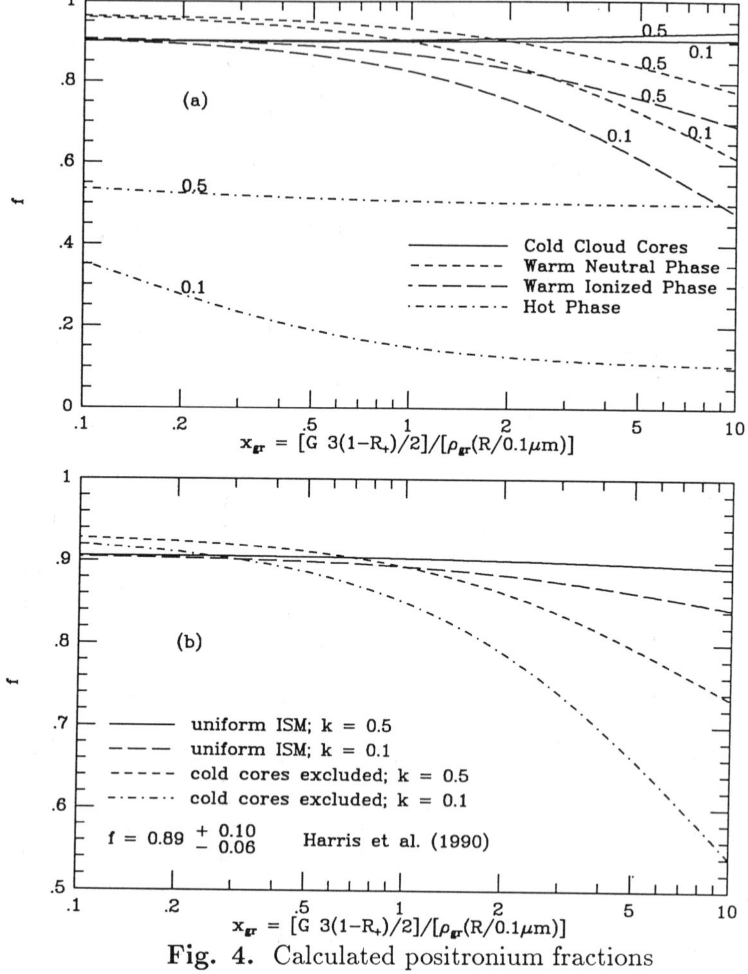

Fig. 4. Calculated positronium fractions

The processes leading to gamma-ray production from positron annihilation are shown schematically in Figure 3 (from ref. 36), where the ambient medium is assumed to be a partially ionized H gas with heavy elements condensed into dust grains. If the initial energy of the positrons is around 1 MeV, as expected for most positron production mechanisms (see below), then the energy loss is almost exclusively due to Coulomb collisions. The positrons then, having lost the bulk of their energy, either form positronium in flight or become thermalized with the electrons in the gas. The thermalized positrons can annihilate directly with both free and atomic electrons, they can form positronium by charge exchange with hydrogen atoms and radiative combination with free electrons, and they can interact with dust grains[35]. (The effects of dust are negligible for positronium formation in flight.) Capture of positrons by grains can lead to the annihilation of the positrons within the grains, or to the formation and subsequent escape of positronium back to the gas. Direct annihilation produces two line photons at 511 keV. Positronium is formed in either the para or the ortho states. Parapositronium decays into two 511 keV line photons, while orthopositronium decays into a 3-photon continuum below 511 keV.

If the ambient gas density, radiation, and magnetic field are sufficiently low, the ionization, excitation, and spin flip of positronium formed in the gas can be ignored so that once positronium is formed in a given state, it will always annihilate from that state. These conditions are well satisfied in the interstellar medium, but not necessarily in the vicinity of a compact object. Assuming that once formed positronium will decay, the ratio of the orthopositronium continuum flux to the flux in the 511 keV line is $2.25f/(2-1.5f)$, where f is the fraction of annihilations occurring via positronium.

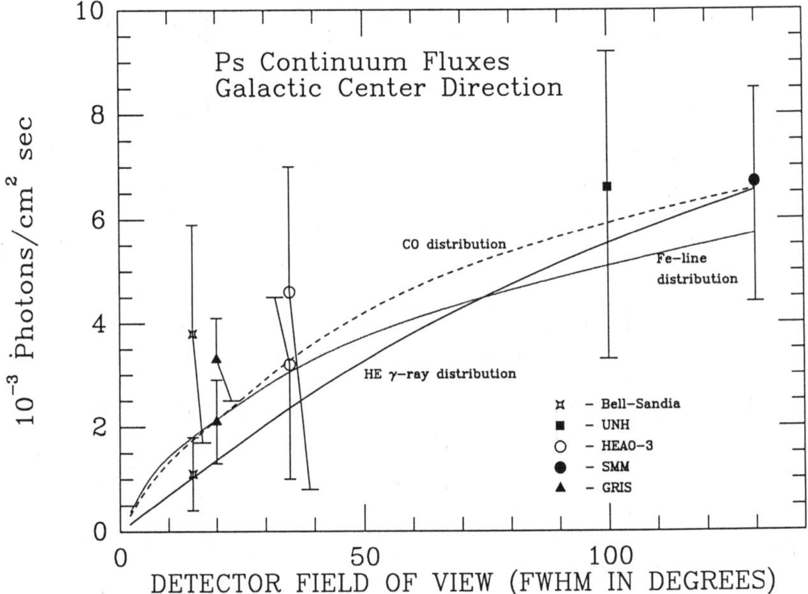

Fig. 5. Orthopositronium continuum observations and calculations

Values of f have been calculated for various assumptions on the ambient medium in which the positrons annihilate and the effects of grains on the annihilation process. The results[36] are shown in Figure 4. Panel (a) pertains to four distinct interstellar medium phases[37]: cold cloud cores, warm neutral and warm ionized cloud envelopes, and hot intercloud gas. In the cores, T = 80K, the gas is neutral and its density is 42 cm^{-3}; in the warm envelopes T = 8000K, the ionization fractions are 0.15 and 0.68, and the densities are 0.37 and 0.25 cm^{-3} for the neutral and ionized cases, respectively; in the hot gas T = 4.5×10^5 K, the gas is fully ionized and the density is 3.5×10^{-3} cm^{-3}. While these parameters are appropriate to a particular model of the interstellar medium[37], they do span a sufficiently broad range of values for the results to be applicable to a large variety of astrophysical conditions. In panel (a), for each phase there are two curves, corresponding to two values of the fraction k of the positrons captured on grains, that escape back to the gas as positronium. This fraction is uncertain, but values of 0.1 to 0.5 span the range of the observational results (see ref. 36). The parameter x_{gr} depends on the ratio G of the average galactic metallicity to its local value, on the probability R_+ that a positron is reflected by a grain, and on the density ρ_{gr} and radius R of the grains. We expect that in the interstellar medium, the value of x_{gr} will be close to 1.

TABLE 4
Positronium Fractions and Orthopositronium Continuum Fluxes (in 10^{-3} Photons cm^{-2} sec^{-1})

Date	f	Ps. Cont.	Calculated Diffuse Ps. Cont. HE	CO	Fe
Direction of the Galactic Center, $l = 0°$					
1977.86	0.9 ± 0.15	3.8 ± 2.1	1.03	1.75	1.81
1977.89	$0.7^{+0.12}_{-0.14}$	$6.6^{+2.6}_{-3.0}$	5.53	5.92	5.09
1979.29	0.5 ± 0.15	1.1 ± 0.7	1.03	1.75	1.81
1979.8	$0.71^{+0.13}_{-0.23}$	$3.2^{+1.3}_{-2.2}$	2.37	3.28	3.06
1980.2	$1.1^{+0.06}_{-0.14}$	$4.6^{+2.4}_{-3.8}$	2.37	3.28	3.06
1981–1987	$0.89^{+0.10}_{-0.06}$	$6.7^{+1.8}_{-2.3}$	6.54	6.57	5.71
1988.33	0.99 ± 0.08	3.3 ± 0.8	1.37	2.14	2.14
1988.83	0.72 ± 0.13	2.1 ± 0.8	1.37	2.14	2.14
Direction $l = -25°$ in the Galactic Plane					
1988.83	1.21 ± 0.08	2.6 ± 0.7	1.25	1.25	1.03

The positronium annihilation fraction shown in panel (b) of Figure 4 has been obtained[36] by combining the annihilation radiation from the four phases under two assumptions: (1) that the positrons are penetrating freely into the cloud cores (uniform ISM), and (2) that they are prevented from penetrating because the effects of interstellar magnetic fields. The relative contributions of the phases were obtained by considering the probabilities of positron annihilation in each one of them, which depend on the ambient densities and the filling factors. In the uniform ISM case, the probabilities are 0.83, 0.09, 0.08 and 0.002 for the cores, the warm ionized and neutral media, and the hot gas, respectively. If the positrons are excluded from the cores, the probabilities are 0.54, 0.45 and 0.01 for the warm neutral, warm ionized and hot phases, respectively. We see that in the uniform ISM case the positrons are expected to annihilate predominantly in the cold cores. In this case, as already mentioned, capture on the grains can be ignored, and f is expected to be close to 0.9 for essentially all values of x_{gr} and k. If the positrons are excluded from the cores but still allowed to penetrate the envelopes, most of them are expected to annihilate in the envelopes. In this case f is still going to be around 0.9 for a broad range of grain parameters. However, for values of x_{gr} larger than about 2, f can be significantly less than 0.9.

The observationally determined orthopositronium continuum fluxes and the corresponding values of f are summarized in Table 4. We have discussed these data in our previous summary[29], except for the 1981-1987, 1988.33 and 1988.83

values. The 1981-1987 values are the average SMM results[15] for these years. The 1988.33, and the 1988.83 $l = 0°$ and $l = -25°$ values have been obtained from the most recent analysis[17] of the GRIS observations. All of these orthopositronium results have been obtained by attributing the excess flux above an assumed (most often power law) continuum at energies $\leq$ 511 keV to orthopositronium annihilation. It was first pointed out by Forrest[38] that the excess could also result from Compton scattering in the source. While Compton scattering in the vicinity of a compact object can be quite important[39], for positron annihilation in the interstellar medium it is more likely that the excess will be produced by orthopositronium annihilation. The orthopositronium continuum fluxes for $l = 0°$ are plotted in Figure 5 versus detector fields of view. The GRIS orthopositronium flux observed from $l = -25°$ is plotted in Figure 6 together with the $l = 0°$ orthopositronium observations with the same detector.

We have calculated the orthopositronium continuum fluxes expected from the diffuse component for the three assumed galactic distributions,

$$\Phi_{3\gamma}(\theta, l) = 3.12\ \Phi_{511}(\theta, l), \tag{2}$$

where $\Phi_{511}(\theta, l)$ is given by Equation (1) and $2.25f/(2 - 1.5f) = 3.12$ is the ratio of the orthopositronium continuum flux to the 511 keV line flux for a positronium fraction $f = 0.9$. The results are given in Figures 5 and 6, and in Table 4.

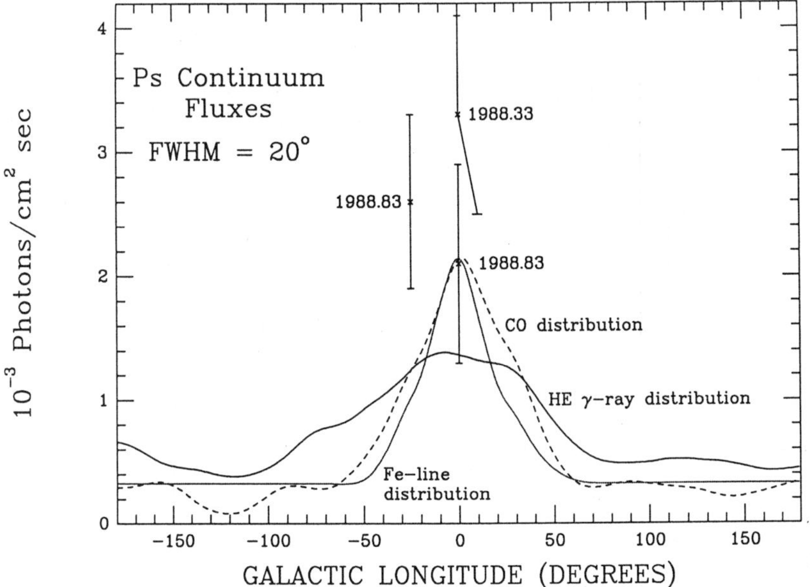

Fig. 6. Orthopositronium continuum fluxes versus galactic longitude

For the Galactic Center $l = 0°$ observations (Figure 5) the calculated fluxes are consistent with the data, indicating that the bulk of the observed orthopositronium continuum could originate from the diffuse component[29] and that the annihilation radiation from the compact source could have a low or even vanishing positronium fraction[36]. However, the uncertainty in the data is large enough (Figure 5) to allow the compact source to also contribute to the observed excess continuum below 511 keV, either by orthopositronium annihilation of Compton scattering. A potential problem is the orthopositronium continuum flux observed at $l = -25°$ (Table 4 and Figure 6), as this flux exceeds the calculated flux for all three models, and the difference is significant at the 2σ level for the assumed diffuse component normalizations, and even more significant if the diffuse emission is lower. It is possible that the discrepancy could be attributed to Compton scattered annihilation radiation from one or several compact sources different from the compact object near the Galactic Center. We are looking forward to new observations to resolve this issue.

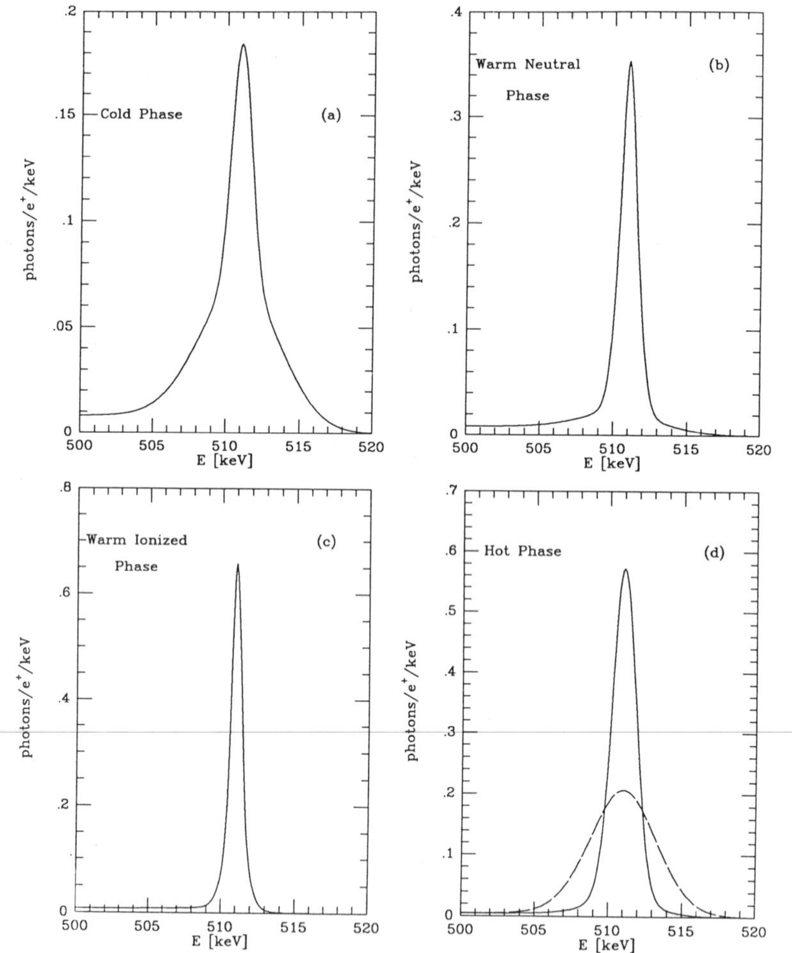

Fig. 7. Annihilation line shapes for the phases of the interstellar medium. Solid curves—x_{gr}=1 k=0.5; dashed curve—x_{gr}=0.

ANNIHILATION LINE SHAPE

The annihilation processes discussed in the previous section also affect the shape of the 511 keV line. Results of detailed calculations[36] for the interstellar medium are shown in Figures 7 and 8, and Table 5. Figure 7 shows annihilation spectra for the cold, warm neutral, warm ionized, and hot phases for various values of x_{gr} and k. All of these spectra are normalized to 1 positron, i.e. the total number of photons contained in the spectrum (extended to zero photon energy) equals $2+0.75f$. We see that in the cold phase (panel a) the effects of the grains are negligible and the spectrum shows the narrow line due to direct annihilation with bound electrons on top of the broader line due to positronium formation and annihilation in flight. For the warm neutral and warm ionized cases (panels b and c) the grains become important only when $k = 0.1$ and $x_{gr} > 1$, but, independent of the values of these parameters, the annihilation line is very narrow. For the hot phase (panel d) the line is quite broad if the effects of the grains are ignored, but it becomes narrower if these effects are included. The full widths at half maximum (FWHM) derived from these spectra are shown in Table 5. The combined spectra of the uniform ISM and the cold cores excluded models are shown in Figure 8, and the derived widths are also given in Table 5. We see that there is a very significant difference between the shape of line for the uniform ISM and cold cores excluded models (compare panels a and b in Figure 8). When the positrons can penetrate the cloud cores, the line shows the broad base resulting from positronium formation in flight, and this broad feature is absent when the positrons are excluded from the cores.

TABLE 5

Values of FWHM and (EFW) (in keV)

x_{gr}	0	0.1	1	2	10
k	–	0.5	0.5	0.1	0.1
Cold Phase	2.17 (5.12)	2.18 (5.12)	2.23 (5.16)	2.24 (5.13)	2.39 (5.16)
Warm Neutral Phase	1.28 (1.91)	1.29 (1.92)	1.36 (1.97)	1.46 (1.94)	1.68 (1.96)
Warm Ionized Phase	0.82 (0.97)	0.83 (0.99)	0.88 (1.14)	0.97 (1.33)	1.35 (1.67)
Hot Phase	5.19 (5.33)	2.32 (3.59)	1.93 (2.13)	1.83 (1.89)	1.82 (1.83)
Uniform ISM	1.60 (4.59)	1.60 (4.58)	1.66 (4.56)	1.72 (4.39)	1.91 (4.08)
Cold Cores Excluded	0.95 (1.46)	0.96 (1.44)	1.02 (1.54)	1.15 (1.63)	1.51 (1.80)

In order to better quantify the difference between the various line shapes, we present an effective full width (EFW), in addition to the more generally used FWHM. In a Gaussian, 76% of the line flux is contained within ±FWHM/2 of the line center. In analogy to this, we define EFW such that the same fraction is contained within ±EFW/2. Clearly, in a Gaussian EFW = FWHM. But for the composite lines resulting from annihilation in the ISM, a smaller fraction of the total flux is contained within ±FWHM/2 than within ±EFW/2, because for these lines the FWHM is determined by a narrow component sitting on top of the broader line from positronium formation in flight. Values of EFW are shown in Table 5 for all four phases of the ISM, as well as for the uniform ISM and cold cores excluded models. We see that EFW is greater than FWHM for all cases, with the difference being particularly significant for the cold phase and the uniform ISM case. Indeed, for these two cases single-Gaussian functions provide poor fits to the calculated spectra. Therefore, the comparison of FWHM's derived from data using Gaussian fits with calculated FWHM's is not the best way to set strong constraints on the models. Ideally, the observed line profile should be compared with the calculations, but in lieu of a detailed fitting, the comparison of the EFW's might also provide useful information.

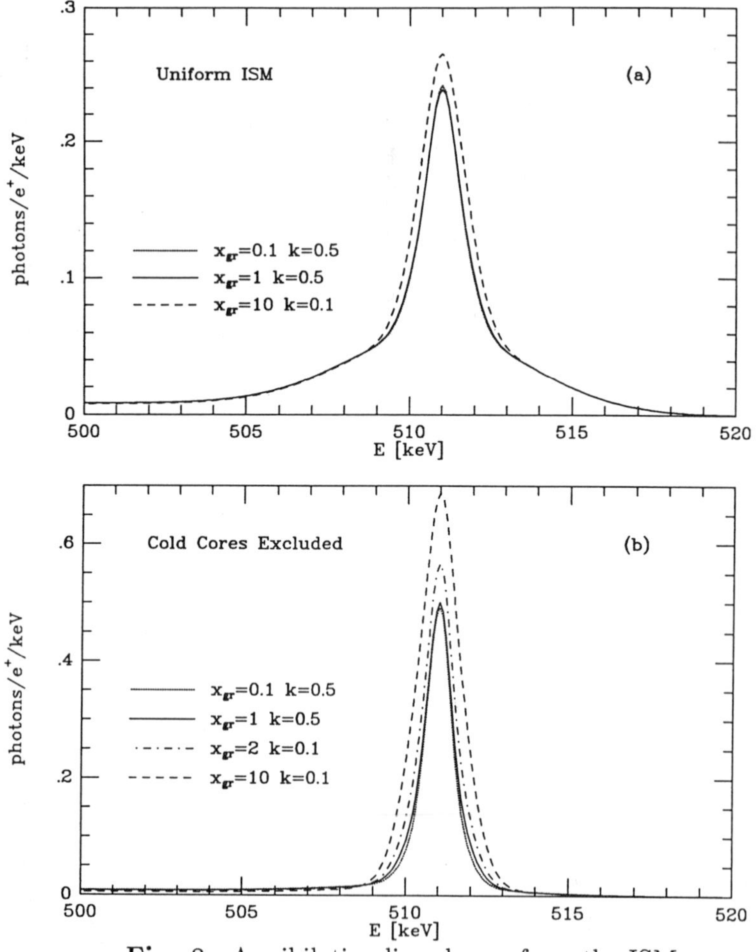

Fig. 8. Annihilation line shapes from the ISM

Currently, there are only upper limits on the widths of the 511 keV line, except for the 1988.83 GRIS observation for which a Gaussian FWHM of 2.9 $\pm$0.6 keV has been obtained[17]. The most stringent limit (obtained by assuming a single Gaussian line shape) is that for the 1979.8 HEAO-3 observations, yielding a width (FWHM) less than 3.4 keV (2σ). By comparing this limit with the FWHM's listed in Table 5, we can only rule out annihilation in the hot phase with no grains. But this is not surprising, since, as we have seen above, annihilation in this phase of the interstellar medium is quite improbable. However, future accurate measurements of the shape of the line from the diffuse component will distinguish between the uniform ISM and the excluded cold core models, as the former predicts a broader line characterized by significantly higher equivalent full width (Table 5). Such an observation can be carried out by observing annihilation radiation in the galactic plane from directions away from the galactic center. The discovery of a narrow line of EFW less than about 4 keV would rule out the uniform ISM model and would imply that the positrons are prevented from penetrating the cloud cores by interstellar magnetic fields.

Future observations should also provide much more information on the nature of the compact source. This source should be studied with detectors combining imaging with high spectral resolution which would determine the shape of the annihilation line without interference from the distributed component. We note that for the 1988.83 observation (the only observation for which the 511 keV line was resolved) only about 50% of the line emission originated from the compact source (Table 3).

TABLE 6

Sources for the Diffuse Galactic Annihilation Radiation

Source	Process	Rate (10^{43} e^+/sec)	Remarks
SN Ia	$^{56}Co(e^+)^{56}Fe$ (19%)	3.25 (ϵ/0.05) (g_I/0.5) $\dot{M}_{56}$	(1)
SN (all)	$^{44}Sc(e^+)^{44}Ca$ (95%)	1.04 $\dot{M}_{56}$	(2)
SN, Novae Wolf-Rayet Others	$^{26}Al(e^+)^{26}Mg$ (82%)	0.40 ±0.15	(3)
" "	$^{22}Na(e^+)^{22}Ne$ (91%)	< 0.4	(4)
Cosmic Rays	pp $\rightarrow \pi^+$...	<0.06	(5)
Low Energy Cosmic Rays	pCNO... $\rightarrow$ β emitters...	<0.2	(6)
Gamma Ray Bursts	pair production	<0.2	(7)
Pulsars	pair production	<0.1	(8)
Required Total		2.1 ±0.6	

(1) $\dot{M}_{56}$ = present rate of ^{56}Fe nucleosynthesis in $M_\odot$/100 years; g_I = fraction of ^{56}Fe produced by SN Ia ($\simeq 0.5$); ϵ = fraction of positrons escaping from SN Ia, ϵ assumed zero for all other SN
(2) Assuming all ^{44}Ca from ^{44}Sc and local $^{44}Ca/^{56}Fe = 1.6 \times 10^{-3}$, and all e^+ escape
(3) From observed[42] 1.809 MeV line flux of $(4.3 \pm 0.8) \times 10^{-4}$ photons cm^{-2} sec^{-1} rad^{-1}, assuming the same spatial distributions and $f = 0.9 \pm 0.1$
(4) From upper limits on the 1.275 MeV line and the analysis of ref. 43
(5) From observed 70–150 MeV gamma rays[30,44]
(6) From upper limit[45] on galactic 4.44 MeV line emission and calculation[46]
(7) From GRB model calculations[47]
(8) From pulsar model calculations[48]

SOURCES OF THE DISTRIBUTED COMPONENT

Although there are a variety of sources of positrons in the galactic disk, we have shown[40,41] that the dominant source should be positrons from the decay of radionuclei, resulting from processes of galactic nucleosynthesis. Our estimates of the average rates of positron production from various possible galactic sources are summarized in Table 6.

As can be seen, the most likely process[40,49,50,51] for producing these diffuse interstellar positrons is β^+ decay of ^{56}Co, ^{44}Sc and ^{26}Al made by various processes of nucleosynthesis. Perhaps the most important source is ^{56}Co produced in Type Ia supernovae. Because of the much greater mass and slower expansion velocity of the nebulae of Type II supernovae, the ^{56}Co decay positrons produced there should nearly all annihilate before the nebulae become transparent with only a negligible fraction escaping. The relative importance of ^{56}Co depends on the fraction ϵ of ^{56}Co decay positrons that can escape into the interstellar medium from a Type Ia supernova. This escape fraction is very uncertain with theoretical estimates[52,53] ranging as high as 0.1. However, the value of ϵ can also be vanishingly small. The contributions of both ^{56}Co and ^{44}Sc depend on the current rate of galactic nucleosynthesis, which is also uncertain. We estimate that $\dot{M}_{56}$ ranges between 0.2 to 0.7 $M_\odot$/100 years, based on the rate of galactic supernovae[54] and supernova model calculations[55,56]. On the other hand, the contributions of ^{26}Al is well determined (Table 6) from direct observations.

In Table 7 we present values for the total diffuse galactic positron production rate for several values of ϵ and $\dot{M}_{56}$. We see that the reasonable, intermediate values of $\epsilon = 0.05$ and $\dot{M}_{56} = 0.4$ can account for the required diffuse positron production rate given in Table 6. But as we have discussed earlier, this rate should be considered as an upper limit, because it was obtained by normalizing the assumed longitude distributions to the observational bounds on the 511 keV line flux (the 1988.83 observation at $l = -25°$ and the BS/G limit at $l = 0°$. Thus, the 2σ upper limit of 3.3 $\times 10^{43}$ positrons sec^{-1} (Table 6) requires that $\dot{M}_{56} < 2.9/(65\epsilon + 1.04)$ $M_\odot$/100 years. Since calculations[53] indicate that ϵ is probably not smaller than 0.01, this implies that $\dot{M}_{56} < 1.7$ $M_\odot$/100 years. This is an interesting limit, since it is less than the average production rate of ^{56}Fe over the lifetime of the Galaxy, ~ 2 $M_\odot$/100 years (obtained for a galactic population I mass of 10^{11} $M_\odot$).

TABLE 7
Estimates of the Total Diffuse Galactic Positron Production Rates

ϵ	$\dot{M}_{56}$ $M_\odot$/100 years	Rate (10^{43} e^+/sec)
0.1	0.7	5.7
0.1	0.2	1.9
0.05	0.4	2.1
0	0.7	1.1
0	0.2	0.6
0	0	0.4

The contribution of ^{26}Al (the case of $\epsilon = 0$ and $\dot{M}_{56} = 0$ in Table 7) provides an absolute lower limit to the total diffuse galactic positron production rate, which is lower than the required rate (Table 6) by about a factor of 5. But since the current rate of nucleosynthesis is not expected to be zero, a more reasonable lower limit is provided by the case $\epsilon = 0$ and $\dot{M}_{56} = 0.2$ (Table 7), which is also lower than the required production rate by about a factor of 3. Such a low contribution of nucleosynthesis would imply the existence of hitherto unknown sources to account for the large 511 keV line fluxes observed with the wide field of view SMM, CESR, UNH and FIGARO detectors (Table 1). A detailed mapping of the galactic disk in 511 keV line emission is needed to resolve these issues.

NATURE OF THE COMPACT SOURCE

We have considered[24,41] the various possible mechanisms for producing the observed annihilating positrons in a compact object and concluded that the most plausible mechanism is pair production by $\gtrsim$ 511 keV gamma rays interacting with each other. Using the > 511 keV continuum spectrum measured[57] with HEAO-3 in the fall of 1979 (1979.8), we obtain a compact source luminosity in photons > 511 keV of $\sim 1.6 \times 10^{38}$ erg sec^{-1}, assuming that the source is at a distance[33] of 7.7 kpc. If the pair production and annihilation are in equilibrium, then the annihilation rate of $\sim 7 \times 10^{42}/(2-1.5f)$ given above, implies that the energy available for pair production at the source is $\sim 14 \times (2 - 1.5f)$ MeV per pair, yielding 7 to 28 MeV/pair, depending on the value of f. Thus, the available energy above the pair production threshold is well in excess of the minimum of 1.02 MeV required if the pairs are produced by photons colliding head-on. In our previous treatment[24,41] we assumed a spherical interaction region in which the interacting photons are isotropic. By extrapolating the observed continuum photon density back to the source, we showed that the required rate of pair production implies that the size of the photon-photon interaction region should not exceed several times 10^8 cm. If the compact object is powered by accretion and releases gravitational energy close to its Schwarzschild radius, then the size of the pair production region should be of the order of several Schwarzschild radii. This implies a compact object mass less than several hundreds solar masses. The most probable candidate is a black hole.

In addition to this model, there have been several suggestions of 10^6 $M_\odot$ black-hole models for the compact annihilation source. We have reviewed these previously[41]. In these models, the pairs are produced either by small-angle photon-photon collisions in a beam[58,59] or by interactions[60] of a photon beam with gas clouds. The efficiency of these models is much lower than that of the isotropic model, as the pairs are produced by photons of energies of tens of MeV or higher. In addition, the beam models cannot directly account for the observed continuum flux > 511 keV and would require variations in the higher energy (> 25 MeV) flux which have not been reported[61].

Returning to our stellar mass black hole model, we pointed out previously that the similarity between the continuum spectra of Cygnus X-1 and the Galactic Center compact source is an additional argument in favor of this model[29,62]. This line of argument has gained support with the report[22] (at the 2σ significance level) of a 511 keV annihilation line from Cygnus X-1. The pairs responsible for the observed 511 keV line from Cygnus X-1 are thought[63] to be produced in photon-photon interactions near the accretion disk of the hole. Furthermore, as we have pointed out before[29], a stellar mass black hole, unlike a 10^6 $M_\odot$ object, need not reside in the nucleus of our Galaxy. The SIGMA discovery that 1E1740.7-2942, at some 120 pc from the nucleus, is probably the compact source of annihilation radiation (at least in the present epoch) is consistent with this argument.

SUMMARY

We have considered a two component model for galactic annihilation radiation: a variable component resulting from one or just a few compact sources at or near the Galactic Center and a steady, diffuse component resulting from positron annihilation in the galactic disk.

We have reviewed the proposed models for the variable, compact source. Here the positrons probably result from photon-photon pair production, most likely around an accreting stellar mass black hole. Our earlier argument, that such a hole need not reside in the nucleus of the Galaxy, is consistent with the SIGMA results (reported elsewhere in this volume) that at the present epoch a promising candidate for the variable source is the hard X-ray source 1E1740.7-2942 located about 120 pc from the Galactic Center.

We have modeled the diffuse component by using observed high energy gamma ray, CO and Fe line emission longitude profiles, and we have normalized these distributions to observational upper limits on the 511 keV line emission from the galactic plane to define the maximum steady diffuse component of that emission that is consistent with all of the observations. We find that for these models the largest consistent diffuse 511 keV line flux is $1.4 \pm 0.2 \times 10^{-3}$ photons cm^{-2} sec^{-1} rad^{-1} from the central radian of the Galaxy, and that the corresponding galactic positron production is $2.1 \pm 0.6 \times 10^{43}$ positrons sec^{-1}. Although this maximum flux was obtained from upper limits on narrow field observations, this range of values is quite consistent with independently determined values obtained for similar distributions from wide field measurements of high statistical significance. We reviewed the sources of diffuse galactic positrons, confirming that the dominant sources are ^{56}Co, ^{44}Sc and ^{26}Al resulting from various processes of nucleosynthesis. If the actual diffuse galactic positron production were much lower, however, hitherto unknown sources of annihilation radiation would be required to account for the large 511 keV line fluxes observed with the wide field of view detectors. A detailed mapping of the galactic disk in 511 keV line emission is needed to clarify the nature of this potentially new population of sources.

We have considered the processes of positron annihilation and we have shown that positron annihilation via positronium with a positron fraction of about 0.9 is expected for a broad range of interstellar parameters. We show that all of the orthopositronium continuum observations from the direction of the Galactic Center are consistent with a diffuse 511 keV line component normalized to $1.4 \pm 0.2 \times 10^{-3}$ photons cm^{-2} sec^{-1} rad^{-1}. However, an orthopositronium detection from the galactic plane at longitude $-25°$ is inconsistent at the 2σ with this normalization. We also discuss the expected width of the 511 keV line and show that high resolution observations could determine whether the positrons are excluded from interstellar clouds or allowed to penetrate into them.

Acknowledgements. We wish to acknowledge K. W. Chan, N. Guessoum, and J. Skibo for various contributions. The research of REL was supported by NASA under Grant NAGW 1970.

REFERENCES

1. W. N. Johnson, F. R. Harnden, and R. C. Haymes, *Astrophys. J.*, **172**, L1 (1972).
2. W. N. Johnson, and R. C. Haymes, *Astrophys. J.*, **184**, 103 (1972).
3. R. C. Haymes *et al., Astrophys. J.*, 201, 593 (1975).
4. M. Leventhal, C. J. MacCallum, and P. D. Stang, *Astrophys. J.*, **225**, L11 (1978).
5. F. Albernhe *et al., Astron. and Astrophys.*, **94**, 214 (1981).
6. B. M. Gardner *et al.,* 1982, in *The Galactic Center,* ed. G. R. Riegler and R. D. Blandford (New York: AIP) p. 144 (1982)
7. M. Leventhal *et al., Astrophys. J.*, **240**, 338 (1980).
8. G. R. Riegler *et al., Astrophys. J.*, **248**, L13 (1981).
9. W. A. Mahoney in *Nuclear Spectroscopy of Astrophysical Sources*, eds. N. Gehrels and G. H. Share, (New York:AIP), p. 149 (1988).
10. G. H. Share *et al., Astrophys. J.*, **326**, 717 (1988).
11. W. S. Paciesas *et al., Astrophys. J.*, **260**, L7 (1982).
12. M. Leventhal *et al., Astrophys. J.*, **260**, L1 (1982).
13. M. Leventhal *et al., Astrophys. J.*, **302**, 459 (1986).
14. G. H. Share *et al., Astrophys. J.*, **358**, L45 (1990).
15. M. J. Harris *et al. Astrophys. J.*, **362**, 135 (1990).
16. M. Leventhal *et al., Nature*, **339**, 36 (1989).
17. N. Gehrels, private communication (1990).
18. M. Niel *et al., Astrophys. J.*, **356**, L21 (1990).
19. J. L. Matteson *et al.,* (this volume).
20. R. J. Murphy, G. H. Share, J. R. Letaw, and D. J. Forrest, *Astrophys. J.*, **358** 290 (1990).
21. E. P. Mazets, S. V. Golenetskii, Yu. A. Guryan, and V. N. Ilyinskii, *Astrophys. Space Sci.*, **84**, 173 (1982).
22. J. C. Ling and W. A. Wheaton, *Astrophys. J.*, **343**, L57 (1989).
23. R. Ramaty, D. Leiter, and R. E. Lingenfelter, *Ann. New York Acad. Sci.*, **375**, p. 338 (1981).
24. R. E. Lingenfelter and R. Ramaty, 1982, in *The Galactic Center,* ed. G. R. Riegler and R. D. Blandford (New York: AIP), p. 148 (1982).
25. G. K. Skinner *et al., Nature*, **330**, 544 (1987).
26. W. R. Cook *et al.*, in *The Center of the Galaxy*, ed. M. Morris, (Dordrecht: Kluwer), p. 581 (1989).
27. P. P. Dunphy, E. L. Chupp. and D. J. Forrest, in *Positron Electron Pairs in Astrophysics*, eds. M. L. Burns, A. K. Harding, and R. Ramaty (New York: AIP), p. 237 (1983).
28. R. Ramaty and R. E. Lingenfelter, in *The Galactic Center,* ed. D. C. Backer (New York: AIP), p. 51 (1987).
29. R. E. Lingenfelter and R. Ramaty, *Astrophys. J.*, **343**, 686 (1989).

30. H. A. Mayer-Hasselwander *et al., Astron. and Astrophys.*, **105**, 164 (1982).
31. T. M. Dame *et al., Astrophys. J.*, **322**, 706 (1987).
32. K. Koyama *et al., Nature*, **339**, 603 (1989).
33. L. J. Reid, *The Center of the Galaxy*, ed. M. Morris, (Dordrecht: Kluwer), p. 37 (1989).
34. R. W. Bussard, R. Ramaty, and R. J. Drachman, *Astrophys. J.*, **228**, 928 (1979).
35. W. H. Zurek, *Astrophys. J.*, **289**, 603 (1985).
36. N. Guessoum, R. Ramaty, and R. E. Lingenfelter, *Astrophys. J.*, in press (1991).
37. C. F. McKee and J. P. Ostriker, *Astrophys. J.*, **218**, 148 (1977).
38. D. J. Forrest, *The Galactic Center,* ed. G. R. Riegler and R. D. Blandford (New York: AIP), p. 160 (1982).
39. R. E. Lingenfelter and X. M. Hua, *Astrophys. J.*, submitted (1990).
40. R. Ramaty and R. E. Lingenfelter, *Nature*, **278**, 127 (1979).
41. R. E. Lingenfelter and R. Ramaty 1989, *Nuclear Physics B, (Proc. Suppl.)*, **10B**, 67 (1989).
42. W. A. Mahoney *et al., Astrophys. J.*, **286**, 578 (1984).
43. J. C. Higdon and W. A. Fowler, *Astrophys. J.*, **317**, 710 (1987).
44. H. Bloemen, *Ann. Rev. Astron. Astrophys.*, **27**, 469 (1989).
45. G. H. Share, private communication (1991).
46. R. Ramaty, B. Kozlovsky, and R. E. Lingenfelter, *Astrophys. J. Suppl.*, **40**, 487 (1979).
47. R. E. Lingenfelter and G. J. Hueter, in *High Energy Transients in Astrophysics*, ed. S. E. Woosley (New York: AIP), p. 558 (1984).
48. P. A. Sturrock, *Astrophys. J.*, **164**, 529 (1971).
49. D. D. Clayton, *Nature, Phys. Sci.*, **244**, 137 (1973).
50. S. E. Woosley and P. A. Pinto, in *Positron Electron Pairs in Astrophysics*, eds. M. L. Burns, A. K. Harding, and R. Ramaty (New York: AIP), p. 98 (1988).
51. M. Signore and G. Vedrenne, *Astron. and Astrophys.*, **201**, 379 (1988).
52. S. A. Colgate, *Astrophys. Space Sci.*, **8**, 457 (1970).
53. K. W. Chan and R. E. Lingenfelter, *21st Internat. Cosmic Ray Conf. Papers*, **3**, 253 (1990).
54. S. van den Bergh, R. D. McClure, and R. Evans, *Astrophys. J.*, **323**, 44.
55. S. E. Woosley and T. A. Weaver, in *Radiation Hydrodynamics in Stars and Compact Objects*, eds. D. Mihalas and K. Winkler (Berlin: Springer Verlag), p. 91 (1986).
56. K. Nomoto, F-K. Thielemann, and K. Yokoi, *Astrophys. J.*, **286**, 644 (1984).
57. G. R. Riegler *et al., Astrophys. J.*, **294**, L13 (1985).
58. R. D. Blandford, in *The Galactic Center,* ed. G. R. Riegler and R. D. Blandford (New York: AIP), p. 177 (1982).
59. M. L. Burns, M. L. in *Positron Electron Pairs in Astrophysics*, eds. M. L. Burns, A. K. Harding, and R. Ramaty (New York: AIP), p. 281 (1983).
60. N. S. Kardashev, in *Positron Electron Pairs in Astrophysics*, eds. M. L. Burns, A. K. Harding, and R. Ramaty (New York: AIP), p. 253 (1983).
61. B. N. Swanenburg *et al., Astrophys. J.*, **243**, L69 (1981).
62. R. Ramaty and R. E. Lingenfelter, *Ann. New York Acad. Sci.*, **571**, 433 (1989).
63. E. P. Liang and C. D. Dermer, *Astrophys. J.*, **325**, L39 (1988).

COMPTON BACKSCATTERED 511 KEV ANNIHILATION LINE AND THE 170 KEV LINE FROM THE GALACTIC CENTER

RICHARD E. LINGENFELTER
Center for Astrophysics and Space Sciences, University of California, San Diego, La Jolla CA 92093 USA
AND
XIN-MIN HUA
Institute for Space and Terrestrial Science, Concord, Ontario, L4K 3C8 Canada

ABSTRACT

We show that Compton scattering of 511 keV electron-positron annihilation radiation produces a line-like reflection feature at $\sim$ 170 keV from backscattered photons. Assuming a simple model of an accretion disk around a compact source, we explore the spectrum of Compton scattered annihilation line emission for a range of disk geometries, opacities, and observing angles, and we find that the line-like feature is produced under a wide variety of conditions. We further show that such Compton backscattering of slightly redshifted annihilation line emission from the inner edge of an accretion disk could account for the 170 keV line emission and higher energy continuum observed[1–3] together with the 511 keV annihilation radiation from the direction of the Galactic center. Identification of the observed 170 keV line as an annihilation line reflection feature provides direct new evidence for the existence of an accreting stellar-mass black-hole source of annihilation radiation at or near the Galactic center.

INTRODUCTION

The role of Compton scattering of x-rays and gamma rays has been studied in a variety of astrophysical sources. The Comptonization of x-rays by plasma clouds has been studied in compact sources[4]. The Compton reflection of the hard x-ray and gamma-ray continuum has been studied in solar flares[5] and in accretion disks around compact objects[6]. The Compton attenuation of gamma-ray line emission has been studied in solar flares[7] and in supernovae[8,9] And the Compton "tails" of gamma-ray lines have been studied as a diagnostic in solar flares[10], and as a continuum source in supernovae[8,11] and an alternative to orthopositronium emission in the variable annihilation radiation source in the direction of the Galactic center[12,13].

Here we point out the astrophysical importance of Compton backscattering of gamma-ray lines that can produce *line-like reflection features* at lower energies. These Compton backscattered features are well-known in the laboratory[14], but their astrophysical significance has not been discussed previously. We consider, by way of illustration, the Compton backscattering of the 511 keV electron-positron annihilation line which produces a line-like feature at $\sim$ 170 keV. We study the Compton scattering of 511 keV photons because they are the most nearly ubiquitous of any astrophysical gamma-ray line photons, and because 170 keV line emission has, in fact, been observed[1–3] to accompany the intense 511 keV line emission from the direction of the Galactic center.

Analysis of the observations of annihilation radiation from the Galactic center region suggests[15] that there are two sources of the emission: a steady, diffuse interstellar source with a broad distribution in galactic longitude, and a variable, compact source at or near the Galactic center. The diffuse galactic emission appears to consist of both 511 keV line emission and orthopositronium annihilation continuum below 511 keV, and it most likely results[16–19] from the annihilation of positrons produced in the decay of radionuclei synthesized by thermonuclear burning in supernovae. The emission from the compact source consists not only of variable 511 keV line emission and possible orthopositronium annihilation continuum below 511 keV, but also a highly variable higher energy continuum just above 511 keV. The origin of this emission is generally attributed[19–22] to annihilating positrons produced around an accreting black hole. The variability may result from episodic accretion or from Compton-scattering obscuration by accreting matter, which could also account[12] for much of the continuum below 511 keV instead of orthopositronium.

We therefore explore the generation of line-like features by Compton backscattering of the 511 keV annihilation line in the context of an accreting black hole source of the variable annihilation line emission from the direction of the Galactic center.

COMPTON BACKSCATTERING OF THE ANNIHILATION LINE

The energy of the Compton-scattered photon relative to the initial photon energy is $r = E'_\gamma/E_\gamma = 1/(1 + \alpha - \alpha cos\phi)$, where $\alpha = E_\gamma/m_ec^2$ and ϕ is the scattering angle. The energy distribution of the Compton-scattered photons is

$$f(r) \propto \left[r - 1 + \frac{1}{r} + \frac{(\alpha r + r - 1)^2}{\alpha^2 r^2}\right]$$

for $1/(2\alpha + 1) \leq r \leq 1$, corresponding to scattering angles $180° \geq \phi \geq 0°$, and $f(r) = 0$ for other values of r. This distribution, which rises steeply at the lowest energies and then cuts off sharply at the 180° backscattered energy of $r = 1/(2\alpha + 1)$, produces a line-like feature just above the minimum energy. This feature can be seen in Figure 1, where we show the energy distribution of Compton scattered 511 keV annihilation radiation photons. For 511 keV photons $\alpha = 1$ and the scattered photons have energies $1/3 \leq r \leq 1$, or 170 keV $\leq E'_\gamma \leq$ 511 keV, so that the scattered photon energy distribution is simply $f(r) \propto (r + 3 - 3/r + 1/r^2)$. As can also be seen in Figure 1, the backscattered photons ($90° \leq \phi \leq 180°$) are compressed into a relatively narrow energy range of 1/6 of 511 kev, while the forward scattered photons are spread over an energy range of 1/2 of 511 keV, three times larger.

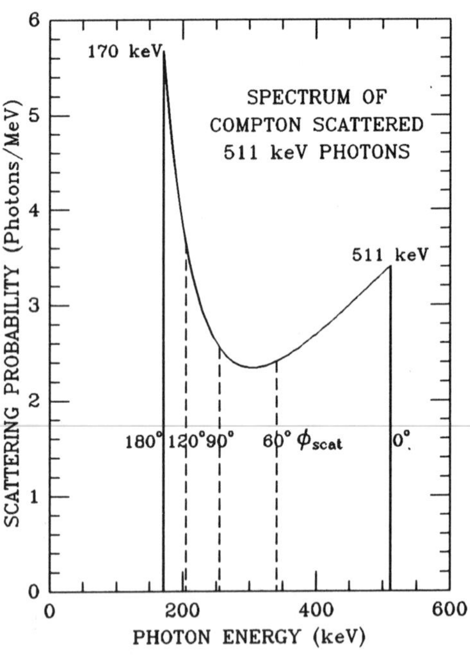

Figure 1. The spectrum of Compton scattered 511 keV photons as a function of the scattered photon energy and scattering angle, showing the line-like backscattered feature at 170 keV.

The Compton scattering distribution shown in Figure 1, of course, is only for singly scattered line photons and multiple scattering produces a broader spectrum extending down to much lower energies. To study the full spectrum resulting from multiple Compton scattering of line photons, we have made a series of Monte Carlo simulations. In all cases we assumed an isotropic, monoenergetic source of 511 keV annihilation line photons, and allowed them to Compton scatter on cold ($kT \ll m_e c^2$) electrons in a simple, uniform density, gaseous disk. In particular, we considered a disk in a cylindrical coordinate system with linear dimensions defined in terms of the Compton scattering optical depth τ at 511 keV, equivalent to 3.5×10^{24} electrons cm^{-2}. The disk was assumed to lie in a plane normal to the z axis with a thicknesses in the z direction equal to τ and a radius $\gg \tau$. In order to approximate the inner edge of an accretion disk around a compact object, we considered a cylindrical hole in the disk centered at the origin with its axis in the z direction normal to the plane of the disk. The size of the hole is defined by an opening angle θ_o which is the zenith angle of the inner edge of the disk measured with respect to the axis of the cylinder. Thus, the radius of the hole is equivalent to $\frac{1}{2}\tau \tan\theta_o$ and the limiting case of $\theta_o = 0°$ is simply the uniform disk without a hole. We have ignored disk rotation in this simple model. We also assumed for simplicity that the source of the 511 keV photons is located at the origin in the center of the hole. Thus the emerging unscattered 511 keV line intensity, as a function of μ the cosine of the observing angle θ, is equal to $1/2\pi$ for $\theta < \theta_o$, and

$$f(\mu) = \frac{1}{2\pi} \exp\left[-\frac{\tau}{2}\left(\frac{1}{\mu} - \frac{\tan\theta_o}{(1-\mu^2)^{1/2}}\right)\right], \text{ for } \theta < \theta_o.$$

The resulting Compton scattered 511 keV line spectra determined for a range of opacities, τ, and disk angles, θ_o, are shown in Figure 2, as a function of μ the cosine of the observing angle. As can be seen, there is a significant singly backscattered line-like feature at 170 keV, or higher energies, in the Compton scattered 511 keV line spectra for each of the disk cases as seen from nearly any angle of observation. The strength, mean energy and shape of this feature all depend on the opacity and the geometry of the disk, i.e. the opening angle of the central hole, and the angle of observation with respect to the disk.

In particular, we see from the upper panels of Figure 2 for large observing angles (small μ), near the plane of the disk, that the 170 keV feature becomes more intense and narrower as the opening angle of the central hole increases. Because the reflecting matter behind the inner edge of the disk subtends a smaller solid angle from the source as the opening angle increases, the scattering angles for observable singly scattered photons are concentrated into a narrower band around 180° and the observable energy band above 170 keV is likewise narrower. Correspondingly at smaller observing angles (larger μ), as θ become less than θ_o, an increasing opening angle concentrates the reflecting matter more nearly perpendicular to the line of sight and there is no reflecting matter directly behind the source. Thus, the peak energy of the reflection feature is shifted up toward 255 keV as scattering angles around 90° dominate the observable singly scattered photon emission, and there is a significant depression in the observable spectrum around 170 keV. This shift in the energy of the peak of the reflection feature can serve as a diagnostic of the observing angle.

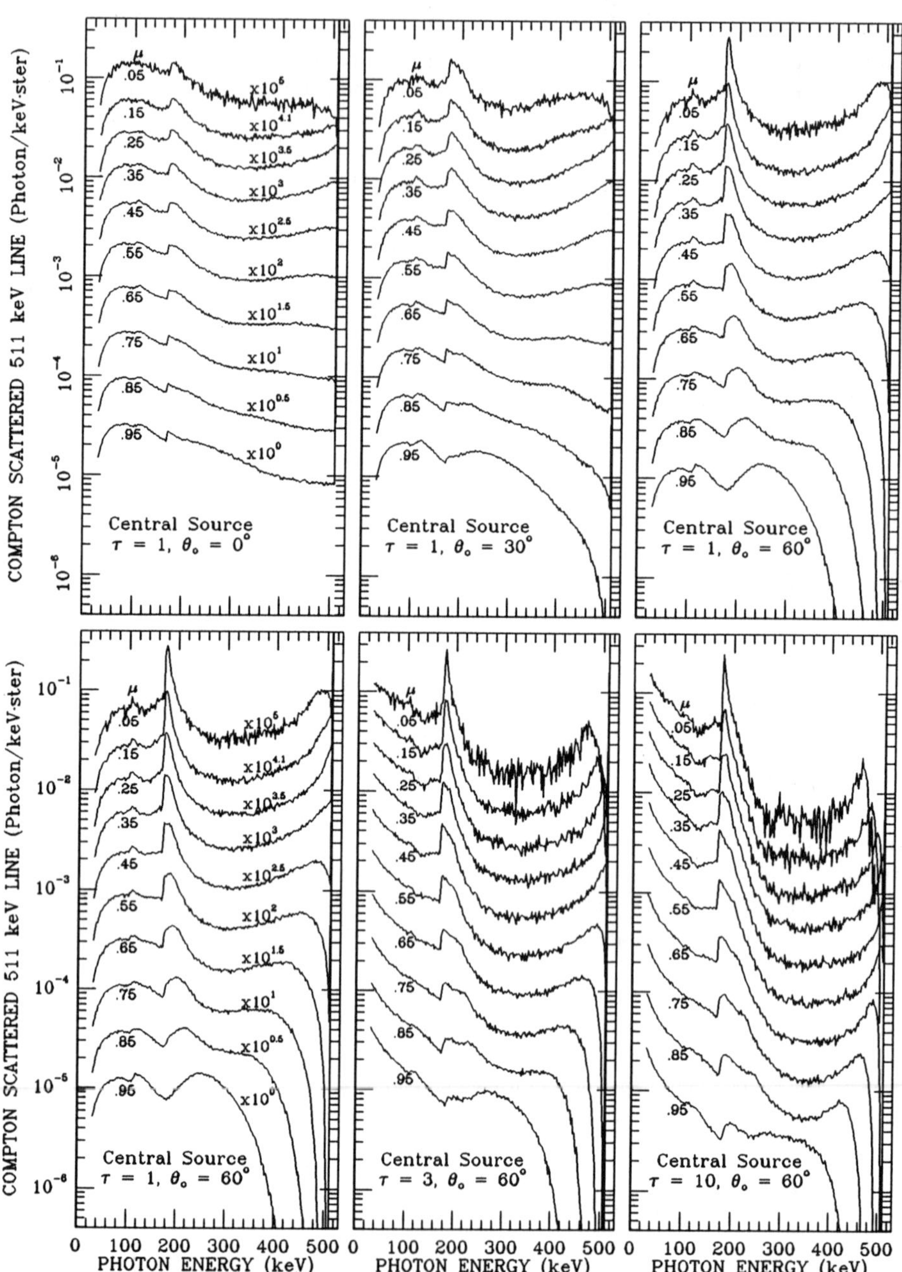

Figure 2. Monte Carlo spectra of the line-like reflection feature at 170 keV and the associated continuum from Compton scattered 511 keV photons in a uniform disk with varying opacities, τ, and central hole opening angles, θ_o, showing the effect of disk opacity and geometry. All of the spectra are normalized to one isotropically emitted 511 keV photon, and individual spectra are displaced from one another by a factor of $10^{0.5}$ or more for clarity.

Considering the effect of disk opacity, we see from the lower panels of Figure 2, that for large observing angles the 170 keV feature also becomes narrower as the opacity of the disk increases, since that further enhances the singly backscattered photons by attenuating the flux of photons with scattering angles around 90° relative to those around 180°. Because the scattering probability for photons moving away from the observer is already close to unity for large observing angles, increasing the disk opacity can not further increase the peak intensity of the 170 keV feature. For small observing angles, however, increasing the disk opacity does attenuate the flux in the singly scattered feature around 255 keV from scattering around 90°.

In addition to the backscattered feature, we see that at higher energies, just below 511 keV, there is a conspicuous forward scattering peak, which is also a characteristic signature of Compton scattering. As we can see, especially in the lower panels of Figure 2, depending upon the angle of observation, the forward scattered emission does not always extend all the way up to 511 keV. This can happen either because there is no matter close to the line of sight to scatter the line photons, as is the situation for observing angles less than the opening angle of the central hole, or because there is so much scattering matter in the line of sight that the source is completely obscured, as is the situation for large observing angles, closest to the plane of the disk. When the forward scattered emission does extend all the way up to 511 keV, it can be observationally confused with the three-photon orthopositronium annihilation emission[12], since the shape of the two spectra are almost identical at energies $>$ 350 keV. However, detection of the Compton backscattered feature at 170 keV can enable us to distinguish between the two processes.

We now compare these calculated spectra with the observations of the annihilation radiation and accompanying 170 keV line from the direction of the Galactic center.

THE 170 KEV LINE FROM THE GALACTIC CENTER

Positron annihilation radiation in the 511 keV line and in the apparent three-photon orthopositronium continuum has been observed from the direction of the Galactic center since 1970. As we discussed above, analysis of these observations suggests[15] that there are two sources of the emission: a steady, diffuse interstellar source with a broad distribution in galactic longitude, and a variable, compact source, quite possibly an accreting black hole at or near the Galactic center.

This emission was first measured by Haymes et al.[23] with low resolution NaI spectrometers in a series of three balloon flights in the early 1970s. However, there was uncertainty as to the identification of the emission, because the observed line appeared to be centered at 476±24 keV during the first two observations in 1970 and 1971 and then at 530±11 keV during the third observation in 1974. It was not until the fall of 1977 that a narrow (FWHM $<$ 3.2 keV) line was clearly identified at an energy of 510.7±0.5 keV together with an apparent orthopositronium continuum by Leventhal, MacCallum and Stang[1] with a balloon-borne high resolution Ge spectrometer. This observation was made during a period $<$ 1977–1979 $>$ when the annihilation line emission from the direction of the Galactic center was significantly higher than average and the variable, compact source was active. It was during this same observation that they[1,2] also detected an unidentified "candidate" line at 170 keV with a width (FWHM) of 12

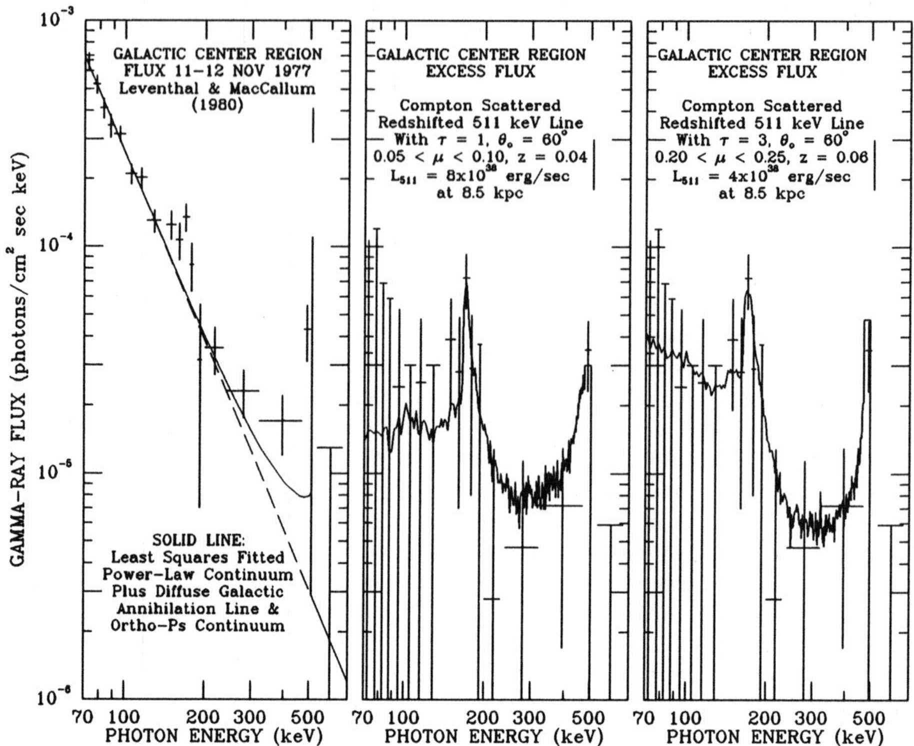

Figure 3. The observed[1,2] spectrum from the Galactic center region on 11-12 November 1977 (left panel), showing the 170 keV and 511 keV lines and other excess flux above a best-fit, low-energy power-law spectrum and the diffuse Galactic annihilation line and orthopositronium continuum flux expected for the detector field of view. This excess flux is also shown separately (central and right panels) with calculated spectra of Compton scattered, slightly-redshifted 511 keV line photons isotropically emitted in a hole of opening angle $\theta_o = 60°$ in a uniform disk. The calculated spectra are for disk opacities τ of 1 and 3, and observing angle cosines μ of 0.05–0.1 and 0.20–0.25, normalized to 511 keV annihilation line luminosities of 8×10^{38} and 4×10^{38} erg s^{-1}, redshifted by 0.04 and 0.06, respectively.

keV and a flux of $(7.4\pm1.8)\times10^{-4}$ photons cm^{-2} s^{-1} (see Figure 3, left panel). This 170 keV line flux was roughly 60% of that in the narrow 511 keV line, $(1.22\pm0.22)\times10^{-3}$ photons cm^{-2} s^{-1}, seen at the same time. Although the 170 keV line was detected at the 4.1σ or 99.998% confidence level, and no similar negative (presumably spurious) features were found in the spectrum at $> 3.5\sigma$, two other positive unidentified "candidate" lines were also detected at 1611 keV with a width of 10 keV and 3700 keV with a width of 500 keV at 4.9 and 3.9σ, respectively. Lacking a "reasonable identification" of any of these lines, Leventhal and MacCallum[2] cautioned against taking them seriously until they were confirmed by other observations.

We now suggest that a reasonable identification of the 170 keV line can be found in Compton backscattering of the 511 keV annihilation line. Moreover, there is now at least one confirming observation of such a line in spectra of the Galactic center region. Matteson et al.[3] observed a line-like feature at 170 keV at the several σ level of significance with a high resolution Ge spectrometer during a balloon flight in May of 1989. There are also suggestions of features in the range from 150 to 250 keV in other published[23–25] spectra. Clearly these and the other spectral observations of the Galactic center region should be reexamined for evidence of such a feature.

In order to identify the 170 keV line observed from the direction of the Galactic center as Compton backscattered annihilation radiation, however, it is necessary to assume a small gravitational redshift. Such a redshift should in fact be expected, if the backscattered feature arises from reflection near the inner edge of an accretion disk around a compact object. Moreover, it shows that the 170 keV line could not originate locally from Compton backscattered annihilation radiation in either the instrument or the earth's atmosphere.

The need to assume a redshift for a Compton backscattering identification of the line arises from the fact that a Gaussian fit to the observed[2] line was centered at 170 keV with a uncertainty of < 0.5 keV and a width (FWHM) of 12 keV, whereas the backscattered feature is cut off sharply below 170.3 keV and such a width would require the effective center of the feature to be shifted higher in energy by $\sim$ FWMH/2, or $\sim$ 6 keV. Thus, if the observed 170 keV line is Compton backscattered 511 keV line emission, the Compton spectrum and the annihilation line must be redshifted by that amount, which would correspond to a $z \sim 6/170 \sim 0.04$. Such a redshift can be produced gravitationally, if the source of the annihilation radiation is located at a distance of $\sim$ 12 Schwarzschild radii from a compact object, which would be consistent with emission and reflection near the inner edge of an accretion disk. This, of course, requires that any observable 511 keV annihilation line photons, coming directly from the source, would be redshifted down by $\sim$ 18 keV to an energy of $\sim$ 493 keV.

Thus, the observed narrow 511 keV line could not be the source of the backscattered photons, but as was previously pointed out[20,26] these photons could not have come from positron annihilation directly in the compact source. That is because the line center of 510.7 ± 0.5 keV determined by Leventhal, MacCallum and Stang[1] only allowed a redshift of < 1.3 keV from the rest energy of 510.9991 keV at the 2σ level, and the subsequent measurement by Riegler et al.[27] in 1979 reduced that to < 0.6 keV. Therefore, even if the diffuse Galactic contribution to the observed line were centered exactly at the rest energy, the annihilation region in which the variable component of the narrow line originates could not be closer than $\sim$ 300 Schwarzschild radii from the compact object. Moreover, if the annihilation region is in a surrounding accretion disk, the observed[1,2] line width, FWHM < 3.2 keV, would not allow it to be closer than $\sim 10^5$ Schwarzschild radii from the compact object, because of Doppler broadening by the Keplerian velocities.

A redshifted annihilation line from the compact source at an energy of $\sim$ 493 keV would lie right in the middle of the data bin just below the 511 keV line in the spectrum of Leventhal, MacCallum and Stang[1,2], shown in Figure 3. The flux of $(1.4 \pm 0.4) \times 10^{-3}$ photons cm^{-2} s^{-1} in this band from 476 to 509 keV is actually larger than that in the narrow 511 keV line and it seems to be quite anomalous, as we shall discuss further below. This flux is also roughly twice that in the 170 keV feature. Such emission may also have contributed to the redshift of the annihilation line observed by Haymes et al.[23] in their first two observations with a low resolution spectrometer.

As can be seen in the left panel of Figure 3, there is a strong power-law continuum underlying all of these features. However, the contribution of the variable source of annihilating positrons to this hard x-ray and gamma-ray continuum is not known, because there are also several highly variable hard x-ray sources[28–29] within the 15° instrumental field of view. But, since we are interested in the non-power-law components of the spectrum, we can simply subtract a best-fit power-law from the observed spectrum. We find that the least squares fit to the spectrum below 135 keV is a power law of $(2.6 \pm 0.1) \times 10^{-4}(E/100\text{keV})^{-2.8\pm0.2}$ photons cm^{-2} s^{-1} keV^{-1}. To this power-law continuum, we also add the diffuse galactic 511 keV line and orthopositronium continuum fluxes expected for the instrumental field of view of 15°, assuming a diffuse galactic 511 keV line flux of $(1.5 \pm 0.3) \times 10^{-3}$ photons cm^{-2} s^{-1} per radian of galactic longitude[15,30–31] and a diffuse galactic positronium annihilation fraction[32] of $0.89^{+0.10}_{-0.06}$.

Comparing the observed flux from the Galactic center region with the sum of the low energy power-law continuum and the diffuse galactic annihilation radiation, we clearly see that there is a significant excess flux not only around 170 keV and in the narrow 511 keV line already noted by Leventhal and MacCallum[2], but also in that seemingly anomalous energy band from 476 to 509 keV, which we suggest includes both redshifted 511 keV line and forward scattered photons from the compact source. The 33 keV width of this data band limits the thermal broadening of the redshifted 511 keV line to a temperature of $< 1.6 \times 10^7$ K, assuming direct annihilation with hot electrons[33].

The excess flux alone is shown in the central and right panels of Figure 3, with the additional uncertainties in both the low energy power-law and the diffuse annihilation radiation included in the error bars. Comparing this excess flux with the calculated Compton scattered 511 keV line spectra from the simple accretion disk model discussed above, we see that such emission can in fact account not just for the 170 keV line-like feature, but for all of the rest of the excess flux, as well. Although we do not attempt to explore the full range of possible spectral fitting here, because that is beyond the scope of this paper, these two examples clearly show that Compton scattered 511 keV line emission can in fact give a good fit to the observations. They also allow us to explore some of the constraints on the disk opacity, opening angle and viewing angle that can be set by the observations, and they give a measure of the gravitational redshift and 511 keV line luminosity that are required to fit the observations. We plan much more detailed studies of the observations with more realistic accretion disk models.

In the calculated spectra, we have spread the redshifted 511 keV line emission over a 33 keV band together with the forward scattered emission for easier comparison with the data bin just below 511 keV. All data upper limits are 2σ. For these comparisons we assume the same redshift ($z = 0.04$ or 0.06) for both the unscattered and scattered 511 keV photons. However, if the annihilation and scattering regions are at different distances from the compact object, as we assumed in the simple accretion disk model, then there is a small difference between the observed redshifts. The observed redshift of the 170 keV feature would be $z_{170} \sim \frac{2}{3}z_s + \frac{1}{3}z_{511}$, where z_s is the redshift of the scattering region and z_{511} is the redshift of the 511 keV line from the annihilation region. The lower energy bound on the 476 to 509 keV data band suggests a lower limit on the redshift of the 511 keV line from the annihilation region of $z_{511} < 0.06 \sim 511/476 - 1$, and the fit to the 170 keV feature requires a redshift of roughly $0.07 > z_{170} > 0.03$. Thus the redshift of the scattering region z_s must be > 0.015, corresponding to a distance of < 30 Schwarzschild radii, while the annihilation region redshift $z_{511} < 0.06$ corresponds to a distance of > 8 Schwarzschild radii, allowing a significant separation between the two.

We also see that no additional orthopositronium continuum is needed to fit the observed spectrum, suggesting that the fraction of positrons annihilating via positronium may be close to zero, even for those positrons which annihilate at distances > 300 Schwarzschild radii from the compact object and produce the variable narrow unshifted 511 keV line emission. However, an additional orthopositronium continuum associated with a positronium annihilation fraction as large as ~ 0.5 could not be excluded for the narrow line alone. This is quite consistent with recent analyses[15,33] of the positronium annihilation fraction observed from the region of the Galactic center, which also suggest a value of 0 to 0.5 for this component.

Assuming a nominal Galactic center distance of 8.5 kpc, a redshifted 511 keV line luminosity of (4 to 8)$\times 10^{38}$ erg s^{-1} is required to fit the observations for disk opacities τ of 1 to 3. Such a luminosity does not seem unreasonably large, since it is of the same order as the luminosity of $\sim 2 \times 10^{38}$ erg s^{-1} observed[25] at ≥ 511 keV by the HEAO-3 Ge spectrometer in the fall of 1979 when the compact source was still active. The redshifted 511 keV line luminosity implies a steady state positron production rate of (2.5 to 5)$\times 10^{44} e^{+}$ s^{-1} in the compact source, assuming a positronium fraction of zero. Such a luminosity also suggests that at least $\sim 1\%$ of those positrons escape to distances > 300 Schwarzschild radii from the source to produce the variable narrow unshifted 511 keV line flux of $(8.4 \pm 2.2) \times 10^{-4}$ photons cm^{-2} s^{-1}, which corresponds to an annihilation rate of $\sim 4 \times 10^{42} e^{+}$ s^{-1} and a 511 keV line luminosity $\sim 6 \times 10^{38}$ erg s^{-1}.

These escaping positrons, as previously suggested[19–20], could also be produced by pair production in photon-photon collisions with photons of energy ≥ 511 keV, close to the compact source, within a distance

$$d \sim \frac{< \sigma_{\gamma\gamma} c > L^2}{24\pi c^3 (m_e c^2)^2 Q} \sim 6.6 \times 10^{-26} L^2/Q,$$

where $\sigma_{\gamma\gamma}$ is the photon-photon pair production cross section and d is in cm for the $\geq$ 511 keV luminosity L in erg s^{-1} and the unshifted narrow line annihilation rate Q in e^{+} s^{-1}. The luminosity L of $\sim 2 \times 10^{38}$ erg s^{-1}, directly observed by HEAO-3, and the annihilation rate Q of $\sim 4 \times 10^{42} e^{+}$ s^{-1} give[19] a radius for the source of the escaping positrons of $d \leq 7 \times 10^{8}$ cm. This radius puts a limit on the mass of the compact source of $M_* = zc^2 d/G \leq 5 \times 10^{3} z M_{\odot}$, or < 200 to 300 $M_{\odot}$, if they escape from a redshift z of 0.04 to 0.06. If we use an L of (4 to 8)$\times 10^{38}$ erg s^{-1}, equal to the redshifted 511 keV line luminosity required to fit the observations for disk opacities $\tau = 1$ and 3, shown in Figure 3, and the corresponding steady state positron production rate of (2.5 to 5)$\times 10^{44} e^{+}$ s^{-1}, we find a source radius $d \leq$ (4 to 8)$\times 10^{7}$ cm for z of 0.06 to 0.04. This radius puts a limit on the compact source mass of only $\sim$ 15 to 20 $M_{\odot}$.

A similar, but separate, estimate of the compact source mass can also be made from the redshift and the annihilation line luminosity required to fit the observations in Figure 3. Taking a positronium fraction of zero, we can define the mass accretion rate $\dot{M} = L_a/\varepsilon_{\pm} c^2$ in terms of the annihilation line luminosity L_a and the fraction $\varepsilon_{\pm}$ of the gravitational energy that is converted into electron-positron pairs. In the simple disk model assumed above, we can also define the mass accretion rate

$$\dot{M} \sim \frac{4\pi G m_p < A/Z > \tau\eta M_*}{\sigma_{511} c (2z)^{1/2}},$$

where η is the ratio of the radial inflow velocity to the free fall velocity, σ_{511} is the Compton scattering cross section at 511 keV, and $< A/Z >$ is the mean atomic mass to charge ratio. The accretion rate does not depend on the angle subtended by the disk, and hence on the opening angle, θ_o, because the opacity of the disk is fixed. Combining these two equations, we see that

$$M_*/M_\odot \sim 0.42 \times 10^{-38} L_a z^{1/2} / \tau\eta\varepsilon_\pm.$$

Thus for an L_a of $(4 \text{ to } 8)\times 10^{38}$ erg s^{-1} required to fit the observations for disk opacities τ of 3 and 1 and redshifts z of 0.06 and 0.04, we find a compact source mass of 0.14 to 0.7 $M_\odot/\eta\varepsilon_\pm$. Nominal values of the order of $\sim$ 0.1 for both η and $\varepsilon_\pm$, give a compact source mass of the order of 20 to 100 $M_\odot$, which is quite consistent with the estimates above. Such estimates strongly suggest an accreting stellar-mass black hole, rather than a massive black hole, as the source of the variable component of the annihilation and its reflected emission down to 170 keV.

SUMMARY

We have shown that Compton scattering of 511 keV electron-positron annihilation radiation produces a line-like reflection feature at $\sim$ 170 keV from backscattered photons. Assuming a simple model of an accretion disk around a compact source, we have explored the spectrum of Compton scattered annihilation line emission for a range of disk geometries, opacities, and observing angles, and we find that the line-like feature is produced under a wide variety of conditions.

We have further shown that such Compton backscattering of slightly redshifted annihilation line emission from the inner edge of an accretion disk can account for the 170 keV line emission and higher energy continuum observed[1,2] together with the 511 keV annihilation radiation from the direction of the Galactic center. Although the identification of the observed 170 keV line as Compton backscattered annihilation radiation requires a small redshift of 0.04 to 0.06, such a redshift should in fact be expected, if the backscattered feature arises from reflection near the inner edge of an accretion disk around a compact object. Moreover, such a redshift shows that the 170 keV line could not originate locally from Compton backscattering of annihilation radiation in either the instrument or the earth's atmosphere.

Identification of the 170 keV line as Compton backscattered 511 keV line emission provides direct new evidence for an accreting compact source in the direction of the Galactic center, because the diffuse Galactic line emission could not generate such a reflection feature at the observed intensity. Moreover, from the redshift and the annihilation line luminosity required to fit the observations, we estimate the compact source mass to be of the order to 10 to 100 $M_\odot$, corresponding to a stellar-mass black hole, rather than a more massive one.

As this example clearly suggests, line-like reflection features from Compton backscattering not only of annihilation line emission, but of other nuclear line emission as well, could also be an important new tool for studying high energy processes in many other astrophysical sources.

Acknowledgements. We thank Michael Briggs, Kai-Wing Chan, Marvin Leventhal, Jim Matteson, Reuven Ramaty, Rick Rothschild and Peter von Ballmoos for helpful discussions. We also thank NASA for financial support under grant NAGW 1970 and the Province of Ontario for support at ISTS. The calculations were carried out on a SUN at ISTS and a VAX at UCSD.

REFERENCES

1. M. Leventhal, C. J. MacCallum and P. D. Stang, *Ap. J. (Letters),* **225,** L11 (1978).
2. M. Leventhal and C. J. MacCallum, *Ann. N. Y. Acad. Sci.,* **336,** 248 (1980).
3. J. L. Matteson, et al. in these proceedings (1991).
4. R. A. Sunyaev, and L. G. Titarchuk, *Astron. Ap.,* **86,** 121 (1980).
5. T. Bai and R. Ramaty, *Ap. J.,* **219,** 705 (1978).
6. T. R. White, A. P. Lightman, and A. A. Zdziarski, *Ap. J.,* **331,** 939 (1988).
7. X.-M. Hua and R. E. Lingenfelter, *Solar Phys.,* **107,** 351 (1987).
8. A. Burrows and L. S. The, *Ap. J.,* **360,** 626 (1990).
9. K. W. Chan and R. E. Lingenfelter, *Ap. J.,* **368,** 515 (1991).
10. W. T. Vestrand, *Ap. J.,* **352,** 353 (1990).
11. P. A. Pinto and S. E. Woosley, *Ap. J.,* **329,** 820 (1988).
12. D. J. Forrest, in *The Galactic Center,* ed. G. R. Riegler and R. D. Blandford (New York: Am. Inst. Phys., 1982) p. 160.
13. L. Bildsten and W. H. Zurek, *Ap. J.,* **329,** 212 (1988).
14. G. F. Knoll, *Radiation Detection and Measurement,* (New York: Wiley 1979) p. 322.
15. R. E. Lingenfelter and R. Ramaty, *Ap. J.,* **343,** 686 (1989).
16. R. Ramaty and R. E. Lingenfelter, in *The Galactic Center,* ed. D. C. Backer (New York: Am. Inst. Phys., 1987) p. 51.
17. S. E. Woosley and P. A. Pinto, in *Nuclear Spectroscopy of Astrophysical Sources,* ed. N. Gehrels and G. H. Share (New York: Am. Inst. Phys., 1988) p. 98.
18. M. Signore and G. Vedrenne, *Astron. Ap.,* **201,** 379 (1988).
19. R. E. Lingenfelter and R. Ramaty, *Nuc. Phys. B (Proc. Supp.)* **10B,** 67 (1989).
20. R. E. Lingenfelter and R. Ramaty, in *The Galactic Center,* ed. G. R. Riegler and R. D. Blandford (New York: Am. Inst. Phys., 1982) p. 148.
21. M. J. Rees, in *The Galactic Center,* ed. G. R. Riegler and R. D. Blandford (New York: Am. Inst. Phys., 1982) p. 166.
22. L. M. Ozernoy, in *The Center of the Galaxy,* ed. M. Morris, Dordrecht: Kluwer Academic, p. 555 (1989).
23. R. C. Haymes, et al., *Ap. J.,* **201,** 593 (1975).
24. W. S. Paciesas, et al., *Ap. J. (Letters),* **260,** L7 (1982).
25. G. R. Riegler, et al., *Ap. J. (Letters),* **294,** L13 (1985).
26. R. E. Lingenfelter and R. Ramaty, in *The Center of the Galaxy,* ed. M. Morris, Dordrecht: Kluwer Academic, p. 587 (1989).
27. G. R. Riegler, et al., *Ap. J. (Letters),* **248,** L13 (1981).
28. J. L. Matteson, in *The Galactic Center,* ed. G. R. Riegler and R. D. Blandford, New York: Am. Inst. Phys., p. 109 (1982).
29. A. M. Levine, et al., *Ap. J., Supp.,* **54,** 581 (1984).
30. G. H. Share, et al., *Ap. J. (Letters),* **358,** L45 (1990).
31. G. H. Share, et al., *Ap. J.,* **326,** 717 (1988).
32. M. J. Harris, et al., *Ap. J.,* **362,** 135 (1990).
33. N. Guessoum, R. Ramaty and R. E. Lingenfelter, *Ap. J.,* in press. (1991).

DIFFUSE 511 keV AND 1.8 MeV EMISSION IN THE GALACTIC PLANE

ON THE ANGULAR EXTENT OF THE GALACTIC 1.8 MEV LINE EMISSION FROM RADIOACTIVE ^{26}AL

V. Schönfelder and M. Varendorff
Max-Planck-Institut für extraterrestrische Physik,
8046 Garching, FRG

ABSTRACT

A measurement of the angular extent of the radioactive ^{26}Al line source along the galactic plane is essential to answer the question of its origin. Supernovae, novae, Wolf-Rayet stars, red giants or energetic explosions at the Galactic Center have been proposed as sources. A summary of previous observations of the gamma-ray line by HEAO-C, SMM, three different balloon borne Germanium spectrometers, and the balloon borne Compton-telescope of MPI is given. The results from the latter experiment are revisited. By applying the maximum entropy method to the data, the "smoothest" distribution is identified which gives the best fit to the data. Fits of other galactic longitude distributions to the data are tested as well. The results from all six experiments are consistent with a diffuse interstellar origin of the ^{26}Al-line with enhanced emission from the Galactic Center region as suggested by various nova and thin hot plasma models in the Galaxy. A pure point source origin of the *entire* line-emission can be excluded at this time. A significant progress in the understanding of the origin of the 1.8 MeV line is expected from GRO. The two instruments COMPTEL and OSSE are expected to perform the first complete survey of the galactic plane in the light of the line.

INTRODUCTION

The observation of the 1.809 MeV gamma-ray line from radioactive ^{26}Al was a milestone for gamma-ray astronomy. It marked the first detection of a cosmic radioactive isotope by means of gamma-ray astronomy. The discovery was made by the high resolution Germanium spectrometer onboard HEAO-C [1,2].

Already in 1977 Ramaty and Lingenfelter [3] had speculated that ^{26}Al might be a promising radioisotope for gamma-ray astronomy. It decays by positron emission (82 %) or electron capture (18 %) into an excited state of ^{26}Mg; the excitation energy is emitted as 1.809 MeV (100 %), and 1.13 MeV (4 %) gamma-ray photons. Whereas Ramaty and Lingenfelter [3] had considered supernovae as the sites of ^{26}Al production, nowadays other objects like novae, red giants, and Wolf-Rayet stars are also discussed. Whatever the actual source of ^{26}Al is, due to its long decay time of $1.04 \cdot 10^6$ years it can still be observed in interstellar space millions of years after its production.

Because most of the candidate source types of the gamma-ray line are distributed differently throughout the galactic plane, the measurement of the galactic longitude profile of the line flux should provide the best means to identify the dominant production site.

Five different longitude distributuons - presenting possible source distributions - will be discussed throughout this paper and are illustrated in Fig. 1.

These are:

(1) A flat supernova distribution (similar to the longitude distribution of 100 MeV gamma rays), which was used by Leising and Clayton [4], and which is based on the radial distribution of molecular clouds in the Galaxy.

(2) A thin hot plasma distribution which was derived from Tenma [5] and Ginga [6] observations of the X-ray iron line at 6.7 keV and a thermal bremsstrahlung continuum in the Galaxy, and which was suggested by v. Ballmoos [7] to be representative of the ^{26}Al-distribution. This distribution is qualitatively similar to the distribution of the observed CO-flux integrated along the line of sight (Dame et al., 1987) [8]. In both cases the distribution is similar to the flat supernova distribution with an additional sharp peak in the central 2 to 3 degrees, which contributes ≈ 20 % to the emission of the central radian. The thin hot plasma distribution is assumed to mark the distribution of all kinds of objects, which produce large stellar winds or shock fronts like massive stars (Wolf-Rayet stars, AGB-stars, and young Type II supernovae) or novae.

(3) A nova distribution, labelled N1, which was derived by Mahoney et al. (1985) [9], and which follows the visual luminosity of the older disc population.

(4) A nova distribution, labelled N2, which is relatively more peaked towards the Galactic Center, and which was derived by Leising and Clayton [4] from the observed radial distribution of novae in M31.

(5) A point source origin at the Galactic Center (not shown in Fig. 1).

To date, observations of the 1.8 MeV line emission have been made by 6 different instruments; these are the HEAO-C [1, 2] and the SMM [10] gamma-ray spectrometers, the balloon borne gamma-ray spectrometer of the Bell-Sandia group [11], the balloon borne Compton telescope of the Max-Planck-Institute [12], and more recently, the two balloon borne Germanium spectrometers of a US/French collaboration [13], and of a GSFC/Bell-New M. collaboration [14]. The results of the 6 observations will be reviewed. All available informations on the line flux and on the angular extent of the line emission will be put together in order to identify the most probable source candidates. At the end of the paper the prospects of an indepth-study of the line-emission by GRO (OSSE and COMPTEL) will be given.

REVIEW OF OBSERVATIONS

1. The HEAO-C Spectrometer

The HEAO-C experiment consists of a cluster of 4 high purity Germanium coaxial detectors inside an active 6 cm thick CsI(Na) anticoincidence shield. The shield defines a collimator with an effective aperture of about 42° FWHM at 1.809 MeV. HEAO-C was a spinning spacecraft with a period of about 20 minutes. For the analysis of the ^{26}Al-results only those periods were used, when the spin axis was oriented towards the galactic pole. During these times the spectrometer scans the galactic equatorial plane, because the axis of its field-of-view points at right angles to the spacecraft spin axis.

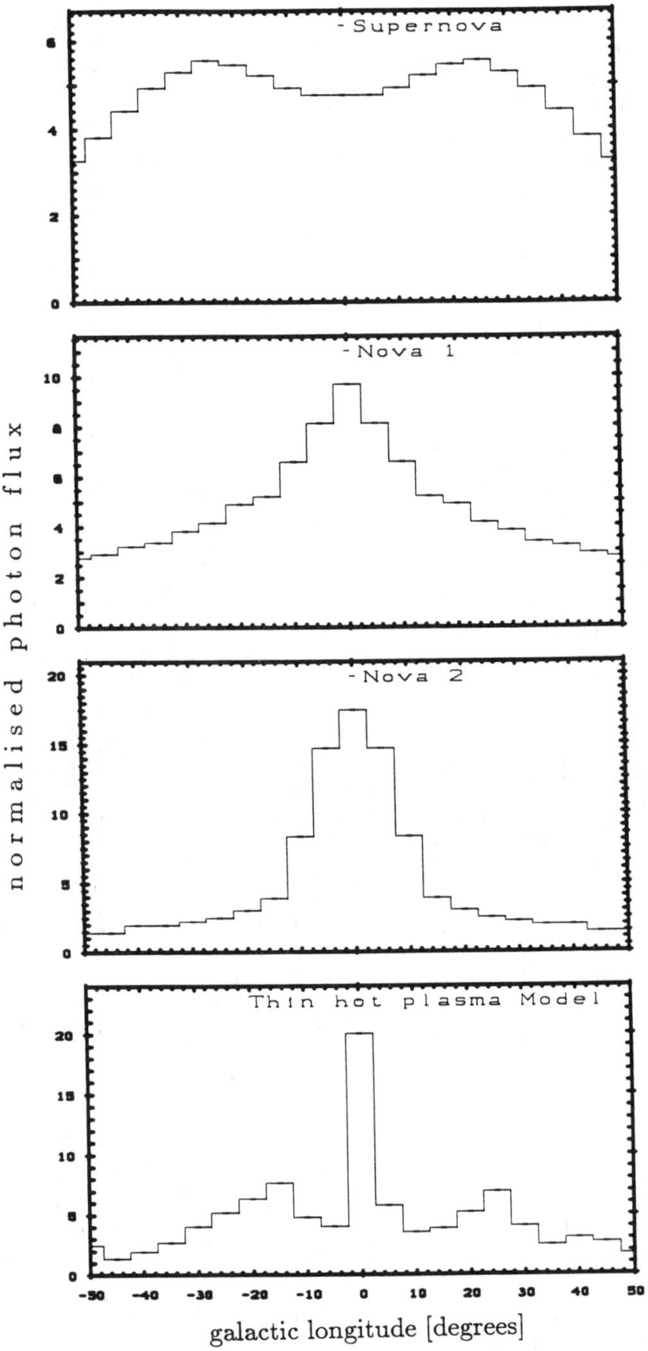

Fig. 1. The ^{26}Al galactic longitude models used throughout this paper (in addition to a pure point source origin at the Galactic Center). For references see text.

One of the major difficulties for analysing the HEAO-C data is due to the fact that at 1.8 MeV the anticoincidence shield is not opaque, but becomes significantly transparent. Due to this behaviour of the instrument, a pure background measurement is never possible: part of the source photons will always penetrate through the shield.

The HEAO-C experimenters have taken the following approach to determine the γ-ray line flux: they assumed a certain spatial distribution for the line emission along the galactic plane. The amplitude of the assumed distribution is obtained by fitting the observed count rate (per 2 keV energy bin) to the model distribution (convolved with the detector response).

The result of the analysis is shown in Fig. 2 over the energy interval 1760 to 1824 keV. The data points represent the net galactic plane emission under the assumption that the latter one has the same shape as the distribution of high energy gamma rays between 70-5000 MeV as measured by COS-B (top distribution of Fig. 1).

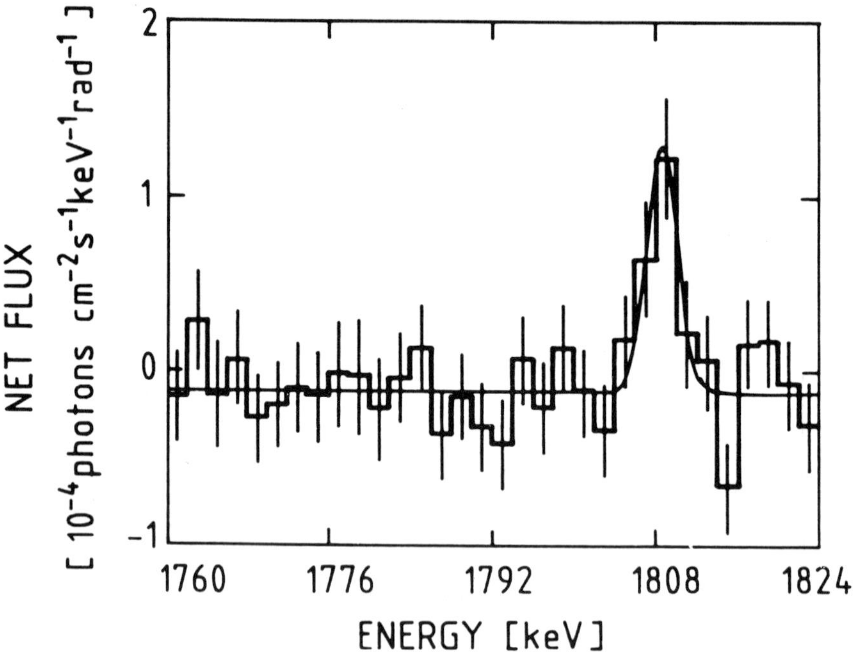

Fig. 2. Net galactic plane emission near 1.809 MeV as measured by HEAO-C.

The line near 1.809 MeV stands out clearly. The peak of the line is at 1808.49 ± 9.41 keV consistent with the 1808.65 ± keV emission from ^{26}Al at rest. The measured line width (smaller than 3 keV) is consistent with the instrument energy resolution - restricting the velocity of the ^{26}Al atoms to less than 250 km/sec. The resulting line intensity is $(4.3 \pm 0.8) \cdot 10^{-4}$ cm^{-2} sec^{-1} rad^{-1} at $l = o$. The statistical significance of the line detection ($\approx 5\sigma$) only weakly depends on the assumed longitude distribution, the value of the line flux, however, is model dependent (see later Table 2 and 3). Only in case of the extreme assumption of a point source origin at the position of the Galactic Center the significance of the detection goes down to 2.2 σ, and then has a value of $(1.4 \pm 0.9) \cdot 10^{-4}$ cm^{-2} sec^{-1} [15].

2. The SMM Observations

A few years after the discovery of the 1.8 MeV γ-ray line by HEAO-C its detection was confirmed by measurements with the SMM γ-ray spectrometer [10].

The SMM gamma-ray spectrometer - operating since 1980 - was designed for solar observations, and not for galactic studies. Still, important measurements of cosmic γ-radiation could be made. The spectrometer consists of seven 7.5 cm x 7.5 cm NaI scintillation detectors. A 2.5 cm thick annulus and a 7.6 cm thick back plate of CsI define an aperture of ≈ 160° FWHM at 1.8 MeV. The axis of the detector has been pointed to within 10° of the Sun throughout the mission. Due to the annual movement of the Earth around the Sun, a scan of the ecliptic plane is obtained within one year. Always in December the Sun crosses the galactic equator within 6° of the Galactic Center. Because of the large aperture of the instrument a source at that position is observable for about 3 months.

The signature of the 1.809 MeV ^{26}Al line in the SMM γ-ray data is visible in Fig. 3. In order to obtain Fig. 3, the data of SMM in the interval 1.6 to 2.0 MeV were fitted by a power law continuum and lines at 1.75 MeV and 1.81 MeV. The variation of the 1.809 MeV line fit during sky viewing periods is shown in Fig. 3 (top) as a function of time from 1980 to 1983.

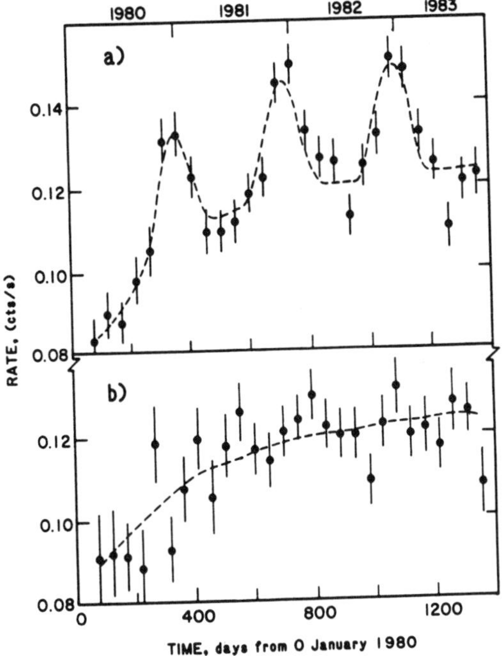

Fig. 3. Temporal variation of a line fit at 1.809 MeV to SMM data between 1.6 - 2.0 MeV. Top: Sky viewing data. Bottom: Earth occulted data (from Share et al., 1985) [10].

The gradual increase is due to the growth of the 1.79 MeV background line from the build up of radioactive ^{22}Na ($t_{1/2}$ = 2.6 years) in the spacecraft. Superimposed are the three increases which always peaked in December, when the Galactic Center was within the field-of-view. This annual modulation disappears, when Earth viewing data were used (i.e. when the Earth filled most of the aperture of the instrument); see Fig. 3 (bottom).

Modelling the galactic component of the observed 1.809 MeV countrate with a spatial distribution which again was assumed to be identical to the one observed by COS-B above 70 MeV yields a flux of (4.0 ± 0.4) 10^{-4} photons/cm^2 sec rad at l = o, which is consistent with the HEAO-C result. The observed line width of 102 ± 10 keV FWHM is consistent with the energy resolution of the SMM-spectrometer, and the significance of the detection of 10 σ ruled out any doubt about the existence of this line.

In a further analysis of the SMM-data [16] limits on the distribution of the 1.809 MeV radiation have been obtained by using the Earth as an occulting disk in the large field-of-view of the instrument. The ^{26}Al-emission from a certain source region along the galactic plane was derived from the difference of two observations: the data when the celestial region was between 1° to 13° below the horizon were subtracted from the data when the same source was 1° to 13° above the horizon. Fig. 4 shows the net ^{26}Al-rate as a function of galactic longitude from observations at different positions of the galactic plane.

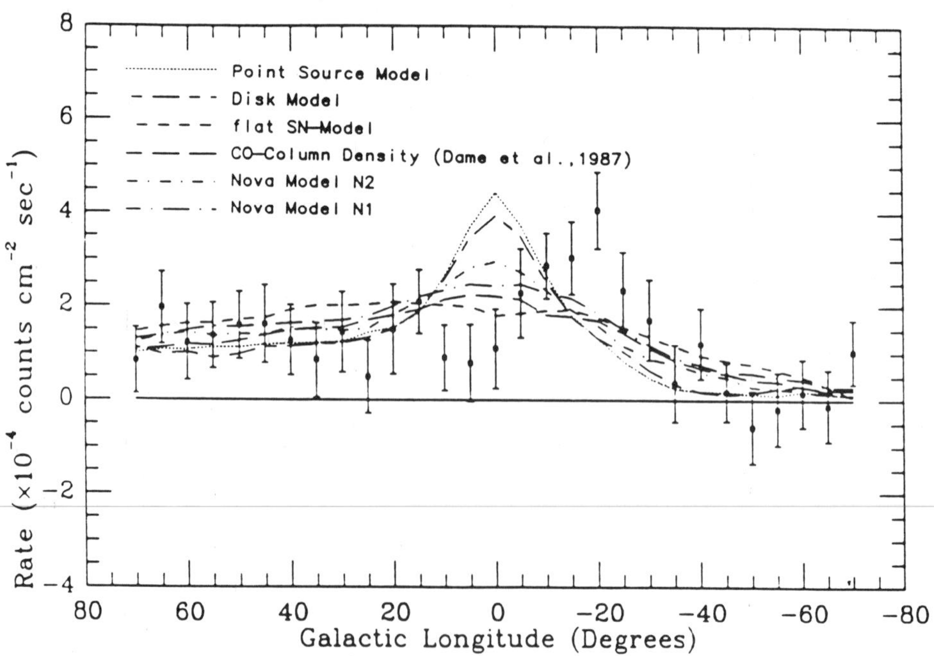

Fig. 4. Background corrected ^{26}Al gamma-ray line countrates from SMM-data using the Earth occultating technique. Also shown are the detector responses to 6 different ^{26}Al distribution models.

Also shown are fits of 6 different ^{26}Al model distributions to the data. Whereas a point source origin at the Galactic Center or an emission centered on the inner 10° around the Galactic Center (diskmodel) can be ruled out at the 4.7 σ or 4.1 σ confidence level, respectively, all the other diffuse emission models presented in Fig. 1 give acceptable fits to the data. The origin of a possible excess at l = -20° remains unsolved. The ^{26}Al gamma-ray line flux within the effective field-of-view of 22° FWHM was found to be $(1.1 \pm 1) \cdot 10^{-4}$ cm^{-2} sec^{-1}.

3. Observations by the Bell-Sandia Group

Constraints on the ^{26}Al distribution along the galactic plane were obtained from four balloon flights with the Bell-Sandia Germanium spectrometer [11]. The instrument consists of a single large Ge-detector surrounded by a thick active anticoincidence shield of NaI. The entrance aperture in the NaI shield defined a field-of-view of 15° FWHM at 1.8 MeV during 3 balloon flights and 87° FWHM during the fourth flight. No statistically significant detection of the ^{26}Al-line was achieved in any of the four flights. By combining all four flights, a point-source flux at the position of the Galactic Center of $(1.3 \pm 0.9) \cdot 10^{-4}$ cm^{-2} sec^{-1} was obtained at the 1.5 σ confidence level. The quoted errors contain statistical, but no systematic uncertainties.

4. Observations with the MPI-Compton Telescope

A first, still crude image of the Galactic Center region in the light of the 1.8 MeV-line was made by the balloon borne Compton telescope of the Max-Planck-Institut für extraterrestrische Physik. This imaging telescope has a wide field-of-view of about 60° FWHM and its angular resolution within the field is ≈ 10° FWHM (the image of a point source is smeared to a disk of ≈ 10° diameter). The telescope consists of an array of 16 cells of liquid scintillator NE213 (15 x 15 x 15 cm^3, each) as upper detector and an array of 32 blocks of NaI (Tl) (15 x 15 x 7.5 cm^3) as lower detector. The balloon flight took place in Brazil in October 1982. During the flight the center region of the galactic plane (a ± 30° broad band around the center) was within the field-of-view for about 2 hours.

During a previous analysis [12], a donought-shaped probability distribution for the arrival direction of each detected photon was projected onto the sky, and the sky image was derived from the multiplication of all these distributions. The result showed most of the emission to come from the Galactic Center: the data were best fitted by a narrow source with an extension smaller than or equal to the angular resolution, but wider distributions could not totally be excluded, e.g. the flat supernova distribution of Leising and Clayton [4] could only be rejected at the 2 σ confidence level. The γ-ray line fluxes obtained from this analysis were significantly (factor of ≈ 3) higher than those from the other experiments.

Since then a re-analysis of the balloon flight data has been performed. The new analysis was based on 2 steps: first, a maximum entropy image was derived from the flight data, and second, maximum likelihood fits between all possible source distributions presented in Fig. 1 and the flight data were performed in order to find out by which of these models the data be fitted best. The γ-ray line fluxes were derived from the results of these model fits.

The maximum entropy method requires the knowledge of the telescope response to any given sky image. The response of a Compton telescope can be described in the simplest way in a 3-dimensional data space defined by the scatter direction (χ, ψ) and the Compton scatter angle $\bar{\varphi}$. In the idealized case, in which the scattered gamma ray is totally absorbed in the lower detector, the pattern of all data points originating from a gamma-ray source with the coordinates (χ_0, ψ_0) lies on a cone in (χ, ψ, $\bar{\varphi}$)-space, where the cone apex is at χ_0, ψ_0) and the cone semi-angle is 45° (see Fig. 5).

The response density along the cone is given by the variation of the Klein-Nishina cross-section for Compton scattering. This idealized "cone mantle" response is blurred by measurement inadequacies in the scintillation detectors, especially by incomplete absorption in the lower detector.

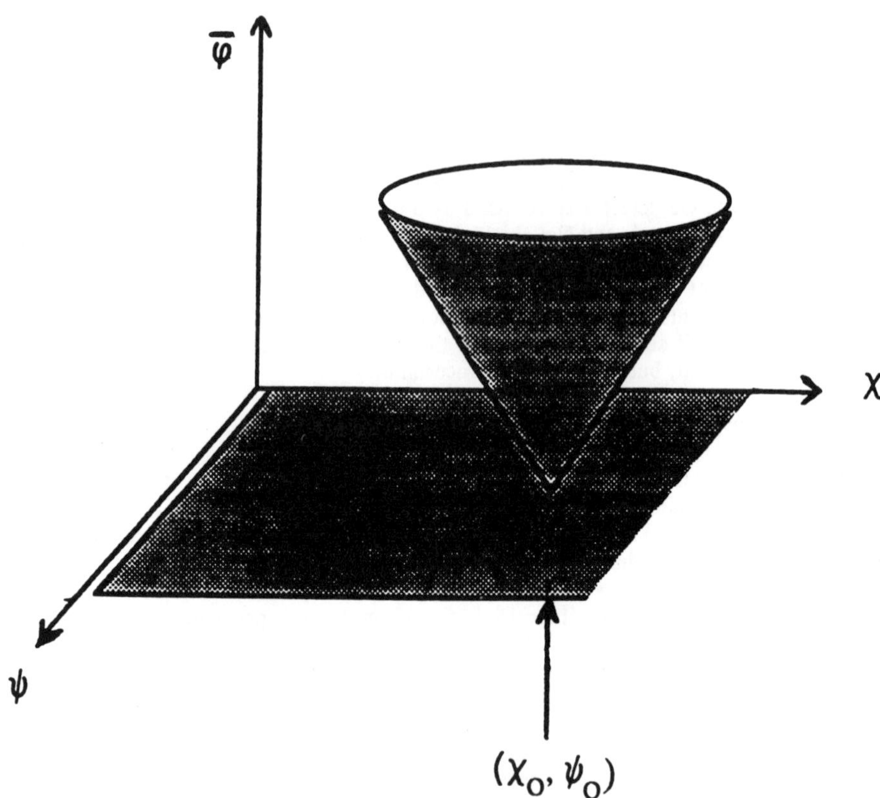

Fig. 5. Illustration of the Compton telescope response of a celestial point source in the 3-dimensional data space $(\chi, \psi, \overline{\varphi})$. The data lie on a cone, the apex of which is at the position of the celestial source (χ_0, ψ_0). The cone semi-angle is 45°. Actually the cone mantle is blurred due to measurement inaccuracies.

Therefore, the event density in data space can be represented by [17]:

$$(1)\quad n(\chi, \psi, \overline{\varphi}) = g(\chi, \psi) \cdot \int\int I(\chi_0, \psi_0)\, A(\chi_0, \psi_0)\, f(\chi - \chi_0, \psi - \psi_0, \overline{\varphi})\, d\chi_0\, d\psi_0$$

where I (χ_0, ψ_0) is the infalling sky intensity distribution. A (χ_0, ψ_0) is the exposure factor (product of effective detection area times observation times), f $(\chi-\chi_0, \psi-\psi_0, \overline{\varphi})$ is the "cone"-function and g (χ_0, ψ_0) is a geometrical function accounting for the incomplete coverage of the upper and lower planes by the detectors.

The maximum entropy image approach is to convolve a (2-dimensional) sky image with the full response of the telescope in the 3-dimensional dataspace (χ, ψ,$\bar{\varphi}$) and to try to match the "mock data" from the trial image to the measured data in the 3-dimensional dataspace. The trial image yielding a statistically acceptable match to the measured data, and at the same time fulfilling the entropy criterion, is defined to be the maximum entropy image.

The exposure factor and the "cone"-function of the Compton telescope were derived from calibration measurements and Monte Carlo simulations. All coordinates in equation (1) are in celestial coordinates, hence, the functions g ((χ, ψ) and A(χ_0, ψ_0) had to be transformed from the telescope system of reference into the celestial coordinate system.

Fig. 6 shows the resulting maximum entropy image of the balloon flight data [18]. The emission is strongly peaked towards the Galactic Center. The extent of the longitude distribution is ≈ 10° FWHM. This image is very similar to the one obtained earlier by the event-projection method [12].

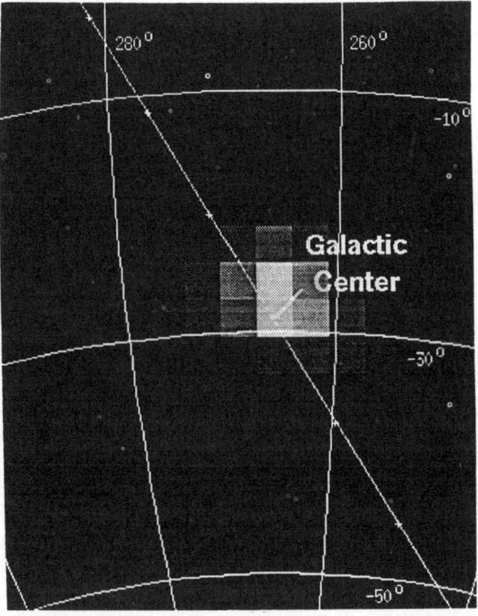

Fig. 6. Maximum entropy image of the central region of the Galaxy obtained from the balloon flight data of the MPI Compton telescope. The grey scale represents fluxes per bin.

In order to investigate the compatibility of the model distributions of Fig. 1 with the balloon flight data a χ^2-test was applied between the data and each of these distributions. For this purpose the model distributions had to be convolved with the response of the telescope so that the comparison with the data could be made in the three-dimensional (χ, ψ, $\bar{\varphi}$)-space. The results of the χ^2-test are listed in Table 1 for all 5 models.

Obviously, the data are better matched by distributions with an enhanced emission towards the Galactic Center than with a wide, flat longitude distribution. The flat supernova model can be excluded at the 2.1 σ confidence level. This conclusion is in agreement with our previous findings using the event-projection method [12].

The line fluxes for the various models were determined by a Monte Carlo simulation. In this simulation the sky was modelled by the longitude distributions. The measurements in the telescope were simulated under actual balloon flight conditions (track and azimuthal orientation of the balloon, dead-time effects). The simulated data were processed in the same way as the observed data, (i.e.maximum entropy images were derived for each of the model distributions). The simulated source strengths were chosen to give a comparable number of events in the image as the real measurements.

Table I: Results of Fitting the MPI Balloon data with Model Longitude Distributions

model	χ^2-probabilities
flat SN-model	7 %
N1 nova model	14 %
N2 nova model	32 %
thin hot plasma model	26 %
point source model	84 %

The fluxes are generally lower than the previously published ones [12]. The reasons are twofold: first, the number of events accepted for the analysis was higher as before. The addition of more events surprisingly lowered the overall flux though the additional events were obtained with comparable exposures; second, for the wider model distributions, especially the flat supernova distribution, the flux determination is now based on all events contained in the image, whereas previously, it was based on the smaller fraction of those events, whose event circles passed through a circle of 4.5° radius around the Galactic Center. This restriction lead to a higher uncertainty of the flux determination for the wider longitude distributions.

The resulting point source flux is $(4.6 \pm 2.3) \cdot 10^{-4}$ cm^{-2} sec^{-1}, whereas the flux from the central radian in the flat supernova model is $(4.9 \pm 2.4) \cdot 10^{-4}$ cm^{-2} sec^{-1}. The fluxes obtained for the other model distributions are given in the later Table III.

5. The GRIS-Experiment of GSFC-BELL-NEW MEXICO

The GRIS experiment [19] is a balloon borne high resolution spectrometer consisting of an array of seven Germanium detectors. Aperture holes in a thick NaI-anticoincidence shield above each Ge-detector define a 24° field-of-view at 1.8 MeV. During a balloon flight from Alice Springs/Australia on October 28 - 30, 1988 the Galactic plane was observed at two positions: at $l = o°$, and $l = 335°$. Background data were taken at the same zenit angle, but at azimuth off-set angles of 200° to 240° from the source observation. The derived ^{26}Al line fluxes for a point source at $l = o$ are $(1.9 \pm 0.8) \cdot 10^{-4}$ and $(2.1\,^{+1.1}_{-1.2}) \cdot 10^{-4}$ cm^{-2} sec^{-1}. For the flat supernova model fluxes at $l = o$ of $(4.2\,^{+1.9}_{-1.7}) \cdot 10^{-4}$ and $(5.4\,^{+2.9}_{-3.1}) \cdot 10^{-4}$ cm^{-2} sec^{-1} rad^{-1} are obtained. Both detections were made at the ≈ 2 σ confidence limit. Details of the observation and analysis are given by Teegarden [14] at this Symposium.

6. The US-French Balloon Borne High Resolution Gamma-Ray Spectrometer

The balloon borne high resolution spectrometer of the US-French collaboration [20] consists of 12 high purity Germanium detectors surrounded at the sides and from the rear by a thick anticoincidence shield of BGO-bars. A field-of-view of 22.5° at 1.8 MeV above each Ge-detector is defined by a 10 cm thick CsI (Na) shield with holes infront of the detectors. Observations of the ^{26}Al gamma-ray line were made during a balloon flight on May 22, 1989 from Alice Springs/Australia. The differential spectrum measured from the direction of the Galactic Center shows a narrow line at 1.805 ± 0.2 keV with a flux of (1.7 ± 0.8) · 10^{-4} cm^{-2} sec^{-1} within the 22.5° field-of-view. The detection was made also at the ≈ 2 σ confidence level. Details of the observations and the analysis are given in a contributed paper by Matteson [13] at this Symposium.

CONCLUSIONS FROM ALL EXISTING ^{26}AL-LINE OBSERVATIONS

All six so far existing observations of the 1.8 MeV γ-ray line have provided a measurement of the line flux. In addition, one of them, the MPI-Compton telescope observation has provided a first - still crude - map of the line emission from the central radian of the galactic plane.

A summary of the flux measurements is given in Table II. Here the fluxes obtained for the two extreme models - a point source and a supernova origin (flat distribution) - are compared with each other. The line flux obtained for the flat supernova model from the central radian of the Galaxy is about 4 · 10^{-4} cm^{-2} sec^{-1}; within the error bars all measurments are consistent with this value. For the point source model HEAO-C, SMM (using Earth occultation), and the three balloon experiments with Germanium detectors derived flux limits to a point source somewhere between (1 to 2) · 10^{-4} cm^{-2} sec^{-1}. If the total flux observed by the MPI-Compton telescope from the central radian is assumed to come from a point source, then a value of (4.6 ± 2.3) · 10^{-4} cm^{-2} sec^{-1} is obtained in contradiction to the other measurements.

Line fluxes for the two nova-models N1 and N2 were derived from the HEAO-C and the MPI balloon data. They are summarized in Table III. Again, the derived fluxes are consistent within the error bars of the measurements.

If the emission from the thin hot plasma model is normalized to a flux of 4.0 · 10^{-4} cm^{-2} sec^{-1} from the central radian, then it predicts an aperture flux of 1.6 · 10^{-4} cm^{-2} sec^{-1} within a 15° narrow field-of-view, which makes this model consistent with the flux results of all six measurements including the MPI-measurement, since this latter observation can be interpreted as superposition of a 20 % contribution from a narrow source component plus a 80 % contribution of the flat supernova model yielding in a total flux of 1.9 · 10^{-4} photons cm^{-2} sec^{-1} within the central 15°.

In summary it can be concluded that a pure point source origin of the *entire* line emission can now be excluded from the flux-measurements of the various experiments with narrow fields-of-view. Though, the image derived by the MPI-Compton telescope is consistent with a point source origin, the flux derived under this hypothesis is inconsistent with the results from the other experiments. Also SMM-analysis using Earth occultations rules out a point source origin at the 4.7 σ confidence level.

The flat supernova model (similar to a COS-B like distribution) leads to consistent line fluxes for all six observations. However, the image derived by the MPI Compton telescope fits this model worst. Still, it cannot be excluded defintitely. At present it is rejected at the 2 σ level, only.

The other three models with enhanced diffuse emission from the central region of the Galaxy, namely the two nova models N1 and N2, and the hot thin plasma model give consistent results with respect to the flux measurements of the six experiments, and also with respect to the image derived from the MPI-data.

Table II: Comparison of 1.8 MeV Line Fluxes Obtained from 6 Different Experiments

Instrument	FOV (FWHM)	Point source origin (Photons/cm^2 · s)	Supernova-Model (Photons/cm^2 · s · rad
MPE	(60°)	$(4.6 \pm 2.3) \cdot 10^{-4}$	$(4.9 \pm 2.4) \cdot 10^{-4}$
HEAO-C	(42°)	$(1.4 \pm 0.9) \cdot 10^{-4}$	$(4.3 \pm 0.8) \cdot 10^{-4}$
SMM	(130°)		$(4.0 \pm 0.4) \cdot 10^{-4}$
SMM	(22°)	$(1.1 \pm 1.0) \cdot 10^{-4}$	
BELL/SANDIA	(15°/87°)	$(1.3 \pm 0.9) \cdot 10^{-4}$	$(3.9 \pm 2.0) \cdot 10^{-4}$
GRIS	(24°)	$(1.9 \pm 0.8) \cdot 10^{-4}$	$(4.2 {}^{+1.9}_{-1.7}) \cdot 10^{-4}$ $(5.4 {}^{+2.9}_{-3.1}) \cdot 10^{-4}$
US/FRENCH	(22°)	$(1.7 \pm 0.8) \cdot 10^{-4}$	--

Table III: Comparison of the 1.8 MeV Line Fluxes from HEAO-C and the MPI Compton Telescope for the two Nova Models (photons/cm^2 sec rad at l = o).

Instrument	Nova-model 1	Nova-model 2
MPE	$(9 \pm 4.5) \cdot 10^{-4}$	$(14 \pm 7) \cdot 10^{-4}$
HEAO-C	$(7.3 \pm 1.5) \cdot 10^{-4}$	$(9.7 \pm 2,0) \cdot 10^{-4}$

OUTLOOK

A complete understanding of the origin of the ^{26}Al-line will only be possible after new, more extended measurements have become available. Such measurements are expected in the near future from the two GRO instruments OSSE [21] and COMPTEL [22]

These two instruments have the capabilities to map the galactic plane in the light of the 1.8 MeV line.

The sensitivity of OSSE for such a survey can be judged from Fig. 8. Here the OSSE 3 σ-sensitivity is compared with the line flux expected from the flat supernova and the N2 nova model after penetration of the γ-radiation through the instrument aperture. It is clear from this diagram that OSSE will mainly concentrate on mapping the galactic plane in the central 2 to 3 radians.

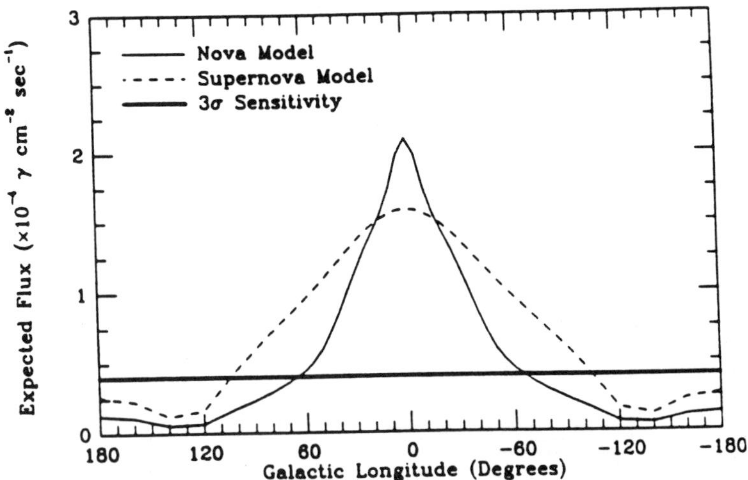

Fig. 8. OSSE 3 σ-sensitivity to map the galactic plane in the light of the 1.8 MeV ^{26}Al-line. From Kurfess et al. (1989) [23].

The COMPTEL sensitivity is illustrated in Fig. 9. Since COMPTEL is an imaging telescope, its sensitivity depends on the longitude bin size over which the emission is averaged. Practically, any bin size within the field-of-view of about 60° can be selected software-wise. In Fig. 9 the COMPTEL 3 σ-sensitivities are compared with the normalized line fluxes expected from the flat supernova and the two nova models N1 and N2 for two bin sizes, 10 and 40 degrees. It is clear from this diagram that COMPTEL will be able to map the entire galactic plane in the light of the ^{26}Al-line. Even in the case of the sharply peaked N2-distribution, the line should still be detectable in the anticenter region, if the bin size is made sufficiently wide.

We are all now impatiently waiting for the launch of the GRO in order to get a better understanding about the origin of the 1.8 MeV ^{26}Al-line.

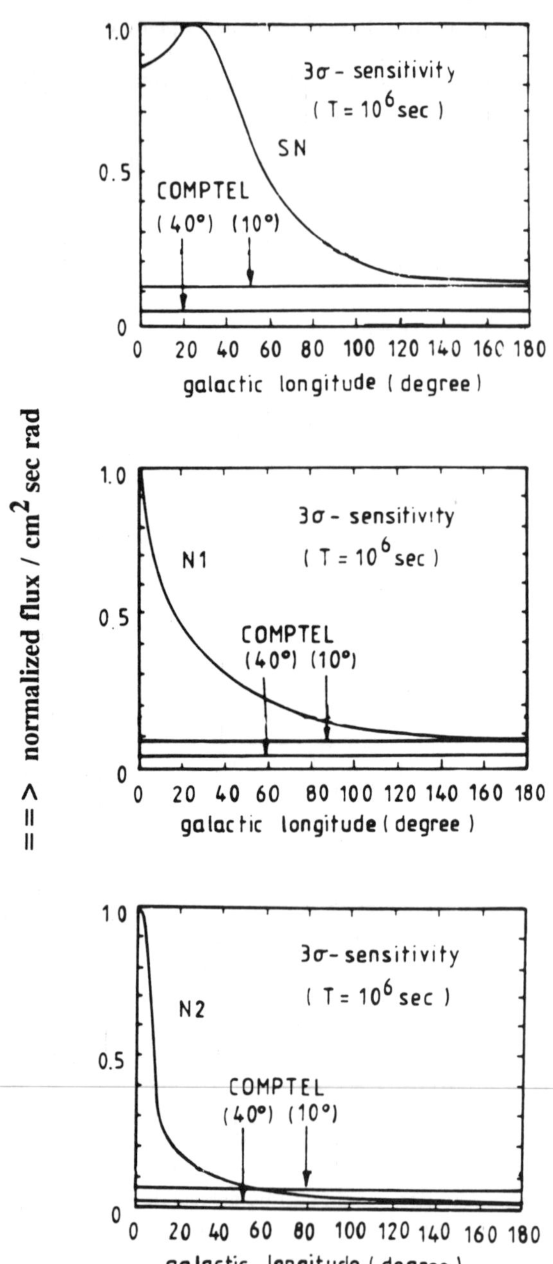

Fig. 9. Illustration of sensitivity of COMPTEL to map the galactic plane in the light of the ^{26}Al-line. The amplitudes of the 3 model distributions are normalized to 1. The sensitivity levels are at $2 \cdot 10^{-5}$ cm^{-2} sec^{-1} rad^{-1} for a 40° bin size and $6 \cdot 10^{-5}$ cm^{-2} sec^{-1} rad^{-1} for a 10° bin size.

REFERENCES

1. W.A. Mahoney, J.C. Ling, A.S. Jacobson, and R.E. Lingenfelter, Ap.J. 262, 742 (1982)
2. W.A. Mahoney, J.C. Ling, W.A. Wheaton, and A.S. Jacobson, Ap.J. 286, 578 (1984)
3. R. Ramaty and R.E. Lingenfelter, Ap.J. 213, L5 (1977)
4. M.D. Leising and D.D. Clayton, Ap.J. 294, 591 (1985)
5. K. Koyama, K. Makishima, Y. Tanaka, Publ. Astr. Soc Japan 38, 121 (1986)
6. K. Koyama, Publ. Astr. Soc Japan 41, 665 (1989)
7. P. von Ballmoos, preprint, Proc. of 28th COSPAR meeting, The Hague (1990)
8. T.M. Dame, H. Ungerechts, R.S. Cohen, E.J. de Geus, I.A. Grenter, J. May, D.C. Murphy, L.-A. Nyman, P. Thaddeus, Ap.J. 322, 706 (1987)
9. W.A. Mahoney, J.C. Higdon, W.A. Wheaton, A.S. Jacobson, Proc. of 19th International Cosmic Ray Conference (La Jolla) 1, 357 (1985)
10. G.H. Share, R.L. Kinzer, J.D. Kurfess, D.J. Forrest, E.L. Chupp, and E. Rieger, Ap.J. 292, L61 (1985)
11. C.J. MacCallum, A.F. Huters, P.D. Stang, and M. Leventhal, Ap.J. 317, 877 (1987)
12. P. von Ballmoos, R. Diehl, and V. Schönfelder, Ap.J. 318, 654 (1987)
13. J. Matteson, these proceedings (1990)
14. B. Teegarden, these proceddings (1990)
15. W.A. Mahoney, BAAS 18, 979 (1986)
16. W.R. Purcell, M.P. Ulmer, G.H. Share, and R.L. Kinzer, Preedings of GRO Workshop, GSFC, page 4-327 (1989)
17. A.W. Strong, K. Bennett, P. Cabeza-Orcel, A. Deerenberg, R. Diehl, J.W. den Herder, W. Hermsen, G. Lichti, J. Lockwood, M. McConnell, J. Macri, D. Morris, R. Much, J. Ryan, V. Schönfelder, G. Simpson, H. Steinle, B.N. Swanenburg, C. Winkler, Proc. of 21th International Cosmic Ray Conference, Adelaide, Vol 4, 154 (1990)
19. N. Gehrels, S. Barthelmy, B.N. Teegarden, J. Tueller, M. Leventhal, and C.J. MacCallum, Proc. of GRO-Workshop, GSFC, p. 4-247 (1989)
18. M. Varendorff, Thesis at Technische Universität München (1991)
20. J. Matteson, M. Pelling, B. Bowman, M. Briggs, R. Lingenfelter, L. Peterson, R. Lin, D. Smith, K. Hurley, R. Pehl, P. von Ballmoos, M. Niel, P. Durouchoux, Proc. of XXI International Cosmic Ray Conference, Adelaide, Vol 2, 174 (1990)
21. J.D. Kurfess, W.N. Johnson, R.L. Kinzer, G.H. Share, M.S. Strickman, M.P. Ulmer, D.D. Clayton, and C.S. Dyer, Adv. in Sp. Res., Vol. 3, no. 4, 109 (1983)
22. V. Schönfelder, R. Diehl, G.G. Lichti, H. Steinle, B.N. Swanenburg, A.J.M. Deerenberg, H. Aarts, J. Lockwood, J. Webber, J. Macri, J. Ryan, G. Simpson, B.G. Taylor, K. Bennett, and M. Snelling, IEEE-Trans. on Nucl. Sc., Vol. NS-31, no. 1, p. 766 (1984)
23. J.D. Kurfess, W.N. Johnson, R.L. Kinzer, M.D. Leising, G.H. Share, M.S. Strickman, D.A. Grabelsky, S.M. Matz, W.R. Purcell, M.P. Ulmer, R.A. Cameron, G.V. Jung, R.J. Murphy, D.D. Clayton, C.S. Dyer, C.M. Jensen, Proc. of GRO Workshop, GSFC, page 3-35 (1989)

GRIS OBSERVATIONS OF ^{26}AL GAMMA-RAY LINE EMISSION FROM THE GALACTIC PLANE

B. J. Teegarden, S. D. Barthelmy, N. Gehrels, J. Tueller
NASA/Goddard Space Flight Center

Marvin Leventhal
AT&T/Bell Laboratories

Crawford MacCallum
University of New Mexico

ABSTRACT

We report on measurements made during a balloon flight of the GRIS (Gamma-Ray Imaging Spectrometer) experiment over Alice Springs, Australia on 28-30 Oct. 1988. Two observations of the galactic plane were made at $l^{II} = 0^o$ and $l^{II} = 335^o$ of durations, 9.9 and 6.9 hr. respectively. ^{26}Al gamma-ray line emission (1809 keV) was detected during both observations, although the latter was statistically marginal. For an assumed galactic longitude distribution similar to high-energy cosmic rays, the inferred fluxes at $l^{II} = 0^o$ are 4.2 (+1.9,-1.7) x 10^{-4} cm^{-1} s^{-1} rad^{-1} (2.5 σ) and 5.4 (+2.9,-3.1) x 10^{-4} cm^{-1} s^{-1} rad^{-1} (1.7 σ) for the two observations. The two observations are therefore consistent with the assumed high energy gamma-ray distribution, but they are consistent with other distributions as well. The marginal statistics of the observations combined with the broad field-of-view (40° effective FWHM) do not permit meaningful tests of the model distributions to be made. The data do, however, suggest that the ^{26}Al emission is distributed over galactic longitude rather than confined to a point source.

1. INTRODUCTION

^{26}Al with its 10^6 yr mean life is a valuable tracer of nucleosynthesis processes in our galaxy. Ramaty and Lingenfelter[1] first suggested that the 1809 keV ^{26}Al line might be detectable in the galactic plane. Such emission was subsequently discovered by HEAO-3[2] and confirmed by Share *et al.*[3], von Ballmoos *et al.*[4], and MacCallum *et al.*[5] Supernovae, novae, AGB stars and Wolf-Rayet stars have been suggested as possible sources for the radioactive ^{26}Al. The galactic longitude distributions for these various types of sources are different. Hence determination of this distribution is one of the most important clues as to the origin of the emission. Thus far our knowledge of the distribution is meager. The results reported herein from the GRIS experiment are a first step towards mapping this galactic longitude distribution, but, as will be seen later, more powerful instruments are required to be able to make meaningful distinctions between the various candidate models.

2. OBSERVATIONS

The GRIS experiment is a balloon-borne high-resolution gamma-ray spectrometer. It consists of an array of seven germanium detectors surrounded by a thick active NaI anticoincidence shield. The results reported herein are from the second flight of the GRIS experiment over Alice Springs, Australia on 28-30 Oct. 1988. The total duration at float of the flight was 44 hr at a typical atmospheric depth of 5 g cm^{-2}. The galactic center ($l^{II} = 0^o$, $b^{II} = 0^o$) was observed for 9.9 hr on 29 Oct. On the following day, a point in the galactic plane 25° off the center ($l^{II} = 335^o$, $b^{II} = 0^o$) was observed for 6.9 hr. During both observations the normal observing mode was to alternate between target and background pointings every 20 min. Background was taken by rotating the experiment in azimuth to that point at which the galactic latitude is a maximum while holding the elevation constant. The one exception to this was when the galactic center was at high elevation near transit. During this period (~ 1.5 hr) the galactic center was tracked with no background pointings.

In deriving the source flux, it is important to take into account the effects of collimator side lobes and shield leakage, particularly if the ^{26}Al emission is distributed over a broad range in galactic longitude. Detailed Monte Carlo calculations were performed and verified with radioactive source calibrations. The nominal field-of-view of the GRIS instrument at 1809 keV is 24° FWHM. The typical shield leakage at 1809 keV is 5%. For the distributions under consideration, the contribution to the flux outside of the main field-of-view is typically 10% - 25% of the flux in the main field-of-view. Strong side-lobes due to collimator cross-hole trajectories significantly broaden the effective width of the aperture. The effective width at 1809 keV (defined as the width of the equivalent triangular aperture that would give the same number of detected counts for a flat longitude distribution) is 40°.

Differences were taken between the 20 min. target pointings and the two adjacent 10 min. background pointings. The final flux values were derived by taking weighted sums of these difference spectra. The analysis was performed under the assumption of different forms for the distribution of the incident ^{26}Al flux in galactic longitude. The assumed distributions were: high energy cosmic-ray (COS-B)[2,6], CO[7], nova[8], and point source. The number of counts/keV in the ^{26}Al line is given by:

$$\frac{\mathbf{dN}}{\mathbf{dE}} = \left[\int_0^{2\pi} \lambda(E)\mathbf{f}(\ell)\mathbf{R}(\alpha(\ell),\beta(\ell))\mathbf{T}(\mathbf{h},\varepsilon(\ell))\mathbf{d}\ell + \mathbf{B}\right]\mathbf{t}$$
$$= [\lambda \mathbf{I} + \mathbf{B}]\mathbf{t} \qquad (1)$$

where

$\ell =$ galactic longitude
$\lambda(E) =$ differential flux at $\ell = 0°$ (ph cm^{-2} s^{-1} kev^{-1} rad^{-1})
(Note: $\lambda(E)$ is broadened by the instrumental energy resolution.)
$\mathbf{f}(\ell) =$ flux distribution (normalized to unity at $\ell =$ 0)
$\alpha,\beta =$ angles in GRIS coordinate frame
$\mathbf{R}(\alpha(\ell),\beta(\ell)) =$ instrument reponse function (effective area vs. angle)
$\mathbf{T}(h,\varepsilon(\ell)) =$ atmospheric transmission
$h =$ atmospheric depth
$\varepsilon =$ elevation angle
$\mathbf{t} =$ duration of observation
$\mathbf{B} =$ instrument background (cts keV^{-1} s^{-1})
$\mathbf{I} \equiv \int_0^{2\pi} \mathbf{f}(\ell)\mathbf{R}(\alpha(\ell),\beta(\ell))\mathbf{T}(h,\varepsilon(\ell))\mathbf{d}\ell$

Then we can write for a target pointing,

$$(dN/dE)_T = [\lambda(E) I_T + B] t_T \tag{2}$$

and similarly for a background pointing,

$$(dN/dE)_B = [\lambda(E) I_B + B] t_B \tag{3}$$

Taking the difference between eqns. (2) and (3) and solving for λ gives:

$$\lambda(E) = \frac{(dN/dE)_T / t_T - (dN/dE)_T / t_B}{I_T - I_B} \tag{4}$$

λ(E) is calculated for each target-background pair and a weighted sum is taken to give the final flux at $l^{II} = 0^o$. This distribution is then fitted with a gaussian folded through the experimental response function to give the total line flux.

The flux spectra, as derived above, are plotted in Fig. 1 for the $l^{II} = 0^o$ and $l^{II} = 335^o$ pointings. Note that what is plotted is the flux at $l^{II} = 0^o$ for both the $l^{II} = 0^o$ and $l^{II} = 335^o$ pointings. The assumed distribution is COS-B (a smoothed version used by Mahoney *et al.*[2] and Share *et al.*[3]). The solid line is the best-fit gaussian folded through the GRIS response function (instrumental resolution = 2.8 keV FWHM at 1809 keV). The values for the integrated 1809 keV line fluxes are given in Table 1. Positive detections of the 1809 keV line were made for both pointings (2.5 σ and 1.7 σ, respectively), although the latter was statistically marginal.

Table 1. 1809 KEV LINE MEASUREMENTS (assuming COS-B distribution)

Measurement	Flux at $l^{II}=0^o$ (10^{-4} cm^{-1} s^{-1} rad^{-1})
GRIS	
$l^{II} = 0^o$	$4.2^{+1.9}_{-1.7}$ (2.5σ)
$l^{II} = 335^o$	$5.4^{+2.9}_{-3.1}$ (1.7σ)
HEAO-3 (Mahoney *et al.*)	4.8 ± 1.0
SMM (Share *et al.*)	4.0 ± 0.4
AT&T/Bell Labs.* (MacCallum *et al.*)	$3.9^{+2.0}_{-1.7}$

* Model-independent analysis.

The GRIS galactic center flux values from the $l^{II} = 0^o$ and $l^{II} = 335^o$ observations agree within errors, which means that the data are consistent with the assumed COS-B distribution. Also given are $l^{II} = 0^o$ flux values from several other experiments. These all appear to be in mutual agreement.

Figs. 2a-c show the two GRIS data points calculated for various assumptions about the galactic longitude distribution and compared with each of the assumed distributions. In each case the original distribution has been folded through the GRIS aperture response function. It is evident from Fig. 2 that no meaningful discrimination can be made between the various assumed distributions. This due to the combination of poor statistics and the broad side-lobe pattern of the GRIS aperture response function. Also, the 25° separation in pointings was chosen for the 511 keV observation and is far from optimum for the 1809 keV observation. A definitive determination of the shape of this distribution must await observation with an instrument such as the NAE/INTEGRAL which has a narrower field-of-view (10° FWHM) and two orders of magnitude greater sensitivity[9].

A similar analysis was carried out assuming a point source at the galactic center. The results are given in Table 2 and compared with other measurements. In a previous analysis of the balloon flight data of the MPI Compton telescope[4] most of the ^{26}Al emission was found to come from the galactic center region: the data were best fitted by a narrow source with an extension smaller than or equal to the angular resolution of the telescope (10° FWHM), but wider distributions could not totally be excluded, e.g. a flat galactic longitude distribution could be excluded only that the 2 σ level. For a point source origin, a flux of $(6.4 \pm 2.6) \times 10^{-4}$ cm^{-2} s^{-1} was obtained. A recent re-analysis of these data[11] has reduced the flux to $(4.6 \pm 2.3) \times 10^{-4}$ cm^{-2} s^{-1}. The derived image of the line emission, however, was found to be very similar to the one obtained earlier:

Table 2. 1809 KEV LINE MEASUREMENTS (assuming point source)

	Flux (10^{-4} cm^{-2} s^{-1})	**Field of View** (FWHM)
GRIS		
$l^{II} = 0^o$	1.9 ± 0.8	24° (see a)
$l^{II} = 335^o$	$2.1^{+1.1}_{-1.2}$	24° (see a)
AT&T/Bell Labs. (MacCallum *et al.* [5])	1.3 ± 0.9	15°
UC/France (Malet *et al.* [10])	1.9 ± 0.9	22.5^o
SMM (Share *et al.* 1985[3])	4.0 ± 0.4	130°
MPI (Schoenfelder and Varendorff 1991[11])	4.6 ± 2.3	60° (see b)

a - Effective FOV = 40°
b - Compton telescope.

the data are better matched by distributions with an enhanced emission towards the Galactic Center than with wide, flat longitude distributions.

We may ask, what is the probability that the five reported $l^{II} = 0^o$ values in Table 2 are consistent with a single value of the flux? A χ^2 calculation yields χ^2/ν = 3.38 for four degrees of freedom giving a probability $P_\chi(\chi^2,\nu) = 8 \times 10^{-3}$. Thus, the point source hypothesis is ruled out at the 0.8% level. The SMM data are also inconsistent with the point source hypothesis at the 4.7 σ level[12].

Best fit results for the 1809 keV line are give in Table 3. The quoted errors for the peak energies include 0.3 keV estimated systematic uncertainty in the energy calibration at 1809 keV. The peak energy for $l^{II} = 0^o$ lies 1.7 σ below the laboratory value. Both values for the FWHM suggest a broadened line, but the $l^{II} = 335^o$ measurement is also consistent with zero line width. To further explore the question of whether our 1809 keV line results are significantly different from the laboratory values, for the $l^{II} = 0^o$ measurement we have determined 2-parameter confidence regions following the procedure of Lampton, Margon, and Bowyer[13]. We find that the laboratory value (E_o = 1808.7 keV, FWHM = 0 keV) has a value $\Delta\chi^2 = \chi^2 - \chi^2_{min}$ = 3.13, which gives, for 2 degrees of freedom, $P_\chi(\chi^2,\nu) = 0.33$. Thus, it is not possible to make a strong statement about any broadening or peak energy shift of the ^{26}Al line.

3. SUMMARY AND CONCLUSIONS

The GRIS experiment has observed the ^{26}Al line at two points in the galactic plane ($l^{II} = 0^o$ and $l^{II} = 335^o$). The results are suggestive of a distributed source, but poor statistics (due to abbreviated observing time) do not permit a definitive statement to be made on the shape of the distribution. Assuming a distributed source that follows the high-energy gamma-rays, the derived GRIS results at $l^{II} = 0^o$ are consistent with most other similar measurements.[2,3,5] These measurements, taken together, are not consistent (at the 0.8% level) with a constant point source at the galactic center. The GRIS data hint at a possible broadening of the 1809 keV line (FWHM = 4.2 ± 2.4 /), but a 2-parameter error analysis reveals that this is significant only at the 33% el. Some broadening is expected due to galactic rotation. A velocity spread of ± km/s (typical of the inner few kpc of the galaxy) would give a ^{26}Al line width of keV (full width).

Table 3. 1809 KEV LINE FIT PARAMETERS (assuming COS-B distribution)

	Flux at l^{II}=0 (10^{-4} cm^{-2} s^{-1} rad^{-1})	**Peak Energy** (keV)	**FWHM** (keV)
= 0°	$4.2^{+1.9}_{-1.7}$	$1806.3^{+1.4}_{-1.6}$	4.2 ± 2.4
= 335°	$5.4^{+2.9}_{-3.1}$	$1808.5^{+2.1}_{-2.0}$	$5.5^{+4.5}_{-5.5}$

ACKNOWLEDGEMENTS

We wish to acknowledge useful and important discussions with Reuven Ramaty, Peter von Ballmoos, and Volker Schoenfelder.

REFERENCES

1. R. Ramaty and R. E. Lingenfelter, Ap. J., **213**, L5 (1977).
2. W. A. Mahoney, J. C.Ling, Wm. A.Wheaton, and A. S. Jacobson, Ap. J., **286**, L578 (1984).
3. G. H. Share, R. L. Kinzer, J. D. Kurfess, D. J. Forrest, E. L. Chupp, and E. Rieger, Ap. J., **292**, L61 (1985).
4. P. von Ballmoos, R. Diehl, and V. Schonfelder, Ap. J., **318**, 654 (1987).
5. C. J. MacCallum, A. F. Huters, P. D. Stang, and M. Leventhal, Ap. J., **317**, 877 (1987).
6. H. A. Mayer-Hasselwander *et al.*, Astr. Ap., **105**, 164 (1982).
7. T. M. Dame *et al.*, Ap. J., **322**, 706 (1987).
8. J. C. Higdon, and W. A. Fowler, Ap. J., **339**, 956 (1989).
9. J. L. Matteson, in High Energy Astrophysics in the 21st Century, ed. P. C. Joss (AIP, New York, 1989), p. 343.
10. M. Malet *et al.*, in Symposium on Gamma-Ray Line Astrophysics, ed. P. Durouchoux and N. Prantzos (AIP, New York,1991) in press.
11. V. Schoenfelder and M. Varendorff, in Symposium on Gamma-Ray Line Astrophysics, ed. P. Durouchoux and N. Prantzos (AIP, New York, 1991) in press.
12. W. R. Purcell, M. P. Ulmer, G. H. Share, and R. L. Kinzer, in Proceedings of the GRO Science Workshop (Greenbelt, MD, 1989), p. 4-327.
13. M. Lampton, B. Margon, and S. Bowyer, Ap. J., **208**, 177 (1976).

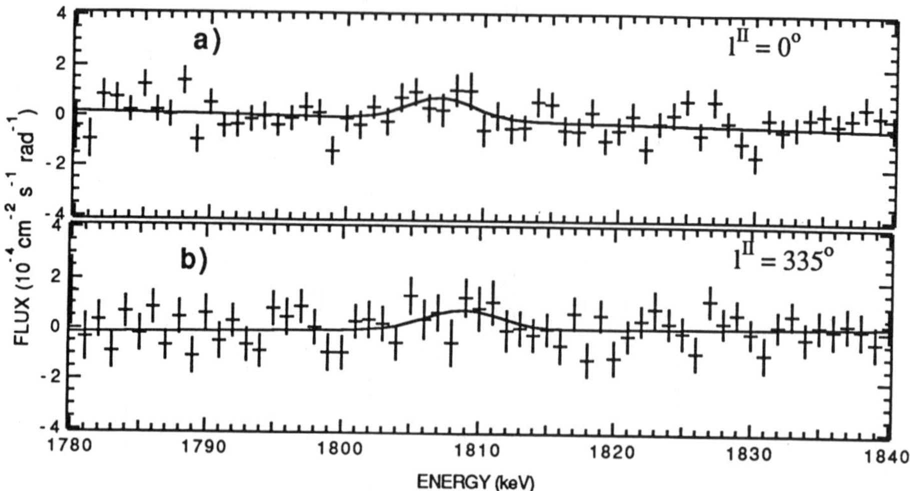

Fig. 1. GRIS spectra of the ^{26}Al line. A high-energy gamma-ray distribution (COS-B[2,6]) is assumed for the flux calculation. Solid lines are best fit gaussians folded through the GRIS instrumental response function. a) $l^{II} = 0°$; b) $l^{II} = 335°$.

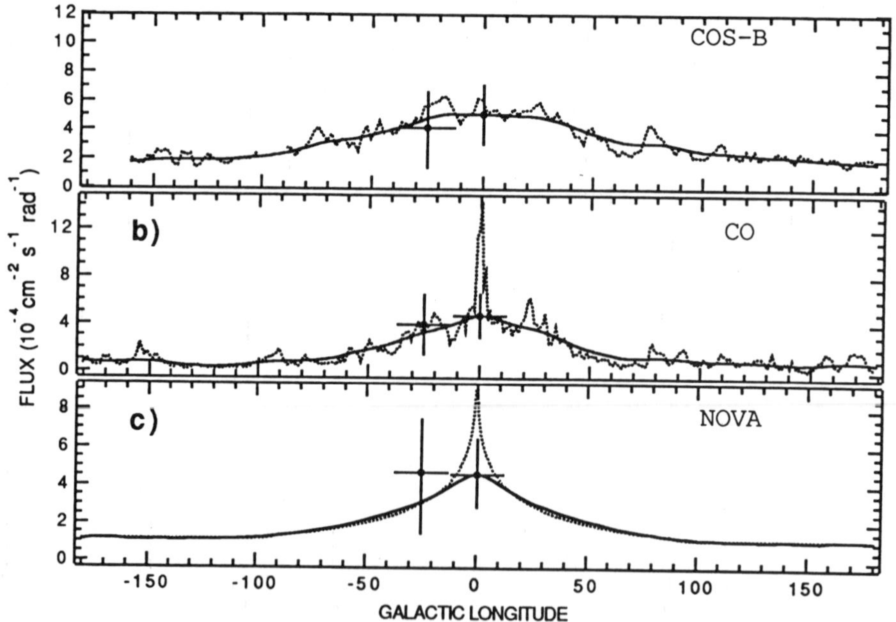

Fig. 2. Comparison of GRIS results with various assumed galactic plane distributions. Dotted line is measured data. Solid line is measured data folded through GRIS aperture response function. a) COS-B distribution[2,6]; b) CO distribution[7]; c) nova distribution[8].

OBSERVATION OF THE GALACTIC 1809 KEV GAMMA-RAY LINE WITH THE HEXAGONE SPECTROMETER

I. Malet, M. Niel, G. Vedrenne, P. von Ballmoos
Centre d'Etude Spatiale des Rayonnements
31029 Toulouse Cedex, France

B. Bowman, M. Briggs, D. Gruber, J. Matteson, M. Pelling, L. Peterson
Center for Astrophysics and Space Sciences
University of California, San Diego
La Jolla, CA 92093 USA

P. Feffer, K. Hurley, R. Lin, D. Smith
Space Science Laboratory
University of California, Berkeley
Berkeley, CA 94720 USA

C. Cork, D.Landis, P. Luke, N. Madden, D. Malone, R. Pehl, M. Pollard
Lawrence Berkeley Laboratory
Berkeley, CA 94720 USA

C. Chapuis, P. Durouchoux
Service d'Astrophysique
Centre d'Etudes Nucleaires de Saclay
91191 Gif sur Yvette Cedex, France

ABSTRACT

We report an observation of the galactic 1809 keV gamma-ray line produced by radioactive ^{26}Al in the interstellar medium. The measurement was performed with our high resolution germanium spectrometer HEXAGONE on a balloon flight in May 1989 from Alice Springs, Australia. Our differential spectrum of the Galactic Center region shows a narrow line at 1809 keV corresponding to a flux of (1.9 +/- 0.9)$\cdot 10^{-4}$ photons$\cdot$cm^{-2}s^{-1} assuming a source at the Galactic Center. We discuss the available observations of the 1809 keV line in the context of models that have been proposed for the origin of the galactic ^{26}Al.

INTRODUCTION

Gamma-ray spectroscopy is a unique tool to study the sites and rates of ongoing nucleosynthesis in- and outside our Galaxy. In 1977 Ramaty and Lingenfelter[1] recognized that a narrow gamma-ray line at 1809 keV should be detectable from the decay of ^{26}Al in the interstellar medium. ^{26}Al is unstable to positron emission with a mean lifetime of $1.0 \cdot 10^{6}$ yr. Its decay feeds the first excited state of ^{26}Mg and the de-excitation photon from this state gives a line at 1809 keV. The first detection of the 1809 keV galactic gamma-ray line was made a couple of years later with the high resolution germanium spectrometer on board HEAO C[2]. The line intensity reported was of $(4.8 \pm 1.0) \cdot 10^{-4}$ ph/cm^{2}·s·rad. A confirmation of the line detection was performed by measurements with the SMM spectrometer [3,4], with a line flux in the central radian of the galactic plane of $(3.5 \ -1.3+0.8) \cdot 10^{-4}$ ph/cm^{2}·s. Both the HEAO C and SMM observations are consistent with a broad galactic origin of the line emission. The intensity in the 1809 keV line corresponds to about 3 M_o of ^{26}Al in the interstellar medium.

In 1982 the 1809 keV line was observed with the imaging Compton telescope of the MPI [5]. Their skymap shows a narrow peak of ^{26}Al emission in the Galactic Center region. Results from the Bell/Sandia spectrometer[6] included data collected on four balloon flights occuring between 1977 and 1984. The equivalent point source flux from the Galactic Center is $(1.3 \pm 0.9) \cdot 10^{-4}$ ph/cm^{2}·s, favoring extended galactic distributions like the HEAO C and SMM observations. The large area germanium spectromenter GRIS has observed the GC region on two balloon flights in 1988. Results are presented in these proceedings[7].

Various attempts have been made to explain the observed γ-ray line emission; different source models have been proposed[2,8,5,9-12], but the question on the origin of ^{26}Al is still unsolved.

In this paper, we report results from the maiden balloon flight of the 12 Detector Germanium Spectrometer HEXAGONE on May 22, 1989 from Alice Springs, Australia. We confirm the presence of a narrow line at 1809 keV in the Galactic Center region and calculate its flux assuming a point source at the GC.

INSTRUMENT AND RESULTS

HEXAGONE was developed by a collaboration of US and French scientists to observe astrophysical gamma-ray line sources. The instrument was designed to have a sensitivity of 10^{-4} ph.cm^{-2}.s^{-1} for nuclear gamma ray lines, and to prove new instrumental techniques that may be applied to future space missions. It consists of 12 Germanium detectors operated at cryogenic temperature. The detectors are surrounded by 240 kg of BGO (rear and sides, thickness 5 cm) and 60 kg of CsI (10 cm thick, front) which are viewed by 61 photomultipliers. Drilled holes in the CsI part of the anticoincidence shield provide collimation to 18° FWHM. At energies above 300 keV, gamma-rays outside the 18° aperture can penetrate the CsI causing a gradual increase in the aperture's FWHM, to 25° at 5 MeV. The twelve 5.5 cm Ø * 5.6 cm long detectors have closed-end coaxial geometry, resulting in a total volume of ~1600 cm^{3}. The relevant characteristics of HEXAGONE are given in table 1; for a more detailed description of this instrument see Matteson et al.[13].

HEXAGONE had its first balloon flight on May 22, 1989 from Alice Springs, Australia. The instrument performed nearly flawlessly. Carried on a Raven 28 million cubic foot balloon, it was at float for 17.4 hours and observed SN1987A for 9.9 hours, the Galactic Center for 6.3 hours (just over half a transit) and the Crab Nebula and the transient X-ray source A0535+26 for 1.3 hours.

TABLE 1. Instrument Parameters

Energy range	:	15 keV - 12 MeV
Energy resolution	:	3.1 keV (FWHM @1.8 MeV)
Field of view	:	22.5° (FWHM @ 1.8 MeV)
Total detector area	:	285 cm^2
Effective area	:	32 cm^2 @1.8 MeV

Four of the detectors had fully operational segmentation and Pulse Shape Discrimination however background reduction from these techniques have not been exploited for the present data analysis. The observing program consisted of a sequence of 20 minute source pointings and 20 minute background pointings. Background was taken by maintaining the elevation angle and changing the azimuth by usually 180° for maximum source modulation. The elevation angle of the Galactic Center varied between a minimum value of 18° at the very beginning of the observations and 83° at transit while the residual atmospheric depth varied from 3.7 to 4.7 g/cm^2.

The data has been energy-calibrated by means of three background lines: a strong line at 198.4 keV from neutron activation of the Ge detectors (^{70}Ge (n, γ) ^{71m}Ge) which escapes our anticoincidence because of its long lifetime $t_{1/2}$ = 21.9 ms; the e^+e^- annihilation line at 511.0 keV and a line of the ^{40}K onboard calibration source at 1460.8 keV. A set of nonlinear calibration curves, parametrized in order to allow for variation with time is used for each detector segment. Both the source and the background spectra show the presence of an instrumental line at 1809 keV. The instrumental component in this line comes from the decay of the first excited state of ^{26}Mg resulting largely from the reaction $^{27}Al(n,np)^{26}Mg^*$. Its energy is the same as for the ^{26}Al line, 1808.7 keV. Spectral fits with a Gaussian and a powerlaw resulted in a line energy of 1808.5 +/- 0.2 keV and a width of 3.9 +/- 0.2 keV FWHM showing that our energy calibration is correct.

Since our shield is not completely opaque to 1809 keV gamma-rays the source flux will depend on the 'on-off' modulation and thus on the assumed source distribution. In the present analysis only one source model for the ^{26}Al is considered: a points source at the GC. The source modulation has been calculated for the geometry of every single observation during the GC transit. The shield transmission for the 'off' pointings varied between 6% and 35% with a mean of 13%. Atmospheric transmission ranged from 56% to 80% during the transit of the Galactic Center.

In a first attempt to deduce a source flux for the galactic ^{26}Al line the "on" and "off" data of the GC observation were subtracted after live time normalization. Two different methods have been used: Subtracting of the 'off' counts (119 counts) from the 'on' counts (140 counts) in a bin 4 keV large centered on the fitted BG line energy, and subtracting the integrals of the 'on-off' Gauss-fits. In both cases, the differential spectrum still shows an excess at 1809 keV, however only with marginal significance consistent with a flux of $(1.8 \pm 1.3)\cdot 10^{-4}$ $ph\cdot cm^{-2}\cdot s^{-1}$.

Background statistics were improved by adding observations of the supernova 1987A with equivalent elevation angles. No measurable source flux at 1809 keV is expected from that position. Variation of the background line is estimated to be unimportant because of the small changes in altitude and geomagnetic cutoff rigidity - possible systematic errors of this will be discussed below. SN background pointings have 100% 'off' modulation since the GC was below horizon. The two methods of BG subtracting have again been applied resulting in nearly equal 'on' - 'off' countrates ($3.4 \pm 1.6 \cdot 10^{-4}$ counts$\cdot$s^{-1}). The source flux remains virtually the same as in the analysis including only the GC Background - $(1.9 \pm 0.9) \cdot 10^{-4}$ ph$\cdot$cm$^{-2}\cdot$s^{-1} - while the significance is enhanced slightly to $> 2\sigma$.

TEST FOR SYSTEMATIC ERRORS

The assistance of SN1987A-observations for better background definition demands an understanding of the degree of neutron activation as a function of time. Since the southern latitude of the balloon decreased by 1.5° during the balloon flight we expect a reduction in the neutron activation with time. This decrease could overestimate the 1809 keV background flux derived from the SN1987A observations which were made early in the flight. We have estimated the variation in neutron activation by studying the behavior of a strong background line at 198 keV resulting from ^{70}Ge(n,γ)^{71m}Ge. Taking into account the overall decrease in this line of less than 6%, the underestimation of the differential flux due to neutron activation should be < 10%. The reliability of our background treatment has been tested with the two instrumental lines at 1764 keV and 1779 keV. While the 1764 keV line comes from the decay of on board ^{238}U (and thus subtracts independently of activation), both the 1779 keV and 1809 keV background lines have their origin in neutron activation of the instrument (see also [2]). The 1779 keV background line results mainly from neutron capture ^{27}Al (n,γ)^{28}Al, like the 1809 keV background line, followed by β-decay to ^{28}Si. It is particularly comforting that the 1779 keV background line is subtracted to within 0.6σ.

DISCUSSION

Source distributions for the 1809 keV emission have been proposed by several authors (see Introduction) however none of the sources seem to satisfy all observational constraints - leaving the question on the origin of the galactic ^{26}Al still unresolved. Supernovae, novae, massive stars (in particular during their Wolf-Rayet phase), red giants and explosion of supermassive stars have been proposed as production site. The total ^{26}Al mass of $3M_{\odot}$ in our Galaxy estimated from the measurements of the 1809 keV line is causing a major problem for nucleosynthesis theory.

Supernovae were the first sites proposed for ^{26}Al production. They have been suggested by Ramaty and Lingenfelter[1] even before the discovery of the 1809 keV line. The anticipated ^{26}Al yield for SN has since gone through several 'ups' and 'downs'. While Woosley and Weaver[14] estimate the ^{26}Al production of supernovae to be only 0.1 $M_{\odot}$, Signore and Dupraz[12] find that SN are more important contributors, producing about four times more ^{26}Al than massive stars in their pre-explosive phases. According to Weiss and Truran[15] who recently have reevaluated the yield of novae, neon-rich novae could represent an important though perhaps not dominant source of ^{26}Al (<1.5 $M_{\odot}$ of ^{26}Al). Prantzos and Cassé[9] found that Wolf-

Rayet stars may contribute significantly to the galactic ^{26}Al in the first stage of their evolution and a very optimistic model involving 8000 WR stars could account for all of the observed ^{26}Al. This suggestion has since been contradicted by Signore and Dupraz[12] who estimate that WR stars contribute at most 0.15 M_o of ^{26}Al.

All authors have major difficulties in explaining the total observed ~3 M_o of ^{26}Al. Walter and Maeder[11] calculate a total amount of ~ 0.4 M_o of ^{26}Al, taking into account the combined yield of all the various sources. Likewise, the present 1809 keV observations do not allow for identification of the source population responsible for the ^{26}Al production. This is primarily due to the poor angular resolution of the present instruments and at least partially due to the uncertain galactic distribution of the sources in question. In the following we try to constrain the extension of the 1809 keV emission with the existing measurements.

Local origin of the ^{26}Al resulting from Wolf-Rayet stars and supernovae activity forming a large bubble in the local ISM has been proposed by Blake and Dearborn [10]. Because of the proximity of the bubble almost uniform 1809 keV emission is predicted within about one steradian centered on the Sco-Cen association (~20° away from the GC). Several observational facts seem to rule out such a model: The geometry of a single bubble in the LISM is incompatible with the measurement of the MPI Compton telescope[5]. Their 4σ detection would not have been possible since almost all of their 'off'-pointings were lying within the proposed bubble therefore canceling out any signal. The Bell/Sandia instrument measured a flux of $1.3 \cdot 10^{-4}$ ph·cm^{-2}·s^{-1} within its 15° field of view (FWHW). Would the ^{26}Al be distributed in the local bubble, this detector should have measured a flux of only ~$0.2 \cdot 10^{-4}$ ph·cm^{-2}·s^{-1} assuming that SMM (FOV 130° FWHM) measured $4 \cdot 10^{-4}$ ph·cm^{-2}·s^{-1} for the entire bubble. Our present analysis of the HEXAGONE data also indicates a flux too high to come from the local bubble. A model-dependent analysis has yet to be accomplished.

The above evidence seems to definitely rule out a *local* bubble as the major contributor to the observed ^{26}Al. However a system of numerous *galactic bubbles or superbubbles* as as the origin of the 1809 keV emission would comply with both observational and theoretical evidence[16,17]. Since all objects producing ^{26}Al have in common the injection of energy into the interstellar medium in the form of explosions and strong winds, high velocity shock fronts are formed with temperatures $>10^7$ °K. Bubble structures of such hot low density gas maintain temperatures of $>10^6$ °K for over 10^4 years. Regardless of the source type that caused these bubbles (massive stars and/or supernovae) thermal bremsstrahlung is emitted revealing the sites of nucleosynthesis. Bremsstrahlung from the galactic ridge and center has been detected by several X-ray telescopes (EXOSAT, TENMA, Einstein IPC, Spacelab 2 and SPARTAN 1) indicating a system of partly overlapping bubbles filling a large fraction of the interstellar space with a strong concentration at the GC. Recently the X-ray satellite GINGA[18] measured a distribution of the galactic ridge thin hot gas with enhanced features in the 5 kpc spiral arms (l=± 25°) and at the Galactic Center where a narrow peak contains a third of the total emission in the central radian. It has been suggested[17] that this distribution reflects the origin of the gamma ray line emission at 1809 keV and 511 keV. Such a model complies with our measurement and the detection of the Bell/Sandia instrument when normalized to the SMM flux ($3.5 \cdot 10^{-4}$ γ/cm^2·s^1 in the central radian). The distribution can reconciliate the skymap of the MPI Compton telescope which shows a large fraction of the 1809 keV emission to come from the GC with the measurements of HEAO C and SMM which require an extended galactic origin of the ^{26}Al.

Further analysis of our data is in course to establish source fluxes for extended galactic distributions; this work will be published elsewhere. NASA's Gamma-ray Observatory (GRO) will soon be able to put rigorous constraints on the source models. With their superior sensitivity and angular resolution GRO's OSSE and COMPTEL experiments will be able to map the large scale distribution of the 1809 keV line emission. Narrow line spectroscopy of the 1809 keV line is currently studied by NASA balloon programs (GRIS and HEXAGONE) - a breakthrough is expected by the planed NAE or INTEGRAL missions.

We are grateful to our colleagues F. Cotin, J. Couteret, J. Coutelier and S. Slassi at CESR, F.Duttweiler, P. James, G. Huszar, G. Beriones and G. Allen at UCSD, H. Primbsch, S. McBride, J. Penegor, and B. Campbell at UCB and J. Poulalion at CEN-Saclay. We thank B. Kubara and his team of NSBF for flying our instrument on a difficult but successful balloon campaign.

REFERENCES

[1] Ramaty R. and Lingenfelter R.E., *Ap.J.*, **213**, L5, 1977
[2] Mahoney W.A., Ling J.C., Wheaton W.A., and Jacobson A.S., *Ap.J.*, **286**,578, 1977
[3] Share G.H., Kinzer R.L., Kurfess J.D., Forrest D.J., Chupp E.L., Rieger E. , *Ap.J.*, **292**, L61,1985
[4] Harris M.J., Share G.H., Leising M.D., Kinzer R.L., Messina D.C., *Ap.J.*, **362**, 1990
[5] von Ballmoos P., Diehl R., and Schönfelder V., *Ap.J.*, **318**, 654, 1987
[6] MacCallum C.J., Huters A.F., Stang P.D., and Leventhal M., *Ap.J.*, **317**, 877, 1987
[7] Teegarden B. et al., these proceedings,1991
[8] Leising M.D., and Clayton D.D., *Ap.J.*, **294**, 103, 1985
[9] Prantzos N. and Cassé M., *Ap.J.*, **307**, 324, 1986
[10] Blake J.B. and Dearborn D.S.P., *Ap.J.*, **338**, L17, 1986
[11] Walter R., and Maeder A., *Astron.Astrophys.*, **218**, 123, 1989
[12] Signore M. and Dupraz C., *Astron.Astrophys.*, in press, 1990
[13] Matteson J. et al., Proc of the 28 COSPAR 1990, The Hague, to be published in *Adv.Space Res,* 1991
[14] Woosley M. and Weaver T., in *Nucleosynthesis and its Implication on Nuclear and Particle Physics*, eds. Audouze, Mathieu; Reidel, Dordrecht, p. 145,1986
[15] Weiss A. and Truran J.W., accepted for publ. in *Astron.Astrophys*, 1990
[16] Montmerle T., in *Advances in Nuclear Astrophysics* ed. E.Vangioni-Flam et al. Edition Frontieres , p.335, 1986
[17] von Ballmoos P., Proc. of the 28 COSPAR 1990, The Hague, to appear in *Adv.Space Res,* 1991
[18] Koyama K., Awaki H., Kuneida H., Takano S., Tawara Y., Yamauchi S., Hatsukade I., Nagase F., *Nature*, **339**, 603, 1989

PRODUCTION AND DISTRIBUTION OF ^{26}Al IN THE GALAXY: THE ROLE OF MASSIVE STARS

Nikos Prantzos

Institut d' Astrophysique de Paris
98bis, Bd. Arago, 75014 Paris
and
Service d' Astrophysique, CEN Saclay
91191 Gif sur Yvette, FRANCE

ABSTRACT

The various sources of production of ^{26}Al in the Galaxy are discussed, in the light of recent observational and theoretical data. It is found that with current stellar and nucleosynthesis models, all the candidate sources still fall short (by factors of >5) from producing the 3 $M_\odot/10^6$ y required by the observations. The corresponding flux distribution in the galactic plane is also discussed; contrary to previous claims, it is argued that this signature cannot easily help to discriminate between the various candidate sources. Possibilities of unambiguous identification of massive stars are discussed, and relevant flux profiles taking into account the spiral structure of the Galaxy are presented.

INTRODUCTION

^{26}Al is unstable to positron emission with a mean lifetime of $\tau_{26} \sim 10^6$ y. Its decay feeds the first excited state of ^{26}Mg at 1.809 MeV, the de-excitation of which gives rise to a gamma-ray photon. The detection of the 1.8 MeV γ-ray line from the galactic center (GC) direction by the HEAO-3 satellite[24], attributed to the decay of $\sim$3 $M_\odot$ of ^{26}Al in the interstellar medium (ISM), is a discovery of paramount importance[7,12]: indeed, ^{26}Al is the first radioactive nucleus ever seen in extrasolar gamma-ray astronomy (the second one being ^{56}Co in SN1987A, detected through its 847 keV and 1238 keV lines). This discovery, along with that of Tc at the surface of S stars (by Merill in 1952) clearly demonstrates that nucleosynthesis is currently active in the Galaxy, since the lifetime of those nuclei (a million years or so) is very short compared to the time-scale of galactic chemical evolution; moreover, it offers an opportunity to confront nucleosynthesis theories with observational data.

That is why ^{26}Al, which can be produced in substantial amounts in various astrophysical sites (like novae, supernovae, massive mass losing stars, red giants, etc.) has been proposed as early as 1977 by Ramaty and Lingenfelter[35] as an interesting candidate for γ-ray line astronomy. Its subsequent detection in the ISM, confirmed by several experiments up to now, came as no surprise, but the origin of the emission discovered by HEAO-3 is not clear as yet. Indeed, the inferred quantity of $\sim$3 $M_\odot$ of galactic ^{26}Al seems difficult to explain with current models of nucleosynthesis[8,32]: the galactic populations of all the proposed sources fall short by factors >5 from reproducing that amount of ^{26}Al in the last $\sim 10^6$ years; however, current uncertainties in the modelling of all the sources, as well as in their frequency of occurence in the Galaxy, do not allow any definite conclusion yet.

In view of that difficulty encountered by theory, it has been expected that observations with good angular resolution could help, mapping the distribution of the 1.8 MeV emission in the Galaxy and thus revealing the distribution and the nature of the underlying sources[4,16,21,33] (making the implicit, and quite plausible, assumption that during its $\sim 10^6$ y lifetime ^{26}Al does not move very far away from its source). Although correct in principle, this expectation faces considerable difficulties in practice: indeed, detectors with angular resolution and sensitivity an order of magnitude better than currently available are needed for a reasonable mapping of the 1.8 MeV emission; and even in that case (which is attainable with next generation gamma-ray experiments), it will be difficult to discriminate between the various classes of proposed sources, since their galactic distributions are poorly known.

The plan of this paper is as follows: In Sect. 1 we summarize the observational data concerning the galactic emission at 1.8 MeV. In Sect. 2 we review the various nucleosynthetic mechanisms and the possible astrophysical sources of ^{26}Al, with an emphasis on the uncertainties affecting the corresponding yields. In Sect. 3 we evaluate the production of the galactic population of each class of those sources (especially the one of massive stars) during the last 10^6 yr, making plausible assumptions about their frequency and distribution in the Galaxy. These assumptions allow a modelling of the distribution of the corresponding 1.8 MeV flux in the Galactic plane, which is also presented in Sect. 3 for the case of massive stars (Wolf-Rayet and supernovae); for the first time, the spiral structure of our Galaxy is taken into account in such a calculation. The possibilities of an unambiguous identification of massive stars, through the profile of their 1.8 MeV emission, are discussed.

OBSERVATIONS OF THE GALACTIC 1.8 MeV EMISSION

The 1.8 MeV line due to the decay of ^{26}Al in the ISM has been detected by six different experiments, up to now: the HEAO-3 and SMM satellites and four balloon borne experiments. Because of poor statistics and angular resolution (FWHM $>15°$), a source distribution in galactic longitude has to be assumed for the deconvolution of the data. Obviously, the derived flux in the GC direction depends on that distribution, which is usually taken to be that of high energy ($>$70 MeV) gamma-rays, detected by the COS-B satellite[27]. The main results concerning the observations of the 1.8 MeV emission are the following (for a more detailed presentation, see Ref. 36 in this volume):

The γ-ray experiment on board the HEAO-3 satellite had an energy resolution of 3.3 keV (FWHM) and a field of view of 42° at 1.8 MeV. The measured width of the 1.8 MeV line (FWHM $<$ 3 keV) implies velocity dispersions of the emitting ^{26}Al atoms less than 250 km s^{-1}, a limit compatible with emission from the ISM. Assuming that the line emission has the same longitudinal profile as the high energy γ-rays detected by COS-B, a γ-ray flux F = 4.8$\pm$0.7 10^{-4}

ph cm^{-2} s^{-1} rad^{-1} toward the GC region is derived[24]; for a point source in the GC, the corresponding flux is F = 1.4 ± 0.9 10^{-4} ph cm^{-2} s^{-1}. A reanalysis of the data with different source distributions leads to similar results[25].

The NaI scintillator detectors of the SMM experiment (130° FWHM field of view) detected the line at the 10σ confidence level and confirmed its observed parameters, i.e. energy and width, as well as flux[38,39]. The line flux is F = 4.±0.4 10^{-4} ph cm^{-2} s^{-1} rad^{-1}, if the COS-B distribution is again assumed. The data is not consistent with a point source centered at the GC, at the 4.7 σ level.

A balloon borne experiment of the Max Planck Institute (MPI), equipped with a Compton telescope, detected the 1.8 MeV line at the 4.5σ level and localized it within 10° of the GC. It was the first detection of the 1.8 MeV emission with an instrument having imaging capabilities. The data does not exclude a diffuse galactic emission, but it is more consistent with a point source located at the GC[49,53]. A recent reanalysis[36] of the data confirms that the emission is peaked towards the GC, the derived flux being lower than the original claims: F = 4.9 ±2.4 10^{-4} ph cm^{-2} s^{-1} rad^{-1} for a COS-B distribution, and F = 4.6±2.3 10^{-4} ph cm^{-2} s^{-1} for a point source in the GC, the latter case being favoured w.r.t. all diffuse source models.

A point source of ^{26}Al in the GC is not favoured, either by the two satellite experiments, or by the data obtained from four balloon observations of AT&T/Bell Labs. between 1977 and 1984 with a 15° field of view (FWHM) instrument[22]. The combined result for all four experiments gives a flux of 3.9 +2.0/-1.7 10^{-4} cm^{-2} s^{-1} rad^{-1} for a COS-B distribution, quite consistent with the satellite experiments; the equivalent point source flux from the GC direction is 1.6±0.9 10^{-4} cm^{-2} s^{-1}, lower than the MPI results. The data favour a distributed source over a point source at the 90% confidence level.

The most recent observations of the Galactic 1.8 MeV line emission were made with balloon borne instruments with similar fields of view: the GRIS experiment in October 1988 (FWHM = 24°), and the HEXAGONE collaboration in May 1989 (FWHM = 22.5°). In the latter case, an emission from the GC direction was detected at the ~2 σ level[26]; the corresponding flux F = 1.9±0.9 10^{-4} ph cm^{-2} s^{-1} is consistent with previous measurements, but the results were analyzed only for the case of a point source in the GC up to now.

The GRIS experiment performed the first measurement of the 1.8 MeV emission *off* the GC direction. Assuming that the source distribution is similar to the COS-B one, the GRIS data give a flux[45] F = 5.4 +2.9/-3.1 10^{-4} ph cm^{-2} s^{-1} rad^{-1} (1.7 σ level) for a galactic longitude l=335°, while towards the GC direction (l=0°) a flux F = 4.2 +1.9/-1.7 10^{-4} ph cm^{-2} s^{-1} rad^{-1} is obtained (2.5 σ level). Although the first step towards a mapping of the galactic distribution of the 1.8 MeV emission, the observations of GRIS do not allow any conclusion on the distribution of the ^{26}Al sources, being consistent with almost all the proposed distributions (except for a point source in the GC direction).

The observational results can be summarized as follows: a) a diffuse galactic emission with F ~ 4 10^{-4} ph cm^{-2} s^{-1} rad^{-1} in the GC direction is favoured by all the experiments but one; b) the longitudinal dependence of this emission is presently unknown, because of the poor angular resolution and sensitivity of the experiments.

A quantitative analysis[15] shows that for an axisymmetric distribution of sources in the Galactic plane (i.e. around the GC), the flux received on a detector from the direction of the GC at a distance of R_{GC} ~10 kpc is numerically given by: F (ph cm^{-2} s^{-1} rad^{-1}) ~ 1. 10^{-46} Q, where Q (in ph s^{-1})is the total galactic γ-ray emissivity (the exact form of the distribution or the distance to the GC affect slightly the numerical coefficient in the above formula, i.e. by less than a factor of two, at least for $7.5<R_{GC}(kpc)<10$). Interpreted through this theoretical relationship, the above observations imply that $N_{26}\sim 10^{56}$ nuclei or $M_{26}\sim 3\ M_\odot$ of ^{26}Al are currently present in the ISM (since $Q = dN_{26}/dt = -\lambda N_{26}$); assuming that we do not live in a privileged epoch, i.e. equilibrium between the production and decay rates of ^{26}Al, one concludes that the average production rate of ^{26}Al in the Galaxy is $\sim 3\ M_\odot/10^6$ y, at least during the last several million years.

ASTROPHYSICAL SOURCES OF ^{26}Al

The main production mechanism of ^{26}Al in astrophysical environments is proton capture on ^{25}Mg [$^{25}Mg(p,\gamma)^{26}Al$]. Thus, the astrophysical regimes of nuclear burning leading to the production of ^{26}Al are (at least in principle): H burning (both hydrostatic and explosive) and C-Ne burning [where substantial quantities of Mg exist, and protons are secondary particles, produced by reactions like e.g. $^{12}C(^{12}C,p)$]. On the other hand, for the gamma-ray photons of its decay to be observable, ^{26}Al has to be ejected in the ISM before destruction. This can be easily achieved in the case of an explosive site, like e.g. a nova or a supernova, as well as in the case of an object suffering extensive mass loss, like a Wolf-Rayet star or a red giant (in the Asymptotic Giant Branch phase or AGB). Obviously, the different regimes of temperature, density, timescale and initial composition in those astrophysical sites imply different modes for the production and destruction of this radioactive nucleus[2,32].

Notice that, for temperatures $T < 4\ 10^8$ K the short-lived ($\tau \sim 7$ s) isomeric state $^{26}Al^m$ (E=226 keV) is not thermalized (e.g. its population with respect to the ground state $^{26}Al^g$ is not given by the equations of statistical equilibrium) and it should be treated as a separate species in detailed nucleosynthesis computations[52].

Hydrostatic nucleosynthesis of ^{26}Al (WR and AGB stars)

Hydrostatic core H burning can lead to the production of significant amounts of ^{26}Al if the central temperature of the star is $T_c >$35-40 10^6 K, i.e. for stars more massive than $\sim 30\ M_\odot$. For temperatures $T<45\ 10^6$ K, ^{26}Al is predominantly destroyed by $^{26}Al(\beta^+)^{26}Mg$, while for higher temperatures $^{26}Al(p,\gamma)$ is the main destruction mechanism in that environment[33]. As the convective stellar core gradually retreats (while the star still burns H on the main sequence) it leaves behind ^{26}Al that has been previously produced and mixed in the core. Thus, some ^{26}Al is found in the stellar enveloppe, where it β^+ decays to ^{26}Mg.

Whether or not it will appear at the stellar surface depends on the mass loss. For stars with M>30-40 $M_\odot$ (depending on metallicity) radiatively driven winds induce such high mass loss rates that the initial envelope of the star is expelled and regions processed by H burning finally emerge at the surface: this is the standard senario for the formation of single *Wolf-Rayet* (WR) stars. Obviously, the higher the mass loss rate and/or the extent of the convective core, the more ^{26}Al will be ejected. During the subsequent phase of central He burning, neutrons released through ^{13}C(α,n) efficiently destroy the remaining ^{26}Al in the stellar core, through (n,α) and (n,p) reactions. However, ^{26}Al continues to be ejected from the stellar envelope, and disappears only when the He burning products appear, in their turn, at the surface[33].

The nucleosynthesis of ^{26}Al in massive mass losing stars has been studied by several authors[9,10,33,34,51]. Despite the use of quite different stellar evolution codes and different prescriptions for convection and mass loss, those calculations lead to rather similar results (to better than 30%): stars in the mass range 40 $< M(M_\odot) <$ 120 eject in the ISM a few 10^{-5} - 10^{-4} $M_\odot$ of ^{26}Al (depending on stellar mass), during the ~2-5 10^5 y of their WR stage; the corresponding 1.8 MeV luminosity (averaged over the duration of the WR stage) is[33]: $L_\gamma \sim 10^{37}$-10^{38} ph s^{-1}.

For this work, we updated our previous[34] calculations using the recently remeasured[18] reaction rate of ^{25}Mg(p,γ)^{26}Al (I90 in the following). This new rate is lower than previously thought[6] (CF88 in the following) by a factor ~5 for temperatures T<80 10^6 K. However, most of the calculations of ^{26}Al production in WR stars were done with an even older value of that rate[5] (lower by ~2-4 than I90 for T>40 10^6 K and higher by ~2-5 than I90 for T<40 10^6 K). In our new calculations the ^{26}Al yield is slightly enhanced (by ~5-15%, depending on the stellar mass) w.r.t. our earlier estimates. These new results and those of a previous calculation[51] with different stellar models (and the older rate) are plotted in Fig. 1. Adopting an Initial Mass function (IMF) of the Salpeter type (i.e. a power law with index α=-2.50), we find an average yield of ~4.5 10^{-5} $M_\odot$ for WR stars with progenitor masses M>40 $M_\odot$, i.e. slightly larger than estimated in previous works. [Notice that the I90 rate for ^{25}Mg(p,γ) is still uncertain by considerable factors for T<20 10^6 K. A substantial increase in that temperature range could lead to an almost complete transformation of ^{25}Mg (solar mass fraction: X_{25}=6.8 10^{-5}) to ^{26}Al in those stars; their yields in that case would still increase by a factor of 2-3, depending on the stellar mass].

Hydrostatic shell H burning takes place at somewhat higher temperatures i.e. T ~70-90 10^6 K. In the case of massive stars (M > 10 $M_\odot$) this site does not seem very promising for the production of ^{26}Al: indeed, for the most massive ones (M>30 $M_\odot$, which could develop extended H burning shells), mass loss considerably reduces the importance of shell burning, while for the less massive ones (M<30 $M_\odot$) the small H shell allows only negligible amounts of ^{26}Al to be produced (and eventually ejected by a SNII explosion). Hydrostatic shell burning in *intermediate mass stars* (1-7 $M_\odot$) during the AGB phase is a different story. During that phase of its evolution, the star consists of a degen-

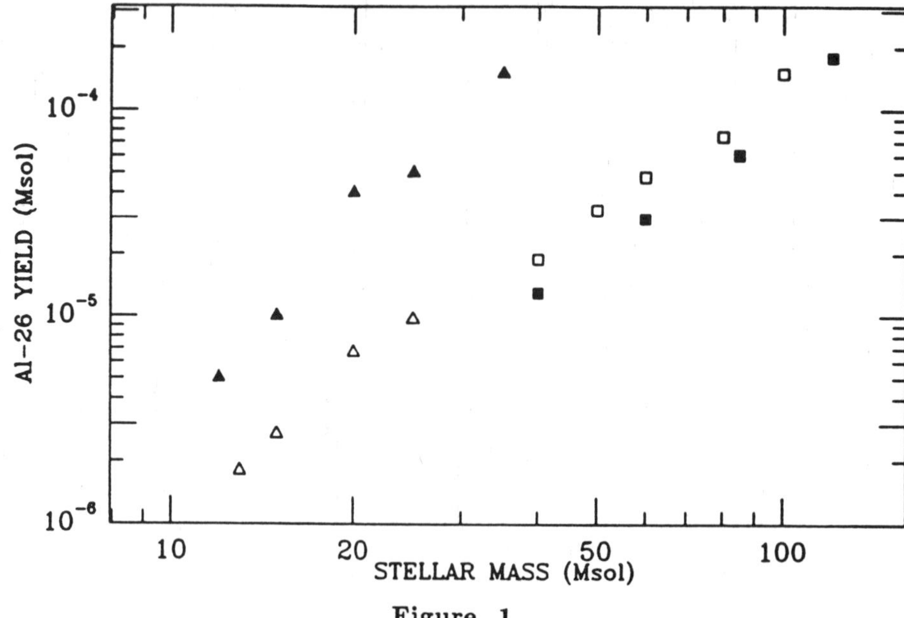

Figure 1

Yields of ^{26}Al from SNII (triangles) and WR stars (squares), according to recent nucleosynthesis calculations. *Open triangles*: Thielemann (1990); *black triangles*: Woosley (1990); *open squares*: Prantzos (1990-this work); *black squares*: Walter and Maeder (1989). The discrepancies between the various calculations are currently of a factor of 4-5 for SNII, and of <30% for WR stars.

erate C-O core, surrounded by a He burning shell, a H burning shell and (part of) the original envelope. This double-shell burning phase is unstable, leading to the developpement of thermal pulses, which progressively expell the stellar envelope on a time-scale of a few 10^5-10^6 y (depending on the stellar mass), forming a planetary nebula. At the base of the He shell, where temperatures are typically T~4 10^8 K, ^{14}N (a product of H burning through the CNO cycle) is converted to $^{25,26}Mg$ through $^{14}N(\alpha,\gamma)^{18}F(\beta^+)^{18}O(\alpha,\gamma)^{22}Ne(\alpha,\gamma)^{26}Mg$ or $^{22}Ne(\alpha,n)^{25}Mg$ (the neutrons of the last reaction being used for the synthesis of s- nuclei). Following the shell He-flash, an extended convective zone develops above the He shell, transporting those nuclei up in the H shell. There, ^{25}Mg is converted to ^{26}Al, at temperatures T~70-90 10^6 K (depending on stellar mass and adopted model). Finally, the convective envelope of the red giant penetrates the underlying H shell and ^{26}Al is brought at the surface and ejected in the ISM. Needless to say that no satisfactory model exists yet for such a complicated scenario, where convection plays a key role.

The nucleosynthesis of ^{26}Al in the envelopes of AGB stars has been originally evaluated[30] on the basis of previous estimates[19] concerning the thermodynamic conditions, the *dredging – up* and the mass loss of stars with M $\sim$ 5 – 7 $M_\odot$, during this evolutionary phase. It was found that nucleosynthesis at the bottom of the convective H envelope may produce abundance ratios $X_{26}/X_{27} \sim$ 0.5 – 1. On the basis of those computations it has been argued[3,13] in a very qualitative way, that red giants could significantly contribute to the production of ^{26}Al at the Galactic level (due to a lack of accurate stellar evolution models, most of those calculations assume that all the initial ^{25}Mg in the H envelope is transformed in ^{26}Al; this is a far from proven conjecture, which leads to an upper limit of 2-3 10^{-5} $M_\odot$ of ^{26}Al per AGB star). The first realistic calculations of nucleosynthesis in that site have been recently performed for stars of 1 and 3 $M_\odot$, and preliminary results are presented in this volume[31]; they suggest that an AGB star may eject up to 3 10^{-8} $M_\odot$ of ^{26}Al (much less than the above qualitative estimates), in a timescale of 7 10^6 y. The average production rate of ^{26}Al by those objects is then $\sim$ 5 10^{-9} $M_\odot/10^6$ y/AGB star. It is, however, very difficult to evaluate the uncertainties due to input physics (e.g. nuclear reaction rates, treatment of convection etc.) to the ^{26}Al yield in that site.

Explosive nucleosynthesis of ^{26}Al (novae and supernovae)

Explosive H burning at peak temperatures T_p $\sim$200-400 10^6 K is thought to occur during *novae* explosions. According to the currently accepted theoretical model for a nova[42], accretion of a critical mass (10^{-4}-10^{-5} $M_\odot$) of H-rich fuel on the surface of a white dwarf leads to a thermal runaway. The accreted material, and a small part of the crust of the white dwarf (which has been mixed in the envelope) are expelled in the ISM after being nuclearly processed. Studies of nucleosynthesis in such conditions show that, depending on the caracteristic timescale of the explosion ($\sim$a few tens to hundreds of seconds), not only ^{25}Mg but also the more abundant ^{24}Mg can be transformed in ^{26}Al, by two successive proton captures (^{24}Mg(p,γ)^{25}Al(β^+)^{25}Mg(p,γ)^{26}Al), provided there is enough time for the β^+ decay of ^{25}Al to take place; ^{26}Al(p,γ) is still the main destruction mechanism for ^{26}Al in that site. On the other hand, some leakage out of the Mg-Al cycle starts occuring at the higher peak temperatures, through ^{26}Al(p,γ)^{27}Si(p,γ)^{28}P etc. (instead of ^{27}Si(β^+)^{27}Al(p,α)^{24}Mg).

The nucleosynthesis of ^{26}Al in novae has been studied mostly in one zone calculations up to now (i.e. with no coupling between the nuclear reaction network and the hydrodynamics). Most of them[2,11,17,20,54,55,56] adopted thermodynamic conditions corresponding to realistic novae models[42,43] (peak temperatures T_p $\sim$150-400 10^6 K and densities ρ_p $\sim$$10^4$ g cm^{-3}) and found a production ratio $X_{26}/X_{27} \sim$ 0.1 - 1; the corresponding mass fraction is X_{26} $\sim$$10^{-4}$, for material with solar initial composition. A major revision[55] in the rates of many relevant reactions, concerning essentially unstable nuclei (in some cases by many orders of magnitude), leads to an important leakage out of the Mg-Al cycle (through ^{27}Si(p,γ)); thus, the production of ^{26}Al is found to be considerably reduced with respect to previous estimates: only a few times 10^{-7} (by mass fraction) for *hot* novae (T_p >200 10^6 K) and a few 10^{-5} for *cold* ones (T_p <200 10^6 K). The most recent calculations of that kind[31], using the I90 rate for ^{25}Mg(p,γ) reaction, find somewhat lower results (X_{26} $\sim$$10^{-5}$ only, for a cold nova with solar composition). Taking into account that a nova ejects on the average $M_{nova} \sim 10^{-4}$ $M_\odot$, the average ^{26}Al yield of a C-O nova turns out to be $\sim 10^{-11}$-10^{-9} $M_\odot$, depending on its peak temperature.

It should be stressed, however, that uncertain reaction rates and thermodynamic conditions are not the only factors affecting the production of ^{26}Al in novae, which is particularly sensitive to the modelling of the site: the basic difficulty comes from the treatment of *convection* (the time-scale of which is comparable to the nuclear one, i.e. 100-400 s), but also from the treatment of the mass loss and the dredging-up mechanisms. Up to now, only results of parametrized, two-zone, computations have been reported[31,57]. ^{26}Al is produced in the lower and hotter zone. Convection brings fresh fuel from the upper zone and removes the fragile nuclei (among them ^{26}Al) from the burning region up to the colder zone, where they can be preserved before being ejected in the ISM. In some cases spectacular production enhancements are found for ^{26}Al (by factors ~10-30, depending on the adopted physical conditions) a fact which clearly illustrates the uncertainties affecting all current (parametrized) nucleosynthesis computations in novae and the need for completely self-consistent computations (i.e. coupling hydrodynamics and nucleosynthesis).

Finally, notice that the initial nova composition greatly affects the ^{26}Al yield. Nucleosynthesis in *O-Ne-Mg rich novae*, coming from white dwarfs with progenitor stars in the 7-11 $M_\odot$ mass range, has been considered by several authors[11,31,54,55]. They find that the ^{26}Al yield is considerably enhanced, since ^{25}Mg *and* part of ^{24}Mg turn into ^{26}Al: ^{26}Al mass fractions up to a few 10^{-3} have been obtained in some parametrized calculations of that kind. Notice, however, that hydrodynamic calculations[43] suggest that the mass ejected by O-Ne-Mg rich novae should be lower than the canonical value of 10^{-4} $M_\odot$, by factors 5-10; the mass of ^{26}Al ejected by such an explosion is then[43] ~a few times 10^{-8} $M_\odot$ (a very generous upper limit). Finally, notice that those O-Ne-Mg rich nova are ~4 times less frequent than the C-O ones, which reduces even more their contribution to the production of ^{26}Al in the Galaxy (see next Section).

Nucleosynthesis of ^{26}Al by *pre-explosive and explosive carbon burning* and *neon burning* in massive stars ending their lives as *Type II supernovae* (SNII) has been considered by many authors[1,28,48,58], especially after the original sugggestion[35] that these sites are interesting candidates for gamma-ray line astronomy.

^{26}Al is produced by $^{25}Mg(p,\gamma)$ during hydrostatic shell carbon- and neon-burning in massive stars, at temperatures $T \sim 1.5\ 10^9$ K, densities $\rho \sim 10^5$ g cm^{-3} and timescales ~a few years. Background protons, coming mainly from reactions on ^{12}C or ^{20}Ne nuclei, are captured on Mg nuclei to produce ^{26}Al: background neutrons, liberated essentially through $^{13}C(\alpha,n)$ and $^{22}Ne(\alpha,n)$ reactions, are now the main destruction agent of ^{26}Al, instead of (p,γ) reactions.

Recent calculations[60] show that the total mass of ^{26}Al produced in those stellar regions *before the explosion* is a few times 10^{-5} $M_\odot$, depending on the stellar mass. Notice that the amount of hydrostatically produced ^{26}Al before the supernova explosion depends crucially on the treatment of time-dependent convection[1,58,60]: if the convective transport of ^{26}Al from the Ne and C burning shells to cooler regions is not rapid enough, ^{26}Al may be considerably destroyed, before the final explosion that ejects it in the ISM (for example, its β decay rate is substantially enhanced at those high temperatures: $\tau_\beta \sim$1400 s at T$\sim 10^9$ K, compared to $\tau_c \sim$1000 s for convection).

The passage of the shock wave heats the stellar regions that have experienced shell C-Ne burning (or were convectively coupled to those shells) to peak temperatures $T_p \sim$2-3 10^9 K, depending on the stellar mass (and the adopted model). *Explosive neon burning* takes place at those temperatures, at peak densities $\rho_p \sim$ a few 10^5 g cm^{-3} and timescales $\tau \sim$1 s. In the hotter regions (T> 2.5 10^9 K) ^{20}Ne is completely exhausted, and ^{26}Al may be destroyed by photodesintegration, while in the colder ones neon burning is incomplete, and ^{26}Al is produced.

The total amount of ^{26}Al produced by a supernova of type II is difficult to estimate, since it depends on several parameters[60], like the stellar mass and initial composition, the treatment of (time-dependent) convection, various nuclear reaction rates (the famous $^{12}C(\alpha,\gamma)$ among others), etc. and certainly requires a realistic model for the explosion.

The explosion of SN1987A, a $\sim$20 $M_\odot$ star in the Large Magellanic Cloud, gave the theoriticians the opportunity to refine considerably their models, comparing them to observations of unprecedented accuracy. Two groups considered the nucleosynthesis of ^{26}Al in that site, with realistic stellar models (the word realistic here is really meaningful!), detailed networks and recent nuclear physics: Woosley *et al.*[59] find that the explosion of that SNII ejected $\sim$2 10^{-5} $M_\odot$ of ^{26}Al, produced essentially by explosive neon burning (and to a smaller extent, by neutrino induced nucleosynthesis, see below); while Thielemann *et al.*[46] find that SN1987A ejected 6.7 10^{-6} $M_\odot$ of ^{26}Al. Notice that in the latter case only explosive production of ^{26}Al has been considered, any hydrostatic contribution being ignored, i.e. the above number is only a lower limit to the real one (not much lower, however, since pre-explosive synthesis of ^{26}Al in stars of that mass is not expected to dominate over the explosive one). Considering the differences in the stellar and nucleosynthesis codes of the two groups, the discrepancy in the obtained ^{26}Al yields (by factor of $\sim$2 for the explosive nucleosynthesis) should be considered as rather satisfactory.

Recently, the same two groups reported preliminary results of a nucleosynthesis programme, aiming to obtain SNII yields of stars with different masses. The first group has completed up to now calculations concerning only the hydrostatic production of ^{26}Al in stars of 12, 15, 20, 25 and 35 $M_\odot$, respectively. A preliminary evaluations of the ^{26}Al yields of the corresponding SNII explosions is reported in this volume[60]: 10^{-5} - 10^{-4} $M_\odot$ of ^{26}Al, somewhat higher than previous estimates (see Fig. 1). In the second case[47], only explosive production of ^{26}Al in realistic SNII models of 13, 15, 20 and 25 $M_\odot$ has been calculated, with results in the range 10^{-6}-10^{-5} $M_\odot$ (plotted also in Fig. 1). Those results show that the SNII yield of ^{26}Al increases rapidly (though not monotonically) with stellar mass, due to the fact that the more massive the star, the larger the zone experiencing appropriate burning conditions for the production of ^{26}Al. Based on those results, and using a standard IMF ($\propto M^{-2.5}$) one finds that the average ^{26}Al yield of SNII in this mass range, i.e. between 12 and 40 $M_\odot$, is ~7. 10^{-6} - 3 10^{-5} $M_\odot$ (it is not obvious how to extrapolate to higher stellar masses, where mass loss considerably affects the evolution of the star, and might affect its structure and nucleosynthesis prior to and during the explosion as well).One should notice that there is a discrepancy by a factor of ~4 between the average SNII yields of ^{26}Al proposed by the two groups; despite the successful modelling of SN1987A, much effort is still needed in order to put the production of ^{26}Al by SNII on more secure grounds and, in particular, to clarify the role of pre-explosive C- and Ne- burning.

A rather exotic mechanism for producing ^{26}Al (and other fragile nuclei) in SNII has been recently proposed: *neutrino induced nucleosynthesis.* The flux of neutrinos from the collapsed stellar core is so high, that some of them (especially the higher energy τ and μ neutrinos) may interact with nuclei in the overlying Si, O, Ne, C, He and H layers; single neutrons and protons are ejected from the de-excitation of those nuclei and subsequently interact with other nuclei present, modifying somehow the classical scheme of explosive nucleosynthesis. A ~50% enhancement to the explosive production of ^{26}Al in a 20 $M_\odot$ star is found in the only detailed calculation[59] of that effect up to now. [Notice: this enhancement is included in the amounts of ^{26}Al reported by Woosley[60] and in Fig. 1]. The major uncertainty in that calculation is the number of neutrinos energetic enough (i.e. with energies E>6-8 MeV) to induce that kind of nucleosynthesis, since the neutrino spectrum emitted from the SNII core collapse is largely unknown. Recent calculations[29] suggest that the neutrino spectrum is truncated in its high and low energy tails, and that the corresponding effect on the explosive nucleosynthesis (and the production of ^{26}Al) in SNII has been overestimated.

Finally, and for sake of completeness, it should be mentioned that a few attempts were made to explain the origin of the observed 1.8 MeV line of ^{26}Al in terms of point (or local) source models. Recent observational evidence suggests, however, that such a possibility should be excluded[50].

The following remarks can be made on the nucleosynthesis of ^{26}Al:

* in general, the reaction rates of stable nuclei are better known in the high temperature regime (i.e. above ~10^9 K), where extrapolation from experimental data is relatively easy, than in the low temperature one, where uknown (or difficult to evaluate) resonances may be involved. However, the higher the temperature, the more unstable (proton rich) nuclei are involved in the nuclear flow. Since no experimental information is available for such nuclei, one has to rely on theoretical (and quite uncertain) estimates for the relevant reaction rates. Thus, the nuclear physics is generally better treated in the case of quiescent nucleosynthesis of ^{26}Al than in the case of explosive nucleosynthesis.

* the production rate of ^{26}Al is proportional to the metallicity of the concerned site, whereas its destruction rate is proportional to metallicity only in the

high temperature regime, where neutrons are the main destroying agent (since neutron production, through (α,n) reactions, is proportional to metallicity). As a result, the *net* production rate of ^{26}Al may be considered as (roughly) proportional to metallicity in the case of massive stars, red giants and novae, but not in the case of supernovae (this holds, of course, under the assumption that the physical conditions of the site do not depend sensitively on metallicity -e.g. through the energy production, the opacity etc.- which can not be exluded). Actually, it has been argued that the SNII yield of ^{26}Al should be inversely proportional to metallicity[58], while that of WR should scale as $z^{1.7}$: indeed, in high metallicity massive stars, the stellar envelope is more efficiently expelled by the radiative pressure on its high opacity layers, so that ^{26}Al has less time to decay before attaining the stellar surface[23,51].

* the time-scale for the ejection of ^{26}Al from the above astrophysical sites is $\tau << \tau_{26}$ for all the explosive sites (novae and supernovae), $\tau \sim \tau_{26}$ for WR stars and $\tau > \tau_{26}$ for AGB stars (for the most massive of them, $\tau \sim \tau_{26}$). In the latter case, the contribution on the galactic ^{26}Al should be treated with care: it is not the total quantity of ^{26}Al ejected by the AGB winds that has to be taken into account, but only an average production over $\sim 10^6$ y.

PRODUCTION AND DISTRIBUTION OF ^{26}Al IN THE GALAXY

In order to evaluate the total amount of ^{26}Al ejected in the ISM by the previously discussed astrophysical sites in the last $\sim 10^6$ y, one needs to know their galactic frequency (for a more accurate evaluation, their galactic distribution is also needed, since the ^{26}Al yield depends in some cases on metallicity, which is a function of position in the Galaxy). The corresponding mass of ^{26}Al can be estimated by the general expression:

$$M_{26}^{G} = \tau_{26}\, M_{26}^{S}\, f_{S},$$

where: $\tau_{26} \sim 10^6$ y, M_{26} is the amount of ^{26}Al ejected by each source, and f_S (y^{-1}) its frequency of occurence in the Galaxy (at least during the last millions of years).

In the case of *novae*, the most prolific producers of ^{26}Al are the O-Ne-Mg rich ones, ejecting up to $M_{26}^{ONeMg} \sim 4\ 10^{-8}\ M_{\odot}$ each (a very generous upper limit). The frequency of novae in the Galaxy is usually estimated to $f_{nova} \sim 40$ y^{-1}, although somewhat higher frequencies cannot be excluded (observations of the spiral Andromeda galaxy show that $f_{nova} \sim 100\ y^{-1}$ there). Taking into account that $\sim 1/4$ of all novae are expected to be O-Ne-Mg rich (i.e. $f_{ONeMg} \sim 10\ y^{-1}$), and neglecting the small contribution of the other novae, we find that the galactic production of ^{26}Al by novae in the last $\sim 10^6$ y is: $M_{26}^{G,nova} \sim 0.4\ M_{\odot}$. A somewhat higher nova frequency would push that number up to 1 $M_{\odot}$, which is really a very generous upper limit on the nova contribution to galactic ^{26}Al (at least, within our current understanding of novae). and still below the observational requirements.

The contribution of massive stars (both SNII and WR stars) to the galactic amount of ^{26}Al has a more complicated story. The fact that SNII are thought to be the major galactic producers of the stable isotope ^{27}Al suggested that the galactic ^{26}Al production of SNII should be correlated to that of ^{27}Al; another motivation for such a correlation was the fact that, at that time, the isotopic production ratio $(X_{26}/X_{27})_{SN} \sim 10^{-3}$ suffered from less uncertainties than the absolute ^{26}Al yield of SNII (due to a lack of detailed models of SNII explosions). Supposing that supernovae produced all the ^{27}Al($X_{27} \sim 6\ 10^{-5}$) in the ISM (that is, in a mass $M_{ISM} \sim 4\ 10^{9}\ M_{\odot}$) during the past $T_G \sim 10^{10}$ years (the age of the Galaxy), the quantity of ^{26}Al produced by supernovae with $(X_{26}/X_{27})_{SN} \sim 6\ 10^{-3}$ during the last $\tau_{26} \sim 10^{6}$ years should be:

$$M_{26}^{G} \sim (X_{26}/X_{27})_{SN}\ (\tau_{26}/T_G)\ X_{27}\ M_{ISM} \sim 0.15\ M_{\odot},$$

if *nucleosynthesis at constant rate* is assumed all over the galactic history [notice that this method allows to circumvent the uncertainties on the galactic rate of SNII]. Thus, it seeemed that supernovae fall short of producing the observed quantity of ^{26}Al by a factor of ~20; stated in a different way, if supernovae were at the origin of ~3 $M_{\odot}$ of ^{26}Al in the ISM, they should have overproduced ^{27}Al by a factor of ~20, as Clayton[7] remarked. This argument was used for sometime as conclusive evidence against the SNII origin of galactic ^{26}Al. However, if the assumption of constant rate nucleosynthesis is dropped, and *galactic chemical evolution* effects are taken into account (as they should), the supernovae contribution may be considerably modified. With simple chemical evolution models it was subsequently shown[8] that the supernova contribution to the ISM may be as high as 0.5 $M_{\odot}$ of ^{26}Al in the past 10^{6} years, still lower than the 3 $M_{\odot}$ required by observations.

On the other hand, early estimates for the contribution of WR stars to the galactic ^{26}Al content were based on rather uncertain evaluation of their total number and distribution in the Galaxy, by extrapolating data concerning the WR density in the solar neighbourhood and low metallicity regions like the Large and Small Magellanic Clouds (this was due to the fact that the WR catalogues are complete only within 2.5 kpc from the Sun). Taking into account metallicity effects on the number of WR stars and their ^{26}Al yield (see below), it was estimated[34] that they could contribute up to 0.5 $M_{\odot}$ of ^{26}Al in the ISM in the last 10^{6} y (an earlier estimate[33] of 2.2 $M_{\odot}$ for this upper limit, requiring ~8000 galactic WR stars, was clearly very optimistic).

Instead of uncertain estimates for WR numbers, it seems more reasonable to evaluate the galactic contribution of *all massive stars* producing ^{26}Al both hydrostatically and explosively, in a unified scheme[41]. This can be done by use of the IMF and the estimated frequency of galactic supernovae. In the following we adopt a power-law IMF ($\Phi(M) \propto M^{-2.5}$), and adopt as lower mass limits M_{LSN}=12 $M_{\odot}$ for core collapse supernova and M_{LWR}=40 $M_{\odot}$ for the formation of WR stars (which are currently thought to explode as Ib supernovae), respectively. With such an IMF, WR stars constitute only ~13 % of the total number of massive stars.

The adopted ^{26}Al yields are those of Fig. 1: in the region 12 to 40 $M_{\odot}$, ^{26}Al is assumed to come from nucleosynthesis in SNII, whereas in the region M>40 $M_{\odot}$ it is assumed to be produced by hydrostatic nucleosynthesis in WR stars. Notice that those numbers are valid for solar metallicity stars ($z_\odot$ ~0.02), and should be modified in view of the observed galactic metallicity gradient[40] of d(logz)/dr = -0.07 dex/kpc for r>4 kpc (r being the galactocentric radius). As discussed in the previous paragraph, the ^{26}Al yield of WR stars is found to be *at least proportional to z* and, according to some estimates, proportional to $z^{1.7}$, while the supernova yield could be either inversely proportional to, or independent of, metallicity; both cases are considered in the following.

There are still considerable uncertainties on the supernova rate and the distribution of massive stars in our Galaxy. In this work we adopt the rate of 1.2 core collapse SN/100 y, predicted by the luminosity function of nearby stars[44] [there is a discrepancy with estimates based on observations of historical supernovae, which give a rate 4-8 times higher, but statistical fluctuations cannot be excluded in that case]. Since massive stars are thought to be formed in molecular clouds, we adopted the distribution of H_2 in the Galaxy (Fig. 2), based on observations of CO emission[37]. Notice that the surface density of H_2 in the innermost galactic regions (R<1 kpc) is difficult to estimate; even more uncertain is the star formation rate there, since there are observational indications for turbulence and/or magnetic fields, which could prevent star formation even in the presence of large H_2 densities[14]. In the following we consider both cases, i.e. a Galaxy with and without massive stars in its innermost regions. [*Notice*: these considerations affect very little the quantity of ^{26}Al ejected by massive stars, which depends essentially on the adopted supernova rate, but are very important for the distribution of the corresponding gamma-ray line emission in the Galaxy, as we shall see below].

The total quantity of ^{26}Al ejected by massive stars in the Galaxy in the last 10^6 y is then calculated as:

$$M_{26} = \int_0^R 2\pi r dr\ \sigma(r) \int_{M_1}^{M_2} \Phi(M) Y(M,r) dM,$$

where r is the galactocentric radius, $\sigma(r)$ the adopted H_2 surface density, R=15 kpc the radius of the Galaxy, M_1=12 $M_\odot$ and M_2=120 $M_\odot$, the adopted IMF is normalized to the adopted SN rate: $\int_0^R 2\pi r dr\ \sigma(r) \int_{M1}^{M2} \Phi(M) dM = 12000/10^6$ y, *Y(M)* are the yields of Fig. 1, $Y(M,r) = Y(M) * (z(r)/z_\odot)^k$, where $z(r)$ is the metallicity [such as $d(log z)/dr = -0.07\ dex/kpc, z(r = 8.5\ kpc) = z_\odot, z(r < 3\ kpc) = 3z_\odot$], and k=1 or 1.7 for WR stars and k=0 or -1 for SN.

Calculations of the galactic ^{26}Al yield of massive stars on the basis of the above assumptions show that: a) nucleosynthesis in supernovae can produce up to 0.15 $M_\odot/10^6$ y, if the yields of Ref. 47 are adopted (extrapolated up to 40 $M_\odot$), and up to 0.50 $M_\odot/10^6$ y if the yields of Ref. 60 are adopted; b) extrapolating the latter yields above 40 $M_\odot$ leads to ~1 $M_\odot$ of ^{26}Al; and c) the winds of WR stars can eject up to 0.20-0.30 $M_\odot/10^6$ y, the larger number being obtained when a massive star population in the innermost galactic regions is considered. Depending on the adopted assumptions, the WR/SN production ratio of ^{26}Al varies from 2/1 to 0.3/1. All massive stars taken together can contribute up to 1.3 $M_\odot$ of ^{26}Al, in the most optimistic case; a SN rate only 2.5 times higher than the standard rate adopted here would obviously allow

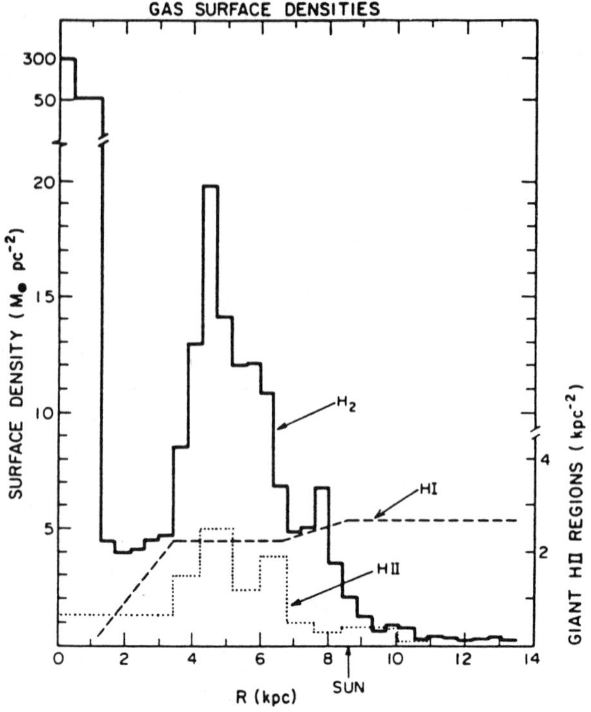

Figure 2

Galactocentric radial distribution of H_2 clouds, acording to Ref. 37. The distribution of massive stars in the Galaxy could follow this one, but considerable uncertainties remain as to the amounts of H_2 and the efficiency of star formation in the inner kpc (Ref. 14).

for massive stars to make all the ^{26}Al inferred from the observations. Notice, however, that this is the most optimistic case conceivable with current SNII and WR models.

It has often been suggested that gamma-ray observations with good angular resolution would help to discriminate between the various proposed sources of ^{26}Al, since the signature of each source (i.e. flux distribution as a function of galactic longitude and/or latitude) is expected to differ from that of the others. Unfortunately, the galactic distributions of all the proposed sources are unknown for the moment; up to know, they have been assumed to follow either the directly observed diffuse galactic emission at some other wavelength (e.g. the CO emission, or the high-energy γ-ray emission of COS B), or some other presumed indicator of young or old stellar populations in the Galaxy (e.g. H_2 clouds for massive stars, progenitors of SNII and WR; surface brightness of spiral galaxies, presumably reflecting the distribution of AGB stars; or the observed nova distribution in the Andromeda galaxy, possibly similar to the one in our own Galaxy).

Calculations performed on the basis of those (more or less plausible) assumptions showed that, contrary to what is usually claimed (and to the original expectations), *none* of the proposed sources has a unique signature, leading to a clear-cut identification. More specifically, it turned out that all the proposed sources can give a flux with a longitudinal profile very peaked in the GC direction; in the case of Population I objects the peak comes from the assumption of an important star formation in the inner 1-2 kpc of the Galaxy[33,34], while in the case of Population II objects the peak is mainly due to the contribution of the Galactic bulge[16,21]. Moreover, it is impossible to discriminate on that basis SNII from WR (both have massive stars as progenitors), or AGB stars from novae (both coming from intermediate mass stars).

The uncertainties affecting the expected flux distribution from massive stars are clearly illustrated in Fig. 3. It can be seen that, depending on the assumptions made on the metallicity dependence of the ^{26}Al yields or the importance of star formation in the inner kpc of the Galaxy, quite different flux profiles are obtained from the *same* radial source distribution (the one of H_2 molecular clouds, adopted in this work). Those results also suggest that only in one case would the identification of a young stellar population be unambiguous: if the observed flux profile is found to be hollow in the GC direction, reflecting a negligible contibution from the innermost 1-2 kpc, and a prominent one from the molecular ring at 3-4 kpc (giving the peaks at l$\sim$ $\pm$30-40°). Indeed, such a profile is not expected from any reasonable distribution of old stars, since the contribution of the galactic bulge is always prominent in that case. Notice that even such a "hollow" distribution cannot be excluded by the GRIS data.

Another way to look for specific signatures of the emission of massive stars is to drop the assumption of radial symmetry for the source distribution in the galactic plane (made in all the previous works) and take into account the *spiral structure* of our Galaxy. Indeed, it is currently thought that star formation in the galactic disk occurs predominently inside the spiral arms, especially in the case of massive stars (observations of SNII in spiral galaxies strongly support this argument). Unfortunately, the spiral structure of our Galaxy is far from being well known, and the results shown in Fig. 4 should be considered only as indicative. They are based on a 2-arm galactic model with logarithmic spirals; massive stars are concentrated inside the 500 pc-wide arms, their azimuthaly averaged radial distribution following again that of H_2 molecular clouds. The *generic feature* in that case (Fig. 4) is the presence of *asymmetries* in the flux profile between the first and the fourth galactic quadrant, and of several spikes, corresponding to the position of the spiral arms (their contribution being enhanced when the line of sight is tangential to them). However, the extent of those asymmetries and features depends crucially on the adopted galactic model, and no prediction can be made on the angular resolution and/or sensitivity necessary to detect them.

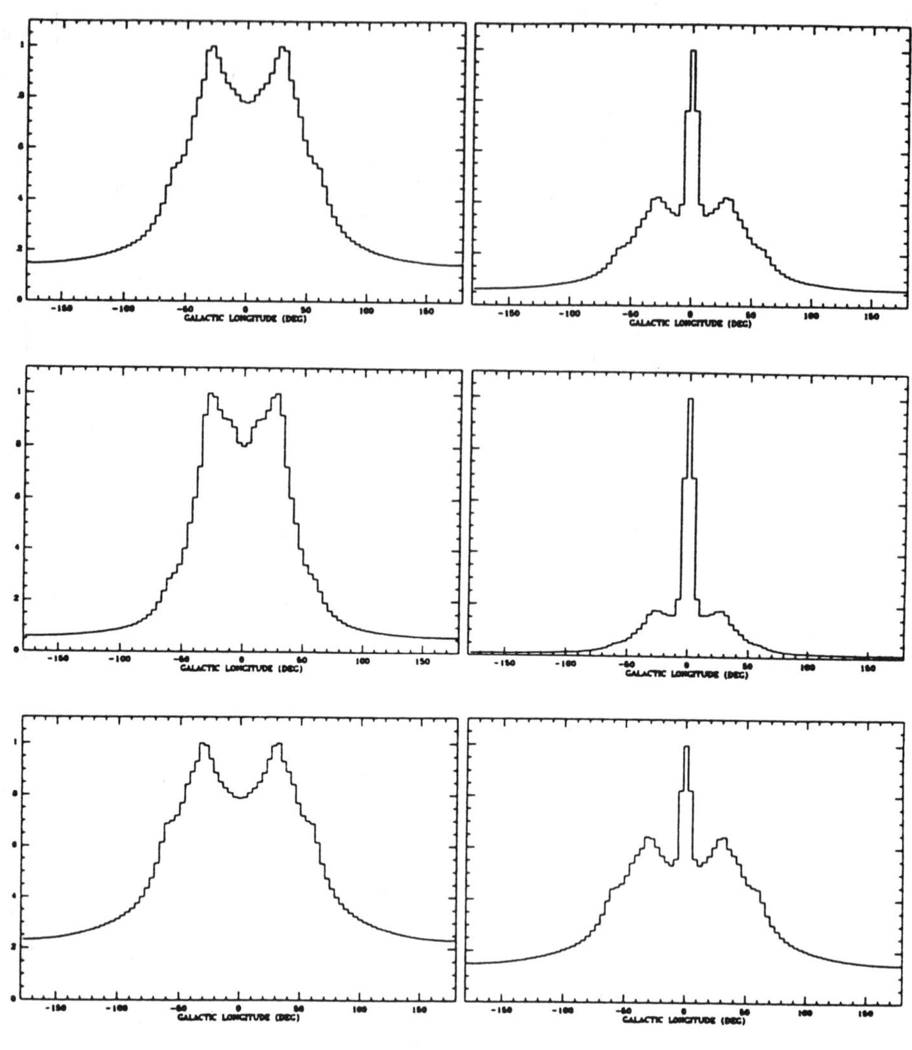

Figure 3

Normalized flux distribution (Max=1) from a population of galactic sources following the radial distribution of Fig. 2. *Left*: star formation in the inner kpc is neglected; *right*: star formation in the inner kpc is included. A radial metallicity gradient of $d(logz)/dr = -0.07\ dex/kpc$ is adopted for the galactic disk (Ref. 40), with z=$z_{\odot}$=0.02 at r=8.5 kpc and $z(r < 3kpc) = 3z_{\odot}$. The yield of each source is assumed to vary with metallicity as $(z/z_{\odot})^k$, where *k=0 (upper panel)*, i.e. the yield is independent of metallicity; *k=1.7 (middle panel)*, probably appropriate for the ^{26}Al yield of WR stars; and *k=-1 (lower panel)*, probably appropriate for the ^{26}Al yield of SNII. Flux is given in bins of 4° in galactic longitude.

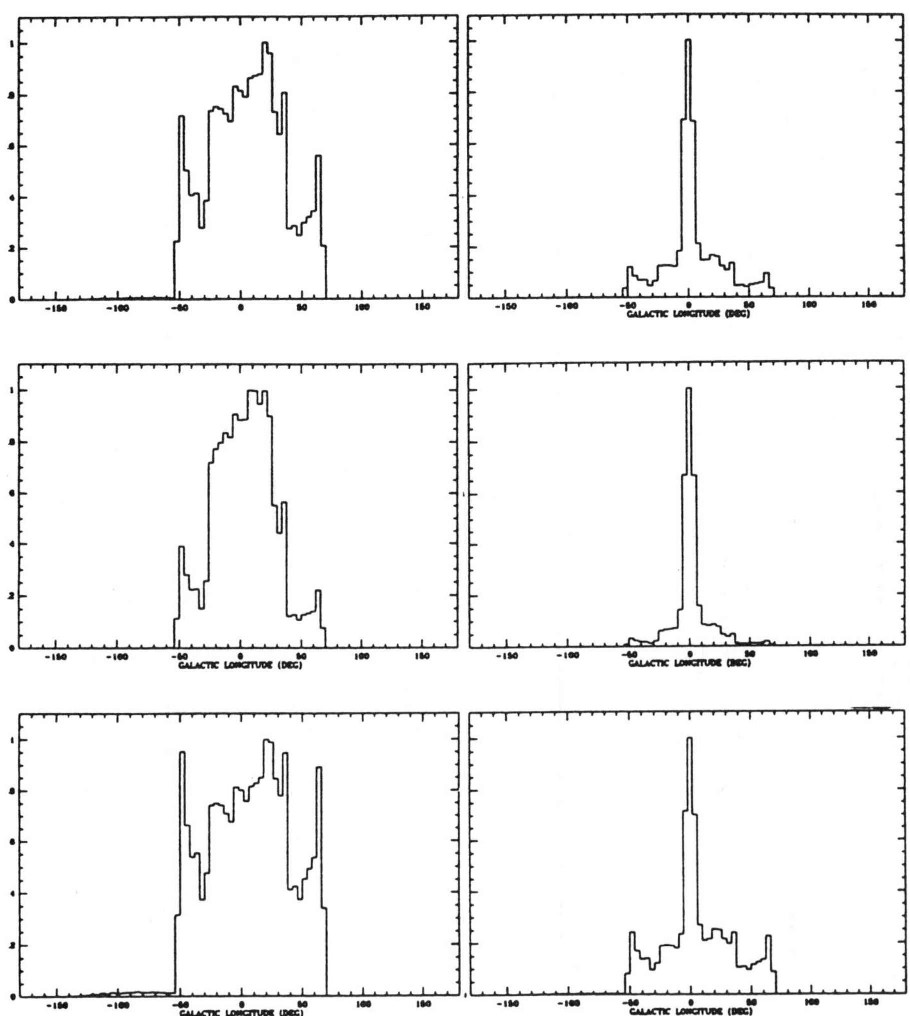

Figure 4

Same as in Fig. 3, but this time a 2-arm spiral structure for the Galaxy is superimposed on the radial H_2 distribution of Fig. 2. Sources are confined inside the 500-pc wide spiral arms.

CONCLUSION

The detection in the galactic plane of a γ-ray line at 1.8 MeV due to the decay of ^{26}Al, boosted theoretical as well as experimental and observational work on the origin of the $\sim$3 $M_{\odot}$ of ^{26}Al currently present in the ISM.

All the basic ingredients (nuclear physics, stellar models, and galactic distribution) of the proposed sources (supernovae, novae, red giants, Wolf-Rayet stars) suffer from (several) serious drawbacks. However, for sites involving low mass stellar progenitors, like novae and AGB stars, the uncertainties on the production of ^{26}Al are currently much larger (factors $\sim$100) than in the case of sites with massive stellar progenitors ($\sim$4 for SNII and $<$2 for WR stars). Following the results reported at this conference[31,60], SNII seem (again!) to be the most promising candidate, but the large uncertainties related to e.g. the treatment of time-dependent convection during C- and Ne- shell burning, or neutrino induced nucleosynthesis etc., do not allow any definite conclusion yet. In our opinion, none of the above proposals can be completely discarded at the present time and much more refined models for the corresponding sites are needed.

Because of the poor angular resolution and the insufficient sensitivity of the experiments, the source distribution is still quite uncertain, although a point source in the GC seems to be excluded by the data[50]. Obviously, further observations, with better angular resolution ($<5^{o}$) and sensitivity ($\sim 10^{-6}$ photons cm^{-2} s^{-1}) are badly needed to clarify the situation. The recently reported results of the GRIS experiment[45], although inconclusive as to the precise form of the distribution, are the first step towards that goal.

Unfortunately, the expected signatures of the various candidate sources (i.e. 1.8 MeV flux as a function of galactic longitude) are not found to be unique, at least at the present stage of our knowledge. More specifically, a peaked distribution towards the GC region may result from the galactic population of either massive stars (WR and SNIb or SNII), or sites with low-mass progenitors (AGB, novae), and cannot be used to unambiguously identify Pop. I or Pop. II objects. It is argued here that a flux distribution hollow in the GC direction (or relatively flat), or the existence of asymmetries between the first and fourth galactic quadrants (even in the case of a peaked distribution), would favour an interpretation in terms of massive stars.

REFERENCES

1. Arnett D. W., Wefel J. P. 1978, *Ap. J.*, **224**, L139
2. Arnould M., Norgaard H., Thielemann F.-K., Hillebrandt W. 1980, *Ap. J.*, **237**, 931
3. Cameron A. G. W. 1984, *Icarus*, **60**, 416
4. Cassé M., Prantzos N. 1986, in *Nucleosynthesis and its implications on Nuclear and Particle Physics*, eds. J. Audouze and N. Mathieu, (Dordrecht : Reidel), p. 339
5. Caughlan G.R., Fowler W.A., Harris M.J., Zimmerman B.A. 1985, *At. Data Nucl. Data Tables*, **32**, 197

6. Caughlan G.R., Fowler W.A., *At. Data Nucl. Data Tables*, **40**, 238
7. Clayton D. D. 1984, *Ap. J.*, **280**, 144
8. Clayton D. D., Leysing M. D. 1987 *Phys. Rep.*, **144**, 1
9. Dearborn D. S. P., Blake J. B. 1984, *Ap. J.*, **277**, 783
10. Dearborn D. S. P., Blake J. B. 1985, *Ap. J.*, **288**, L21
11. Delbourgo-Salvador P., Mochkovitch R., Vangioni-Flam E. 1985, *Proceedings of the ESA Workshop : Recent results on Cataclysmic Variables*, Bamberg, p. 229
12. Fowler W. A. 1984, *Rev. Mod. Phys.*, **56**, 149
13. Frantsman Yu. 1989, *Sov. Astr.*, **33**, 565
14. Gusten R. 1989, in *The Center of the Galaxy*, Ed. M. Morris (IAU Symposium), p. 89
15. Higdon J. C., Lingenfelter R. E. 1976, *Ap. J.*, **208**, L107
16. Higdon J. C., Fowler W. A. 1989, *Ap. J.*, **339**, 956
17. Hillebrandt W., Thielemann F.-K. 1982, *Ap. J.*, **255**, 657
18. Iliadis Ch. *et al.* 1990, *Nucl. Phys.*, **A512**, 509
19. Iben I., Truran J. 1978, *Ap. J.*, **220**, 980
20. Lazareff B, Audouze J., Starrfield S., Truran J. 1979, *Ap. J.*, **228**, 875
21. Leising M. D., Clayton D. D. 1985, *Ap. J.*, **294**, 591
22. McCallum C.J., Huters A.F., Stang P.D., Leventhal M. 1987, *Ap. J.*, **317**, 877
23. Maeder A. 1984, *Adv. Sp. Res.*, **4**, 55
24. Mahoney W. A., Ling J. C., Wheaton W. A., Jacobson A. S. 1984, *Ap. J.*, **286**, 578
25. Mahoney W. A., Higdon J. C., Ling J. C., Wheaton W. A., Jacobson A. S. 1985, *Proceedings of the 19th ICRC*, La Jolla, USA, OG 3.2-3 (p. 357)
26. Mallet I., *et al.*, 1990, *this volume*
27. Mayer-Hasselwander et al. 1982, *9th Texas Symp. Relat. Ap.*, 211
28. Morgan J. A. 1980, *Ap. J.*, **238**, 674
29. Myra E. S., Burrows A. 1990, *Ap. J*, **364**, 222
30. Norgaard H. 1980, *Ap. J.*, **236**, 95
31. Paulus G., Forestini M., 1990, *this volume*
32. Prantzos N. 1987, in *Nuclear Astrophysics*, Eds. W. Hillebrandt, R. Kuhfuss, E. Mueller, J. Truran (Springer-Verlag), p. 250
33. Prantzos N., Cassé M. 1986 *Ap. J.*, **307**, 324
34. Prantzos N, Cassé M., Arnould M. 1987, *Proceedings of the 20th ICRC*, Moscow, OG:2.3-4 (**2**, 152)
35. Ramaty R., Lingenfelter R. E. 1977, *Ap. J.*, **213**, L5
36. Schoenfelder V. 1990, *this volume*
37. Scoville N. Z., Sanders D. B. 1987, in *Interstellar Processes*, eds. H. Thronson and D. Hollenbach (Reidel), p. 21
38. Share G. H., Kinzer R. L., Kurfess J. D., Forrest J. D., Chupp F. L., Rieger E. 1985, *Ap. J.*, **292**, L61
39. Share G. H., Kinzer R. L., Kurfess J. D., Forrest J. D., Chupp F. L., Rieger E. 1985, in *Proceedings of the 19th ICRC*, La Jolla, USA, OG:3.2-1, 353

40. Shaver P. A., Mc Gee R. X., Newton L. M., Danks A. C., Pottasch S. R. 1983, *M. N. R. A. S.*, **204**, 53
41. Signore M., Dupraz C. 1990, *A. A.*, **234**, L15
42. Starrfield S., Truran J. W., Sparks K. 1978, *Ap. J.*, **226**, 186
43. Starrfield S., Sparks K., Truran J. W. 1986, *Ap. J.*, **303**, L5
44. Tammann G., Van den Bergh S. 1990, preprint (to appear in *Ann. Rev. Astr. Ap.*)
45. Teegarden B., *et al.*, 1990, *this volume*
46. Thieleman F.-K., Hashimoto M., Nomoto K. 1990, *Ap. J.*, **349**, 222
47. Thieleman F.-K. 1990, *private communication*
48. Truran J. W., Cameron A. G. W. 1978, *Ap. J.*, **219**, 236
49. von Ballmoos P., Diehl R., Schoenfelder V. 1987 *Ap. J.*, **318**, 654
50. von Ballmoos P. 1990, *this volume*
51. Walter R., Maeder A. 1989, *A. A.*, **218**, 123
52. Ward R. A., Fowler W. A. 1980, *Ap. J.*, **238**, 266
53. Weber W. R., Schonfelder V., Diehl R. 1986, *Nature*, **323**, 692
54. Weiss A., Truran J. W. 1990, *A. A.*, **238**, 178
55. Wiescher M., Gorres, J., Thielemann, F.-K., Ritter, H. 1986, *A. A.*, **160**, 56
56. Wolf M. T., Leising M. D. 1988 in *Gamma-Ray Spectroscopy in Astrophysics*, Eds. N. Gehrels and G. Share, AIP, p. 136
57. Woosley S. E. 1986, in *Saas-Fe Lectures : Nucleosynthesis and Chemical Evolution*, Eds. B. Hauck, A. Maeder and G. Meynet (Geneva Observatory), p.78
58. Woosley S. E., Weaver T. A. 1980, *Ap. J.*, **238**, 1017
59. Woosley S. E., Hartmann D. H., Hoffman R. D., Haxton W. C. 1990, *Ap. J.*, **356**, 272
60. Woosley S. E. 1990, *this volume*

A DISTRIBUTION FOR THE GALACTIC ^{26}AL AND e^+e^- LINE EMISSIONS

P. von Ballmoos
Centre d'Etude Spatiale des Rayonnements
31029 Toulouse, France

ABSTRACT

A distribution for the galactic gamma-ray line emissions at 1809 keV and 511 keV is proposed. It is suggested that both the diffuse ^{26}Al and the e^+e^- emission are reflected by the distribution of thin hot plasma as derived from X-ray observations of the galactic ridge. The distribution consists of a flat galactic distribution with enhanced features at l=± 25° and a dominant narrow peak at the Galactic Center containing a third of the total emission in the central radian. This hypothesis is substantiated by the existing ^{26}Al detections and most of the observations at 511 keV. Evidence for a connection between thin hot plasma and gamma-ray line emission is presented : The relevant nucleosynthesis-processes infer explosive escape of the ^{26}Al from the dense high temperature regions of fast production and decay - as in disruption of stars (supernovae) or strong stellar winds (eg. Wolf-Rayet stars). In either of the cases the mechanical energy of the emanating shock front is transformed into thermal energy, leading to gas temperatures of $\geq 10^7$ °K at the shock. X-ray telescopes who have now mapped hot tenuous plasma at different scales can thus indicate the site of recent nucleosynthesis. In these sites the decaying radionuclei (^{56}Co, ^{26}Al and others) will produce positrons that annihilate with lifetimes comparable to ^{26}Al; the rate of annihilation was found to be sufficient to explain all the observed 511 keV radiation.

INTRODUCTION

During the last decade γ-ray line spectroscopy provided direct evidence for ongoing synthesis of intermediate and heavy elements in the universe. The identification of the 1809 keV line from radioactive ^{26}Al (lifetime: $7 \cdot 10^5$ y) in the ISM of our Galaxy has been the first major breakthrough, followed by nuclear lines at 847 keV and 1230 keV from freshly synthesized ^{56}Co (lifetime: 70 d) detected in SN1987A. The lifetime of the identified nuclides together with the measured fluxes, put constraints on the rate of nucleosynthesis while line-widths and -positions characterize the region where these elements decay in.

In order to identify the sites of nucleosynthesis in our Galaxy several attempts have been made analyzing existing balloon- and satellite-data with respect to the angular extent of the galactic ^{26}Al emission. Apparently contradictory results were obtained by the different experimentators. Theoretists fueled the controversy even further by alternatingly favoring various source candidates (SN, novae, WR stars and other massive stars). At present none of the proposed source populations seem to be able to produce the observed ^{26}Al - not even by taking into account the combined yield of all sources together[1]. Likewise, a controversy is under way on the distribution of the galactic 511 keV annihilation radiation. Here the best way to interpret the apparently conflicting data seems to be a recent interpretation of Lingenfelter and Ramaty[2] where the 511 keV emission consists of two components: A variable point-like positron source at the Galactic Center and an extended galactic source following the distribution of the COS-B high energy gamma-ray emission.

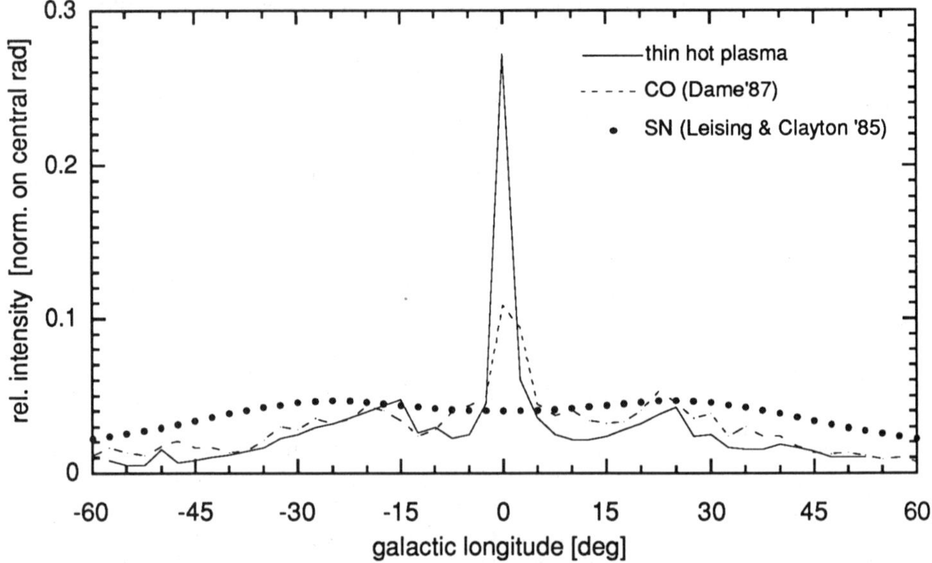

Figure 1. Comparison of the proposed distribution for the gamma ray emissions at 1809 keV and 511 keV (thin hot plasma, adapted from GINGA data [3]) with the Supernova(CO) distribution of Leising and Clayton [4] and a more recent CO distribution [5]. Distributions are normalized to have equal integrals in the central rad of the Galaxy.

In this paper we suggest that the diffuse ^{26}Al and the e^+e^- emissions are both reflected by the distribution of the galactic ridge thin hot gas (Fig 1). Massive stars are thought to synthesize ^{26}Al and other radionuclides during normal and final phases of their life. The positrons produced when those radioactive elements decay are sufficient to explain the observed 511 keV line emission. Sources capable of synthesizing intermediate and heavy elements (SN, WR stars etc.) are the same objects that have been proposed to shock-heat vast bubble structures of thin hot plasma. Recently the X-ray satellite GINGA[3] has mapped the galactic hot tenuous gas. Its distribution is suggested to reflect the two galactic gamma ray lines *regardless* of the source types producing both the radionuclides and the hot plasma. The proposed distribution resembles the CO distribution[5] throughout the galactic ridge. At the Galactic Center however a dominant narrow peak (~2° FWHM) contains a third of the total emission in the central radian.

THE DISTRIBUTION OF THE GALACTIC ^{26}AL

So far two satellite experiments, HEAO-C[6] and SMM[7,8] and two balloon detectors, one from Bell/Sandia[9] and one of the Max Planck Institute[10,11] have detected a gamma-ray line at 1809 keV in the ISM. All experiments support a galactic origin of the line emission. In general, they converge with respect to the intensity of the line; that is to a flux of the order of $4 \cdot 10^{-4}$ $\gamma/cm^2 \cdot s \cdot rad$. This corresponds to about 2-4 M_o of ^{26}Al in the ISM, depending on the assumed solar galactocentric distance and the galactic distribution of the radioisotope. In order to identify the class(es) of astrophysical objects involved in ^{26}Al nucleosynthesis it is crucial to deduce the angular distribution of the 1.8 MeV emission. Yet both satellite instruments had large fields of view (HEAO C : 42° FWHM,

SMM 130° FWHM) with no directional information within them. The Bell/Sandia spectrometer had no imaging either but a narrower aperture; however the low sensitivity of the balloon measurements did not allow for scans. The only instrument with imaging capabilities is the balloon borne MPI Compton telescope (Field of view: ~55° FWHM, angular profile for a point source ~10°). The 1.8 MeV skymap[10] shows most of the emission to come from the Galactic Center, the data being best fitted with a narrow source of < 15° FWHM. Because of the still modest angular resolution and sensitivity of the balloon telescope wider distributions could not be excluded definitely: The broad 'SN' model of Leising and Clayton[4] (FWHM≅50°, based on earlier CO measurements, Fig 1) is rejected on a 2 σ level. On the other hand both HEAO C[6] and SMM[12] data seem to be inconsistent with a single point source at the Galactic Center. The SMM data rule out a point source origin at a level of 4.7σ while HEAO C would detect such a source at only 2σ.

The distribution of the galactic thin hot gas complies with the narrow concentration indicated by the MPE skymap as well as with the diffuse emission required by HEAO C and SMM. All presently available 1809 keV measurements are consistent with such a model. A test for the distribution is the modeled flux for narrow apertures when normalized with the SMM flux in the central radian. An aperture flux of $1.5 \cdot 10^{-4}$ $\gamma/cm^2 \cdot s$ is predicted for narrow field detectors (15° FWHM). The observation of the Bell/Sandia instrument (FOV: 15° FWHM) of $1.3 \pm 0.9 \cdot 10^{-4}$ $\gamma/cm^2 \cdot s$ is in agreement with this value. In these proceedings two new observations of the ^{26}Al line with the GRIS[13] and HEXAGONE[14] spectrometer are reported. Both measurements are consistent with the proposed distribution, but most of the extended emissions (such as CO-, WR-, Novae-models) also fit the data acceptably. Because of the modest statistical significance of the derived fluxes, only the point source model at the GC and a bubble in the local ISM seems to be excluded by the ensemble of the 1809 keV data[14].

THE DISTRIBUTION OF THE GALACTIC e^+e^- EMISSION

Intense gamma ray line emission at 511 keV from the Galactic Center region has been observed since the early seventies with balloon and satellite experiments (see LR, Lingenfelter and Ramaty (LR)[2] for a review). In 1979/80 the positron annihilation source supposedly disappeared within half a year. Until 1988 instruments with a narrow field of view (F.O.V) measured only upper limits for its flux while the gamma ray spectrometer on board the SMM spacecraft (130° F.o.V.) observed a steady flux from this source.
It has been suggested[2] that the seemingly conflicting results can be explained by a two component model: a variable, compact source ($<10^{18}$ cm) at or near the Galactic Center, and a continuous extended disk emission.

The scenario presented here offers an alternative interpretation of the 511 keV data: An extended disk emission similar to that presented in LR is now associated with a *narrow but diffuse* component at the Galactic Center. In Figure 2 all available measurements of the annihilation line are plotted against the field of view of the detectors they were observed with. A curve reflecting the expected aperture flux has been calculated by integrating the proposed model distribution folded with a triangular response function for the detector aperture. The curve has been normalized to a line flux of $2.3 \cdot 10^{-3}$ $\gamma/cm^2 \cdot s$ from the central radian of the Galaxy - a figure deduced by eight years of SMM data[8]. Because of SMM's wide FOV (130°) the model dependence of this normalization is rather weak - both peaked and extended distributions within this FOV fit the data acceptably. Most of the data comply with the curve in Fig. 2 - in particular the satellite experiments and all the new generation balloon experiments. While the null results of the Bell/Sandia and GSFC/CEN experiments in the early eighties have rather large upper limits, the high fluxes of the broad aperture measurements of CESR and UNH can not be explained.

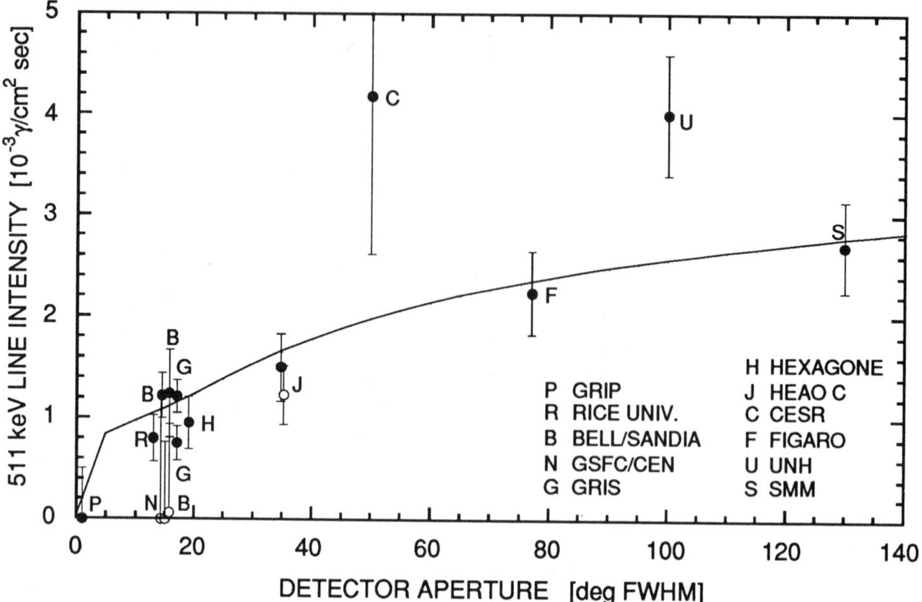

Figure 2. The various measurements of the 511 keV annihilation line versus the aperture of the detectors. Full circles display data taken before 1980 or after 1987 were the compact source at the Galactic Center was supposedly in a "on" state, empty circles show data from the "off" period 1980 -1987. Error bars are 1σ except for the null measurements (2σ upper limits). The solid line represents the expected flux from a distribution following the galactic thin hot plasma normalized on the SMM flux from eight years of data. The UNH data point has been corrected for orhopositronium (as in LR); note that the revised HEAO C figures [15] are used. References: GRIS [16], HEXAGONE [17], FIGARO [18], SMM[8], others see LR [2] and references therein.

Compact Objects in the Galactic Center Region

Hard X-ray variability of one or several objects at the GC has been observed by various experiments. HEAO C[20], Spacelab 2[21], GRIP[22], ART-P[23] show that the GC region is 'flaring' as different objects turn on and off on short time scales. Recently the SIGMA collaboration[24,25] has reported on observations of the GC source 1E1740.7-2942 performed in fall 1990. A strong spectral feature in the energy interval 300-700 keV was detected that emanated and vanished within days. Whereas none of the above instruments has yet detected the narrow 511 keV line itself, the SIGMA result indicates the existence of a pair plasma near 1E1740.7-2942.

Variability of the narrow 511 keV line itself is not vigorously required for the proposed distribution when keeping in mind the uncertainties in many of the data points (eg. error bars showing statistical uncertainties only). Yet, variability of one or several point sources within the central (and the ridge) component could possibly take place on a level of $\sim 5 \cdot 10^{-4}$ γ/cm^2·s. A position similar to this is supported by the limits on a variable 511 keV source set by SMM data[19]. No compelling evidence for variability is found during eight years of SMM observations. In 1988 when the compact annihilation source presumably has reappeared[26] the SMM data were consistent with a transient 511 keV source emitting $\leq 4 \cdot 10^{-4}$ γ/cm^2·s.

In the LR scenario variability of a central compact object is supported by the similarity between the spectra of the Galactic Center (observed in fall'79 by HEAO C) and

Cyg X-1, which is thought to be a black hole. The similarity of the spectra is best when both 511 keV line and positronium continuum are subtracted from the GC spectrum. Despite of being four times closer, a 511 keV line from Cyg X-1 has only once[27] been detected marginally ($4.4{\cdot}10^{-4}$ $\gamma/cm^2{\cdot}s$ on a $<2\sigma$ level). The two objects may therefore not have in common the production of positrons.

An alternative possibility to explain variability of a narrow 511 keV flux at the GC has been presented by Webber et al.[28]. Amplitudes in the 511 keV intensity of the above order could be caused by local temperature-changes in the annihilation region. In SNR and wind-generated bubbles temperature-phases ranging from a few tens of °K to $>10^7$°K can coexist in pressure equilibrium[29]. Positrons from decayed radionuclids (^{56}Co, ^{26}Al etc) are stored in the warm and hot phases with lifetimes $\sim 1.5\text{-}9{\cdot}10^5$ years depending on whether or not positrons are prevented by magnetic fields from penetrating 'cold cloud cores'. Since temperature changes in the 'positron cloud' affect mainly the formation of positronium - the 3-photon continuum below 511 keV would be affected most in such a scenario. This could explain variability of the continuum associated with the annihilation line.

NUCLEOSYNTHESIS AND THE THIN HOT PLASMA

^{26}Al is thought to be produced in supernovae, novae, explosions of supermassive stars. Massive stars synthesize ^{26}Al during their normal life, in particular during their Wolf-Rayet phase. All these objects have in common the injection of energy into the interstellar medium in the form of explosions and strong winds. The ejected gas contains the synthesized ^{26}Al which can escape destruction in the high temperature burning region thanks to the fast removal by stellar winds or disruptions. High velocity shock fronts are formed when the kinetic energy of the ejecta is dissipating in the ISM. The transformation into thermal energy is extremely efficient so that temperatures $>10^7$ °K occur, particularly in the thin gas region lying behind the shock front. Bubbles of such hot low density gas maintain temperatures of $>10^6$ °K for over 10^4 years when the adiabatic phase ends (for a review see McCray[29]). As a result a system of partly overlapping bubbles filling a large fraction of the interstellar space is expected to emit detectable thermal bremsstrahlung at X-ray energies.

X-ray observations of the galactic ridge thin hot plasma have been obtained by EXOSAT[30], TENMA[31] and GINGA[3] An intense narrow but diffuse hot gas component at the Galactic Center has been observed with Einstein IPC[32], Spacelab 2[21], SPARTAN 1[33], GINGA[3,34]. Temperatures of 10^7-10^8 °K have been obtained by thermal bremsstrahlung fits to the data over energy ranges within 1-15 keV. The thermal character of the X-ray emission is certified by the 6.7 keV iron line detected with TENMA and GINGA. Assuming a solar galactocentric distance of R_o=10 kpc the total luminosity of the thin hot plasma in our Galaxy is of the order of 10^{38} erg/s. We would like to point out the distinctive shape of the plasma distribution which is best shown by the GINGA observation: The galactic distribution of the thin hot plasma is enhanced in the 5 kpc spiral arms ($l = \pm 25°$) and at the Galactic Center where a narrow peak contains a third of the total emission in the central radian. This distribution is suggested to reflect the ^{26}Al in our Galaxy.

For Wolf-Rayet stars a link between ^{26}Al production and the disk component (TENMA) of the X-ray emitting plasma has been discussed by Montmerle[35]. The positron yield from radionuclids produced in WR stars has not yet been studied in depth. Should the remnants of supernovae and supermassive stars dominate the heating of the galactic bubbles and superbubbles (as is generally believed) the synthesized radionuclei will produce positrons that are distributed similarly to the X-ray plasma and the 1809 KeV radiation. Recent calculations of Chan and Lingenfelter[36] demonstrate that sufficient nucleosynthesis positrons should escape from supernovae (mainly type I) to produce the observed 511 keV line intensity. The positrons escaping into the ISM annihilate with a lifetime of ~10^5-10^6 years (comparable to the lifetime of ^{26}Al but much longer than the average supernova occurring time) result in a steady diffuse annihilation radiation.

The widths of the observed 1809 keV and 511 keV lines indicate that the decay of ^{26}Al and annihilation does not take place in the shock-heated shells themselves. According to calculations of Guessoum et al.[37] positrons annihilating in the warm partly ionized gas (T~8000 °K) result in a line width of <2 keV, less than the width expected for the cold cloud cores (80°K) and the hot phase (>10^5 °K). Although only a few measurements resolved the 511 keV line beyond the intrinsic detector-widths of ≥2 keV FWHM, annihilation in the warm phase within the bubbles seems to fit the available data best.

Further tests of the presented hypothesis are needed: Fitting the model to existing SMM and HEAO data and, (re)analysis of Galactic Center images with CALTECH's GRIP and the Franco-Soviet SIGMA using a few large resolution elements. A sensitive test will come from NASA's Gamma-ray Observatory (GRO). With their superior sensitivity and angular resolution the GRO OSSE and COMPTEL will be able to map the large scale distribution of the line emission at 511 keV and 1809 keV.

The author appreciated constructive discussions with Mark Leising and acknowledges support from the Institute for Astronomy of the ETH Zürich where a part of this paper has been compiled.

REFERENCES

[1] Walter R., and Maeder A., *Astron.Astrophys.*, **218**, 123, 1989
[2] Lingenfelter R.E., and Ramaty R., *Ap.J.*, **343**, 686 (LR), 1989
[3] Koyama K., et al., *Nature*, **339**, 603, 1989
[4] Leising M.D., and Clayton D.D., *Ap.J.*, **294**, 103 (LC), 1985
[5] Dame T.M. et al., *Ap.J.*, **322**, 706, 1987
[6] Mahoney W.A., Ling J.C., Wheaton W.A., and Jacobson A.S., *Ap.J.*, **286**, 578, 1984
[7] Share G.H., Kinzer R.L., Kurfess J.D., Forrest D.J., Chupp E.L., Rieger E., *Ap.J.*, **292**, L61,1985
[8] Harris M.J., Share G.H., Leising M.D., Kinzer R.L., Messina D.C., *Ap.J.*, **362**, 1990
[9] MacCallum C.J., Huters A.F., Stang P.D., and Leventhal M., *Ap.J.*, **317**, 877, 1987
[10] von Ballmoos P., Diehl R., and Schönfelder V. , *Ap.J.*, **318**, 654, 1987
[11] Schönfelder V., these proceedings, 1991
[12] Purcell W.R., et al., in *GRO-Science Workshop*, p.4-327, 1989
[13] Teegarden B.. *et al.*, these proceedings , 1991
[14] Malet I. *et al.* , these proceedings, 1991
[15] Mahoney W.A., in *Nuc.Spectroscopy of Astrophys. Sources*, (AIP Conf. Proc. 170), p.149, 1988
[16] Leventhal M., Proc of the 28 COSPAR, The Hague, to be published in *Adv.Space Res*, 1991
[17] Matteson J. et al., Proc of the 28 COSPAR, The Hague, to be published in *Adv.Space Res*, 1991
[18] Niel M. et al., *Ap.J*, **356**, L21, 1990
[19] Share G.H., Leising M.D., Messina D.C., Purcell W.R., *Ap.J.*, **385**, L45, 1990
[20] Riegler G.R., Ling J.C., Mahoney W.A., Wheaton W.A., Jacobson A.S., *Ap.J.*, **294**, L13, 1985
[21] Skinner et al., *Nature*, **330**, 544, 1987
[22] Prince T., Proc of the 28 COSPAR, The Hague, to be published in *Adv.Space Res*, 1991
[23] Sunyaev R. and the GRANAT team, *IAU Circilar* No. 5104, 1990
[24] Mandrou *et al.*, IAU circular 5140, 1990
[25] Paul et al., these proceedings, 1991
[26] Leventhal M. et al., *Nature*, **339**, 36, 1989
[27] Ling J.C. and Wheaton Wm.A., *Ap.J.*, **343**, L57, 1989
[28] Webber W, Schönfelder V., Diehl R., *Nature*, 323 , 692, 1986
[29] McCray R., in *Spectroscopy of Astrophys.Plasma* , Cambridge Astrophysics Series, p. 255, 1987
[30] Warwick R.S., Turner M.J.L., Watson M.G., Willingale R., *Nature*, **317**, 218, 1985
[31] Koyama K., *Publ. Astr. Soc. Japan* **41**, 665, 1989
[32] Watson M.G., Willingale R., Grindlay J.E., Hertz P., *Ap.J.*, **250**, 142, 1981
[33] Kawai N., et al., *Ap.J.*, **330**, 130, 1988
[34] Yamauchi S., Kawada M., Koyama K., Kuneida H., Tawara Y. , Hatsukade I., *Ap.J.*, **365**, 532,1990
[35] Montmerle T., in *Adv in Nuclear Astroph* ed. E.Vangioni-Flam et al. Ed. Frontieres , p.335, 1986
[36] Chan K-w. and Lingenfelter R.E., Proc. 21st ICRC, OG7.2-4, 253, 1990
[37] Guessoum N., Ramaty R. and Lingenfelter R.E., preprint, 1990

POSSIBLE ORIGIN OF THE DIFFUSE GALACTIC 6.7 keV IRON LINE, 511 keV ANNIHILATION LINE AND OF 1809 keV AL26 LINE.

C.Chapuis[1], P.Wallyn[1], Z.He[1,2], Ph.Durouchoux[1]

1. Service d'Astrophysique
Centre d'Etudes de Saclay
91191 Gif sur Yvette Cedex
2. NO.3 department
High Energy Physics Institute
Academia Sinica
Beijing, CHINA

ABSTRACT

A recent measurement of the distribution of the 6.7 keV iron line flux along the galactic plane gives the opportunity to compare a model with the experimental results. The model is based on the assumption that supernova ejecta are responsible of the excitation of the interstellar iron and so of the 6.7 keV iron line. For that, we used the current supernova remnant distribution[1] and a recent metallicity gradient[2]. The result of the computation is well correlated with the observed iron flux versus the galactic longitude[3]. The supernovae are progenitors of radioactive nuclei, some of which can decay via β^+ emission, for example ^{56}Co, ^{44}Ti and ^{26}Al. We assume that the Galactic source of 511 keV observed[4,9] in the direction of the Galactic Center[4] are from two different origins: a steady diffuse component, and a variable point source at or near the Galactic Center. We will consider here the diffuse source of 511 keV only. The 1809 keV gamma-ray line was discovered in 1982 and confirmed[5,6]. We propose possible distributions for both 511 keV and 1809 keV galactic lines which seem to fit the data quite well.

INTRODUCTION

Iron is distributed along the galactic plane in the interstellar medium. The 6.7 keV line is a fluorescence line of the iron and could come from the excitation of interstellar iron by supernova ejecta[7].

A part of the Galactic Plane is a source of 511 keV photons coming from the annihilation of positrons. The positrons can annihilate with electrons, or neutral matter like hydrogen or grains. Two photons of 511 keV can be produced directly after the annihilation. A bound state (positronium) can be formed by radiative recombination with free electrons or by charge exchange with a gas atom or molecule. Two kinds of positronium can be produced: orthopositronium (triplet state) and parapositronium (singlet state). The annihilation of this bound state gives different spectra. The decay of parapositronium produces two photons of 511 keV energy, whereas the orthopositronium decay produces in addition a three photon continuum below 511keV.

This 511 keV line was observed by numerous experiments[9,(4 for a review)] and the study of the time variation of this line seems to show there are two different components for the 511keV line: a point source at or near the Galactic Center, and a diffuse source.The positrons could be, to a significant extent, produced in supernovae, especially by β^+ decay of radioactive nuclei synthesized in such stars. The half-life of ^{56}Co is 77.3 days and so, most of the positrons, 19% of which are produced in ^{56}Co beta decays, annihilate inside the supernova ejecta before it becomes thin enough for them to escape. The ^{44}Ti seems to be a good candidate for the Galactic emission of positrons (half-life 47 yr).

The aluminum 26 decays to an excited state of magnesium-26 (β^+ decay), which returns to the fundamental level by emission of a 1809 keV photon (half-life 7.4×10^5yr).

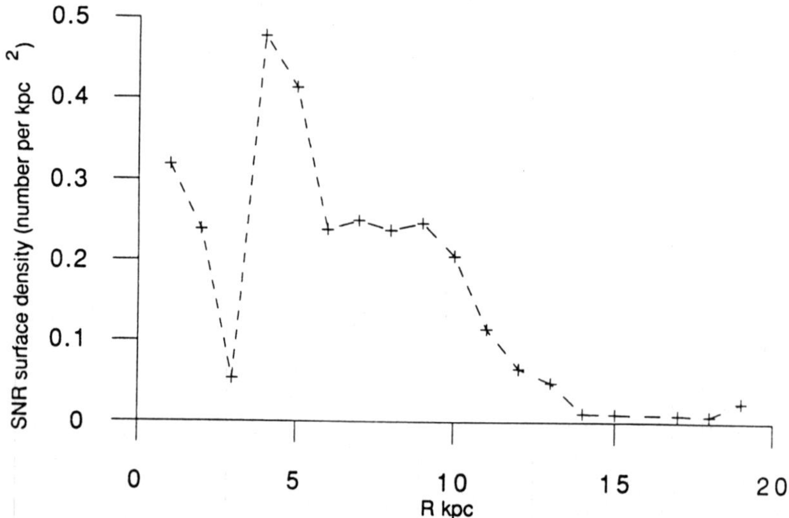

figure 1. SNR surface density in the Galaxy versus the galactocentric distance, assuming an axisymmetric distribution.

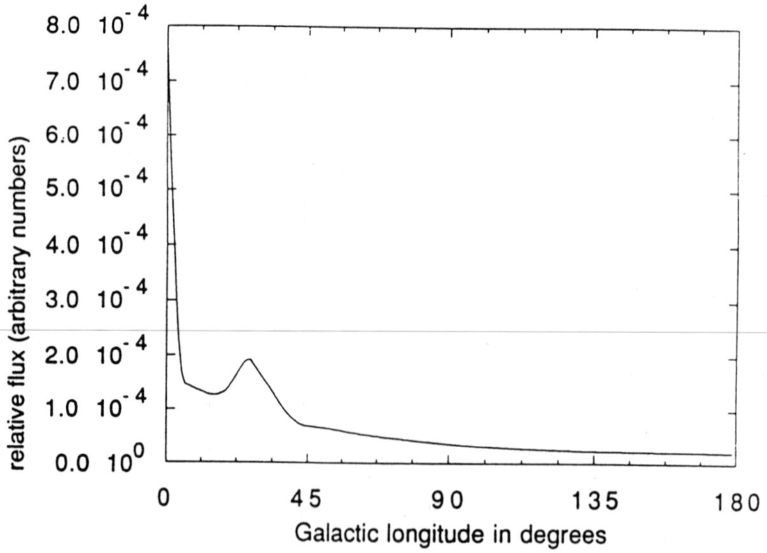

figure 2. relative flux of the 6.7 keV function of the Galactic longitude

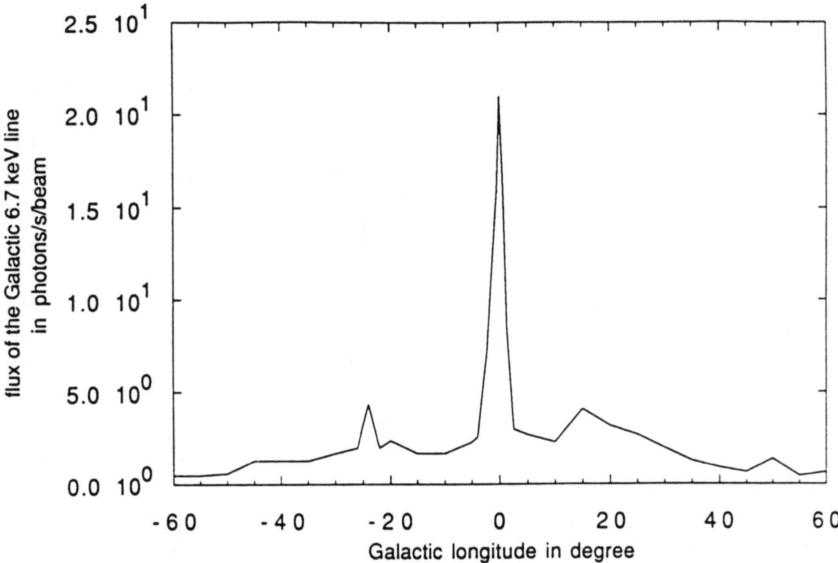

figure 3: flux of the 6.7 keV line versus the Galactic longitude (from Koyama et al. 1989)

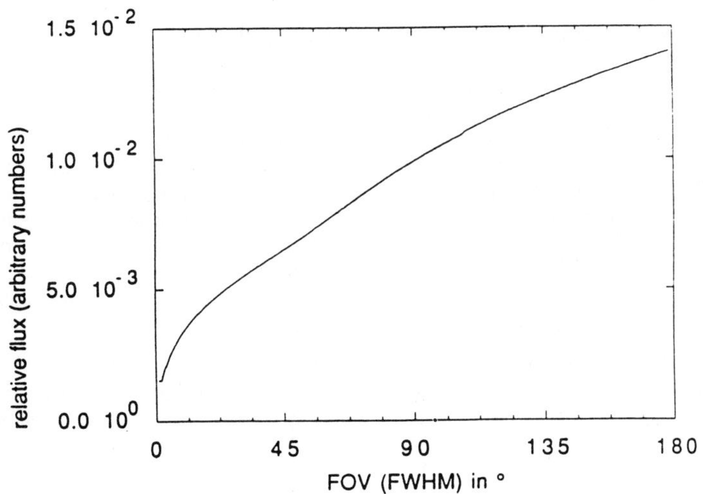

figure 4. relative flux versus the FOV in degree for the 511 keV line

HYPOTHESIS AND RESULTS

Study of the predicted Galactic distribution of 6.7 keV

We assume that the probability of the 6.7 keV iron line excitation is proportional to the local axisymmetric SNR surface density we derived from the data of Green[8] (figure 1) considering the distance to the Galactic Center R_0=8.5kpc. The profile of the SNR surface density (figure 1) follows the data of the CO surface density, so that we expect the SNR surface density follows that of CO in the central kpc. We have normalized the CO axisymmetric[13] distribution to the SNR one. We then fixed the local density of matter to be the SNR surface density and determined the local mass of iron to be proportional to the local metallicity. The local metallicity at x_r is based on the value of Chiosi[2], -0.05dex.kpc^{-1}, which was normalised to the metallicity of the Sun. The metallicity in the Galactic Bulge is not known. The maximum of metallicity in the vicinity of the Galactic Center was taken to be 5xZ_0, while the minimum of the metallicity was taken to be 0.5xZ_0 for high Galactocentric distance.

The differential flux in the 6.7 keV line emitted at a distance varying between x_r and x_r+dx_r from the Earth and located at the galactic longitude varying between l and l+dl is then given by the formula

$$df \,\alpha\, (\sigma_{SNR}(R)^2 \;\; Z(R)\; dx_r dl)/(4\Pi\; x_r)$$

We sum this along a line of sight and obtain the relative distribution of the 6.7 keV line versus the Galactic longitude. The figure 2 indicates the same profile that was observed (figure 3)[3]: a sharp central peak and two symetric features are seen at about +25° and -25°. The GINGA experiment[2] observed the 6.7 keV iron line in the galactic plane and associated to it a spatial distribution of the flux along the galactic longitude (figure 3). There is a consistency between the predicted distribution and the experimental one (one sharp central peak and two features at ±25°). Note that the experimental curve seems to be asymmetric!

Study of the predicted Galactic distribution of 511 keV line

Galactic positrons should be principally produced in supernovae by ^{56}Co, ^{44}Ti and ^{26}Al decays and should be responsible, to a significant extent for the galactic diffuse emission of 511 keV. If we then assume that the surface density of positrons is proportional to the SNR distribution, this distribution could be a good tracer of these positrons. The probability of annihilation of positrons should be dependent on the local density of matter and on the metallicity because of a higher interaction cross section: the positrons will preferentially annihilate in a medium of high matter density and high metallicity. We take into account these facts in fixing this probability to be proportional to the SNR surface density and to the metallicity. The final curve giving the relative flux versus the galactic longitude is then the same as the predicted 6.7 keV line, the 511 keV surface density of flux being proportionnal to $\sigma_{SNR}(R)^2$x$Z(R)$. The relative flux curve versus longitude was then used to monitor the flux versus FOV curve (figure 4) assuming a triangular response. This curve was normalised to the steady flux of SMM. The measurements are consistent with a diffuse source if we skip the fluxes when the point source was supposed to be ON (figure 5).

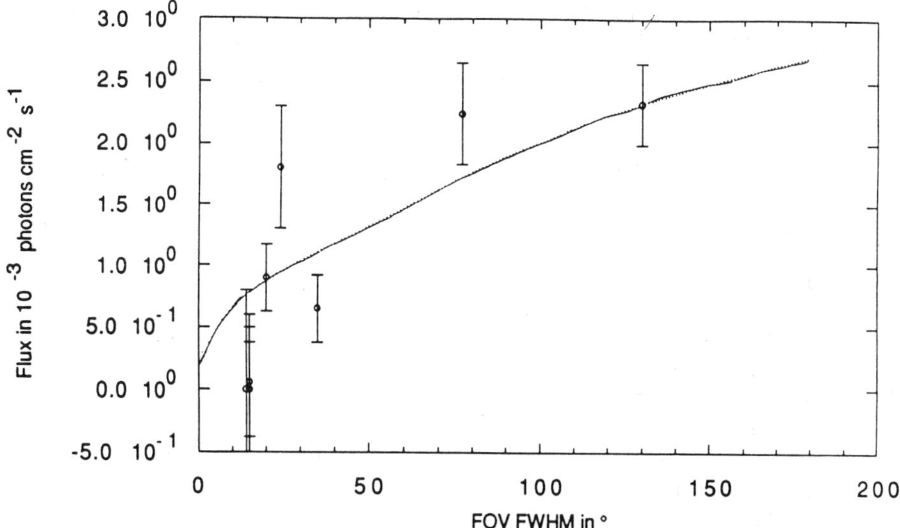

Figure 5. Measured fluxes of the 511 keV line superposed with the predicted one normalized to SMM results versus the FOV of the experiments

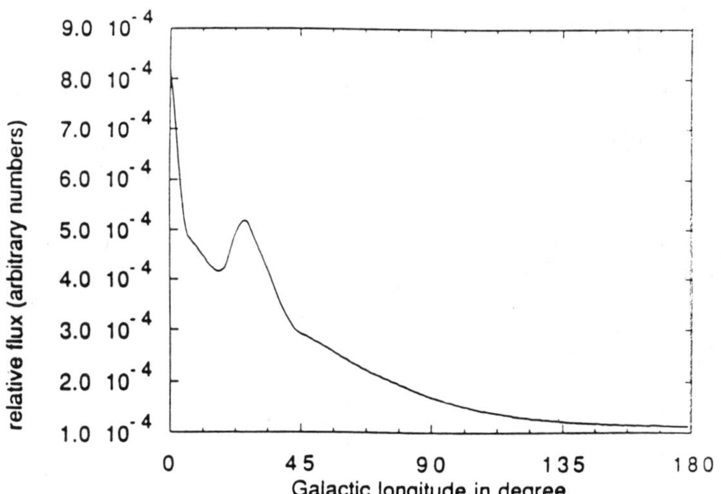

figure 6. Aluminum 26 1809 keV flux distribution versus the Galactic longitude

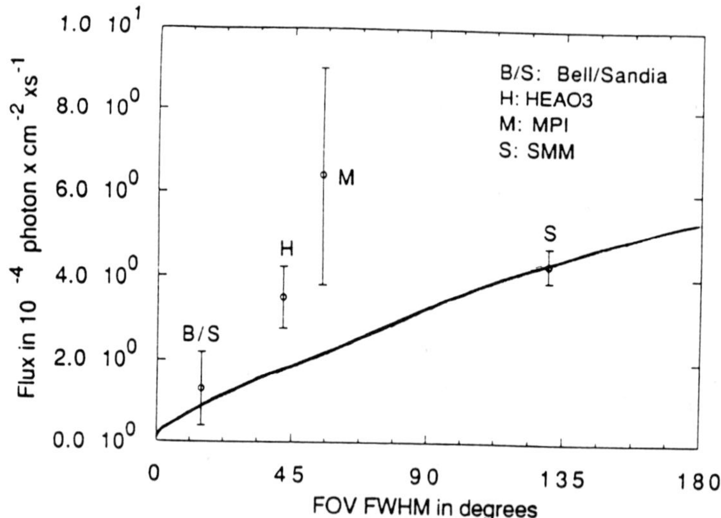

figure 7. predicted distribution and measurements of the 1.809MeV flux with the FOV

Study of the predicted Galactic distribution of the 1809 keV line

We therefore assume that supernovae produce the major part of the aluminum 26 in the Galaxy and are responsible of the 1809 keV gamma-ray line. We predict the relative distribution of the flux of this line along the galactic plane (figure 6) assuming that the surface density of aluminum is proportionnal to that of SNR and to the metallicity. The curve giving the relative flux versus the longitude is then smoother than the 511 keV predicted one or than the 6.7 keV predicted one. The comparison between the secondary peaks intensities at about 25° of the curves of the figures 2 and 6 put in light that there is no proportionnality between the 511keV and the 1809 keV distribution. So, there is no correlation between the 6.7keV and the 1809 keV distribution. We then predict the 1809 keV relative flux for different FOV assuming a triangular response of the collimator of the experiment and normalized it to the SMM[12] value (figure 7). The measurements[10,11,12] are in good agreement with the predicted curve.

CONCLUSION

Our predicted 6.7 keV flux is in good agreement with the measurements[3]. There is a proportionality between the 6.7 keV flux distribution along the galactic longitude and the 511keV one but the model shows no proportionality between the 6.7keV and the 1809keV distribution but show no correlation between the 511 keV distribution and that of 1809 keV. The results of the predictions point out that the 511keV distribution has the same profile as the 6.7keV one while the Aluminum 26 distribution is not the same as the 6.7keV one. All the predicted relative distribution are in good agreement with the results of the measurements.

ACKNOWLEDGEMENTS

We thank N. Prantzos, J.P. Meyer and T.Montmerle for helpful and constructive discussions.

REFERENCES

1.Clark, D., Caswell, J., *Mon.Not.R.Astr.Soc.*, **174**, 267-305, 1976
2.Chiosi, C., "Nucleosynthesis and chemical evolution" 16th Advanced Course Swiss Society of Astrophysics and Astronomy 1986
3.Koyama et al., *Nature*, 339, **603**, june 1989
4.Lingelfelter, Ramaty, *Ap.J.*,**343**, 686, 1989 August 15
8.Green, D.A., *Mon.Not.R.Astr.Soc.*, **209**, 449-478, 1984
5.Mahoney et al., *Ap.J.*, **262**, 742, 1982
6.Mahoney et al., *Ap.J.*, **286**, 578, 1984
7.Phinney, S., *Nature*, **348**, 1990 November 1
8.Green, D.A., *Mon.Not.R.Astr.Soc.*, **209**, 449-478, 1984
9.Chapuis *et al.* this symposium 1990
10.v.Ballmoos,P., Diehl, R., Schoenfelder, V., *Ap.J.*, **318**, 654, 1987 July 15
11.MacCallum, C., J., Huters, A., F., Stang, P., D., Leventhal, M., *Ap.J.*, **317**, 877, 1987 June 15
12.Share,G.,H., Kinzer,R.L., Kurfess,J.D., Forrest, D.,J., Chupp, E.L., Rieger,E., *Ap.J.*, **292**, L61, 1985 May 15
13.Clemens, D.,P., Sanders, D.,B., Scoville, N., Z., *Ap.J.*, **327**, 139, 1988 April 1

A possible correlation between the 511 keV annihilation radiation and the 6.7 keV iron line emission in the galactic plane.

Z.He, Ph.Durouchoux, C.Chapuis, P.Wallyn
DPhG/SAP, CEN-Saclay, 91191 GIF-SUR-YVETTE, CEDEX, FRANCE

Abstract

Since the 1970's, more than twenty observations of the annihilation radiation from the Galactic Plane have been performed. Previous analysis showed that there could be two components to this emission: (1) a variable, compact source of 511 keV annihilation radiation within the Galactic Centre region, (2) a steady, diffuse interstellar 511 keV annihilation source. In order to evaluate the contributions of these two components, several models assuming extended sources have been used for the diffuse source emission. For example, the distribution of radiation is consistent with that of high energy gamma rays or molecular (CO) gas. We suppose here that the diffuse 511 keV line intensity distribution follows the iron line (6.7 keV) radiation along the galactic longitude since they could have a common progenitor: supernovae remnants. The result shows that our hypothesis can account for the observations and is consistent with an extended source having an intensity of 2.0*10(-3) ph/cm2/s/rad near the Galactic Centre. However, there is no strong evidence for a variable compact source.

Analysis and results

Many scientists have tried to compare the 511 keV distribution with the distribution of different types of objects along the Galactic Plane.(eg. Lingenfelter & Ramaty, CO molecular gas and high-energy gamma ray. P.von ballmoos,(Ref.23) thin hot plasma, etc.) Here we use a different approach; we took the distribution of the 6.7 keV line emission, and we have analysed 25 observations performed from 1970 to 1989, 24 of which pointed toward the galactic centre. The data are listed in Tab.3.

We obtained the 6.7 keV iron line emission distribution for galactic longitudes from -60° to 60° from K.Koyama et al.(Ref.6) In order to extrapolate the radiation to angles which fit the SMM observations(F.O.V=130°), we suppose that the radiation from those angles with absolute values ranging from 60° to 130° is proportional to the radiation emitted by the supernovae remnants along the galactic longitude, assuming they have a same emissivity. Using the supernovae remnant distribution in the galactic plane from D.H.Clark & J.L.Caswell,(Ref.20) we calculated the distribution versus galactic radius (assuming a Galactic Centre distance of 8.5 kpc)and then transformed it (a rotational symmetry is assumed) to the SNR radiation distribution along the galactic longitude. We made the calculation in this way since it might avoid some bias in the SN observations. The extrapolated 6.7 keV iron line emission radiation is formed by normalising the SN radiation distribution to the known values at ± 60° and is shown in Fig.1.

We assumed the diffuse 511 keV annihilation radiation along the galactic longitude follows this iron line emission in the galactic plane. Applying a simple triangular response for each instrument, we obtain the assumed contribution from the extended source to detectors having a given field of view (FWHM). These values are normalised to a detector with a F.O.V of 130°(SMM) and are shown in Fig.2.

We fitted the curve(shown in Fig.2) to the observational data by the least squares method. Five groups of data were selected to be the samples. The result shows that the best fit values are consistent with each other and almost independent of the selection of the samples. The results are shown as follows in Tab.1.

Tab.1. Best fit results

Samples	Flux of detec. with FWHM=130° *10(-3)ph/cm2/s	Flux around the Galactic Centre *10(-3)ph/cm2/s/rad	Chi-Square value/d.o.f
1. All observations except those from 1977 -- 1980	2.4	1.9	1.79
2. All SMM data	2.4	1.9	1.01
3. All observations	2.5	2.0	2.44
4. Observation with NaI detectors	2.5	2.0	1.65
5. Observation with Ge detectors	2.6	2.1	3.63

Assuming that the flux around the Galactic Centre is 2.0*10(-3) ph/cm2/s/rad, Fig.3 shows the observational data with 1 sigma error bars and the estimate of the contribution of diffuse emission (solid curve). The excess flux (which should be the flux of a variable compact source if it exists) is shown in Fig.4 and listed in Tab.3. From Fig.4 and Tab.3, one can see that more recent observations are more consistent with the estimated diffuse emission. A variable compact source is not required to account for our calculation. (statistical significance is less than 3.0 sigma)

In addition, we took into account an observation of the galactic plane 25° west of the Galactic Centre made with the GRIS instrument (ref.17) on October 30,1988. Our calculation predicts a flux measured by this experiment: 4.8*10(-4) ph/cm2/s, which is consistent with the observational result: < 5.0*10(-4) ph/cm2/s.

A possible astrophysical explanation

In the theory of supernovae, the core collapse would occur under certain critical conditions. Explosive nuclear reactions or possibly electromagnetic wave pressure from the collapsing object generate an outward moving shock in the envelope and may lead to a powerful explosion of supernovae. The main nucleosynthesis products formed during a SN explosion and decay chains are shown in Tab.2. These have sufficiently long half-lives to produce gamma rays in optically thin regions.

Tab.2. The most important nucleosynthesis products and their decay chains in supernovae explosion. (from Matteson (1982).)

Decay (Parent-Daughter)	Gamma-ray energy(keV) and Branching(%)	half Life	Yield/SN event A.	B.
Ni(57) -> Co(57)	127(14),1370(86)	30 hr	2(53),	5(52)
Co(57) -> Fe(57)	14.4(89),122(80),136(11)	270 d	2(53),	5(52)
Ni(56) -> Co(56)	163(85),276(34),427(34) 748(51),812(85)	6.1 d	4(54),	2(54)
Co(56) -> Fe(56)	847(100),1030(16),1240(67) 1760(14),2600(17),e+(20)	77 d	4(54),	2(54)
Na(22) -> Ne(22)	1275(100),e+(90)	2.6 y	2(51),	1(52)
Ti(44) -> Sc(44)	68(100),78(100)	46 y	3(51),	4(51)
Sc(44) -> Ca(44)	1156(100),e+(94)	3.9 y	3(51)	
Fe(60) -> Co(60)	58.6(100)	3(5) y	4(50),	4(50-51)
Co(60) -> Ni(60)	1170(100),1330(100)	5.3 y	4(50),	4(50-51)
Al(26) -> Mg(26)	1809(100),1130(4),e+(85)	7.4(5)y	5(40),	5(50)

From Tab.2 we see that positrons are generated in several processes. Furthermore, theoretical predictions show that a fraction of the positrons produced in the decay process are expected to escape into the interstellar medium. In a tenous interstellar gas with a density of 1H/cm3, the positron life time versus annihilation can be as long as 10(5) years.(Ref.11) If we assume a galactic supernovae explosion rate about once every several tens of years, positrons should accumulate from several thousand supernovae. Their annihilation should thus produce diffuse galactic gamma-ray line emission at 511 keV.

On the other hand, the iron produced in SN explosions might be an important contribution of the total amount in the interstellar medium. By some powerful interactions (shocks, strong radioactive radiations, etc), the iron is heated to the temperature of several keV and forms an optically thin hot plasma. Helium-like iron is the most abundant ionization state in this kind of plasma and will emit 6.7 keV X-rays when it captures electrons.

With this hypothesis, the 511 keV annihilation line radiation distribution should be similar to the 6.7 keV iron line emission.

Conclusion

From the results shown above, the 'iron-like' diffuse 511 keV radiation model can account for the observations quite well and a time-variable compact source is not required. We believe that the real diffuse positron annihilation distribution is close to that of the 6.7 keV line emission. It gives us a clue that 511 keV annihilation radiation and 6.7 keV emission might come from a common origin. Observations with better angular resolution are clearly needed to provide a final anwser to this very important astrophysical problem.

We thank R.E.Lingenfelter, J.M.Bonnet-Bidaud, T.Montmerle and N.Prantzos for their helpful discussions.

Tab.3. Data and Results

No.	Time	Type	F.O.V	511 KeV Flux		Experiment	Excess	Sign.
	(Year)		FWHM(°)	Flux*	Error*		*,**	Sigma
1	70.9	NaI	24.0	1.8	0.5	Rice Univ.(15)	0.70	1.4
2	71.89	NaI	24.0	1.8	0.5	Rice Univ.(15)	0.70	1.4
3	74.25	NaI	13.0	0.80	0.23	Rice Univ.(15)	0.01	0.0
4	77.13	Ge	50.0	4.18	1.56	CESR/CENS(F)(15)	2.47	1.6
5	77.86	Ge	15.0	1.22	0.22	Bell/Sandia(15)	0.37	1.7
6	77.89	NaI	100.0	4.0	0.6	Univ. NH(15)	1.72	2.9
7	79.29	Ge	15.0	2.35	0.71	Bell/Sandia(15)	1.50	2.1
8	79.8	Ge	35.0	1.85	0.21	JPL/HEAO-3(15)	0.45	2.1
9	80.2	Ge	35.0	0.65	0.27	JPL/HEAO-3(15)	-0.75	-2.8
10	81.0	NaI	130.0	1.95	0.45	SMM(15)	-0.55	-1.2
11	81.89	Ge	15.0	0.0	0.6	Bell/Sandia(15)	-0.85	-1.4
12	81.89	Ge	15.0	0.0	0.38	GSFC/CENS(F)(15)	-0.85	-2.2
13	82.0	NaI	130.0	2.05	0.45	SMM(15)	-0.45	-1.0
14	83.0	NaI	130.0	2.30	0.45	SMM(15)	-0.20	-0.4
15	84.89	Ge	15.0	0.0	0.5	Bell/Sandia(15)	-0.85	-1.7
16	85.0	NaI	130.0	2.55	0.45	SMM(15)	0.05	0.1
17	86.0	NaI	130.0	1.75	0.45	SMM(15)	-0.75	-1.7
18	87.0	NaI	130.0	2.32	0.33	SMM(17)	-0.18	-0.5
19	88.0	NaI	130.0	2.67	0.37	SMM(17)	0.17	0.5
20	88.33	Ge	18.0	0.75	0.17	GRIS(17)	-0.18	-1.1
21	88.83	Ge	18.0	1.21	0.16	GRIS(17)	0.28	1.7
22	88.9	NaI	77.0	2.24	0.41	FIGARO 2(17)	0.17	0.4
23	89.0	NaI	130.0	2.84	0.27	SMM(17)	0.34	1.3
24	89.39	Ge	20.0	0.89	0.27	UC/FRANCE ***	-0.10	-0.4
25	88.83	Ge	18.0	<0.5	\	GRIS		

(It pointed 25° west of GC in the galactic plane.)

*: The unity is: *10(-3) ph/cm2/s
**: Assumed flux for 130° detector(SMM) is 2.5*10(-3) ph/cm2/s
**: Assumed flux near Galactic Centre is 2.0*10(-3) ph/cm2/s/rad
***: See also C.Chapuis et al. presented at this symposium.

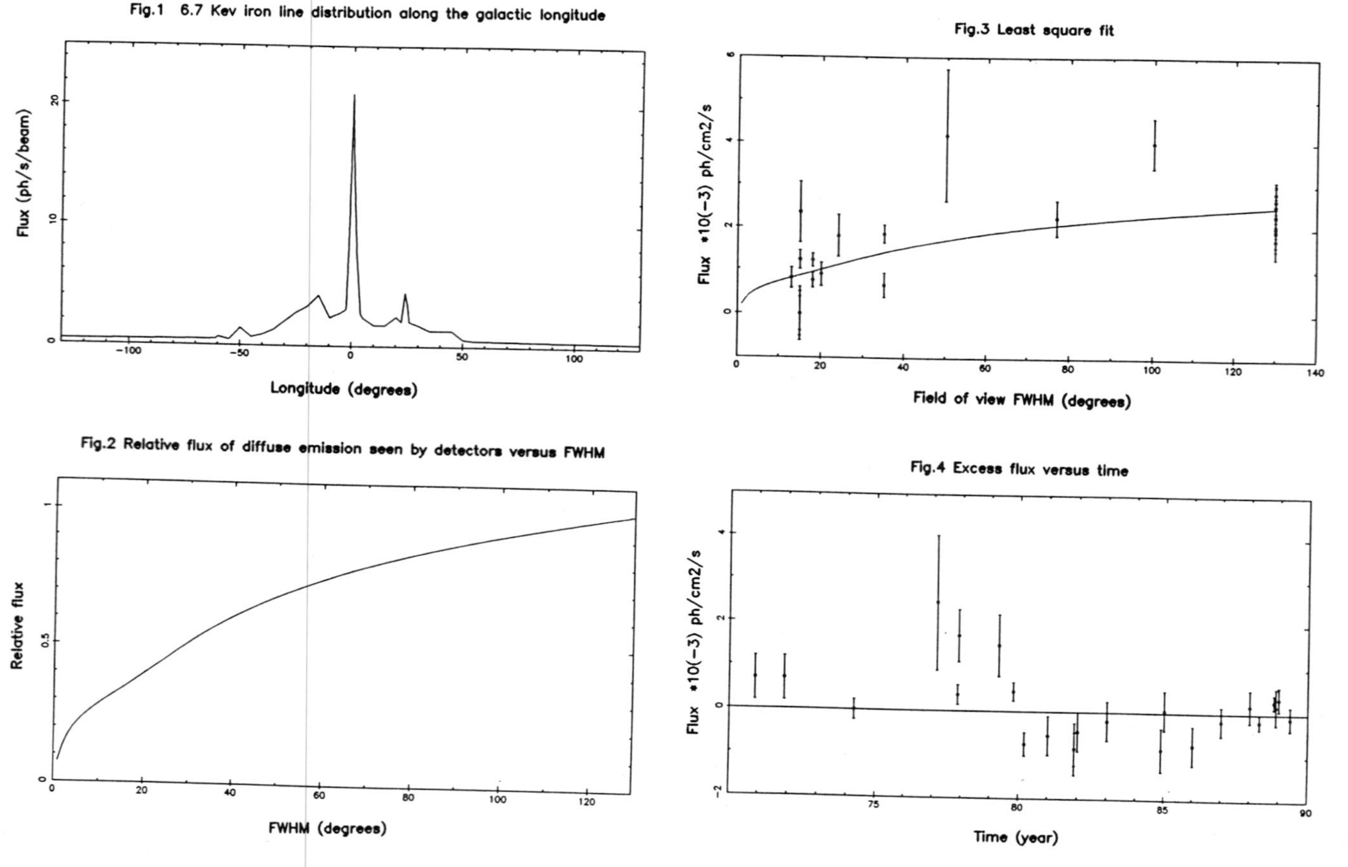
Fig.1 6.7 Kev iron line distribution along the galactic longitude
Flux (ph/s/beam)
Longitude (degrees)
Fig.3 Least square fit
Flux *10(−3) ph/cm2/s
Field of view FWHM (degrees)
Fig.2 Relative flux of diffuse emission seen by detectors versus FWHM
Relative flux
FWHM (degrees)
Fig.4 Excess flux versus time
Flux *10(−3) ph/cm2/s
Time (year)

Figure captions.

Figure 1. 6.7 keV line radiation distribution along the galactic longitude which is extrapolated to 130°.(with the supernovae remnant assumption).

Figure 2. Contribution from the 'iron-like' diffuse 511 keV radiation versus the field of view of the detectors.

Figure 3. If the diffuse 511 keV flux near galactic centre is 2.0 ph/cm2/s/rad., the expected flux(solid curve) seen by a detector with a given field of view together with observational data with 1 sigma error bars.

Figure 4. The excess fluxes detected from 1970 to 1989 versus time.

References

1. R.Ramaty & R.E.Lingenfelter. Nature Vol.278 127, 8 March 1979.
2. R.E.Lingenfelter & R.Ramaty. Nuclear Physics B,Proc.Suppl.,1989.
3. J.C.Ling & Wm.A.Wheaton. Proceedings of gamma-ray Observatory Science Workshop,Goddard Space Flight Center, April 1989.
4. Gerald H.Share & Mark D.Leising et al. Ap.J 1990 Aug.1 358:L45-L48
5. Thierry Montmerle. Proc.IAU Symp.143, Indonesia, June 18-22,1990.
6. K.Koyama, H.Awaki et al. Nature. Vol.339. No.6226. pp.603-605. 22nd June,1989.
7. W.David Arnett. Ap.J.1979 May 15. 230:L37-L40.
8. W.R.Webber et al. Nature. May 22,1986.
9. D.A.Green. Mon.Not.R.astr.Soc.(1984)209 449-478.
10. G.H.Share et al. Proc. of the 20th international cosmic ray conference.
11. R.E.Lingenfelter & R.Ramaty. 19th international cosmic ray conference. 1985, V.9, 19-41.
12. S.M.Matz & G.H.Share. Ap.J October,1990.
13. N.Guessoum & R.Ramaty et al. In press. NAS/NRC Resident Research Associate.
14. R.Ramaty & R.E.Lingenfelter. In press, Physics World, 1990.
15. R.E.Lingenfelter & R.Ramaty. Ap.J. 343:686-695, Aug.15,1989.
16. M.Leventhal et al. Ap.J. 302:459-461, Mar.1,1986.
17. M.Leventhal. Paper of COSPAR, June,1990
18. R.Ramaty & R.E.Lingenfelter. Ap.J, 213:L5-L7, April 1,1977.
19. S.Yamauchi,M.Kawada et al. Optically thin hot plasma near the galactic center.(Mapping observations of 6.7 Kev iron line) Accepted to the Ap.J.
20. D.H.Clark & J.L.Caswell. Mon.Not.R.Astr.Soc.(1976)174:267-305.
21. Ph.Durouchoux. Gamma-ray Line Astronomy. Private communication.
22. Ramaty,R & Lingenfelter,R.E., (1981),Philos. Trans.R.Soc. London, Ser.A.,301,671
23. P.von Ballmoos. Paper of COSPAR, June,1990

SHIFTS OF THE ^{26}AL LINE DUE TO GALACTIC ROTATION

Jeff SKIBO and Reuven RAMATY
Laboratory for High Energy Astrophysics
Goddard Space Flight Center
Greenbelt, MD 20771, USA

ABSTRACT

We have evaluated the shape of the ^{26}Al line using a galactic disk rotation curve and assuming various galactic distributions of radioactive aluminum. The line shape depends on galactic longitude, detector opening angle, and the assumed aluminum distribution. For a detector of 20° aperture and 2.5 keV resolution, the shift of the line peak can be as large as 0.4 keV. Such a shift could be detected with the proposed Nuclear Astrophysics Explorer.

DISCUSSION

Gamma ray line emission at 1.809 MeV from the decay of interstellar ^{26}Al was observed[1,2] with HEAO-3 and SMM, as well as with balloon borne experiments[3,4]. While these observations clearly indicate the presence of several solar masses of radioactive Al in the interstellar medium, they are not sufficiently accurate to map out the spatial distribution of the ^{26}Al. Past studies[5,6,7] have emphasized the distribution of the 1.809 MeV line intensity as a function of galactic longitude. We have investigated the expected shape of the 1.809 MeV line for various assumed galactic spatial distributions of ^{26}Al assuming a galactic rotation curve[8].

As two limiting cases, we take the radial distributions of molecular clouds and novae given in ref. (5). These distributions, as well as the galactic rotation curve, are displayed in Figure 1. The distribution of molecular clouds, with its annular structure, is in sharp contrast with the nova distribution, which is strongly peaked towards the galactic center. In both cases we assume an exponential z-dependence with a scale height of 70 pc for molecular clouds and 180 pc for novae. We normalize the distributions to yield 3 $M_\odot$ of ^{26}Al in the galaxy. We note that the nova distribution of ref. (5), which we use, is much more peaked toward the galactic center than the nova distributions employed in ref. (7). In the present work we have chosen the former to provide a strong contrast to the broad annular molecular distribution.

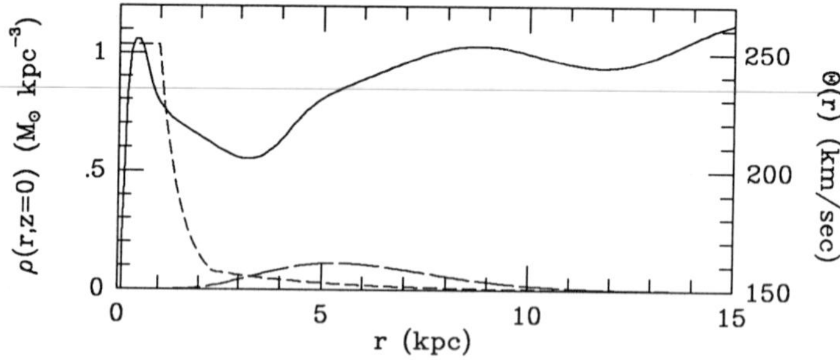

Figure 1. The galactic rotation curve - solid line, CO distribution - long dash, and the nova distribution - short dash. The CO and nova distributions are normalized to yield 3 $M_\odot$ of ^{26}Al.

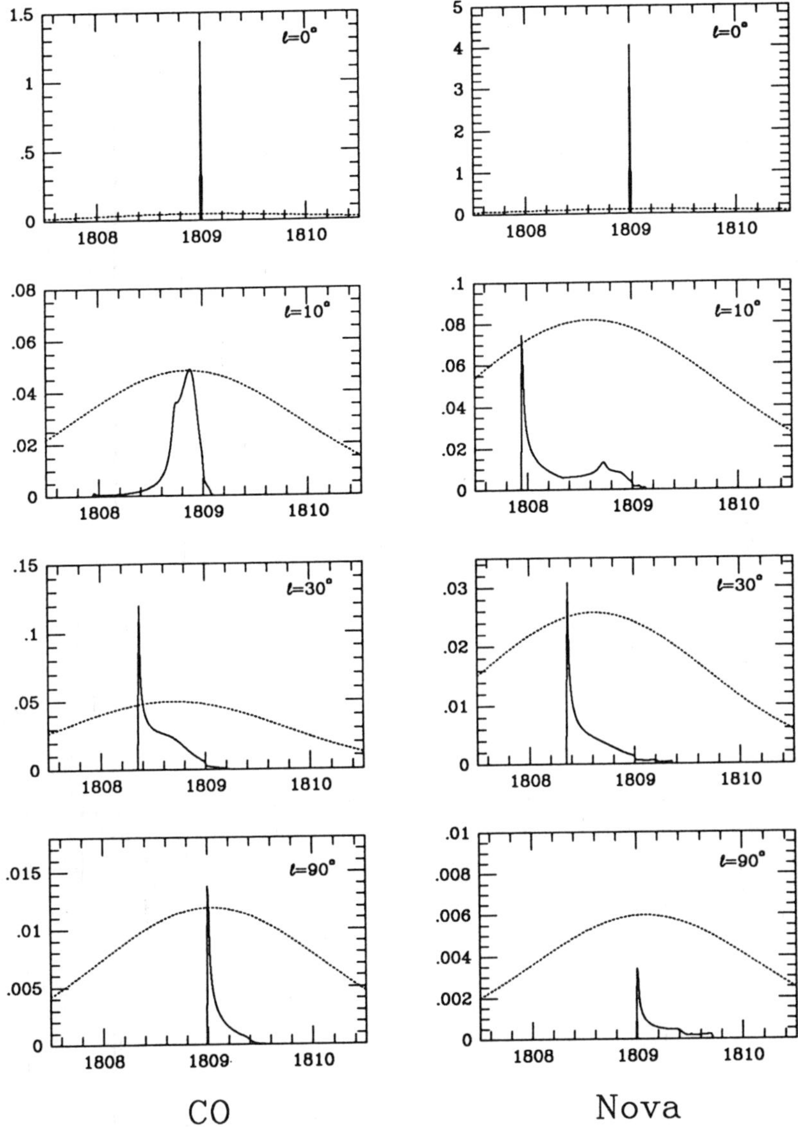

Figure 2. The spectra obtained by computing line of sight integrals through the galactic plane at latitude 0° and longitudes 0°, 10°, 30° and 90° are depicted by the solid curves and are in units of (photons sec^{-1} cm^{-2} $ster^{-1}$ keV^{-1}). The dotted curves result from the convolution of these spectra with a detector response as follows: triangular in longitude with FWHM 20°, theta function in latitude out to ±10° and Gaussian in energy with FWHM 2.5 keV. For the dotted curves the ordinates in units of (photons sec^{-1} cm^{-2} keV^{-1}) should be multiplied by 0.001.

Table 1.

l (degrees)	CO peak (keV)	CO FWHM (keV)	NOVA peak (keV)	NOVA FWHM (keV)
0	1809.00	2.53	1809.00	3.11
10	1808.85	2.54	1808.61	2.91
30	1808.70	2.54	1808.62	2.55
90	1809.03	2.51	1809.08	2.52

Using the distributions and rotation curve of figure 1, we compute specific intensities by carrying out line of sight integrals through the galactic plane. The results for longitudes 0°, 10°, 30° and 90° and latitude 0° are shown in figure 2 by the solid curves. Because of symmetry these spectra are only given for positive longitudes where they are mostly red shifted. The corresponding negative longitudes would simply display the same spectra reflected about 1809 keV. We see that the spectra exhibit a large amount of structure due to the convolution along the line of sight of the assumed emissivity distribution and galactic rotation. The spike like feature at extreme redshifts (e.g. at $l = 30°$) arises from emission from the subcentral point, i.e. the point along the line of sight with smallest galactocentric radius. For the rotation curve of figure 1, this point yields the largest heliocentric radial velocity along the line of sight. The difference between the CO and nova cases for $l = 10°$ is due the different aluminum distributions. For the CO case, the subcentral region resides where there is a paucity of gas, hence the spike at maximum redshift is not so pronounced. For the nova case the aluminum distribution increases rapidly towards the galactic center, ensuring that the emission is maximal at the subcentral point. For $l = 90°$ the blueshift is due to the motion of the Sun towards the outer galaxy.

Also shown in figure 2 (by the dotted curves) are fluxes resulting from the convolution of the specific intensities with detector characteristics as given in the caption. We see that much of the structure mentioned above disappears. The convoluted spectra can be characterized by a shift of the peak energy and a broadening of the line. These are given in Table 1. The maximum shift of about 0.4 keV should be detectable with next generation detectors such as NAE. Galactic rotation also broadens the line. For our assumed rotation curve and detector opening angle (20°), the line is broadened from 2.5 keV to a maximal value of 3.11 keV. We note, however, that the intrinsic line shape is definitely not Gaussian, as can be inferred from figure 2.

From a dissertation to be submitted to the graduate school, University of Maryland, by J. S. in partial fulfillment of the requirements for the Ph.D. degree in physics.

REFERENCES

1. Mahoney, W. A., Ling, J. C., and Jacobson, A. S. 1982, Ap. J., **262**, 742.
2. Share, G. H., *et al.* 1985, Ap. J., **292**, L61.
3. von Ballmoos, P., Diehl, R., and schonfelder, V. 1987, Ap. J., **318**, 654.
4. MacCallum, C. J., *et al.* 1987, Ap. J., **317**, 877.
5. Leising, M. D., Clayton, D. D. 1985, Ap. J., **294**, 591.
6. Pantzos, N., Cassé, M. 1986, Ap. J., **307**, 324.
7. Higdon, J. C., Fowler, W. A. 1989, Ap. J., **339**, 956.
8. Clemens, D. P. 1985, Ap. J., **295**, 422.

NOVAE

GAMMA-RAY LINES FROM CLASSICAL NOVAE

Mark D. Leising
E.O. Hulburt Center for Space Research
Naval Research Lab, Washington DC 20375

ABSTRACT

There is now a wealth of evidence supporting the idea that classical novae are thermonuclear runaways on white dwarfs. Explosive hydrogen burning produces substantial quantities of unstable nuclei. These nuclei emit γ-rays and β^+'s which might be detectable to present and future γ-ray spectrometers. Here we describe the production and observability of several unstable nuclei, review the observations made to date, evaluate the prospects for future observations, and discuss the importance of detection of γ-rays for improving our understanding of the nova phenomenon.

INTRODUCTION

Both theory and observation implicate a thermonuclear runaway in the degenerate envelope of a white dwarf accreted from a companion star as the cause of the classical nova outburst. See Reference 1 for a review of many aspects of classical novae. How the white dwarf and its envelope arrive in the condition required for this runaway is not yet clear. There remain many details regarding the interface of the dwarf and envelope, the nuclear burning conditions, and the transport of material in and out of the burning regions near the time of peak temperature which are uncertain. Gamma-ray lines offer a unique opportunity to probe the conditions within the burning regions and the dynamic processes initiated by the violent runaway. In general, the γ-ray luminosities depend only on the quantities of the emitting nuclei synthesized and the characteristics of any attenuating medium, not on the ambient temperatures or densities after the runaway (although β^+ annihilation is sensitive to the physical conditions). Detection of γ-ray lines would provide firm observational constraints on some aspects of theoretical models of the outbursts and force us to confront physical phenomena in novae which we which cannot currently model (e.g., convection).

THERMONUCLEAR NOVAE

When a white dwarf in a binary accretes matter slowly enough, a substantial envelope, 10^{-5} to 10^{-4} $M_{\odot}$, depending on the white dwarf mass, can be accumulated. The base of the envelope, which is supported by electron degeneracy pressure, is heated by the luminosity of the white dwarf and steady hydrogen burning, first p-p reactions, then CNO cycle reactions, until a runaway ensues. The temperature then rises rapidly, exceeds the Fermi temperature which is near 10^8 K (when the increased thermal pressure causes the envelope to begin expanding), and peaks shortly thereafter. The peak temperature, and indeed the time scale of these and all subsequent developments is determined by the mass of the white dwarf and its envelope, and the available energy. For the most massive dwarfs and envelopes the temperature might reach 5×10^8 K, but the typical value is probably $\lesssim 2 \times 10^8$ K. The high temperatures induce proton captures on most nuclei in the burning region, transforming stable seed nuclei into unstable proton-rich nuclei. Further processing and energy generation must then await the weak decays of those nuclei. The total available energy thus depends on the abundances of the catalysts such as C, O, and Ne. If enough of these nuclei from the white dwarf can be exposed to the hot bath of protons from the envelope, the energy generated can exceed the gravitational binding of the envelope, which can then be ejected. These ideas have been largely developed by Starrfield, Sparks, and Truran, and coworkers (see the chapter by Starrfield in reference 1 and references therein).

The extreme temperature gradient across the envelope at the peak of the burning produces rapid convective energy transport which can thoroughly mix the envelope material. Very abundant unstable nuclei with lifetimes longer than the convective time scale could appear at the surface where they are in principle detectable from their nuclear decay or positron annihilation γ-rays. Unstable nuclei with even longer lifetimes (greater than a few days) could survive the ejection and thinning of the envelope. Then their decay could be observed in γ-rays even if their abundances are relatively small. Those with lifetimes exceeding the time between nova events which eject them could accumulate in the interstellar medium and be detected as (nearly) constant sources of γ-rays. We discuss the potentially interesting nuclei in each lifetime regime separately.

^{26}Al (mean lifetime = 10^6 years)

This is the only isotope discussed here which has been definitely detected in an astrophysical source other than the Sun, namely the Galactic plane, with an apparent concentration toward the Galactic center direction. Aluminum-26 was first detected by the HEAO 3 γ-ray spectrometer[2,3] and its presence was confirmed by that on the Solar Maximum Mission[4]. Novae represent a plausible source of the observed ^{26}Al, but there are several other viable sources (see Reference 5 and references therein). No single nova would alone be detectable in the 1.809 MeV line of ^{26}Al, because of its slow decay. However the 10^7 to 10^8 novae which are thought to occur in the Galaxy during the ^{26}Al lifetime could produce the observed flux of 4×10^{-4} cm^{-2} s^{-1} if on the average they eject 10^{-7} $M_\odot$ of ^{26}Al.

Calculations of nova nucleosynthesis suggest a production of mass fraction X(^{26}Al) ~ 10^{-4} [6,7,8] and observed novae (and calculated nova models) typically eject a total mass of ~ 10^{-4} $M_\odot$[1], but these numbers can both vary by a factor of 10 in either direction. Some novae have been observed to eject material extremely rich in neon and heavier elements. If these enrichments simply reflect enhanced pre-outburst abundances of nuclei in this mass range, the production of ^{26}Al should also be increased proportionately. This conclusion, and the many factors affecting ^{26}Al in novae even when transport into and out of the nuclear burning regions is neglected, are explicitly pointed out in Reference 8. Together, the uncertainties in total Galactic ^{26}Al production of novae allow for their being the sole source, but the same can be said for other potential sources. This might well be a case where the observations teach us how novae do in fact work.

The determination of the origin of the ^{26}Al might await a definitive measurement of its distribution in the Galaxy. Novae in M31 are predominantly a spheroidal population, with a very small (if any) additional disk component[9]. If Galactic novae are distributed like novae in M31, the γ-ray distribution would be strongly peaked in the direction of the Galactic center. However, it appears that the local Galactic nova frequency is substantially higher than that at the corresponding position in M31, so the two nova distributions are probably not identical.

Two subsequent detections of Galactic ^{26}Al by detectors flown on balloons have not yet clarified this situation. One experiment indicated a distribution of ^{26}Al strongly concentrated toward the Galactic center direction[10], while another suggested a more broadly distributed source[11], although neither measured the distribution with high degree of confidence. It is possible that all ^{26}Al measurements are now consistent with each other, e.g., Schönfelder, this volume. The Gamma Ray Observatory will include two instruments which can measure the actual Galactic distribution of novae, COMPTEL and OSSE. Their measurements of the ^{26}Al distribution might finally settle the question of its source, but it is by no means certain that they can. This problem has by now been exhaustively (maybe exhaustingly is a better word) discussed by many authors and dozens of plausible Galactic distributions have been put forth. It is entirely possible that the measured distribution will be ambiguous. Prantzos (this volume) demonstrates how one will be able to explain almost any observed distribution with some variation on almost any proposed source.

The ^{26}Al observations and their interpretations are discussed elsewhere in these proceedings, but here we will just add a few relevant points. There is evidence in meteoritic samples that there was live ^{26}Al in the forming solar system in a proportion to ^{27}Al at least 10 times greater than what we infer that ratio to be in the present-day local ISM. Perhaps some event associated with the solar birth injected this radioactivity into the collapsing cloud. This is in itself an interesting, but long story. There are other pieces of meteorites (namely SiC grains) in which the inferred ratio ^{26}Al/^{27}Al at the time of condensation is much higher, the largest such ratio being 0.2[12]. These are probably fossils where the composition of matter ejected from nucleosynthesis sites is relatively well preserved. This large ratio can be naturally explained as the result of condensation of classical nova ejecta, but again they are not the only source of such a large ratio, and this fact says nothing about which source currently ejects the bulk of the ^{26}Al into the ISM.

^{22}Na (3.75 years)

With its 3.75 year lifetime, ^{22}Na decays fast enough to be detected in the ejecta of individual novae and yet lives long enough to accumulate from the outbursts of many novae. This nucleus was suggested by Clayton and Hoyle[13] as a potentially interesting diagnostic of novae. Upper limits to the γ-ray flux at 1.275 MeV from ^{22}Na have been obtained for Nova Cygni 1975[14], for a diffuse Galactic center component[2,15], and for several individual novae with ejecta rich in neon[15]. If 40 novae occur in the central Galaxy per year and eject 10^{-4} $M_{\odot}$ on average, the production of ^{22}Na is limited to $X(^{22}Na) \lesssim 2x10^{-4}$ in those novae, based on the lowest upper limit for the Galactic center 1.275 MeV line flux, $1.2x10^{-4}$ cm^{-2} s^{-1} [15]. This mass fraction is about the largest predicted by calculations of nucleosynthesis in material which starts with solar abundances. This upper limit does constrain the frequency of novae near the Galactic center which eject material 100 times overabundant in neon (such as N CrA 81 [16] and N Aql 82 [17]), and convert neon to ^{22}Na with the highest calculated efficiency, to about six per year[15]. This limit also implies that ^{22}Na contributes $\lesssim$5% of the β^+'s currently observed in the ISM. It has been suggested, based on rather uncertain statistical arguments, that several previously undiscovered neon-rich novae in the nearby Galactic disk should contribute significantly to the Galactic plane 1.275 MeV flux[18]. Scaling these arguments by the SMM upper limit, the average ^{22}Na mass ejected per neon-rich disk nova is limited to roughly 1.5x10^{-7} $M_\odot$.

Table I

SMM Limits on 1.275 Mev γ-ray Line Emission from Novae

	Galactic Center	N Cyg 1975	N CrA 1981	N Aql 1982	N Vul 1984 #2	Sum of Four Novae
Flux[a] Limit (10^{-4} photons cm^{-2} s^{-1})	1.2	25	3.2	2.5	3.0	...
^{22}Na Mass Limit (10^{-6} $M_\odot$)	3.1	1.5	8.6	1.6	0.7	0.5

[a] 99% confidence; at outburst for individual novae, assumed constant for GC.
[b] assuming equal ^{22}Na mass produced in each.

A summary of the lowest limits to 1.275 MeV fluxes from novae is given in Table I. The upper limits to the mass of ^{22}Na ejected by each of four observed individual neon-rich novae are ~ 10^{-6} $M_\odot$[15]. These novae typically eject only ~10^{-5} $M_\odot$ of material so this does not greatly constrain theoretical models. Nucleosynthesis in this mass range in novae has in general been investigated with fairly crude simulations. There are indications that the net production of very fragile species (such as ^{22}Na and, to a lesser extent, ^{26}Al) might be dramatically increased over those of standard parameterized (isothermal, in mass) calculations because of their removal from the burning regions by rapid convection[7]. One hydrodynamic calculation of a nova outburst on a Ne-O white dwarf, including nuclear reactions beyond ^{22}Na, ejected 1.6x10^{-7} $M_\odot$ of that isotope[7]. However the only other hydrodynamic nova calculation (on a C-O white dwarf) including ^{22}Na synthesis known to this author, ejected only 10^{-13} $M_{\odot}$ of ^{22}Na (Prialnik 1987, private communication). Improved theoretical estimates of the synthesis of species such as this are clearly needed, but detection of ^{22}Na in novae would go a long way toward establishing the validity of theoretical models of the entire nova phenomenon.

^{7}Be (77 days)

The synthesis of ^{7}Li (the daughter of ^{7}Be) is important for more than confirmation of nova theory. Calculations of big bang nucleosynthesis suggest a production of (Li/H) ~ 10^{-10} by number[19], in very good agreement with observed atmospheric abundances in hot, low mass, metal-poor halo stars[20]. Population I objects show striking uniformity in lithium abundance, near (Li/H) ~ 10^{-9}, except in some stars which presumably have convective zones deep enough to destroy surface lithium. Possible explanations are then either: 1) Big Bang estimates of Li production are correct and there is a Galactic source which has produced 90% of the Population I Li, or 2) The big bang somehow produced (Li/H) ~ 10^{-9} which has been uniformly destroyed in hot halo subdwarfs, and there has been no Galactic production of Li.

As pointed out by Arnould and Nørgaard[21], and Starrfield et al.[22], novae are possibly significant contributors to the Galactic Li abundance. It would be ejected as ^{7}Be which, as Clayton[23] suggested, could possibly be observed from its 478 keV β^- capture γ-ray line (which is emitted in 10% of its captures). The net production of ^{7}Be depends sensitively on the thermal history of the ejected material, and therefore on the treatment of convection (it is easily destroyed at high temperatures), and on the pre-outburst abundance of ^{3}He, which would presumably be accreted from the giant companion of the white dwarf. A nova at 1 kpc would have to eject ~4×10^{-9} $M_{\odot}$ of ^{7}Be to produce a flux of 10^{-4} cm^{-2} s^{-1} initially. According to calculations[22], this mass could be achieved in a total ejected mass of 10^{-4} $M_{\odot}$ if the red giant atmosphere accreted onto the white dwarf contained 20 times the solar abundance of ^{3}He.

The only γ-ray limit on ^{7}Be also comes from SMM[24]. Upper limits for line emission at 478 keV near 10^{-3} cm^{-2} s^{-1} at outburst were determined for three recent novae, and a limit of 1.7×10^{-4} cm^{-2} s^{-1} was determined for steady state emission (which is not quite accurate) from the direction of the Galactic center. These do not yet significantly constrain theoretical models, but instruments with better sensitivity and the luck of a very close nova could produce a most interesting detection.

^{13}N (862 s), ^{14}O (102 s), ^{15}O (176 s), ^{18}F (158 m)

As pointed out by Clayton and Hoyle[13], short-lived positron emitters might be carried quickly to the nova surface where the decay positrons would annihilate with electrons. The resulting 511 keV photons might be detectable from nearby Galactic novae, yielding important information on the burning conditions and convective transport. These ideas were refined somewhat and the ^{18}F nucleus was added as a possible candidate[25].

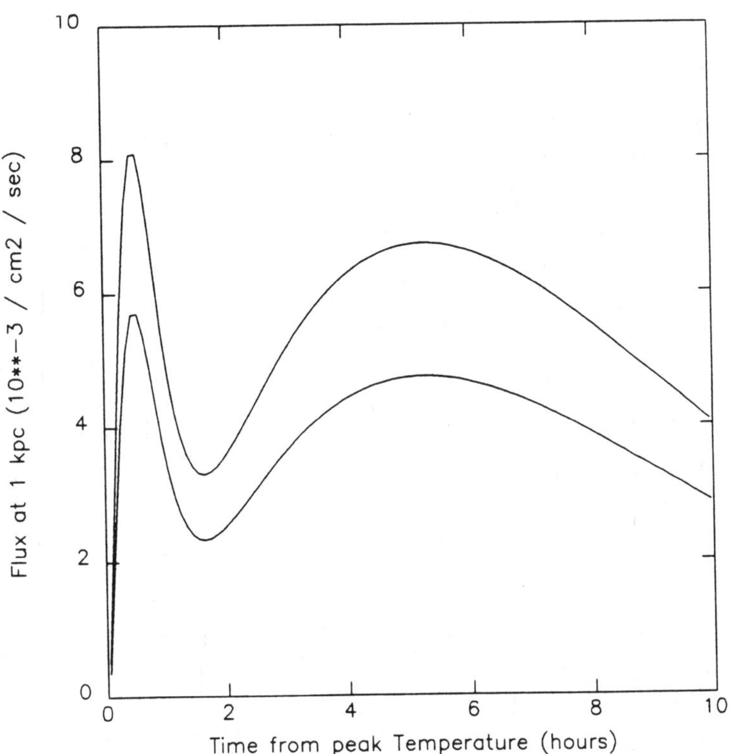

Figure 1. The flux from a model nova[25] in the band 50-100 keV (upper curve) and in the 511 keV line (lower curve).

In the hot hydrogen burning, many of the CNO nuclei initially present are transformed into these unstable species. The relative abundances among them depend on the peak temperature reached. Clearly the longer-lived nuclei offer the best prospects for detection, because the nova envelope can expand further before they decay and their surface abundances depend less on the rapidity of convective transport from the burning region to the surface. It is not clear that these unstable nuclei can reach relatively thin regions at all before decaying. Self-consistent calculations of their transport are not yet available. However, it may be that significant energy production very near the photosphere is required to explain the apparent super-Eddington luminosities of some novae. Beta-unstable nuclei carried there by rapid convection are a natural explanation in the context of the thermonuclear runaway model.

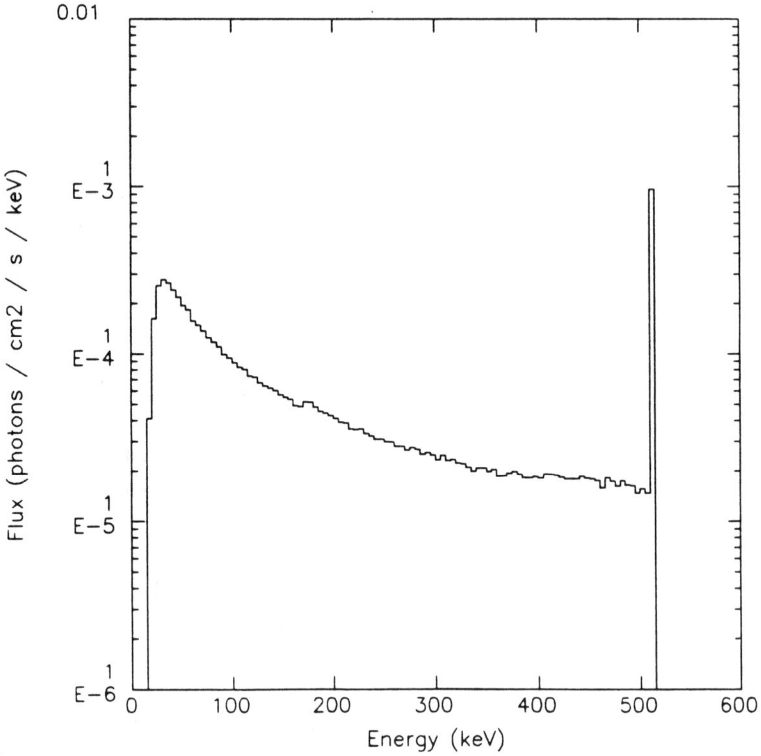

Figure 2. The emergent spectrum from the same model as Figure 1 at 5 hours after peak temperature.

For typical nova models with ^{13}N abundances of a few percent by mass, fluxes at 1 kpc at 511 keV from that nucleus might reach 10^{-2} cm^{-2} s^{-1} over about 30 minutes following runaway[25]. Some models with ten times more ^{13}N suggest fluxes exceeding 0.1 cm^{-2} s^{-1}. Abundances of ^{18}F of 10^{-3} by mass in the outer envelope lead to 511 keV fluxes well over 10^{-3} cm^{-2} s^{-1} for periods of several hours (at 1 kpc). Shown in Figures 1 and 2 are the light curves and spectra of these emissions for model 1 of Reference 25. Explosive hydrogen burning, possibly in novae, is a likely source of the bulk of Galactic ^{18}O[26], which results from the decay of ^{18}F. For novae to be significant contributors, ejected mass fractions of at least $X(^{18}O) = 2x10^{-3}$ (after decay of ^{18}F) are required, and then the later time fluxes of Figure 1 are quite reasonable for a very nearby nova.

As one has no advance knowledge that novae, even relatively nearby ones, are about to occur, detecting them requires large field of view detectors or extremely good luck. Also, we may wait for a long time for a nova as close as 1 kpc. Still, these fluxes are potentially detectable in existing and planned detectors (such as the SMM γ-ray spectrometer and GRO OSSE and BATSE) to distances of a few kiloparsecs. A search of the SMM data for this signature is underway. The most extreme estimated fluxes from ^{13}N positrons might be detectable to OSSE even from novae up to the distance of the Galactic center, where that detector may be pointed for long periods and where novae are thought to be very frequent. Gamma rays from the positrons emitted by shorter-lived nuclei do not appear to be currently detectable, except from very nearby novae. Note that the ^{14}O decay also produces a line at 2.31 MeV at a comparable intensity to its annihilation line. Also a significant continuum from Compton scattering of the lines would be produced and might be detectable in lower energy detectors. BATSE will provide the best hope of measuring this distinct continuum from nearby novae and the prospects are discussed by Fishman et al. (this volume).

CONCLUSION

Clearly, there is the potential for learning a tremendous amount about a fascinating physical situation by observing γ-ray lines from novae. The theory has not yet evolved to the point where upper limits on γ-ray fluxes are terribly interesting, but it soon should. The next generation of γ-ray spectrometers will greatly improve the possibility of detecting emission from novae, especially in the fortuitous event of a very close nova. A detection would place important constraints on models of novae and on general theories of convection, and possibly even address larger issues.

REFERENCES

1. M.F. Bode and A. Evans, *Classical Novae* (New York:Wiley) 1989.
2. Mahoney, W. A., Ling, J. C., Jacobson, A. S., and Lingenfelter, R. E., Ap. J., **262**, 742 (1982).

3. Mahoney, W. A., Ling, J. C., Wheaton, W. A., and Jacobson, A. S., Ap. J., **286**, 578 (1984).
4. Share, G. H., Kinzer, R. L., Kurfess, J. D., Forrest, D. J., Chupp, E. L., and Rieger, E. 1985, Ap. J. (Letters), **292**, L61.
5. Clayton, D. D., and Leising, M. D., Physics Reports, **144**, 1 (1987).
6. Wiescher, M., Görres, J., Thielemann, F.-K., and Ritter, H., Astr. Ap., **160**, 56 (1986).
7. Woosley, S. E., in Nucleosynthesis and Chemical Evolution, ed. B. Hauck, A. Maeder, and G Meynet (Geneva Observatory: Geneva) p. 85 (1986).
8. I. Nofar, G. Shaviv, and S. Starrfield, Ap. J. in press (1991).
9. Ciardullo, R., Ford, H. C., Neill, J. D., Jacoby, g. H., and Shafter, A. W. Ap. J., **318**, 520 (1987).
10. von Ballmoos, P., Diehl, R., and Schönfelder, V., Ap. J., **318**, 654 (1987).
11. MacCallum, C. J., Huters, A. F., Stang, P. D., and Leventhal, M., Ap. J., **317**, 877 (1987).
12. Zinner, E., Amari, S., Anders, E., and Lewis, R., Nature, **349**, 51 (1991).
13. Clayton, D. D., and Hoyle, F., Ap. J. (Letters), **187**, L101 (1974).
14. Leventhal, M., MacCallum, C., and Watts, A., Ap. J., **216**, 491 (1977).
15. Leising, M. D., Share, G. H., Chupp, E. L., and Kanbach, G., Ap. J., **328**, 755 (1988).
16. Williams, R. E., Ney, E. P., Sparks, W. M., Starrfield, S. G., Wyckoff, S., and Truran, J. W., M. N. R. A. S., **212**, 753 (1985).
17. Snijders, M. A. J., Batt, T. J., Roche, P. F., Seaton, M. J., Morton, D. C., Spoelstra, T. A. T., and Blades, J. C., M. N. R. A. S., **228**, 329 (1987).
18. Higdon, J. C., and Fowler, W. A., Ap. J., **317**, 710 (1987).
19. Yang, J. et al., Ap J., **281**, 493 (1984).
20. Spite, F., and Spite, M., Astr. Ap., **115**, 357 (1982).
21. Arnould, M., and Norgaard, H., Astr. Ap., **42**, 55 (1975).
22. Starrfield, S., Truran, J. W., Sparks, W. M., and Arnould, M., Ap. J., **222**, 600 (1978).
23. Clayton, D. D., Ap. J. (Letters), **294**, L97 (1981).
24. Harris, M. J., Leising, M. D., and Share, G. H., Ap. J. in press (1991).
25. Leising, M. D., and Clayton, D. D., Ap. J., **323**, 159 (1987).
26. Wiescher, M., Görres, J., and Thielemann, F.-K., Ap. J., **326**, 384 (1988).

RADIONUCLIDES OF INTEREST FOR γ-RAY LINE ASTRONOMY FROM NOVAE AND RED GIANTS

G. Paulus and M. Forestini
Institut d'Astronomie et d'Astrophysique
Université Libre de Bruxelles, B-1050 Bruxelles

ABSTRACT

The contribution of asymptotic giant branch (AGB) stars to the ^{26}Al content of the galactic disk is discussed, as well as the role of novae in the nucleosynthesis of ^{22}Na and ^{26}Al. In particular, the effects of various uncertainties will be stressed in some detail.

INTRODUCTION

The radionuclides ^{22}Na ($t_{1/2} \approx 2.6\,$yr) and ^{26}Al ($t_{1/2} \approx 7 \times 10^5\,$yr) are crucial to γ-ray astronomy and cosmochemistry. An emission of characteristic γ-ray lines (at 1275 keV for ^{22}Na and 1809 keV in the case of ^{26}Al) always accompanies the ^{22}Na and ^{26}Al β-decays. The ^{26}Al γ-ray line has been observed in the interstellar medium by various satellite and balloon-borne γ-ray detectors.[1,2] From the line intensity, a (model dependent) estimate of $\sim 3\,M_\odot$ of live ^{26}Al in the ISM today has been inferred. However, only an upper limit has been placed up to now on the amount of interstellar ^{22}Na.[3]

Primitive meteorites bear the signature of these two radionuclides. More specifically, ^{22}Na is thought to be the progenitor of an extraordinary meteoritic Ne component made virtually of pure ^{22}Ne and referred to as Ne-E. On the other hand, it has been discovered that ^{26}Al has decayed in situ in various meteoritic inclusions leading to the observation of a ^{26}Mg excess.

^{22}Na and ^{26}Al could be significantly produced by the hot ($T \gtrsim 10^8$ K) NeNa and the cold ($T \gtrsim (2-3) \times 10^7$ K) or hot MgAl chains of reactions respectively. These various H burning modes take place either in explosive environments (novae, supernovae) or in hydrostatic conditions (AGB and Wolf-Rayet stars).[4]

However, the conclusions one can derive about the ability of these processes to account for the meteoritic and γ-ray observations are sensitive not only to the details of the astrophysical modelling of the appropriate sites, but also to the still uncertain rates of several key reactions on unstable nuclei.

These two aspects have motivated the present work and are dealt with in the following. On one hand, we will present results of detailed calculations of ^{26}Al production in AGB stars, relying on better astrophysical and nuclear physics ingredients, and following the nucleosynthesis with the aid of a fairly extended reaction network. On the other hand, we will show to what extent do the predicted ^{22}Na and ^{26}Al yields depend on the uncertainties still plaguing current nova outburst models. The impact of new nuclear data will also be quantitatively assessed in this case.

ASYMPTOTIC GIANT BRANCH STARS AND ^{26}Al

Stars of initially 1 to $\sim 6-8\,M_\odot$ on the main sequence evolve onto the AGB of the Hertzsprung-Russell diagram. The structure of those objects is basically a double shell-burning (H and He) configuration on top of a dense C-O core, and capped by an extended H envelope.

Observations have unveiled the presence, in AGB atmospheres, of H- and He-burning products[5] as well as s-process elements.[6] These nuclides are expected to be manufactured inside the star and transported to the stellar surface by the third dredge-up.[7] This dredge-up mechanism is however still poorly understood and its efficiency largely depends on such unknowns as convection modelling, mass loss rates or metallicity.[8]

Nørgaard[9] and Cameron[10] have suggested that ^{26}Al could be synthesized by H burning at the bottom of the convective envelope. This so-called "hot bottom burning" (HBB) process is neither generally obtained in AGB models[11] nor definitely ascertained observationally,[12] despite some claims to the contrary.[13]

The contribution of lower mass AGB stars to the interstellar ^{26}Al has now been quantitatively estimated. Detailed calculations of the evolution of 1 and $3\,M_\odot$ stars, from the T-Tauri phase to the AGB stage, have been undertaken.[14] In these models, the nucleosynthesis of various species (among which ^{26}Al) is followed through a reaction network comprising 38 isotopes linked by 120 reactions. The models rely on better physical input (e.g. opacities, latest nuclear reaction rates, $T(\tau)$ profiles suited to red giant atmospheres).[14] We hereunder present the results for the $3\,M_\odot$, $Z = 0.03$ model.

Figure 1 illustrates the Al profiles as a function of the fractional mass coordinate of the star. At this point in its evolution, the AGB model has experienced 13 thermal pulses. We also emphasize that our calculations have not shown the occurence of HBB. $X(^{26}\mathrm{Al})$ reaches a maximum value of $\sim 10^{-4}$ in the H-burning shell. This value corresponds to the conversion of the initial amount of ^{25}Mg via $^{25}\mathrm{Mg}(\mathrm{p},\gamma)^{26}\mathrm{Al}$, exemplified by the dip of $X(^{25}\mathrm{Mg})$ around $M(r)/M \approx 0.207$. In the convective thermal pulse $X(^{26}\mathrm{Al}) \sim 10^{-5}$, resulting from the dilution of ^{26}Al swallowed by the previous pulses. Deeper inside the He-burning shell ($M(r)/M \lesssim 0.195$), the ^{26}Al concentration decreases mainly through $^{26}\mathrm{Al}(\beta^+)^{26}\mathrm{Mg}$ and $^{26}\mathrm{Al}(\mathrm{n},\mathrm{p})^{26}\mathrm{Mg}$. The reacting neutrons have been produced by the reaction $^{13}\mathrm{C}(\alpha,\mathrm{n})^{16}\mathrm{O}$. Figure 1 also shows that $X(^{27}\mathrm{Al})$ has remained essentially unaffected at its initial value.

In this model, as is generally the case for other computations, the third dredge-up is not self-consistently found. The dredge-up depth is thus a free parameter. However, Fig. 2 shows that $X(^{26}\mathrm{Al})$ in the envelope quickly becomes rather insensitive to the amount of dredged-up mass, approaching a kind

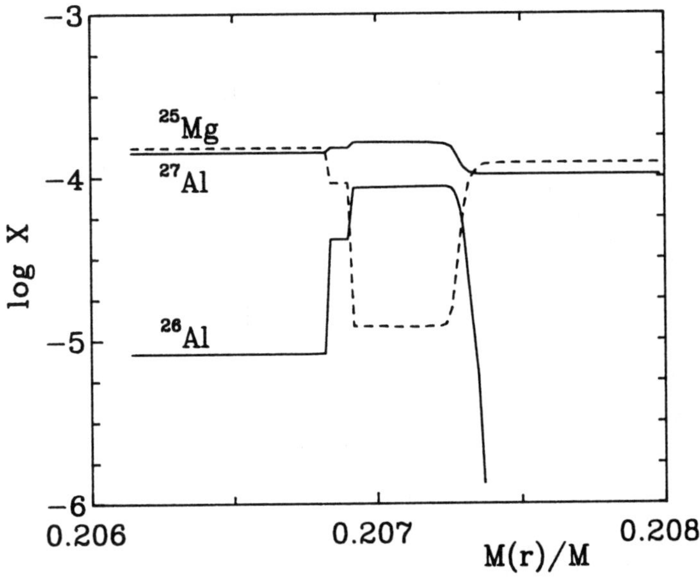

Fig. 1. ^{25}Mg and Al profiles in the $3\,M_{\odot}$, $Z = 0.03$ AGB model after the 13th thermal pulse.

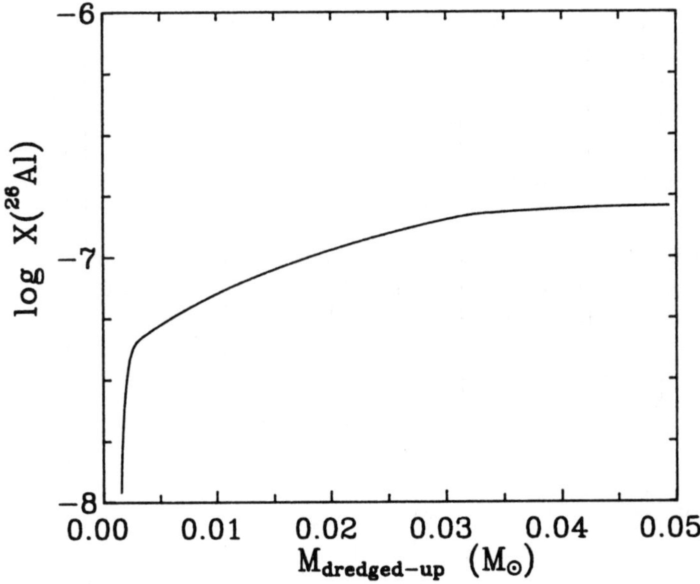

Fig. 2. ^{26}Al concentration in the AGB envelope as a function of the dredged-up mass after 13 thermal pulses.

of saturation value of $\sim 10^{-7}$. This leveling off is a consequence of the fact that relatively little ^{26}Al is brought to the envelope for dredge-ups of reasonable depth. With $X(^{26}\mathrm{Al}) \sim 10^{-7}$ and $X(^{27}\mathrm{Al}) \sim 10^{-4}$, the H envelope is characterized by a ratio $^{26}\mathrm{Al}/^{27}\mathrm{Al} \sim 10^{-3}$, quite close to the values deduced from recent measurements of ^{26}Mg excesses in primitive meteorite grains.[15]

The 3 $M_\odot$ AGB model gives a yield which depends on the number of pulses experienced by the star. Our calculations predict that after 13 pulses, $\sim 10^{-8}$ $M_\odot$ of ^{26}Al has been added to the ISM. At this point, the remaining envelope still contains 2 $M_\odot$ of gas. On grounds of our data, an extrapolation at constant $\dot{M}$ required 32 extra pulses to strip the star completely of its envelope. As a result, we estimate that $\sim 3 \times 10^{-8}$ $M_\odot$ of ^{26}Al are blown in the ISM over a $\sim$ 7 million yr period.

The data of the 3 $M_\odot$ model have been used to estimate the production of ^{26}Al by AGB stars of main sequence masses in the range $1 - 3\,M_\odot$. To this end, we have assumed that (i) the interpulse duration, and (ii) the mean dredged-up ^{26}Al mass per pulse are the same for all the objects in the above mass interval. This hypothesis appears reasonable in the light of envelope models[16] for a 1 $M_\odot$ AGB star which indicate that the results of complete models (currently being calculated) should not differ too much from the 3 $M_\odot$ ones. The mass loss rates for low-mass AGB stars are computed according to ref. 17. Integrating over an IMF,[18] and halting the calculation when the envelope of the less massive star vanishes, we find an average ^{26}Al yield of $\sim 0.1\,M_\odot$ injected in the ISM every 10^6 yr.

We conclude that AGB stars of main sequence masses in the range $1 - 3\,M_\odot$ seem unable to account for the present ISM ^{26}Al content. Other ^{26}Al sources probably should be invoked, and we now turn to another low mass candidate.

PRODUCTION OF ^{22}Na AND ^{26}Al IN NOVAE

The classical nova outburst is nowadays believed to result from a thermonuclear runaway occuring in a hydrogen-rich shell accreted by the CO or ONeMg white dwarf component of a binary system.[19]

At the temperatures ($T \approx (2-4) \times 10^8$ K) and densities ($\rho \sim 10^4$ gcm^{-3}) thought to prevail during the runaway, H burning proceeds via the hot modes of the CNO, NeNa and MgAl chains in which repeated proton captures produce many unstable nuclei. Those with half-lives long enough to survive until the expanding ejecta become transparent to γ rays might be directly observed via their decay photons.Observations of these $\gamma-$ray lines would provide important constraints on models of nova dynamics and nucleosynthesis.

No detailed reaction network calculations directly coupled to the hydrodynamics have been performed to date. Production of the important radionuclides ^{22}Na and ^{26}Al has been studied in the framework of parametrized calculations.[20] In these computations, the accreted H layer is exposed to a peak temperature

T_0 and peak density ρ_0, and allowed to cool exponentially on a given timescale τ ($T = T_0 \exp(-t/\tau)$). The density declines also exponentially with a characteristic timescale 3 times shorter ($\rho \propto T^3$).

The calculations reported here were motivated by the availability of new nuclear reaction cross-sections involving ^{22}Na[21] and ^{26}Al.[22] In the temperature range of interest for nova models ($1 \lesssim T_8 \lesssim 4$), the new ^{22}Na(p, γ)^{23}Mg rate is at least 10 times lower than the Caughlan and Fowler (CF) rate.[23] On the other hand the new ^{25}Mg(p, γ)^{26}Al varies between 1/10 and 9/10 the CF rate over the same temperature interval. By the same token, other sources of uncertainty have been explored.

Table 1A illustrates the effect of the new reaction rates. The calculations were conducted for $T_0 = 1.46 \times 10^8$ K, $\rho_0 = 2.4 \times 10^3$ gcm^{-3}, and initial ^{12}C and ^{16}O mass fractions of 0.45 and 7.2×10^{-3} respectively.[24] The lower ^{22}Na(p, γ) rate leads to $X(^{22}\mathrm{Na})$ in the ejecta 7 − 50 times greater than the value derived from the CF rate. On the other hand, the lower ^{25}Mg(p, γ) rate has the detrimental effect of producing less ^{26}Al by a factor of $\sim 2 - 3$.

As was already noticed previously[25,20] the description of the convective mixing during the nova runaway greatly influences the nucleosynthetic yields. To evaluate this uncertainty, a crude approximation to convection has been used in which the burning shell is divided in two zones convectively coupled on a timescale τ_{con}. The calculations were carried out with an inner zone of thickness $p = 0.5$ and $T_0 = 2.52 \times 10^8$ K, and an outer zone of thickness $1 - p$ whose temperature started and remained 1/2 as great. The peak density of the inner zone was $\rho_0 = 1.4 \times 10^4$ gcm^{-3}, while that of the outer zone was 1/8 as great and remained so. The mixing timescale was fixed at a typical value $\tau_{con} = 300$ s. The results are displayed in Table 1B and compared to the predictions of a 1 zone model (i.e. $p = 1$). Due to convection, fresh fuel is continuously provided to the hot zone from the cold one, and the fragile products are removed from the burning region before being destroyed. As a result, ^{22}Na and ^{26}Al can be enhanced in the ejecta by factors of 25 and 4, respectively.

Table 1A. Effect of new reaction rates on the ^{22}Na and ^{26}Al mass fractions.

	$X(^{22}\mathrm{Na})$	$X(^{26}\mathrm{Al})$
CF88	4.6(−7)	3.3(−5)
"High"	3.2(−6)	1.2(−5)
"Low"	2.2(−5)	1.1(−5)

Entries $a.b(c)$ stand for $a.b \times 10^c$

Table 1B. Influence of convective mixing on the ^{22}Na and ^{26}Al production levels.

	p	$X(^{22}\mathrm{Na})$	$X(^{26}\mathrm{Al})$
1 zone	1	4.0(−7)	1.6(−6)
2 zones	0.5	1.0(−5)	5.5(−6)

^{22}Na and ^{26}Al can be synthesized in significant amounts for accreted shells enriched in seed ONeMg nuclei, as was found previously.[20, 26] This case has been reinvestigated here in the light of the new reaction rates. The results are listed in Table 1C. A H shell enriched in ONeMg at the level of 25% by mass could lead to ejecta in which $X(^{22}\mathrm{Na})$ and $X(^{26}\mathrm{Al})$ are increased by factors of 65 and 100 respectively compared to a shell of initial solar composition.

The thermodynamic conditions (T_0, ρ_0) achieved during the outburst are still quite model dependent. Table 1D shows the effect of changing T_0 on the concentrations of ^{22}Na and ^{26}Al in the ejecta. For a case of initial ONeMg = 25% by mass and $\rho_0 = 10^4$ gcm^{-3}, lowering T_0 from 2.5×10^8 to 1.5×10^8 K increases $X(^{22}\mathrm{Na})$ fivefold and $X(^{26}\mathrm{Al})$ by a factor of 12.

Table 1C. Production of ^{22}Na and ^{26}Al for two different initial compositions.

Composition	$X(^{22}\mathrm{Na})$	$X(^{26}\mathrm{Al})$
$\odot$	5.1(−5)	1.1(−5)
ONeMg = 0.25	3.3(−3)	1.1(−3)

Table 1D. Impact of thermo–dynamic conditions on ^{22}Na and ^{26}Al production.

T_0 (K)	$X(^{22}\mathrm{Na})$	$X(^{26}\mathrm{Al})$
2.5(+8)	6.5(−4)	9.7(−5)
1.5(+8)	3.3(−3)	1.2(−3)

To summarize, the explored cases lead to a wide range of values for the predicted ^{22}Na and ^{26}Al yields because of still uncertain nuclear reaction rates, convection modelling, thermodynamic conditions and composition of the burning shell.

The calculations reported here have been compared to those of ref. 20 using the conditions of ref. 24. For a 2 zone model with $p = 0.1$ and $\tau_{con} = 1000$ s, the predicted $X(^{22}\mathrm{Na})$ in the ejecta is twice Woosley's value, while ^{26}Al is $\sim$ 70 times more abundant in Woosley's computations. The reason for these differences lies solely in revised nuclear data. In particular, the $^{26}\mathrm{Al(p,\gamma)}^{27}\mathrm{Si}$ reaction rate used in ref. 20 is much lower than the one used here for $T < 3 \times 10^8$ K.

At the galactic level, assuming a frequency of 40 novae per year and an ejected mass of 10^{-4} $M_\odot$ per event, we estimate that no more than 0.5 $M_\odot$ of ^{26}Al has been added to the ISM over the last million yr. As far as ^{22}Na is concerned, its short half-life prevents any significant accumulation effect in the ISM, so that the observation of nearby ($\lesssim$ few kpc) bright (ONeMg−rich) novae would offer the best opportunity to detect it.[3]

The authors are grateful to M. Arnould for helpful discussions.

REFERENCES

1. G. Vedrenne, Rep. Prog. Phys. 51, 393 (1988).
2. B.J. Teegarden, S.D. Barthelmy, N. Gehrels, J. Tueller, M. Leventhal, C.J. MacCallum, this volume.
3. M.D. Leising, G.H. Share, E.L. Chupp, G. Kanbach, Astrophys. J. 328, 755 (1988).
4. M. Arnould, Phil. Trans. R. Soc. Lond. A323, 251 (1987).
5. A. Jorissen, V.V. Smith, D.L. Lambert, in preparation (1991).
6. V.V. Smith, D.L. Lambert, Astrophys. J. Supp. 72, 387 (1990).
7. I. Iben, Jr., A. Renzini, Ann. Rev. Astron. Astrophys. 21, 271 (1983).
8. J. C. Lattanzio, in *Origin and Distribution of the Elements*, ed. G. J. Mathews (World Scientific, Singapore, 1987), p. 398.
9. H. Nørgaard, Astrophys. J. 236, 895 (1980).
10. A.G.W. Cameron, Icarus 60, 416 (1984).
11. I.J. Sackmann, Astrophys. J. 235, 554 (1980).
12. J. Mould, N. Reid, Astrophys. J. 321, 156 (1987).
13. V.V. Smith, D.L. Lambert, Astrophys. J. 345, L75 (1989).
14. M. Forestini, M. Arnould, M. El Eid, in preparation (1991).
15. E. Zinner, S. Amari, E. Anders, R. Lewis, Nature 349, 51 (1991).
16. M. Forestini, G. Paulus, M. Arnould, in preparation (1991).
17. D. Reimers, Mem. Soc. Roy. Sci. Liège 6th series, 6, 369 (1975).
18. R. Wielen, Highlights of Astronomy 3, ed. G. Contopoulos (Reidel, Dordrecht, 1974), p. 39.
19. S. Starrfield, in *Classical Novae*, eds. M.F. Bode and A. Evans (Wiley, N.Y., 1989), p. 39.
20. S.E. Woosley, in *Nucleosynthesis and Chemical Evolution*, ed. B. Hauck, A. Maeder and G. Meynet (Geneva Observatory, Geneva, 1986), p. 1.
21. S. Seuthe, C. Rolfs, U. Schröder, W.H. Schulte, E. Somorjai, H.P. Trautvetter, F.B. Waanders, R.W. Kavanagh, H. Ravn, M. Arnould, G. Paulus, Nucl. Phys. A514, 471 (1990).
22. Ch. Iliadis, Th. Schange, C. Rolfs, U. Schröder, E. Somorjai, H.P. Trautvetter, K. Wolke, P.M. Endt, S.W. Kikstra, A.E. Champagne, M. Arnould, G. Paulus, Nucl. Phys. A512, 509 (1990).
23. G.R. Caughlan, W.A. Fowler, At. Data Nucl. Data Tables 40, 283 (1988).
24. M. Wiescher, J. Görres, F.-K. Thielemann, H. Ritter, Astron. Astrophys. 160, 56 (1986).
25. B. Lazareff, J. Audouze, S. Starrfield, J.W. Truran, Astrophys. J. 228, 875 (1979).
26. A. Weiss, J.W. Truran, Astron. Astrophys. 238, 178 (1990).

DETECTABILITY OF EARLY, LOW-ENERGY GAMMA RAYS FROM NEARBY NOVAE BY BATSE/GRO

G.J. Fishman, R.B. Wilson, C.A. Meegan
M.N. Brock, J. M. Horack
NASA/Marshall Space Flight Center, AL 35812

W.S. Paciesas, G.N. Pendleton,
Physics Dept., University of Alabama in Huntsville

B.A. Harmon
Universities Space Research Association

M. Leising
Naval Research Laboratory, Washington D.C.

ABSTRACT

The Burst and Transient Source Experiment (BATSE) on the Gamma Ray Observatory, having an all-sky observing capability and high sensitivity in the low energy gamma-ray region, has a reasonable chance of detecting emission from a nearby nova within the first few hours of its outburst.

INTRODUCTION

The energy source of classical novae is well-understood as a thermonuclear runaway in CNO material on an accreting white dwarf in a binary system. Short-lived, positron-emitting nuclei are produced in large quantities. Their subsequent decay helps power the ejection of the envelope and produces large fluxes of annihilation radiation in the burning region. However, there is still great uncertainty in the hydrodynamics of the highly unstable, expanding envelope. The emerging fluxes of annihilation radiation and the Comptonized spectra from several possible models of novae have been calculated[1,2]; figure 1 shows the fluxes from key isotopes based on one of these models.

The amount of mixing and density fluctuations in the envelope significantly affect the magnitude and evolution of the emerging annihiation radiation. Also, since the envelope is still very thick during the decay of most of the radioactive nuclei considered here, many of the 511 keV annihilation photons are Compton scattered to lower energies before they emerge. This results in a hard spectrum, peaking in the energy region between 20 to 100 keV. (It should be noted that the highly Comptonized spectrum from SN1987A was detected much sooner than expected, now thought to be due to deep mixing in the expanding envelope.)

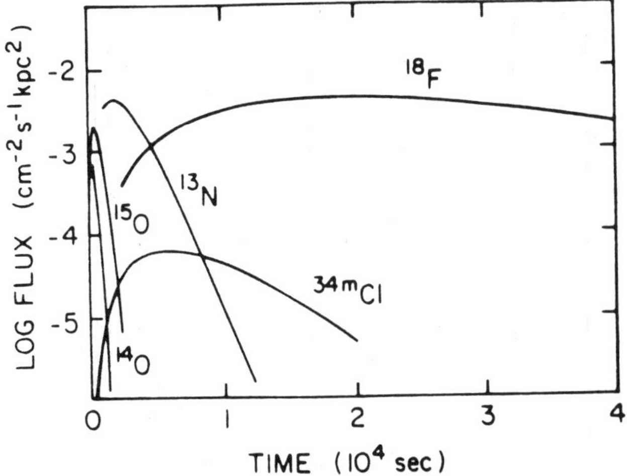

Fig. 1. Emerging flux of 511 keV photons from a nova, based on Model 1 in Ref. 1.

DETECTABILITY

BATSE is designed primarily for the observation of gamma-ray bursts. It can also detect and study longer-lived sources such as that described here. The experiment consists of eight independent detector modules located at the corners of the GRO spacecraft. Each module contains a large area detector (LAD) and a spectroscopy detector (SD). The LADs are 1.27 cm thick and have an uncollimated area of 2025 cm^2 each. The LADs cover a nominal energy range from 25 keV to 1.7 MeV. A transient source in this range, such as that expected from a nearby nova, can be detected by BATSE in two different ways. The first method uses earth occultation; a description of the sensitivity of BATSE using this technique has been described previously[3,4]. We assume that the nova will be observed in the energy bands shown in Table 1 and have estimated the expected backgrounds in those bands. For two detectors with axes 35 deg from the direction of the nova, each occultation step would produce a signal of the indicated statistical significance. There will be two occultation steps observed during each 90 minute orbit, providing a good sample of the build-up and decay of the emission from the nova. Approximately 10% of the sky, near the GRO orbit poles, will not be observable by the occultation technique.

The second observational technique would be to use background samples from the same portion of the GRO orbit as the nova observation, but with a time difference of a day or more. This technique is being used successfully in several studies with data from the gamma-ray spectrometer on the SMM. Since a longer time base can be used for both source and background observations, the statistical uncertainties can be reduced considerably. The signals from multiple detectors can be compared for consistency with that expected from a distant point source. However, the technique is subject to potentially larger systematic errors which are instrument-unique and are difficult to estimate prior to launch.

The rate of nearby novae is uncertain due to selection effects and unknown amounts of obscuration. A reasonable estimate of the rate of novae within 1 kpc of the earth is one per five years, with an uncertainty of about a factor of two. This is of the same order as the GRO lifetime, which has a similar uncertainty.

Table I - Calculated occultation signal from a nova @ 1 kpc, observed by two BATSE detectors, 35 deg off-axis and assuming 300 s backgrounds.

E (keV)	Signal (sigma)
30- 50	2.8
50-100	11.0
100-200	10.5
200-500	8.3
511 keV line	2.2

CONCLUSION

The promising observation of evolving gamma-ray spectra from a nearby nova would yield important new information on the early hydrodynamics of novae as well as the abundances of several radioactive isotopes. The short-lived emission requires wide-field or all-sky detectors, such as those of the BATSE/GRO.

REFERENCES

1. M.D. Leising and D. Clayton, Ap.J. 323, 159, (1987).
2. M.D. Leising, invited paper, these proceedings.
3. G.J. Fishman et al., GRO Sci. Wkshp., GSFC (1989).
4. W.S. Paciesas et al, V.3, 351, XIX ICRC, (1985).

INTERSTELLAR ^{22}Na AND THE EXCESS ^{22}Ne IN THE GALACTIC COSMIC RAY SOURCE

Shohei Yanagita
Department of Earth Sciences, Ibaraki University, Mito 310, Japan

Ken'ichi Nomoto
Department of Astronomy, University of Tokyo, Tokyo 113, Japan

ABSTRACT

The role of novae outburst as a source of the excess ^{22}Ne at the Galactic Cosmic Ray source is reexamined quantitatively. Isotopes ^{22}Ne which are the daughter nuclei of radioactive ^{22}Na synthesized in novae outbursts get excess energy over thermal particles due to nuclear recoil effects by beta-decay and can preferentially be injected as cosmic ray seeds into the shock wave at nova explosion where particles are accelerated. The excess ^{22}Ne can be explained if a fraction of 4.5×10^{-4} of the total production of ^{22}Ne via ^{22}Na in novae is accelerated. The implications of this scenario are discussed with possible observations of the 1.275 MeV gamma ray line from novae.

INTRODUCTION

Studies of elemental and isotopic composition in the Galactic Cosmic Ray (GCR) have been important clues in the investigation of cosmic ray origin, acceleration, and transport in the galaxy. Derived elemental composition at the GCR source relative to the solar system abundance of elements shows a general pattern which seems to be organized by atomic properties such as first ionization potential (FIP) of elements[1] in the sense that high FIP elements are depleted relative to low FIP elements. Similar trend is also seen in the abundance of solar flare energetic particles. In this context, the bulk of seed material of GCR seems to be derived from normal stars[1] rather than special sources which reflect imprints of specific nucleosynthetic processes. The dominant elements in the GCR source, H, He, and N, however, do not follow this orderly dependence on FIP[1] and are depleted further by a factor of 10 to 3.

On the other hand, it has been discovered that the isotopic ratio of some neutron rich isotopes at the GCR source is higher than that of the solar system material[2]. The firmly established anomaly among them is the excess of ^{22}Ne/^{20}Ne which is higher by a factor of 3.5 to 5.5 than the solar system standard[2].

All viable model of the origin of GCR must be able to reproduce the elemental and isotopic composition at the source and also must accommodate appropriate acceleration mechanism which can boost cosmic ray seed particles to relativistic energies and can maintain the energy density of GCR in the galaxy.

Recently we have proposed a new model of GCR origin[3] which assume that the GCR comes from Type I SN ejecta (SNI), Type II SN ejecta (SN II), and the Interstellar Medium (ISM) swept up by the ejecta. This model has succeeded to reproduce the GCR source abundance of elements from H to Ni within a factor of ~2.5 using recent predictions of the composition in SN ejecta.[4,5] The problem of the deficiencies in H, He, and N at the source disappears in our model. In addition to this remarkable result, there are some other reasons favouring our model of GCR. Supernovae can supply enough energy to maintain the energy density of GCR in the galaxy and also can provide an efficient particle acceleration mechanism by the first order Fermi process in the shock waves associated with its explosion.[6]

So far, so good. In this report, we will examine if we can understand by our new model of the origin of the GCR the isotopic anomalies seen in the source focussed on the ^{22}Ne anomaly.

NOVA AS A SOURCE OF THE EXCESS ^{22}Ne

Our new model of SN origin of GCR predicts the isotopic ratio of ^{22}Ne/^{20}Ne as ~0.14, using the mixing ratio of SNI:SNII:ISM=0.20:0.62:0.18 which can reproduce best the source elemental composition.[3] This value is short to explain the ratio of 0.42 at the GCR source.[2] Our model fails in explaining the excess ^{22}Ne as other models of SN origin of GCR unless we invoke other ingredients such as supermetallicity[7] in the progenitor of SN. Therefore, we must invoke an additional source of ^{22}Ne other than SN.

It has been proposed a model which attributes ^{22}Na ($t_{1/2}$=2.6 yr) synthesized at nova explosion to the origin of the excess ^{22}Ne. Recapitulating the model, the idea is that at the moment of beta-decay, the daughter nucleus gets both ionized and selectively heated, hence injected, by the recoil energy of the electron emission. This mechanism may not be expected to work for other nuclides than ^{22}Ne, because either radioactive progenitors other than ^{22}Na are too short-lived inappropriate for acceleration or they are rapidly locked in grains.[8]

Recent radio observations[9,10,11] have shown some evidence for acceleration of relativistic electrons in nova outbursts, a few tens of days or several tens of years after its explosion, presumably due to shock waves which might be attributed to either collisions of nova ejecta with circumstellar material or nova wind. Physical processes that accelerate relativistic particles in young SN remnants should also operate, on a smaller scale, in nova ejecta. If nuclear component is accelerated simultaneously with electrons by the first order Fermi process, the shocks will accelerate suprathermal ions more efficiently than thermal ions by a factor of[12]

$$[(E_{th}(E_{th}+2))/(E_{spt}(E_{spt}+2))]^{(1-s)/2} , \qquad (1)$$

here E is kinetic energy of injected particles expressed in rest mass energy unit and s=(r+1)/(r-1), where r is the shock compression ratio. The kinetic energy of ^{22}Ne which originate from ^{22}Na is higher by ~60 eV than that of ambient thermal particles.[8] Therefore, preferential acceleration of these ^{22}Ne can be expected. Actual enhancement factor of ^{22}Ne due to this preferential acceleration depends on the nature of shocks and the temperature of thermal pool of ions which eventually will be picked up as cosmic rays.

The production rate of ^{22}Na which novae can potentially supply[13] as the source of the excess ^{22}Ne in the GCR source is estimated as

$$N^{22}_{prod}=6.6\times10^{41}\ (R_{nova}/40yr^{-1})\ (M_{ej}/10^{-4}M_\odot)\ (X_{22}/10^{-4})\ \text{atoms/sec} \qquad (2)$$

where R_{nova} is the rate of nova events in the galaxy, M_{ej} is the total mass of nova ejecta and X_{22} is the mass fraction of ^{22}Na in the ejecta. This rate should be compared with an estimation of the production rate of the excess ^{22}Ne of $\sim3\times10^{38}$ atoms/sec in the whole galaxy[8]. Novae can be a source of the excess ^{22}Ne if it can accelerate ^{22}Ne which has originated from ^{22}Na with a moderate efficiency of $\sim4.5\times10^{-4}$. If only ONeMg white dwarfs rather than CO white dwarfs involve[13] in producing interstellar ^{22}Na, a higher, though not impossible, efficiency of $\sim1\times10^{-2}$ may be required. The power required to sustain the excess ^{22}Ne ($\sim5\times10^{-4}$ of the CR flux) is $\sim1.5\times10^{37}$ ergs/sec in the whole galaxy. The mechanical power injected by nova outbursts is $\sim1.3\times10^{39}$ ergs/sec. Consequently, only a small fraction (<2 %) of this power would suffice to energize the excess ^{22}Ne.

CONCLUDING REMARKS

The elemental abundance of the bulk of the GCR at the source can be understood if we assume the GCR originate from SNI, SNII, and the ISM which is swept up by the SN ejecta. The significant, though small in absolute flux, excess ^{22}Ne in the GCR may originate in nova explosions. An energy dependence of ^{22}Ne/^{20}Ne might be observed, because particle acceleration efficiencies may be different between SN and novae explosions.

The novae which are the source of the excess ^{22}Ne are expected to produce detectable flux levels of 1.275 MeV gamma rays of 1.6×10^{-4}/cm^2/sec from ^{22}Na decay[13] and are also expected to be a significant source of the interstellar ^{26}Al.[13] Simultaneous observations of both gamma rays and radio from novae are highly awaited to testify that novae are not only major contributors to galactic nucleosynthesis but also are sources of uncommon isotopes in the GCR.

REFERENCES

1. J.-P. Meyer, Proc. 19th Internat. Cosmic Ray Conf. (La Jolla), 9, 141 (1985).
2. R.A. Mewaldt, Cosmic Abundance of Matter (AIP Conf. Proc. 183, 1989), p. 124.
3. S. Yanagita, K. Nomoto, and S. Hayakawa, Proc. 21th Internat. Cosmic Ray Conf. (Adelaide), 4, 44 (1990).
4. M. Hashimoto et al., Astron. Astrophys. 210, L5 (1989).
5. K. Nomoto et al., Ap. J. 286, 644 (1984).
6. R. Blandford and D. Eichler, Phys. Report 154, 1 (1987).
7. S.E. Woosley and T.A. Weaver, Ap. J. 243, 561 (1981).
8. S. Yanagita, Proc. 19th Internat. Cosmic Ray Conf. (La Jolaa), 3, 175 (1985).
9. S. Padin et al., Nature 315, 306 (1985).
10. S.P. Reynolds and R.A. Chevalier, Ap. J. 281, L33 (1984).
11. A.R. Taylor et al., Astron. Astrophys. 183, 38 (1987).
12. D.C. Ellison et al., Ap. J. 360, 702 (1990).
13. A. Weiss and J.W. Truran, Astron. Astrophys., to appear (1990).

SN1987A AND SUPERNOVAE

MEASUREMENTS OF GAMMA-RAY LINE PROFILES FROM SN1987A AND THE STRUCTURE OF THE EJECTA

J. Tueller
Laboratory for High Energy Astrophysics
NASA/Goddard Space Flight Center, Code 661
Greenbelt, MD 20771 USA

ABSTRACT

Only a few measurements of the crucial gamma-ray line profiles from SN1987A were possible because no satellite instrument had the required high-resolution germanium detectors. These profiles provide the most direct measure of the distribution of freshly synthesized radioactive material (^{56}Co) in the supernova ejecta. All of these observations were made by balloon-borne instruments that were rushed into service after the supernova occurred. The preponderance of the measured profiles show line shapes that are not consistent with the standard spherically-symmetric homogeneous mixed models. The lines are broader (show higher velocities) than would be predicted from an initial distribution concentrated near the center of the ejecta and they are not blueshifted as would be expected for an optically thick source. The gamma-ray line profiles are remarkably similar to the line profiles observed for the infrared (IR) iron lines before day 500. In particular, the redshift of 710 ±270 km s^{-1} from the LMC rest energy observed in the GRIS gamma-ray line profiles is also present in both the near and far IR lines of iron. The consistency of the line profile from gamma-ray to IR energies indicates that this profile probably reflects the distribution of freshly synthesized material in the expanding supernova shell.

INTRODUCTION

Supernova 1987A in the Large Magellanic Cloud provided a once-in-a-lifetime opportunity to directly confirm nucleosynthesis in a Type II supernova by the observation of lines from the decay of ^{56}Co. These lines were first detected by the SMM satellite [1] and confirmed by several balloon experiments [2,3,4,5]. Although the detection of freshly synthesized radioactive material is a historic confirmation of widely held theories on the origin of elements, the pre-existing models for the structure of the supernova remnant [6,7,8] did not accurately predict the observed gamma-ray and hard x-ray light curves. Significant emission of gamma rays and hard x-rays begin well before the predicted time of approximately 300 days after the neutrino burst. The light curves for the lines and Compton scattered continuum were flatter than predicted. The models were quickly modified by introducing mixing to allow the escape of gamma rays at a time when the remnant was still optically thick [9,10,11]. The mixing in these models was adjusted to fit the early hard x-ray and gamma-ray light curves. The mass of ^{56}Co could be determined from the bolometric light curve (~0.075 $M_\odot$ [12]) because the radioactive decay of this one isotope is the dominant source of energy in the expanding shell after the first few weeks. The predictions of all these spherically-symmetric homogeneous mixed models are qualitatively similar: gamma rays from the back side of the supernova would be more attenuated by the optically thick ejecta than those from the front and the resulting observed lines would be blueshifted. Since gamma rays interact predominantly by Compton scattering, which usually removes the photon from the region of the line, the profiles yield a relatively direct measure of the distribution of radioactive material in the supernova without the complications of ionization state and resonant scattering

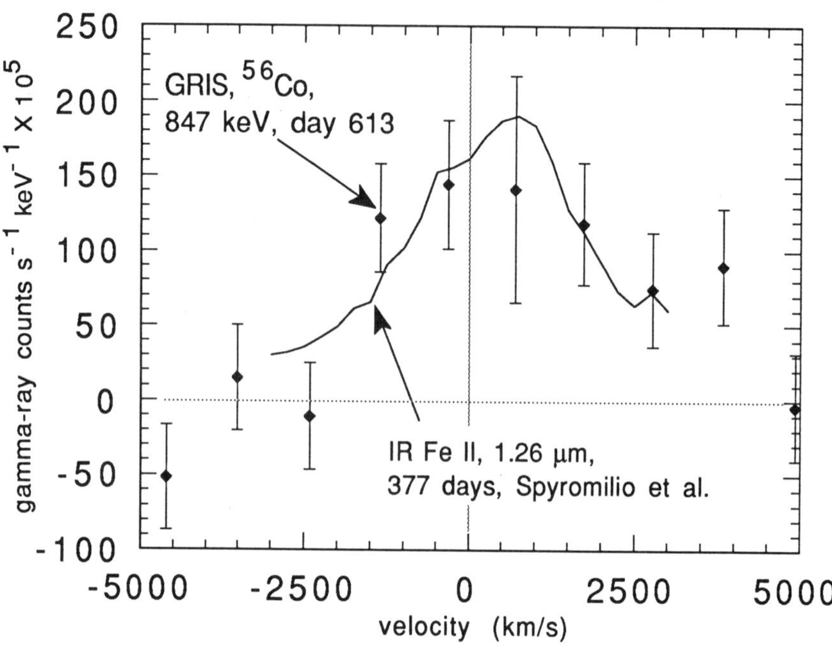

Figure 1. The near IR Fe line profile has been normalized to the gamma-ray counts and plotted over the GRIS line profile. The excellent correspondence over a large energy range is illustrated, suggesting that these profiles are optically thin and represent the velocity distribution of freshly synthesized material.

effects that commonly distort the IR and optical line shapes. High-resolution measurements made prior to ours by other balloon-borne germanium instruments from Lockheed/MSFC [2], Florida /GSFC [3], and JPL [4] did not provide an unequivocal result for the gamma-ray line profile from SN1987A (see Table I). The Gamma-Ray Imaging Spectrometer (GRIS) was the largest and most sensitive of the instruments available in time to observe the supernova lines before they had decayed away and provided the definitive measurement of line profiles.

RESULTS

Several high-resolution measurements of the gamma-ray line profiles from SN1987A were made before the first flight of the GRIS instrument in May 1988. The most sensitive of these measurements was made by the Lockheed/MSFC group. Their instrument contained an array of nine 5.5 cm germanium detectors in a 10 cm NaI shield. Because of their relatively thin shield and small detector size, they did not achieve as high sensitivity as GRIS, but at the 847 keV line their sensitivity was respectable and the ^{56}Co line was detected. Their line profile was quite unexpected, showing a broad feature with no apparent blueshift. The Lockheed group did not perform fits to Gaussian line profiles and their results were not available in a form which was easy to compare with models. Since the feature was very broad in comparison to the instrument resolution, a simple fit to their counts spectrum, ignoring the details of the instrument response, could determine the line profile without introducing significant systematic errors. Fits to this feature with a Gaussian line profile were performed and they confirmed our qualitative impressions of the Lockheed group's spectral feature (see Table I). Shortly after the Lockheed/MSFC flight, measurements were made by two smaller instruments with single germanium detectors. The results of the JPL group, which detected only the 1238 keV line, are completely consistent with the standard models (blueshifted and narrow). However, the line flux obtained by JPL is high (~2σ) when compared with other instruments or the standard models, and this suggests that it may have been effected by a statistical fluctuation. The Florida/GSFC group (not related to GRIS) in a flight from Antarctica made marginally significant detections of the two strongest lines from ^{56}Co. Their analysis produced very strange confirmation of the previous results, agreeing with the Lockheed line profile at the 847 keV line and with the JPL profile at the 1238 keV line. It is very difficult to imagine a model which would allow such radically different line profiles for lines which come from the decay of a single isotope. It must be noted that their analysis procedure gives errors on the line energy that are significantly smaller than the width of the line divided by its significance. This suggests that a more conservative analysis might produce larger estimates of the errors and better consistency. In the Spring of 1989, the UC/France group observed the supernova with an instrument similar to GRIS but could only place upper limits on the ^{56}Co lines [13].

GRIS has an array of 7 large (6.0 to 7.5 cm) n-type germanium detectors in a very heavy (15 cm) NaI active shield (for further description of the instrument see 14, 15). The gamma-ray line results from the two GRIS observations of SN1987A in May and October of 1988 have been published [16, 17] and the gamma-ray continuum results are presented in these proceedings [15]. The two strongest lines of ^{56}Co were detected with a combined significance of 7.8σ and all the line profiles show a consistent shape. The lines have been characterized by performing a least-squares fit to a Gaussian line profile (see Table I). The derived line fluxes are in good agreement with previous measurements and with the spherically-symmetric homogeneous mixed models. All of the line centroids are redshifted from the rest energies at the LMC (846.0 and 1237.2 keV) and the blueshifted model predictions. The line widths are consistently broader than predicted by these models. These results are completely consistent with the other measurements of the 847 keV line and are of greater significance than the other measurements of the 1238 keV line. The agreement of the line flux values with the models is not surprising as these models have had their mixing adjusted to fit the observed values at early times. The failure to predict the line profiles indicates that these models do not correctly describe the structure of the ejecta.

DISCUSSION

The gamma-ray line profiles observed from SN1987A are hard to reconcile with any simple model. The broad lines require acceleration of the freshly synthesized material from the inner layers of the ejecta into the faster moving hydrogen envelope. The observed bolometric light curve and the line fluxes both suggest that most of the gamma rays are still being absorbed in an optically thick source at the times of the line profile measurements. Using a number of 0.075 $M_\odot$ of ^{56}Ni produced in the supernova explosion, the GRIS lines correspond to 13% of the gamma rays on day 433 and 27% on day 613. In contradiction to these results, the gamma-ray line profiles have the qualitative appearance of an optically thin source with as many redshifted as blueshifted photons observed. In fact, the line profiles show a slight but

Table I SN1987A: High Resolution Gamma-Ray Line Profiles

LINE FLUX (ph cm^{-2} s^{-1} x 10^{-4})

INSTRUMENT	DAY	847 keV Line		1238 keV Line	
Lockheed/MSFC	248-250	10 ± 2.8	(3.6σ)	—	
JPL	287	—		21 ± 7	(3.0σ)
Florida/GSFC	319-322	23 ± 20	(1.2σ)	21 ±10	(2.1σ)
GRIS	433	21 +12,-9 [a]	(2.3σ)	9.1 ± 2.1	(4.3σ)
	613	6.5 ±1.4	(4.6σ)	3.1 ±0.9	(3.4σ)

LINE CENTROIDS (keV)

INSTRUMENT	DAY	847 keV Line	1238 keV Line
Lockheed/MSFC	248-250	843.7 ± 0.9 [b]	—
JPL	287	—	1240.8 ± 1.7
Florida/GSFC	319-322	844.1 ± 1.0	1239 ± 1.7
GRIS	433	844.6 ± 1.2 [a]	1235 ± 2.2
	613	844.2 ± 1.3	1233.5 ± 1.8

LINE WIDTHS FWHM (keV)

INSTRUMENT	DAY	847 keV Line	1238 keV Line
Lockheed/MSFC	248-250	9.2 ± 2.0 [b]	—
JPL	287	—	8.2 ± 3.4
Florida/GSFC	319-322	6.0 ± 2.6	10.4 ± 3.4
GRIS	433	8.7 +4.9,-3.3 [a]	14.8 ± 5.2
	613	10.8 +2.9,-2.4	11.0 +3.7,-3.1

a. Possible systematic errors are not included in the error estimates.
b. Fit performed by N. Gehrels.

consistent redshift as shown in Table II for the weighted average of the GRIS measurements. The average of all the measurements yields a lower value of 370 ± 140 km/s (2.6σ), but restricting the average to measurements with >3σ significance yields 760 ± 210 km/s (3.7σ). Without some confirmation such an unusual and difficult to understand result might be ignored.

Confirmation of this result could only come from the IR, where the supernova was nearly transparent and lines from the freshly synthesized material (Co and Fe) have been observed. We were surprised and pleased to find that the IR line profiles are identical to the gamma-ray line profiles within the uncertainties. Figure 1 shows the profile of a near IR Fe II line [18] (1.26 μm) plotted over the GRIS 847 keV line data. This high resolution profile is very precise and the IR line shape is apparently an excellent fit to the gamma-ray line data. Table II summarizes the gamma-ray and IR redshifts and line widths observed and illustrates the good agreement in these two energy regions. The fraction of the freshly synthesized Fe observed in the far IR is calculated to be 33% in good agreement with the gamma-ray optical depths and it has been interpreted as emission form a separate optically thin component [19]. Although the IR line shapes are notoriously subject to distortion, the excellent agreement over a wide energy range from far IR to gamma rays, where scattering processes are distinctly different, suggests that this profile represents the underlying distribution of visible radioactive material in the supernova. We must note that after day 500 the IR line profile undergoes a dramatic change as the supernova becomes optically thick due to the formation of dust and possibly other effects. The dust does not effect the gamma rays observed at day 613 by GRIS, but this is exactly as would be predicted for Compton scattering.

CONCLUSIONS

To resolve the apparent contradictions in the interpretation of the SN1987A line profiles, it is necessary to drop the assumption of spherical symmetry or homogeneity, perhaps both. The good agreement between gamma-ray and IR line profiles suggests that the profiles accurately reflect the velocity distribution of radioactive material in the supernova ejecta. Material from the inner edge of the ejecta must be accelerated to velocities in excess of 3000 km/s. The constant optical depth from gamma rays to IR suggests a two component model with the observed lines coming from an optically-thin region and the rest of the radioactive material hidden in small optically-thick clumps. Calculations using hydrodynamic codes show the development of fragmentation due to instabilities in the ejecta [20, 21]. These calculations produce many fingers of high Z material that extend into the H envelope at high velocities, however, such large numbers of fragments tend to average out the macroscopic non-spheroidal geometries required to give a redshifted line. Models with a toroidal distribution of material or a few very large optically-thick fragments could probably match the line profile but, at present, we are not familiar with a model of the supernova explosion that supports such geometries. Spectroscopic observations show that the remnants of the progenitor's red giant wind are at rest in the LMC and they eliminate any possibility of appealing to an enormous anomalous velocity for the progenitor star to explain the redshift. We conclude that present theories of supernova explosions do not provide an adequate explanation of the observed line profiles from SN1987A and suggest that significantly more complex descriptions will be required.

Table II Gamma-Ray and IR Line Profiles of SN1987A

INSTRUMENT/ WAVELENGTH	DAY (after supernova)	CENTROID REDSHIFT (km/sec)		LINE WIDTH FWHM (km/sec)
GRIS/gamma-ray 847 and 1238 keV lines of ^{56}Co [a]	433 & 613	710 ± 270 [a]	(2.6σ)	3200 ± 500 [a]
AAO/near IR 1.26 μm of Fe II	494	700		2700
KAO/far IR	407 & 409			
17.9 μm of Fe II		450 ± 200	(2.3σ)	2900 ± 80
26.0 μm of Fe II		680 ± 200	(3.4σ)	2500 ± 150

a. The 847 keV line on day 433 is not included due to possible systematic uncertainties.

REFERENCES

1. Matz, S. M. *et al.* 1988, *Nature*, **331**, 416.
2. Sandie, W. G. *et al.* 1988, *Ap. J. Let.*, **342**, L91.
3. Mahoney, W. A. *et al.* 1988, *Ap. J. Let.*, **334**, L81.
4. Rester, A. C. *et al.* 1988, *Ap. J. Let.*, **342**, L71.
5. Teegarden, B. J. *et al.* 1989, *Nature*, **339**, 122.
6. Gehrels, N., MacCallum, C. J., and Leventhal, M. 1987, *Ap. J. Let,.* **320**, L19.
7. Chan, R. W., and Lingenfelter R. E. 1988, in *Nuclear Spectroscopy of Astrophysical Sources* (AIP Conf Proc 170), ed. N. Gehrels and G. H. Share (New York:AIP), p. 110.
8. Pinto, P., and Woosley, S. E. 1988, *Ap. J.*, **329**, 820.
9. Bussard, R. W., Burrows , A., and The, L.S. 1989,*Ap. J.*,**341**, 401.
10. Pinto, P., and Woosley, S. E. 1988, *Nature*, **333**, 534.
11. Grebenev, S. A., and Sunyaev, R. A., 1989, *Physics of Neutron Stars*, ed. D. A.Varshalovich, A. D Kaminker, G. G. Pavlov, and D. G. Yakovlev (New York: Nova Science Publishers).
12. Woosley, S. E. *et al.* 1987, *Ap. J.*, **318**, 664.
13. Mattesson, J. *et al.* 1991,in *Supernovae*, ed. S. E. Woosley (New York: Springer-Verlang), p. 283.
14. Tueller, J. *et al.* 1988, in *Nuclear Spectroscopy of Astrophysical Sources* (AIP Conf. Proc. 170), ed. N. Gehrels and G. H. Share (New York:AIP), p. 439.
15. Barthelmy, S. in these proceedings.
16. Tueller, J. *et al.* 1990, *Ap. J. Let.*, **351**, L41.
17. Tueller, J. *et al.* 1989, in *Supernovae* , ed. S. E. Woosley (New York: Springer-Verlag), p. 278.
18. Spyromilio, J. *et al.* 1989, submitted to MNRAS.
19. Haas, M. R. *et al.* 1990, *Ap. J.*, **360**, 257.
20. Arnett, D. in these proceedings.
21. Nomoto, K. in these proceedings.

GRIS INSTRUMENT CONTINUUM RESULTS FROM SN1987A

S. Barthelmy[1], N. Gehrels, B. J. Teegarden, J. Tueller
Laboratory for High Energy Astrophysics, Code 661
NASA/Goddard Space Flight Center,
Greenbelt, MD 20771, USA

M. Leventhal
Physical Research Laboratory, 1E-349
AT&T/Bell Laboratories
Murray Hill, NJ 07974, USA

C. J. MacCallum
Department of Physics and Astronomy
University of New Mexico
800 Yale Blvd.
Albuquerque, NM 87131, USA

ABSTRACT

We present results on the continuum emission from SN1987A from two balloon flights of GRIS (Gamma-Ray Imaging Spectrometer) over Alice Springs, Australia, in the spring and fall of 1988. The durations of the two observations were 11.2 and 11.7 hours, respectively. The measured continuum spectra (95-2560 keV for the spring and 55-2560 keV for the fall) are consistent with Compton scattering of the ^{56}Co decay lines. The 10HMM model agrees well with our continuum data in the energy dependence of the continuum both in the slope of the spectra and the step increases below the 2599, 1238, and 847 keV ^{56}Co decay lines.

I. INTRODUCTION

The early appearance of the X- and gamma-rays (lines and continuum) from the supernova SN1987A caused the early models of Pinto and Woosley[1,2] to be adjusted to include more mixing of the ^{56}Ni (and hence ^{56}Co) into the outer and faster moving layers of the expanding shell. Thus the 10HMM model was derived from the 10HM model. The continuum emission is due to the Compton down-scattered gamma-ray photons from the radioactive decay lines of ^{56}Co (at 847, 1238, and 2599 keV) escaping through the overlying envelope material. As the envelope thins, exposing more and more of the ^{56}Co to direct observation, the Comptonization, and hence the continuum, decreases.

II. OBSERVATIONS

The GRIS instrument is a balloon-borne high resolution gamma ray spectrometer. The basic instrument consists of seven of the world's largest high purity n-type co-axial Germanium crystals. The detector active volume is ~1530 cm^3 with a total effective

[1]Universities Space Research Association Resident Associate

area of 155 cm^2 at 200 keV (and 74 & 46 cm^2 at 600 & 1200 keV, respectively). The seven detectors are surrounded by a thick (15 cm minimum) active anticoincidence shield of NaI viewed by 36 photomultiplier tubes. Seven collimator holes in the top of the anticoincidence shield define the 17° FWHM (at 500 keV) field of view. Considerable effort was made in the design of the anticoincidence shield housing and the detector cryostats to minimize the amount of passive material inside the shield and thus the background. See Tueller *et al.*[3,4] for more details of the GRIS instrument.

The linearization of the seven 32000 channel pulse height analyzers and energy calibration of the instrument was derived from the three strong and narrow background lines at 198, 511, and 1460 keV. The 1460 keV line came from an onboard ^{40}K source. Additional events in the 511 keV line came from analyzing shield-vetoed events in a special instrument mode which operated continuously during normal observations. Below 1460 keV a time-dependent linear-energy correction was made using these three background lines. Above 1460 keV, a flight-averaged parabolic correction was made in addition to the linear extrapolation from below 1460 keV. The useful energy range of the instrument is 20 to 8000 keV. Because of an electronics problem in the detector preamps, saturating events from charge particle interactions in the germanium caused the degradation of the resolution in the maiden Spring 1988 flight to ~1%. In the Fall 1988 flight, the energy resolution was 1.8 keV FWHM at 500 keV.

The flights described here were made from Alice Springs, Australia. During the Spring 1988 flight (30 April 1988, day 433 since the supernova explosion), lasting 25.5 hours at float, the supernova SN1987A was observed for 10.2 hours at a average atmospheric depth of 5.0 $g\text{-}cm^{-2}$. In the Fall flight (28 October 1988, day 613) SN1987A was observed for 11.7 hours out of 43.5 hours at an average depth of 4.7 $g\text{-}cm^{-2}$. Because of the low declination of SN1987A (RA = 84°, DEC = -69°) the elevation angle never rose above 44.5°. Atmospheric attenuation and instrumental reasons limited the minimum elevation angle to 20°. During periods when SN1987A was not observable, other targets were observed: Galactic Center and Galactic Plane (see Gehrels and Teegarden this conference), and the Crab nebula. Observations of the Crab for the two flights verified the proper operation of the instrument.

The normal mode of operation is to divide the observing time into alternating 20 minute segments of on-source target and off-source background pointings. Background pointings are made at identical elevation angles to the matching source pointing, but at azimuth angles offset from the target, chosen to avoid known gamma ray sources. During the Fall 1988 flight we chose not intersperse background pointings during the second transit of SN1987A. This was done to maximize the statistics for the line analysis, however it prevents those 12 hours of data from being included in the continuum emission analysis. In this paper, only observations containing good matching background observations are reported.

III. ANALYSIS and RESULTS

After channel compression (~4 to 1) and energy calibration, spectra are formed for each target and background observing segment pair. The background segment spectra are formed from the last 10 minutes of the preceding background pointing and the first 10 minutes of the following background pointing. After correcting for live-time, the spectra are subtracted to yield proper source-only flux spectra. These source flux spectra are then added together to form a total source spectrum for the entire transit of the target. This summed source count rate spectrum is then used in the model fitting process along with the instrument response matrix. A trial model photon flux spectrum is then multiplied by the instrument response matrix to yield a trial count rate spectrum. The residuals between this trial model count spectrum and the observed count spectrum are minimized to yield the best fit by the reduced χ^2 method. The fitting process also yields the observed photon spectrum with errors extrapolated to the top of the atmosphere. It is this top-of-atmosphere photon spectrum which is displayed in the figures. Further details of the fitting procedure and the method of deriving the errors are described in Tueller *et al.*[3] The instrument response matrix contains the total effective area of the seven detectors and the atmospheric attenuation plus the combined-detector resolution function (the diagonal and nearly diagonal elements) as well as the atmospheric and instrumental Compton scattering response (the off-diagonal elements). The sum of the off-diagonal elements at 500 keV is 50% of the full-energy peak diagonal elements and when convolved with an $E^{-1.22}$ power law, the count rate contribution to the assumed full-energy peak bin from Compton down-scattered photons is less than 10%. GRIS is a very clean instrument.

Because of the extremely low fluxes involved, the data were binned into eleven energy intervals. Counts falling in windows around the ^{56}Co decay lines 847, 1238, and 2599 keV with widths equal to twice the instrument's FWHM resolution at each line were rejected. Counts were also rejected in windows around the instrumental background lines (138, 198, and 511 keV), because these regions have proportionally much higher error contributions to the result than source emission in the window.

The Spring and Fall flight results of the above analysis are shown in Figure 2. Because the overall flux is lower in the Fall flight, the poorer statistics do not allow all eleven energy bins to be plotted and larger energy bins are used. Discriminator and interference problems in the Spring flight prevented use of the data below 95 keV, whereas we were able to go down to 50 keV in the Fall data. These lower energy limits may be lowered in the final analysis. For each flight the highest energy points are one sigma upper limits.

IV. DISCUSSION

In Figure 1, the results of the Spring flight of GRIS are compared to the measurements of JPL: (Mahoney *et al.*[5], day 286); Frascati: (Ubertini *et al.*[6], day 407); MSFC: (Fishman *et al.*[7], day 411); CalTech: (Cook *et al.*[8], day 414); HEXE (Sunyaev *et al.*[9], day 425). The solid squares are the GRIS results. In general there is good agreement with the other measurements. However, the spread in flux values at any given energy seems larger than can be explained by temporal variations alone due to different day numbers for each balloon flight. The JPL high energy point, while high at only the two sigma level, could be explained by the five month earlier observation. The CalTech points are low which can be explained by instrument pointing errors[10]. The MSFC points are consistently low. It is unlikely that time variations can be the explanation as the two flights were only 22 days apart. The explanation is most likely due to their lack of using a full-blown detector response matrix in the spectral deconvolution analysis[7]. The Frascati and HEXE points agree well in the limited energy range of overlap.

Along with the GRIS Spring and Fall 1988 observations in Figure 2, the continuum curves for the 10HMM model[2,11] for the days since explosion 433 and 613 are shown. These days were interpolated from the model run for days 400, 600, and 800. In general the 10HMM model agrees well with our data both in flux level and energy dependence; although this is not too surprizing, because the "mixing" in this Compton scattering based model was adjusted to get the ^{56}Co line fluxes correct.

For the Spring 1988 GRIS data, the best fit power law using the data between 105 and 830 keV is

$$\text{Flux} = (4.9 \pm 1.2) \times (E/100\ \text{keV})^{-1.22\pm0.24}\ [10^{-5}\ \text{ph-cm}^{-2}\text{-sec}^{-1}\text{-keV}^{-1}]$$

where E is in units of keV. There is insufficient data to quote a fit to the Fall data. Included in Figure 2, as the long-dashed line, is the power law fit to the Spring data described above. There is clear evidence for a break in the energy dependence in the data above 847 keV -- the data fall below the extrapolated power law. The significance of the differentials between the three high energy data points above 847 keV and the best fit power law is 3.1 sigma. This break in the spectrum above 847 keV plus the agreement of the flux levels with the Compton scattering based 10HMM model is evidence that the continuum emission from SN9187A is in fact due to the Compton down-scattering of the radioactive decay lines of ^{56}Co produced in the supernova.

REFERENCES

1) P. Pinto and S.E. Woosley, Ap. J., **329**, p 820, 1988.

2) P. Pinto and S.E. Woosley, Nature, **333**, p 534, 1988.

3) J. Tueller, *et al.*, Ap. J. Let., **351**, L41, 1990.

4) J. Tueller, *et al.*, AIP conf. on Nuclear Spectroscopy of Astrophysical Sources, ed. N. Gehrels and G.H. Share (New York: AIP) p 439, 1988.

5) W.A. Mahoney, *et al.*, Ap. J. Let, **334**, L81, 1988.

6) P. Ubertini, *et al.*, Ap. J. Let., **337**, L19, 1989.

7) G.J. Fishman, *et al.*, Adv. Space Res., **10**, No 2,p 55, 1990.

8) W.R. Cook, *et al.*, IAUC 4584, 1988.

9) R.A. Sunyaev, *et al.*, Soviet Astronomy Letters, **16**, No 5, pp 403, 1990.

10) W.R. Cook, *et al.*, Ap. J. Let, **334**, L87, 1988.

11) P. Pinto and S.E. Woosley, private communication, 1988.

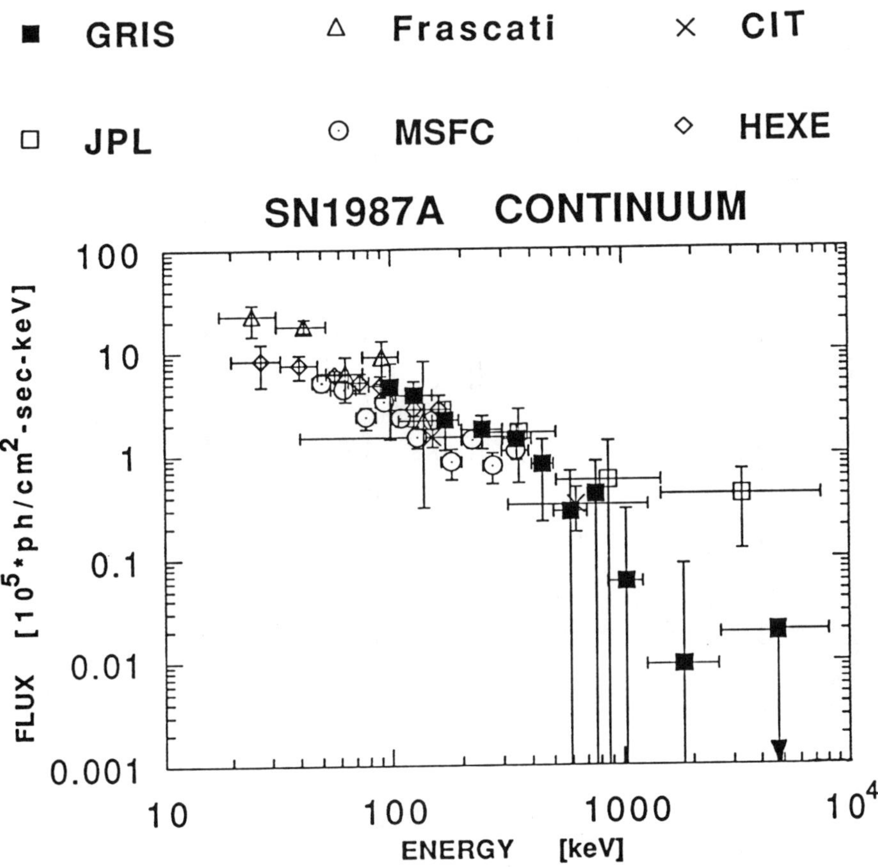

Fig. 1) Comparison of the continuum flux measurements of SN1987A in the Spring of 1988 by GRIS (day 433), JPL (286), Frascati (407), MSFC (411), and Cal Tech (414), HEXE (425).

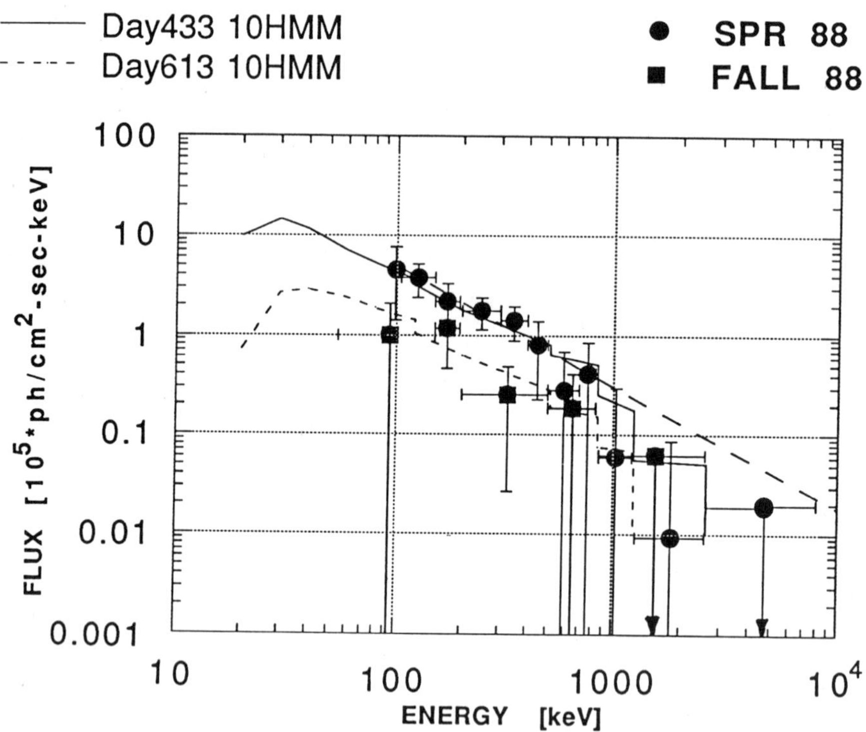

Fig. 2) Comparison of the GRIS continuum flux measurements (filled circles and squares) and the 10HMM model (solid and dashed lines) for days 433 and 613 since the SN1987A explosion. The long-dashed line is the best fit power law to the Spring data in the 105 to 830 keV region. See the text for a discussion of the significance of the discrepancy between this power law and the data points above 1238 keV.

HARD X-RAYS FROM SUPERNOVA 1987A: RESULTS OF MIR-KVANT AND GRANAT IN 1987-1990 AND EXPECTATIONS

R.Sunyaev, S.Grebenev, A.Kaniovsky, V.Efremov, A.Kuznetsov, M.Pavlinsky, N.Yamburenko
Space Research Institute, Academy of Sciences, Moscow, USSR

J.Englhauser, S.Doebereiner, W.Pietsch, C.Reppin, J.Truemper
Max-Planck-Institut fuer Extraterrestrische Physik, Garching, FRG

E.Kendziorra, M.Maisack, B.Mony, R.Staubert
Astronomisches Institut der Universitat Tubingen, FRG

ABSTRACT

Results of the Supernova 1987A hard X-ray observations in 1987-1990 with MIR-KVANT and GRANAT are presented. They allow to estimate degree of ^{56}Co mixing over the envelope and set hard restrictions on abundances of radioactive ^{57}Co, ^{44}Ti and ^{22}Na produced during the outburst. The upper limit at three standard deviation level on the ^{57}Co/^{56}Co ratio inside the envelope is equal to 1.5 of the Earth's ^{57}Fe/^{56}Fe ratio. Future evolution of the SN1987A hard X-ray luminosity connected mainly with central source is discussed.

MIR-KVANT OBSERVATIONS OF HARD X-RAYS FROM SN1987A

In 1987 August the HEXE instrument aboard MIR-KVANT module discovered hard 20-200 keV X-rays from SN1987A[1,2]. They were identified with ^{56}Co decay photons experienced multiple Compton scatterings in the supernova opaque envelope[3-5]. Simultaneously GINGA satellite detected X-rays from SN1987A in the 6-30 keV energy band[6]. Since that time SN1987A has become a main target of the MIR-KVANT observations.

During two years from June 1987 up to June 1989 seven series of intense observations of SN1987A were carried out[7]. X-ray spectra obtained during these observation are presented in Fig.1. Their spectral shape is consistent with predictions of the radioactive model. Diffusion of the ^{56}Co decay gamma-photons in the envelope accompanied by multiple recoil effect moves majority of photons towards $h\nu \leq 30$ keV where they are absorbed due to the photoabsorption beginning to prevail over Compton scattering.

The spectra in Fig.1 demonstrate an increase of the flux from 1987 August to 1988 January[8] due to rapid increase of the envelope transparence. From 1988 January to 1989 June a continuous decline[9,10] of the flux was observed which was mainly connected with decreasing of ^{56}Co amount in the envelope. Already in 1988 September a strong change of the spectral shape was detected. This is explained by decreasing of the envelope optical thickness with respect to Thomson scattering. The number of successive scatterings which ^{56}Co decay gamma-photons experience in the envelope become insufficient to decrease photon energy due to multiple recoil effect up to value $h\nu \leq$ 50 keV. Similar sharp cutoff of the flux at energies below 20 keV in

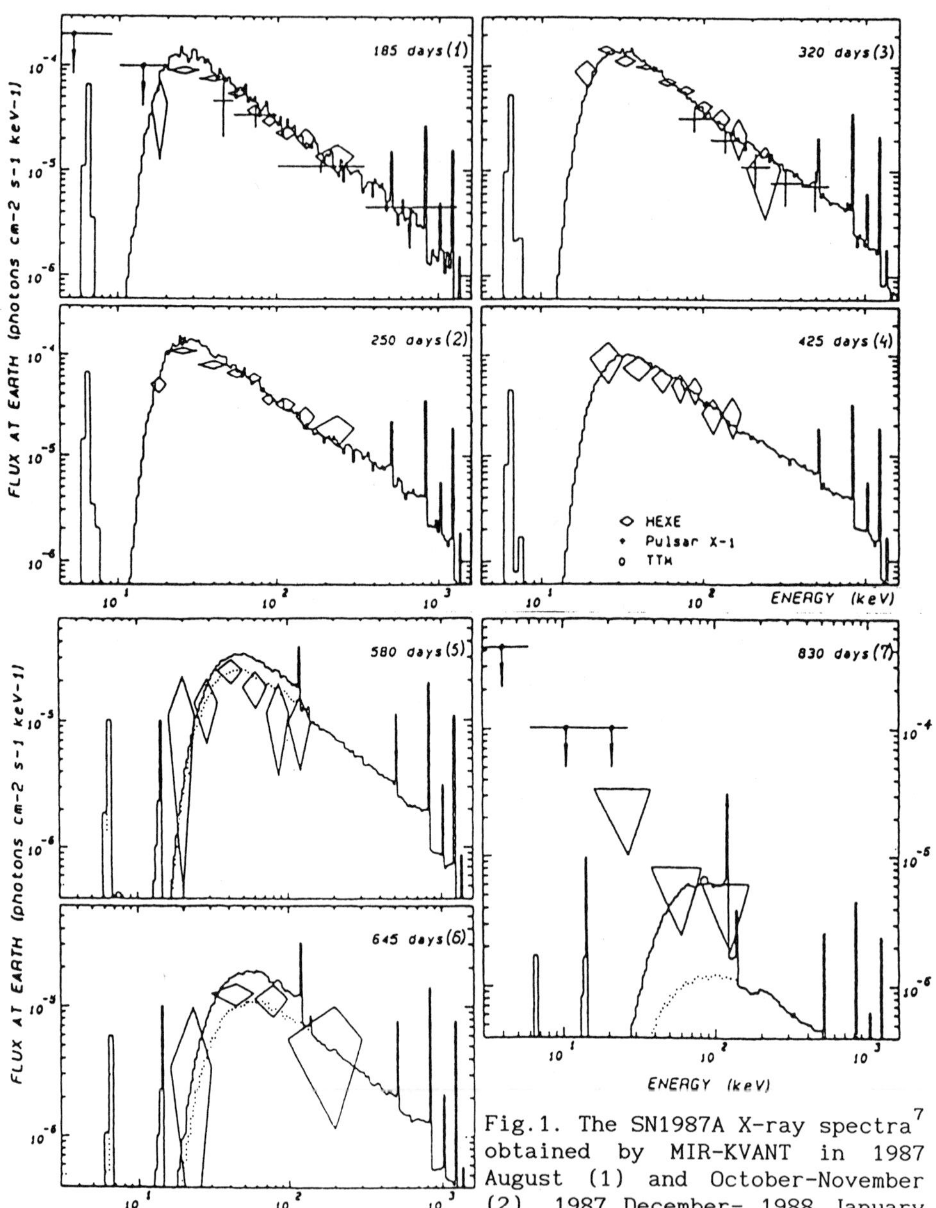

Fig.1. The SN1987A X-ray spectra[7] obtained by MIR-KVANT in 1987 August (1) and October-November (2), 1987 December- 1988 January (3), 1988 April (4), September-October (5) and November (6), 1989 May-June (7) (diamonds and crosses - the HEXE and Pulsar X-1 data, crosses marked by circles - the TTM upper limits). The errors correspond to one standard deviation, the upper limits - to three standard deviations. Results of the Monte-Carlo calculations are presented by solid lines. In graphs 5-7 a ^{56}Co portion in the total ^{56}Co and ^{57}Co emission is shown by dotted lines. The ratio ^{57}Co/^{56}Co is equal to double ratio ^{57}Fe/^{56}Fe at the Earth.

1987 August and 1988 January was connected with photoabsorption by heavy elements.

MIXING OF ^{56}Co AND DIRECT ESCAPE GAMMA-LINES

Early detection of the SN1987A hard X-rays by GINGA satellite[6] and KVANT module[1,2] was the first evidence of the radioactive ^{56}Co strong mixing over the envelope volume. Recently this conclusion was confirmed by measurements of velocity dispersion for infrared lines of iron and cobalt ions[11] and also by broad width of the ^{56}Co direct escape gamma-lines.[12,13] These independent observations testify to presence of the radioactive cobalt in outer envelope layers having high expansion velocities.

Spectral evolution of the supernova X-ray radiation (Fig.1) allow to estimate the degree of real cobalt mixing over the envelope. In Fig.2 the results of such analysis[7] are shown. About 60% of cobalt occupies central region of the envelope. The rest 40% is mixed over the whole envelope volume. In Fig.1 the supernova X-ray spectra calculated using the Monte-Carlo technique for this model of ^{56}Co distribution[7,14] are shown by histograms. The figure demonstrates coincidence of the observational data and results of the calculations.

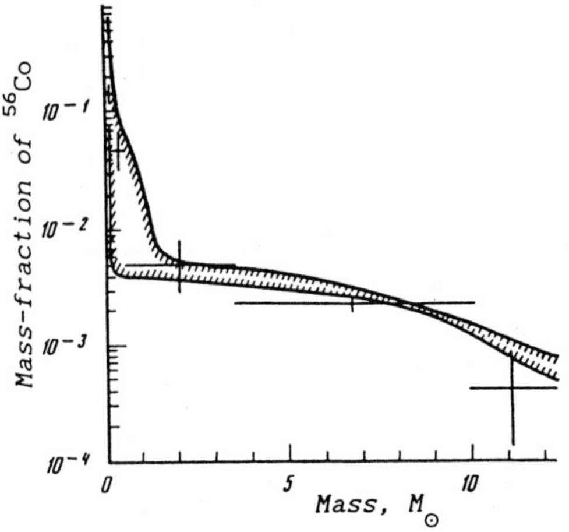

Fig.2. The ^{56}Co distribution (mass-fraction) over the envelope[7] which allows to simulate the SN1987A X-ray light curves observed by HEXE. The results of two independent approaches are presented. In the first approach it is assumed that cobalt is uniformly distributed over five spherical layers of the envelope (crosses). In the second approach the cobalt distribution is described by superposition of two Gaussians, the 67% confidence level region is shown. It is clear that both approaches give similar results.

In Fig.3 the fluxes in two bright direct escape gamma-lines, 847 and 1238 keV, are presented vs. time since the outburst. Both theoretical predictions obtained on basis of the adopted model of ^{56}Co distribution and total combination of the existing observational data are presented. The radiation in these lines was detected by SMM orbital station[15] and during several[12,13,16-18] balloon borne experiments. A small amount of cobalt being moved into outer envelope layers leads to earlier appearance of gamma-rays.

ABUNDANCES OF ^{57}Co, ^{44}Ti AND ^{22}Na

During the whole period of observations up to June 1989 the instruments of MIR-KVANT module has not detected a statistically significant enhancement of X-ray luminosity in the 45-105 keV energy

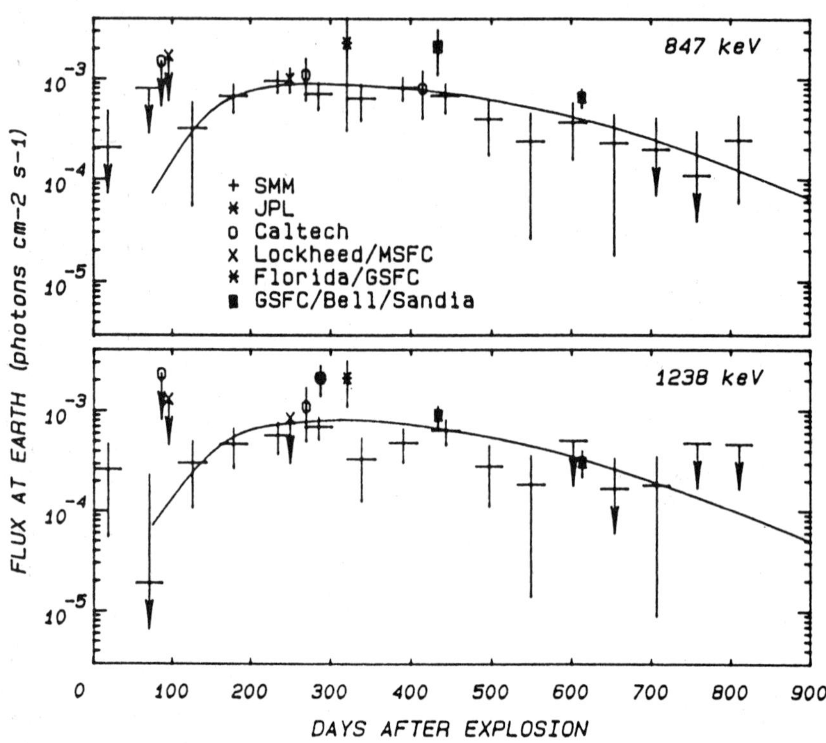

Fig.3. Fluxes in the ^{56}Co gamma-lines 847 and 1238 keV vs. time since the outburst. The results of Monte-Carlo calculations for the model of ^{56}Co mixing given in Fig.2 and the results of observations.[12-13,15-18]

band over predictions of the ^{56}Co decay model. The upper limits at three standard deviation level for the abundance ratio ^{57}Co/^{56}Co in the supernova envelope were equal to 2.4 of the Earth's ratio ^{57}Fe/^{56}Fe in 1988 September, 3.3 in 1988 December and 1.8 in 1989 June. Note that the ratio ^{57}Fe/^{56}Fe at the Earth is 0.024[19] and the SN1987A envelope contained 0.075 $M_\odot$ of ^{56}Co (^{56}Ni) at the moment of outburst[20]. All the data obtained from 1988 September to 1989 June allowed us to obtain a 3σ limit for portion of ^{57}Co decay photons in the SN1987A hard X-ray light curve. This limit corresponds to ratio of abundances ^{57}Co/^{56}Co (see Table) 1.5 times exceeding the Earth's ^{57}Fe/^{56}Fe ratio. The obtained limits depend on the used envelope model[7] and on the degree of cobalt mixing (it is assumed that ^{57}Co is distributed similarly to ^{56}Co, see Fig.2). The observations in 1989 May-June gave an upper limit on flux in the cobalt 122 keV narrow line $3.9\cdot10^{-4}$ photons$\cdot$cm^{-2}s^{-1} (at 3σ level). This corresponds to the ^{57}Co/^{56}Co ratio 6 times exceeding the Earth's ^{57}Fe/^{56}Fe ratio. The expected fluxes in the ^{57}Co direct escape lines 122 and 136 keV are presented in Fig.4. vs. time. The ^{57}Co/^{56}Co ratio was taken equal to 1.5 of the Earth's ^{57}Fe/^{56}Fe ratio.

Table. Abundances of radioactive species in the envelope

isotope	HEXE 3σ upper limit	predictions[21,22]
^{56}Co (^{57}Ni)	$2.8\cdot10^{-3}M_\odot$	$4.3\cdot10^{-3}M_\odot$
^{44}Ti	$9\cdot10^{-3}M_\odot$	$1.2\cdot10^{-4}M_\odot$
^{22}Na	$1.3\cdot10^{-3}M_\odot$	$2\cdot10^{-6}M_\odot$

The data obtained in 1989 May-June gave also a possibility to estimate fractions of the ^{22}Na and ^{44}Ti radioactive photons in the X-ray 45-105 keV flux from SN1987A in 830 days after the explosion. Corresponding upper limits at three standard deviation level on mass of ^{22}Na and ^{44}Ti contained in the envelope at the moment of explosion are given in Table. The predictions[21,22] for amount of ^{57}Co, ^{44}Ti and ^{22}Na obtained on basis of the explosive nucleosynthesis calculations are also given in the Table.

FUTURE SN1987A X-RAY EVOLUTION

The data presented in Fig.1 demonstrate the strong decrease of the SN1987A X-ray flux in the fall 1988. In 1989 May-June (830 days since the outburst) for the first time during two years of the supernova observations the instruments aboard MIR-KVANT were not able to detect its hard X-rays[7]. The upper limits at three standard deviation level for the supernova X-ray luminosity were $L(1\text{-}6\ \text{keV}) \leq 3.6\cdot10^{36}$, $L(6\text{-}15\ \text{keV}) \leq 5.4\cdot10^{36}$ and $L(15\text{-}105\ \text{keV}) \leq 1.35\cdot10^{37}$ erg/s where the distance to SN1987A was assumed to be 55 kpc.

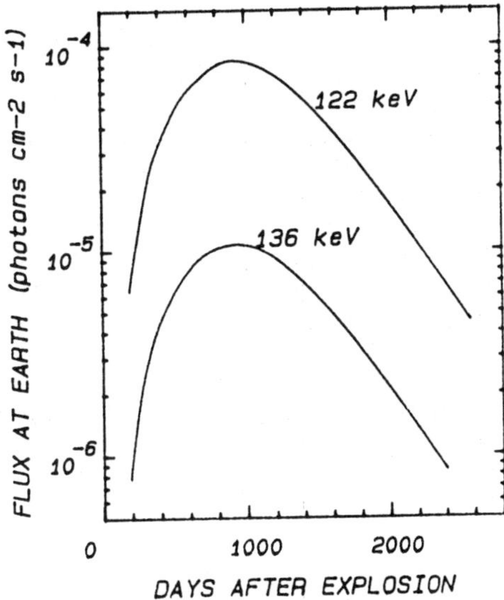

Fig.4. Fluxes in the ^{57}Co gamma-lines 122 and 136 keV vs. time since the outburst. The predictions on basis of the Monte-Carlo calculations for the model of ^{56}Co mixing given in Fig.2. The ratio $^{57}Co/^{56}Co$ is equal to 1.5 of the ratio $^{57}Fe/^{56}Fe$ at the Earth.

In 1990 ART-P and SIGMA imaging telescopes on board GRANAT satellite continued X-ray observations of SN1987A. During the session on September 23 (on the 1309th day since the outburst) hard upper limits were obtained by ART-P[23] for its X-ray luminosity $L(3\text{-}6\ \text{keV}) \leq 3.1\cdot10^{36}$, $L(6\text{-}15\ \text{keV}) \leq 5.5\cdot10^{36}$ and $L(15\text{-}30\ \text{keV}) \leq 2.5\cdot10^{37}$ erg/s.

One of the most intriguing problems of future SN1987A hard X-ray evolution is connected with the nature of the developed star remnant. It may be a young pulsar similar to the Crab pulsar, an accreting neutron star (either with a strong magnetic field or without it) or an accreting black hole. In any of these cases the remnant hidden inside the supernova expanding envelope must produce X-rays. In Fig.5 one of possible intrinsic X-ray spectra of the remnant is shown by dashed lines. We assumed that the remnant is a black hole with the Cyg X-1 like spectrum and with the same X-ray ($h\nu \geq 1$ keV) luminosity. Note the X-ray luminosity of Cyg X-1 is equal to[24] $2.3 \cdot 10^{37} (d/2.5\ \mathrm{kpc})^2$ erg/s.

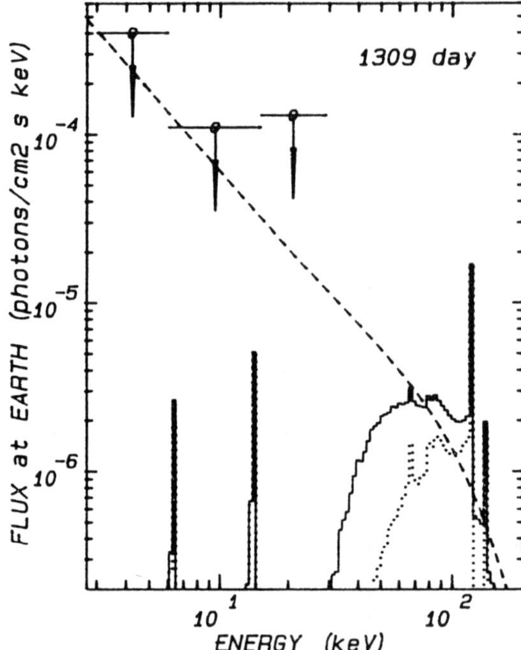

Fig.5. The X-ray spectrum emerging from the SN1987A envelope on the 1309th day since the outburst (a solid line). It was calculated assuming that a source with the Cyg X-1 like spectrum and with the same X-ray luminosity ($L_{\mathrm{in}}(>1\ \mathrm{keV}) = 2.3 \cdot 10^{37}$ erg/s) was hidden inside the envelope (a dashed line). The radiation connected with radioactive decay of ^{56}Co, ^{57}Co and ^{44}Ti was also taken into account (a dotted line). The ratio $^{57}Co/^{56}Co$ equal to 1.5 of the Earth's $^{57}Fe/^{56}Fe$ ratio and $1.2 \cdot 10^{-4} M_\odot$ of ^{44}Ti were adopted.[21] The ART-P upper limits (at three standard deviations level) are presented.

Unfortunately the supernova envelope is still opaque in standard X-ray band. An initial X-ray spectrum of the remnant is strongly distorted by photoabsorption and Compton scattering (Fig.5, a solid line). Today, three years after the outburst, the Thomson depth of the envelope has decreased up to 1-2 and photoabsorption dominates in the envelope opacity. The Monte-Carlo calculations of X-ray transfer through the envelope show a very strong absorption at energies below 30 keV. In the case of a power law $I_\nu \sim \nu^{-\alpha}$ spectrum of the remnant the spectrum of radiation escaping the envelope has maximum near[25]

$$h\nu_m \simeq 50 \sqrt[3]{3\tau_*/\alpha} \left(\frac{1300\ \mathrm{day}}{t}\right)^{2/3} \mathrm{keV}, \text{ where } \tau_* = \tau_a(\nu, t)\left(\frac{h\nu}{50\ \mathrm{keV}}\right)^3 \left(\frac{t}{1300\ \mathrm{day}}\right)^2$$

and $\tau_a(\nu, t)$ is the envelope optical depth with respect to photoabsorption. The isotope ^{56}Fe, daughter element of ^{56}Co, plays the decisive role in process of photoabsorption. Therefore the optical depth $\tau_a(\nu, t)$ is strongly dependent on degree of cobalt mixing. For the model of cobalt distribution given in Fig.2 $\tau_* \simeq 1$. Strong clumpiness of the envelope matter developing due to the Rayleigh-Taylor instability may be another reason why $\tau_a(\nu, t)$ will be decreased.

The data of MIR-KVANT in 1989 May-June and GRANAT in 1990 September set strong restrictions on possible X-ray luminosity of the stellar remnant. So assuming that the remnant is a black hole with the Cyg X-1 like spectrum and using the 3σ HEXE upper limit for the 15-105 keV flux escaping the envelope on the 830th day we obtain the 3σ upper limit on the intrinsic $h\nu \geq 1$ keV remnant luminosity, $L_{in} \leq 3.5 \cdot 10^{37}$ erg/s. The ratio $^{57}Co/^{56}Co$ was adopted to be equal to 1.5 of the Earth's ratio $^{57}Fe/^{56}Fe$. If the ratio $^{57}Co/^{56}Co$ in the envelope is equal to the Earth's ratio then $L_{in} \leq 9 \cdot 10^{37}$ erg/s.

The energy absorbed in the opaque envelope (about 70% of the initial remnant X-ray luminosity 3 years after the outburst) goes on its heating and is reemitted in the infrared, submillimeter and optical bands. Therefore the optical observations[26] give another possibility to restrict the intrinsic luminosity of X-ray source hidden inside the envelope $L_{in}(>1 \text{ keV}) \leq 3 \cdot 10^{37}$ erg/s.

REFERENCES

1. R.Sunyaev et al., Nature, 330, 227 (1987).
2. R.Sunyaev et al., Soviet Astron. Lett., 13, 1027 (1987).
3. S.Grebenev, R.Sunyaev, Soviet Astron. Lett., 13, 945 (1987).
4. P.A.Pinto, S.E.Woosley, Astrophys. J., 329, 820 (1988).
5. S.Kumagai et al., Astrophys. J., 345, 412 (1989).
6. T.Dotani et al., Nature, 330, 230 (1987).
7. R.Sunyaev et al., Soviet Astron. Lett., 16, 403 (1990).
8. R.Sunyaev et al., Soviet Astron. Lett., 14, 579 (1988).
9. R.Sunyaev et al., Soviet Astron. Lett., 15, 291 (1989).
10. J.Englhauser et al., 23d ESLAB Symp. Proc., ed. J.Hunt and B.Battrick (Noordwijk, ESA Publ. Divis., 1989), p.397.
11. E.F.Erickson et al., Astrophys. J. (Letters), 330, L39 (1988).
12. J.Tueller et al., Astrophys. J. (Letters), 351, L41 (1990).
13. A.C.Rester et al., Astrophys. J. (Letters), 342, L71 (1989).
14. S.Grebenev, R.Sunyaev, Soviet Astron. Lett., 14, 675 (1988).
15. M.D.Leising, G.H.Share, Astrophys. J., 357, 638 (1990).
16. W.R.Cooke et al., Astrophys. J. (Letters), 334, L87 (1988).
17. W.A.Mahoney et al., Astrophys.J. (Letters), 334, L81 (1988).
18. W.G.Sandie et al., Astrophys. J.(Letters), 334, L91 (1988).
19. A.J.W.Cameron, Essays in Nuclear Astrophysics, ed. C.A.Barnes, D.D.Clayton and D.N.Schramm (Cambridge: Cambridge Univ. Press., 1982), p.33.
20. S.E.Woosley et al., Astrophys. J., 318, 664 (1987).
21. M.Hashimoto, K.Nomoto, T.Shigeyama, Astron. Astrophys., 210, L5 (1989).
22. S.E.Woosley, P.A.Pinto, D.Hartmann, Astrophys.J., 346, 395 (1989)
23. R.Sunyaev et al., IAU Circular, 5122 (1990).
24. R.Sunyaev, J.Truemper, Nature, 279, 506 (1979).
25. S.Grebenev, R.Sunyaev, Soviet Astron. Lett., 14, 1066 (1988).
26. N.B.Suntzeff et al., Astron. J. (1991), in press

SN 1987A : FROM GAMMA-RAY LINES TO INFRARED CONTINUUM

M. Cassé[1], R. Lehoucq[1,2] and C.J. Cesarsky[1]

1. Service d'Astrophysique, CEA Saclay, France
2. Institut d'Astrophysique de Paris, France

ABSTRACT

On the basis of similar Monte–Carlo algorithms of γ–ray line transfer applied to the ejected envelope of SN 1987A, as modelled by various authors, explanations and predictions concerning the electromagnetic manifestations of the supernova at various wavelengths are presented and commented. The evolution of the 847 and 1238 keV line intensity, of the hard X–ray flux and of the bolometric luminosity are reasonably accounted for in terms of pure radioactivity up to day $\simeq$ 1000. Complications arise beyond. We stress the importance of observations at low γ–ray energies (68 and 78 keV) to clarify the role of ^{44}Ti.

1. INTRODUCTION

Before 1987 astrophysicists were already convinced that unique informations are encoded in energy, shapes and intensities of γ–ray lines from radioactive nuclei produced by explosive nucleosynthesis in supernovae[1–6]. Supernova 1987A, with its close distance and its alleged envelope, provided a perfect opportunity to observe theses lines. They are thought to be sensitive to the details of nucleosynthesis just before and during the explosion, as well as to the position of the frontier between the infalling material and the matter ejected[7–9].

Up to day $\simeq$ 500, the bolometric light curve of the supernova indicated that it was powered by ^{56}Co, decay product of ^{56}Ni synthesized during the explosion[10–15]. Moreover, the most prominent γ–ray lines, a year after the explosion or so, were predicted to be the two decay lines of ^{56}Co at 847 keV and 1238 keV, and they have been indeed detected by the SMM satellite and by various balloon borne detectors[16–21]. The intensity of these lines has been observed to remain approximately constant between 200 and 400 days from the arrival of neutrinos and has declined afterwards. The early appearance of these lines and that of the hard (> 30 keV) X–rays[22,23] is in desagreement with the prediction of models of comptonization of the ^{56}Co γ–ray lines featuring a spherical expansion of the ejecta, without strong inhomogeneities and without mixing of the nucleosynthesis products of the dead star throughout the envelope. However, various authors have shown that the high energy observations of the object can be accounted for by partially mixed models[24–29]. Independent evidences in favour of the mixing of ^{56}Ni in the envelope has been brought by infrared spectrometry: Witteborn et al[30] have found that the FWHM of the Ni II 6.63 μ line is $\simeq$ 3600 km s^{-1}, much wider than predicted by models without mixing. Since strong Rayleigh-Taylor instabilities are expected to develop in the early stages of the expansion[7,8,31–36], a certain amount of mixing throughout the envelope is likely to have occured.

Our aim is to look in more details at the radioactive isotopes ^{57}Co and ^{44}Ti, which have longer lifetime than ^{56}Co (table 1), and which therefore may be important at the present time (winter 1991) and in the next future, giving rise to observable γ–ray lines and partaking noticeably or dominantly in the powering of the supernova at late stages.

Decay	τ	E_γ (keV)	%	ΣE_γ	$\overline{E_e}$ (keV)	%	E_e (keV)	$M/M_\odot$
^{56}Co	113.7 d	847	100	3670	β^+ 660	19	125	0.070
↓								
^{56}Fe		1238	68					
^{57}Co	391 d	122	89	136				4.3 10^{-3}
↓								
^{57}Fe		136	11					
^{44}Ti	78 yr	62	100	2263	β^+ 597	94	561	1.2 10^{-4}
↓								
^{44}Sc		78	100					
↓								
^{44}Ca		1156	100					
^{60}Co	7.6 yr	1173	100	2505	β^- 96	100	96	10^{-5}
↓								
^{60}Ni		1132	100					

Table 1. Characteristics of the γ–ray line emittors (mean lifetime, energies of the main γ–ray photons, branching ratio, total γ–ray energy, mean energy of the positron/electron emitted, branching ratio of β–decay, energy deposited by the positron/electron and ejected mass of the radioactive nucleus of interest[53]

We are particularly interested in the results expected from SIGMA and GRO and future space missions, like NAE or INTEGRAL (see also Pinto and Nomoto et al, this conference). We do not discuss the possible effect on the light curve of a hidden energy source, as for example a pulsar or an accreting object in the center of SN 1987A (see Ref. 37–40, and Nomoto et al, this conference).

2. PORTRAIT AND BIRTH OF THE RADIOACTIVE NUCLEI

Among the radioactive species synthesized and ejected by supernovae, the most important, both by mass and by its consequences, by far, is ^{56}Ni, the progenitor of iron through the decay chain $^{56}\mathrm{Ni} \rightarrow {}^{56}\mathrm{Co} \rightarrow {}^{56}\mathrm{Fe}$. Truran, Arnett and Cameron[41], in 1967, and Bodansky, Clayton and Fowler[42], in 1968, demonstrated theoretically that under explosive conditions close to nuclear statistical equilibrium, ^{56}Ni is the primeval nucleus to be synthesized. Parametrized models of explosive nuclear reprocessing were developed to follow the genesis of the iron peak. The influential parameters were singled out, namely the temperature, the initial composition, through the neutron excess of the shocked material, and the density of the burning zones, through the freeze out of nuclear reactions which can occur in a dense bath of α particles or not[43,44]. Finally, nuclear networks were incorporated into hydrodynamical codes, principally in the US by Woosley and Weaver and in Japan by Nomoto and their collaborators. The subject of explosive nucleosynthesis has been reviewed by Woosley and Weaver[45,46] and Thielemann and Nomoto[47]. Nowadays explosive nucleosynthesis is under systematic study (Ref. 48–50, and Nomoto, this meeting).

More specifically, a detailed nuclear network including more than two hundred species has been coupled to a hydrodynamical model of SN 1987A[51,48]. The mass of the hydrogen envelope (6.7 $M_\odot$) and the energy of the explosion were adjusted to fit the early optical light curve. The initial conditions were derived from a spherical 6 $M_\odot$ helium core evolved up to core collapse[52]. The pre and post explosive composition are shown on fig. 1 (Ref. 49).

It can be seen that all Ni isotopes are generated within a narrow mass–shell of the exploding star, between $\simeq$ 1.5 and 1.7 $M_\odot$. The ^{56}Ni production peaks at 1.65–1.7 $M_\odot$, while that of ^{57}Ni begins to fall off slowly at 1.65 $M_\odot$, and becomes negligible at 1.7 $M_\odot$. Imposing a mass–cut corresponding to 0.07 $M_\odot$ of ^{56}Ni produced, as indicated by the light curve after day $\simeq$ 100, the total amount of ^{57}Ni and ^{44}Ti ejected are 4.3 10^{-3} and 1.2 10^{-4} $M_\odot$, respectively. Accordingly, the mass of matter involved in the collapse (1.6 $M_\odot$) leads to a neutron star mass of 1.45 $M_\odot$, a plausible value. The ratio $^{57}\mathrm{Fe}/^{56}\mathrm{Fe} \simeq 0.056$ (products of the decay of ^{57}Ni and ^{56}Ni), is about twice the value in the solar system (0.024). Similar calculations have been performed by Woosley et al[53] leading to 2.6 10^{-3} $M_\odot$ of ^{57}Co and 5.7 10^{-5} $M_\odot$ of ^{44}Ti in the 18 $M_\odot$ case, which is remarkably close to Nomoto's values given the complexity of the calculation. However unanimity has never been a proof of truth. Many uncertainties plague these estimates (mass of the progenitor star, explosion energy, and above all convection prescriptions, rate of the $^{12}\mathrm{C}(\alpha,\gamma)^{16}\mathrm{O}$ reaction and electron capture rates).

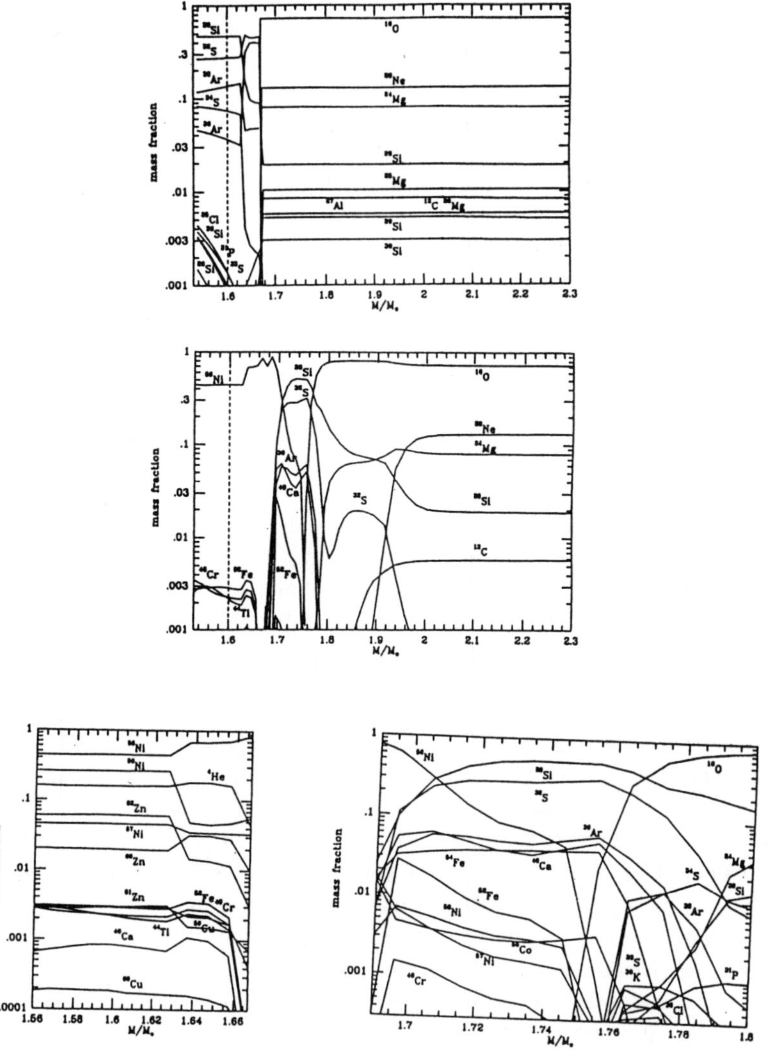

Fig. 1 <u>Preexplosive and postexplosive compositions</u>[48] The mass fraction of the different nuclei produced in quiet nucleosynthesis (upper panel) and in the wake of the shock wave in the innermost shells close to the mass–cut (lower panels) have been calculated consistently. A mass–cut is imposed at 1.63 $M_\odot$ to get 0.07 $M_\odot$ of ^{56}Ni. Nuclear statistical equilibrium is reached at $T > 5\ 10^9$ K.α–rich freeze–out (density $< 10^8$ gcm^{-3}) prevails up to 1.66 $M_\odot$ and then incomplete Si–burning ($T = 4$–$5\ 10^9$ K) up to 1.74 $M_\odot$.

Coming back to our heros, ^{56}Ni decays with a lifetime of 8.8 days into an excited state of ^{56}Co. The second desintegration step, with a mean lifetime of 111.3 days is perfectly suited to leave its imprint on the electromagnetic manifestations of supernovae, and indeed it does. Type Ia supernovae (SN Ia) show radioactive tails, as some SN II (fig. 2). Among them is SN 1987A, which is thought to originate from the explosion of a massive star (18 to 20 $M_\odot$ initially) stripped of a part of its hydrogen-rich envelope (e.g. Ref. 15), and, as said before, an armada of flying objects has observed the 847 and 1238 keV line of cobalt decay accompanied by the hard X-rays coming from their comptonization.

The role of ^{57}Co as γ-ray producer was pointed out by Clayton[55] in 1974. Originating from ^{57}Ni, it decays with a lifetime of 391 days into an excited state of ^{57}Fe by pure electron capture. The daughter nucleus deexcites by photon emission at 122 keV and 14 keV. Another set of important lines is provided by ^{44}Ti. This isotope decays with a mean lifetime of 78 years into an excited state of ^{44}Sc, which cascades to its ground state emitting lines at 68 and 78 keV. ^{44}Sc shortly decays (τ = 2.9 h) into the 1156 keV state of ^{44}Ca. Furthermore, positrons escaping from ^{56}Co (0.19 positron per decay) and/or ^{44}Ti decays (0.94 positron per decay), even in small fraction, would explain a substantial part, if not all, of the diffuse 511 keV line emission of the galaxy (Ramaty, this conference). Concerning ^{44}Ti, it is worth mentioning that its positron emission would induce long lasting effects on the electromagnetic display of the supernova (see below), provided it is produced in sufficient quantity, for, even when the remnant becomes transparent to γ-rays, positrons, in all likelihood, leave their kinetic energy to matter due to their efficient magnetic confinement related to their small gyroradii. This nuclide could take turn with cobalt isotopes in the shaping of the late bolometric (mostly infrared) light curve. Its yield seems to be moderatly sensitive to the mass-cut, according to recent calculations[49,50] and to the method used to artificially generate the shock wave at the base of the ejected envelope[50]. However, in absence of a clear and detailed model of supernova explosion a certain scepticism is healthy concerning the yield of minor isotopes. Nevertheless, all experts of the domain, on the strength of their calculations, maintain that the 44/56 ratio is limited at production to twice the solar system ratio (see below). Observers will judge.

Beyond day $\simeq$ 1000 it is an another story.

3. TITANIUM-44 AND THE LATE LIGHT CURVE OF SN 1987A

Up to day $\simeq$ 1000, if one excludes a few difficulties concerning the width of the 847 line[21] and the fit of the GINGA data[28], the radioactive model armed with only cobalt isotopes yields gratifying results (for reviews see Ref. 58–60). The γ-ray light curves (fig. 3), the hard X-ray spectra at different epochs (fig. 4) are well explained (at the expense of introducing an ad-hoc mixing, however) and the bolometric light curve as well (fig. 5), at least up to day $\sim$ 800.

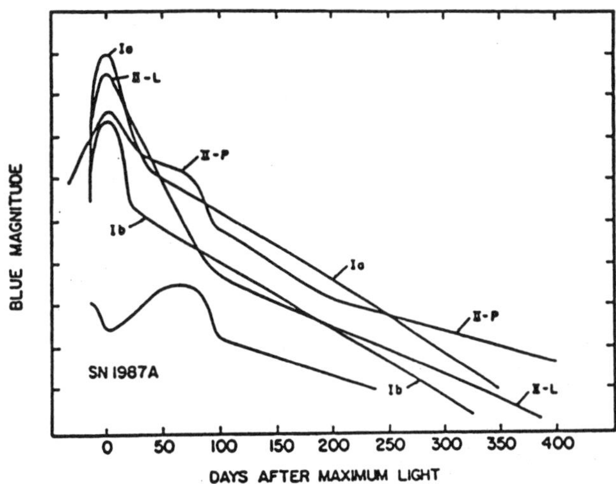

Fig. 2. Light curves of supernova of different types [54] It seems plausible that radioactive decay from the Ni–Co–Fe chain powers all supernovae at late time, excluding, may be the pair creation type. The difference between the mean lifetime of ^{56}Co and the rate of decline of the blue light curve is, in some cases attributed to the leakage of X and γ rays, though the role of bolometric correction could be significant.

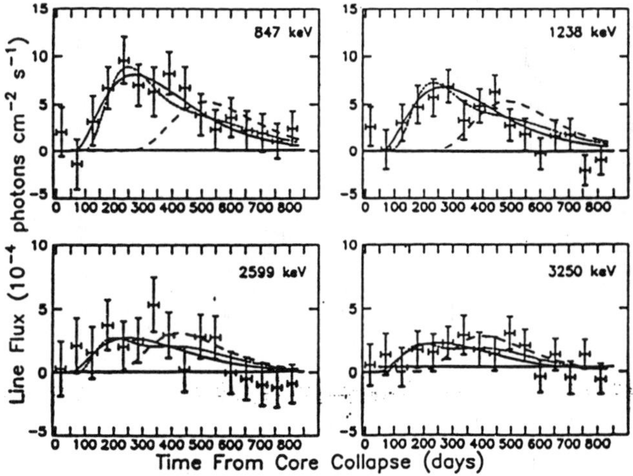

Fig. 3. γ-ray light curves [56]. SMM has delivered a consistent set of data from mid–1987 through mid–1988. Other more episodic measurements compare well with these. A reasonable fit is obtained analytically with a bimodal distribution of ^{56}Co (see Leising and Share for details). Numerical fits are however required to explain consistently the hard X–ray and γ–ray fluxes (see text).

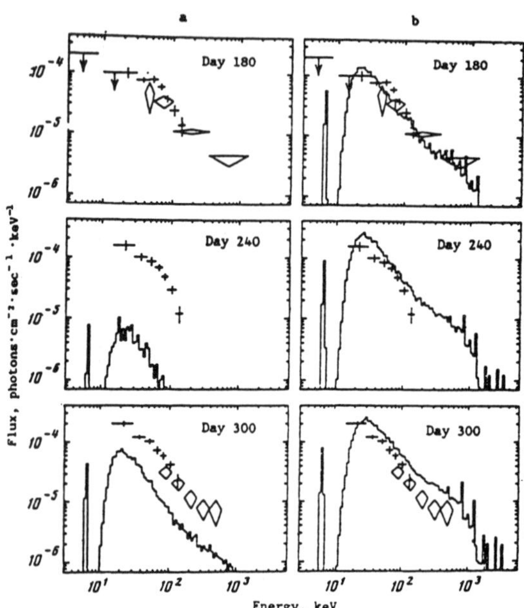

Fig. 4. Early X–ray spectra [57]. This figure clearly illustrates the need of internal mixing. a) without, b) with.

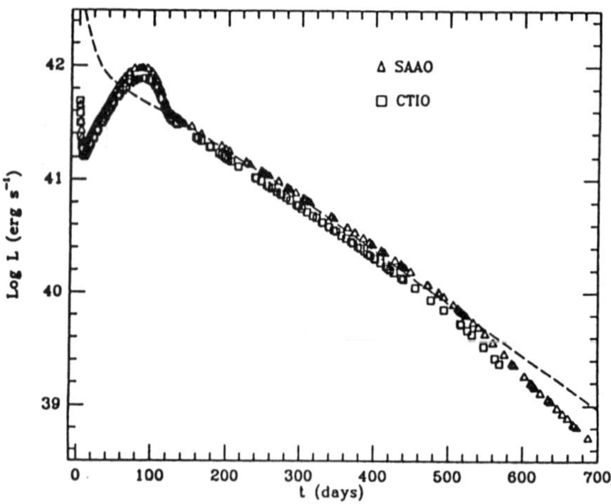

Fig. 5 Early light curve [15]. After day $\simeq$ 500 the ejecta ceases to be opaque to hard photons, more and more radioactive energy leaks out and the light curve steepens.

In this chapter we will concentrate on the recent stagnation of the bolometric luminosity of the magellanic supernova[61,62]. More specifically, we will focus on a single isotope, ^{44}Ti, to try to answer a unique question: can we exclude that the late flattening of the light curve is due to persistent radioactive energy injection by ^{44}Ti alone ? Another reflection on this theme is presented by Pinto, in the same volume.

In the wake of the shock wave triggered by core bounce, the innermost layers are the sites of a rebirth of nuclear activity. Radioactive species are generated in amounts that have been calculated by various authors[8,9,41,51,63]. Due to their long lifetime ^{57}Co (391 d) and ^{44}Ti (78 yr), become dominant at late times. The low energy of their decay lines (122 keV and 78 and 68 keV respectively) and above all the positron emission of the second, make them efficient heating agents of the ejecta (Table 2).

Table 2. Compton cross sections vs Thomson's ($\sigma_T = 6.65\ 10^{-25}$ cm^2)

E_γ [kev]	σ_C/σ_T
847	0.34
511	0.43
122	0.70
78	0.79
68	0.80

Severe limits have been set by Woosley (this conference, see also Woosley and Hoffman 1990, Ref. 63) on the yield of ^{44}Ti (less than twice the solar value). If he is right — this is worth a bottle of Veuve Cliquot, our best champagne — if the ESO data on the infrared light curve beyond day $\simeq$ 1000 are good and well extrapolated to longer wavelengths, and if, finally, the abundance ratio 57/56 is less than 2 times solar, as indicated by infrared line spectrometry[64,65], therefore the electromagnetic emission of a central object is probably overtaking radioactivity as the main energy source of the young supernova remnant.

Indeed, the nucleosynthetic arguments sound right, but in these speculative matters theoretical prudence is advantageous; thus, free from theoretical prejudice we would like to explore the possibility that ^{44}Ti, by itself, dominates the energetics of the whole supernova debris. We see at the moment no observation that could go against this idea, but in the future, undoubtly, γ-ray lines originating from the decay of this isotope, at 68 and 78 keV in particular, should bring, by their presence or their absence, the decisive proof or disproof this scenario. Already, upper limits on the hard X-ray flux have been used to limit the amount of ^{44}Ti to 9 10^{-3} $M_\odot$ (Ref. 68). It is worth verifying this result.

To achieve this goal we proceed step by step.

1. Starting from the model of Lehoucq et al.[27] (i.e. a Monte–Carlo algorithm adapted to the description of the photon transfer in a post–supernova structure adopted from Nomoto et al.[7]), we evaluate the mass of ^{56}Co required to fit the first part of the light curve after maximum (from day $\simeq$ 100 to day 600) i.e. $M_{56} = 0.075\ M_\odot$, and we fix the expansion velocity at the edge of the envelope. We get $v_{exp} = 11500$ km s^{-1}

2. Adopting the upper limit on the 57/56 ratio imposed by IR spectrometry, namely 57/56 $<$ 2 times solar, we try to fit the second part of the light curve (from day 600 to day 800) by fine adjustment of M_{57}, without violating the X–ray constraints (fig 8 and 9). We get 57/56 $<$ 1.5 times solar, in agreement with the value given by Sunyaev et al (1989) (Ref. 66,67).

3. All other parameters being settled, we adjust M_{44} to obtain a reasonable fit of the third part of the ligth curve (beyond day 900). We get $M_{44} = 2.7\ 10^{-3}$ $M_\odot$ (fig. 6, Table 3). The CTIO data[61], superimposed on the figure are also reasonably accounted for.

4. We check the compatibility of M_{44} with the upper limits of the hard X–ray flux at day 830 released by the MIR–KVANT satellite. Other measurements are less constraining including that of the SIGMA satellite. We get $M'_{44} < 2$ to $3\ 10^{-3}\ M_\odot$, still marginably compatible with M_{44} (Fig. 7 to 9).

Thus, there is at least a purely radioactive solution to the problem of the light curve beyond day $\simeq$ 100.

Our results differ significantly from that of Sunyaev et al. (1990) (Ref. 66), who obtained $9\ 10^{-3}\ M_\odot$ for both M_{44} and M'_{44}. Since our methods are comparable, the sources of differences should be sought in the initial conditions and in the ^{44}Ti distribution influenced by the artificial mixing procedure (Table 3).

More precisely, Arnett's[10] (Sunyaev) model and Nomoto's (this work) differ by the mass of hydrogen, which is an efficient comptonizing agent since it offers 1 electron per a.m.u (10 $M_\odot$ against only 6.7 $M_\odot$ in our case), and also by the density and velocity profiles. However both models admit the same E/M ratio. We also note that the zoning of Sunyaev et al is very coarse (only 3 zones) compared to ours. The Thomson optical depths to the center, τ_T, scaling at t^{-2}, are quite similar at the fiducial time 100 days (close to 240, Table 3). It is not surprising that both models and others as well, fit the γ–ray (847 keV) light curve and the hard X–ray spectra at different epochs up to day 645 equally well (e.g. Sunyaev et al, this conference and reference 60) and thus deliver results

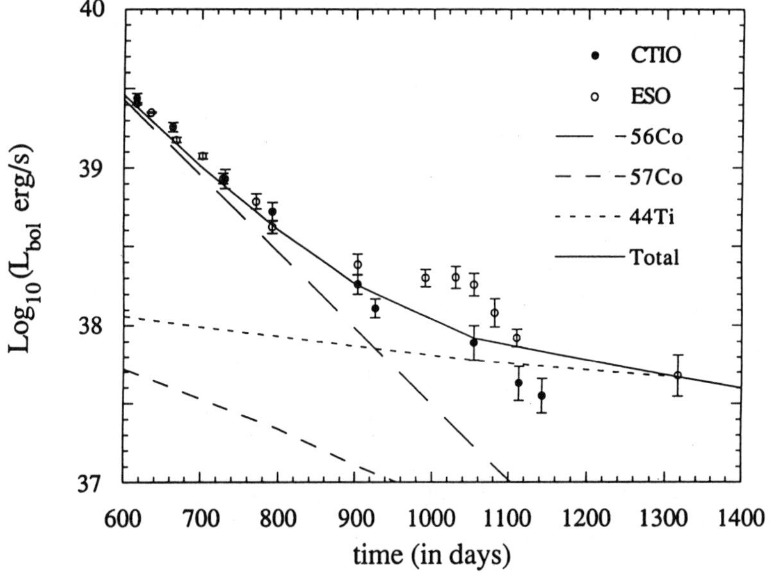

Fig. 6. Fit of the late light curve

The late light curve is infrared dominated, and the ESO[62] and CTIO[61] data are discrepant.

Our fit is based on the following parameters : $M_{56} = 0.075\ M_{\odot}$, $M_{57} = 2.7\ 10^{-3}\ M_{\odot}$, $M_{44} = 2.8\ 10^{-3}\ M_{\odot}$ and $v_{exp} = 11500$ km s^{-1}. The amount of ^{44}Ti is barely consistent with the hard X–ray upper limits delivered by the KVANT experiment of day 830 (Ref. 66) as shown in fig. 9.

The ESO fit, based on Woosley's model, is obtained by adding a central source of $9.5\ 10^{37}$ erg s^{-1} or $1.8\ 10^{38}$ erg s^{-1} to the radioactive light curve generated by $M_{56} = 0.069\ M_{\odot}$, $M_{57} = 1.7\ 10^{-3}\ M_{\odot}$, $M_{44} = 10^{-4}\ M_{\odot}$ (Ref. 62).

Another fit, of similar quality, is obtained with $M_{56} = 0.069\ M_{\odot}$, $M_{57} = 2.4\ 10^{-3}\ M_{\odot}$ and $M_{44} = 2.10^{-3}\ M_{\odot}$. The adjusted column depth ($9.5\ 10^{4}$ g cm^{-2} at 10^{6} s) is slightly higher than ours (7.10^{4} g cm^{-2}) (Ref. 62).

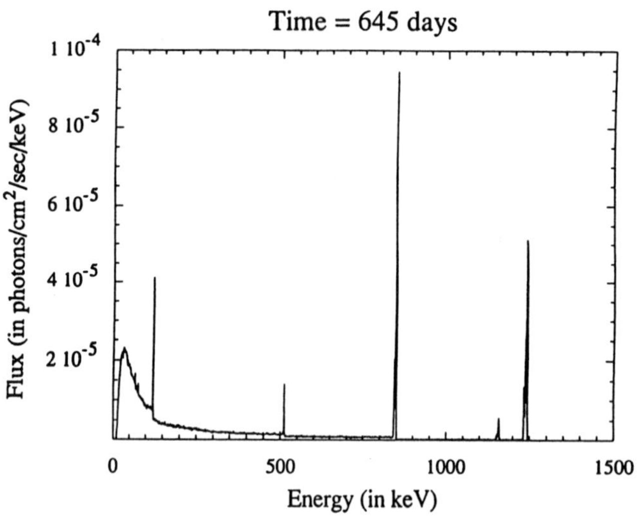

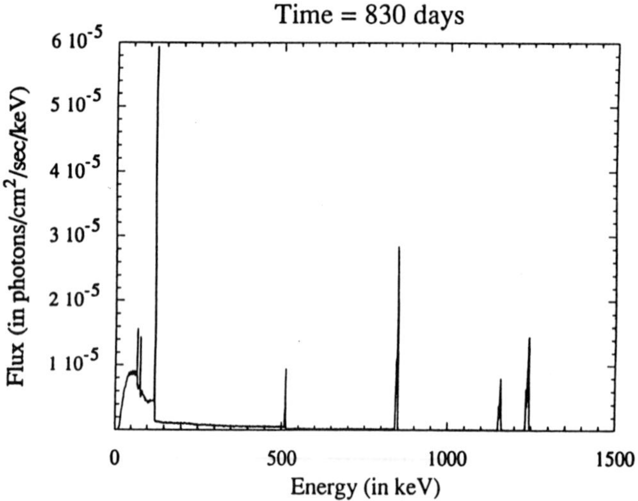

Fig. 7 Change of the spectrum of hard photons between 645 and 830 days. The most intense lines are respectively at 847 keV (^{56}Co) and 122 keV (^{57}Co). At 830 d, the fraction of photons below $\sim$ 70 keV coming from the comptonization of the 68 and 78 keV lines (^{44}Ti) is significant. The parameters are the same as in Fig. 6.

t[d]	L_{56}[erg s^{-1}]	L_{57}[erg s^{-1}]	L_{44}[erg s^{-1}]	log L_{Tot}[erg s^{-1}]
600	2.7 10^{39}	5.3 10^{37}	7.0 10^{37}	39.46
700	9.0 10^{38}	3.4 10^{37}	9.9 10^{37}	39.01
800	2.9 10^{38}	2.2 10^{37}	8.6 10^{37}	38.60
900	9.6 10^{37}	1.3 10^{37}	7.4 10^{37}	38.26
1054	1.7 10^{37}	6.4 10^{36}	6.1 10^{37}	37.92
1316	1.0 10^{36}	1.9 10^{36}	4.7 10^{37}	37.67

Table 3. Contribution of the different radionuclides to the light curve.

The amounts of ^{56}Co, ^{57}Co and ^{44}Ti are respectively 0.075, 2.7 10^{-3} and 2.8 10^{-3} $M_\odot$, the velocity at the edge of the ejecta is 11500 km s^{-1}, corresponding to a column depth of 7.10^4 g cm^{-2} to the center at 10^6 s (against 9.5 10^4 g cm^{-2} in Bouchet et al.[62]), or, equivalently to a Thomson optical depth to the center of 243.7 at 100 d (against $\simeq$ 240 in Grebenev and Sunyaev, 1988 (Ref. 25), their fig. 1). For comparison, the optical depths at other energies are $\tau_{847} = 84$, $\tau_{122} = 172$, $\tau_{78} = 191$ and $\tau_{68} = 196$ at the same fiducial time. At 830 days τ_{78} and τ_{68} are still as high as 2.8. Thus the predicted hard X-ray flux is somewhat dependent on the extent of radial mixing of the radioisotopes adopted, and on the probable clumping of the ejecta. A more concentrated distribution leads to a more efficient comptonization, thus less ^{44}Ti is required to produce the same photon intensity below $\sim$ 70 keV.

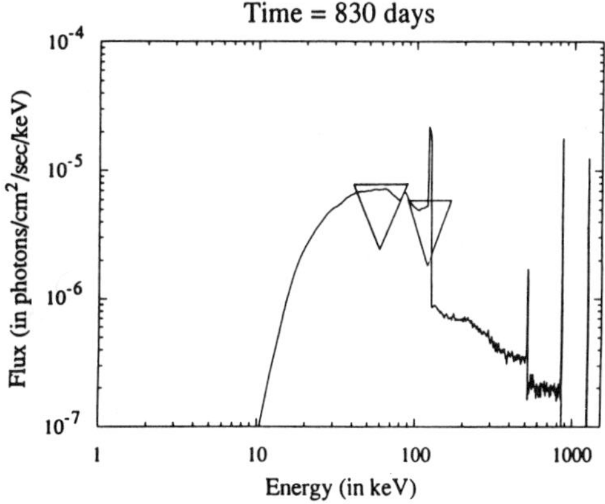

Fig. 8 Sensitivity test : v_{exp} = 11500 km s^{-1} and 57/56 = twice solar. There is little room for ^{44}Ti, if any.

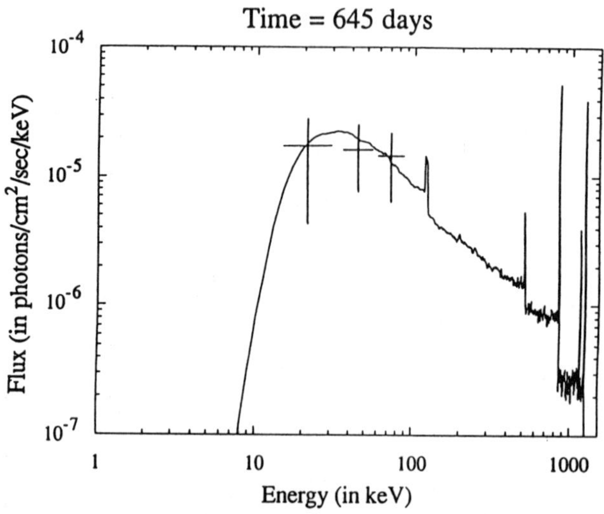

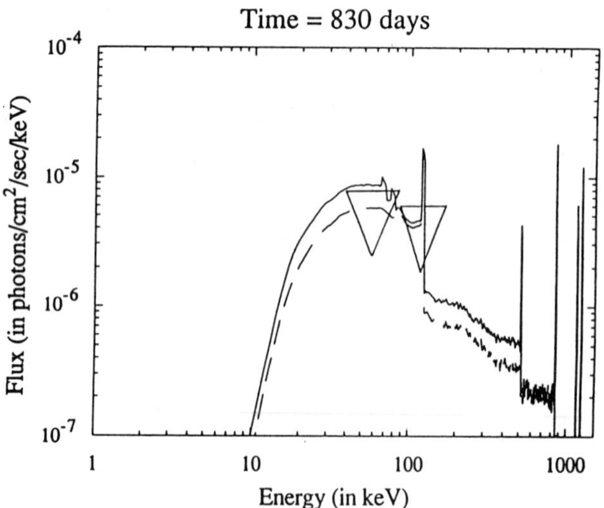

Fig. 9 Consistency test : v_{exp} = 11500 km s^{-1}, 57/56 = 1. 5 times solar (i.e. $M_{57} = 2.7\ 10^{-3} M_{\odot}$) and $M_{44} = 2.8\ 10^{-3} M_{\odot}$. The contribution of ^{57}Co to the hard X–ray flux is indicated by the doted line on the right panel.

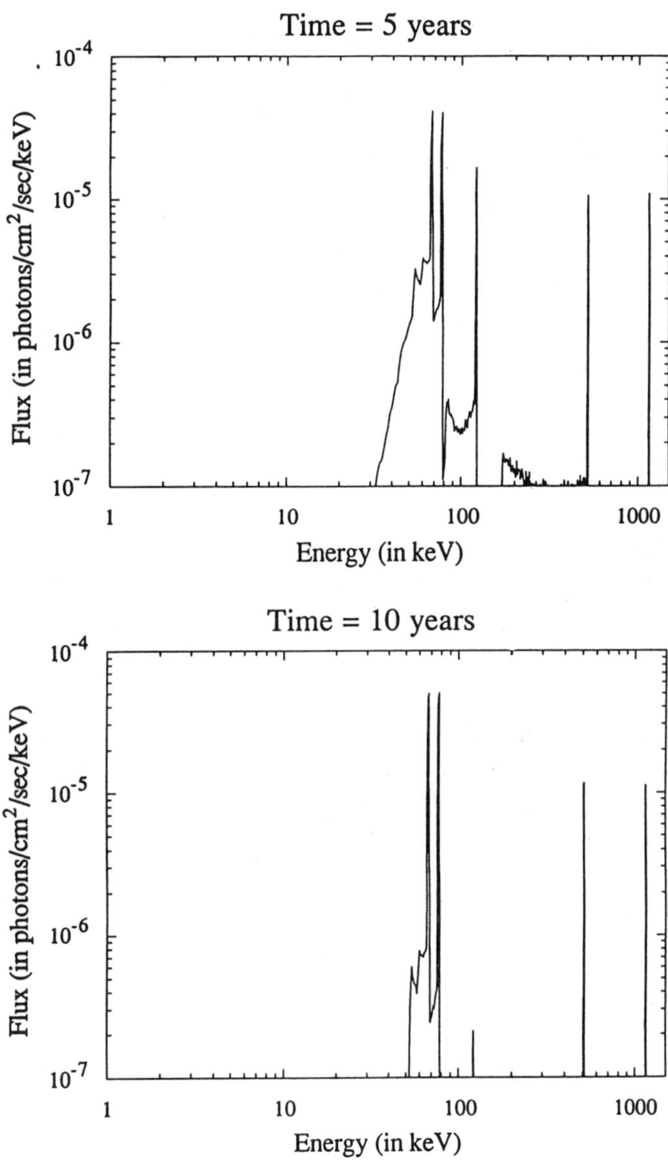

Fig. 10. Predicted high–energy spectra 5 and 10 years after explosion.

that are not very model dependent. This is not anymore the case for the hard X–ray spectrum on day 830. In absence of a clear understanding of the origin of the discrepancy on M'_{44} (see however Table 3), we urge other modelists to make themselves the calculation. Another remark, in passing : Woosley, Pinto and Hartmann (1990) (Ref. 38) lacking a detailed Kurie plot for ^{44}Ti, estimate the average kinetic energy to 1/2 of the maximum energy, i.e. 735 keV. From the informations we have gleaned from the authorized services in Saclay we get 597 keV. Thus a slight correction is necessary to compare our calculations.

4. ASTROPHYSICAL CONSEQUENCES

Let us examine the consequences of the proposed solution. Besides being at variance with the best theoretical predictions, the high–titanium scenario has another disadvantage : it forces us to admit that SN 1987 A is not typical among SN II supernovae[68]. This is philosophically unpleasant, but cannot be discarded on pure logical grounds. The lack of detection of 68 and 78 keV point sources in the galactic plane indicates that SN 1987A–like events, if equally generous in ^{44}Ti, are rare, according to the diagram presented by Mahoney et al. at this conference, giving the probability of detection versus the recurrence time and the amount of ^{44}Ti synthesized. Indeed this diagram strictly deals with type Ia events, which are believed to be rather old objects. The same work remains to be done on the basis of the galactic distribution of pop I objects. We do not think, however, that the conclusions would be strongly alterated.

Another difficulty concerns galactic evolution up to the birth of Sun. To avoid an overabundance of ^{44}Ca (the end product of 44 decay chain) with respect to ^{56}Fe, one can assume, once again, that SN 1987 A is atypical. However, one can alleviate the problem considering that SN Ia, which are generous iron producers, dilute the contribution of SN II and SN Ib.

5. CONCLUSION

A definite test of the ^{44}Ti predominence is the direct observation of its characteristic decay lines in the spectrum of SN 1987 A which would be conspicuous. The flux, close to 10^{-4} photon cm^{-2} s^{-1}, (fig. 10) would be quite easily detectable by GRO, and future missions like NAE or INTEGRAL. (For lower energy prospects see Pinto, this conference). The lines are predicted to be rather sharp since the bulk of Ti is concentrated in the metal core, at velocities less than say 2000 to 3000 km s^{-1}.

REFERENCES

1. Clayton, D.D., Colgate, S.A., and Fishman, G., 1969, Ap.J., 155, 75.
2. Clayton, D.D., 1973, in "Gamma-Ray Astrophysics", ed : F.W. Stecker and J.I. Trombka (Washington : NASA), p. 263.
3. Clayton, D.D., 1982, in "essays in Nuclear Astrophysics", ed : C.A. Barnes, D.D. Clayton, and D.N. Schramm (Cambridge University Press), p. 401.
4. Lingenfelter, R.E., and Ramaty, R., 1978, Physics Today, 31, 40.
5. Ramaty, R., and Lingenfelter, R.E., 1981, Phil. Trans. R. Soc. Lond. A301, 671.
6. Woosley, S.E., Axelrod, T., and Weaver, T.A., 1981, Comments in Nucl. Part. Phys. 9, 185.
7. Nomoto, K., Shigeyama, T., Kumagai, S., and Hashimoto, M., 1988, Proc. Astr. Soc. Australia, 7, 490.
8. Woosley, S.E., 1988, Ap.J., 330, 218.
9. Aufderheide, M., Thielemann, F.K., and Baron, E., 1990, Ap.J., in press.
10. Arnett, W.D., 1988, Ap.J., 331, 377.
11. Shigeyama, T., Nomoto, K. and Hashimoto, M., 1988, Astron. Astrophys., 196, 141.
12. Woosley, S.E., 1988b in "Supernova 1987A in the Large Magellanic Cloud" (Cambridge University Press), p. 289.
13. Woosley, S.E., Pinto, P.A., and Ensman, L., 1988, Ap.J., 324, 446.
14. Nomoto, K., Shigeyama, T., and Hashimoto, M., 1988 in "Atmospheric Diagnostics of Stellar Evolution", Proc. IAU Colloq. n° 108. Berlin : Springer-Verlag, in press.
15. Arnett, W.D., Bahcall, J.N., Kirshner, R.P., and Woosley, S.E., 1989, Ann. Rev. Astron. Astrophys., 27, 629.
16. Matz, S.M. et al., 1988, Nature, 331, 416.
17. Sandie, W.G. et al., 1988, Ap.J. (Letters), 334, L91.
18. Cook, W.R. et al., 1988, Ap.J. (Letters), 334, L87.
19. Rester, A.C. et al., 1989, Ap.J. (Letters), 342, L71.
20. Gehrels, N., MacCallum, C.J., and Leventhal, M., 1987, Ap.J. (Letters), 320, L19.
21. Tueller, J. et al., 1990, Ap.J. (Letters), 351, L41.
22. Donati, T. et al., 1987, Nature, 330, 230.
23. Sunyaev, R.A. et al., 1990, Sov. Astron. Lett., 16, 403.
24. Pinto, P.A., and Woosley, S.E., 1988, Ap.J., 329, 820.
25. Grebenev, S.A., and Sunyaev, R.A., 1988, Sov. Astron. Lett., 14, 288.
26. Kumagai, S., Itoh, M., Shigeyama, T. Nomoto, K., and Nishimura, J., 1988, Astron. Astrophys., 197, L7.

27. Lehoucq, R., Cassé, M., and Cesarsky, C.J., 1989, Astron. Astrophys., 224, 117.
28. The, L.S., Burrows, A., and Bussard, R.W., 1990, Ap.J., 352, 731.
29. Ebisuzaki, T., and Shibazaki, N., 1988, Ap.J. (Letters), 327, L5.
30. Witteborn, F. et al., 1989, Ap.J. (Letters), 338, L9.
31. Chevalier, R.A., 1976, Ap.J., 207, 872.
32. Müller, E., et al., 1989, Astron. Astrophys., 220, 167.
33. Arnett, D., Fryxell, B.A., and Müller, E., 1989, Ap.J. (Letters), 341, L63.
34. Hashisu, I., Matsuda, T., Nomoto, K., and Shigemaya, T., 1990, Ap.J. (Letters), 358, 57.
35. Benz, W., and Thielemann, F.K., 1990, Ap.J. (Letters), 348, L17.
36. Ebisuzaki, T., Shigemaya, T., and Nomoto, K., Ap.J. (Letters), 344, L65.
37. Shigemaya, T., and Nomoto, K., 1990, Ap.J., 360, 242.
38. Woosley, S.E., Pinto, P.A., and Hartmann, D., 1989, Ap. J., 346, 395.
39. Grebenev, S.A., and Sunyaev, R.A., 1988, Sov. Astron. Lett., 14, 452.
40. Xu, Y., Sutherland. P., and McCray, R., 1988, Ap.J., 327, 197.
41. Truran, J.W., Arnett, W.D., and Cameron, A.G.W., 1967, Can. J. Phys., 45, 2315.
42. Bodansky, D., Clayton, D.D., and Fowler, W.A., 1968, Pys. Rev. Lett., 20, 161.
43. Woosley, S.E., Arnett, W.D., and Clayton, D.D., 1973, Ap. J. Suppl., 231, 26.
44. Nomoto, K., Thielemann, F.K., and Yokoi, K., 1984, Ap.J., 286, 644.
45. Woosley, S.E., 1986, in "Nucleosynthesis and Chemical Evolution", 16th Advanced Course of the Swiss Soc. of Astron. and Ap. "Saas–Fee Lecture Notes", ed. B. Hanck, A. Maeder, and G. Meynet, (Geneva Observatory : Geneva), p. 1.
46. Woosley, S.E., and Weaver, T.A., 1991, Ecole de Physique Théorique des Houches, Session LIV : Supernovae, ed : J. Audouze, S. Bludman, R. Mochkowitch and J. Zinn–Justin, in press.
47. Thielemann, F.K., and Nomoto, K., 1991, Ecole de Physique Théorique des Houches, Session LIV : Supernovae, ed. J. Audouze, S. Bludman, R. Mochkovitch and J. Zinn–Justin, in press.
48. Thielemann, F.K., Hashimoto, M., and Nomoto, K., 1990b, Ap.J., 158, 17.
49. Nomoto, K. et al., 1990, in "Chemical and Dynamical Evolution of Galaxies" ed : F. Ferrini, J. Franco, and F. Matteuci, to appear.
50. Arnould, M., and Prantzos, N., 1991, Ecole Joliot–Curie de Physique Nucléaire 1990, ed : P. Aguer, P. Bouche, P.H. Heenen, S. Kox and P. Quentin, (IN2P3, France) p. 1.
51. Hashimoto, M., Nomoto, K., and Shigeyama, T., 1989, Astron., Astrophys., 210, L5.

52. Nomoto, K., and Hashimoto, M., 1988, Physics Report, 163, 13.
53. Woosley, S.E., Pinto, P.A., and Weaver, T.A., 1988, Proc. Astron. Soc. Australia, 7, 390.
54. Wheeler, J.C., and Harkness, R.P., 1990, Rep. Prog. Phys., 53, 1467.
55. Clayton, D.D., 1974, Ap.J., 188, 155.
56. Leising, M.D., and Share, G.H., 1990, Ap.J., 357, 638.
57. Sunyaev, R.A. et al., 1988, Sov. Astron. Lett., 14, 247.
58. Burrows, A., 1990, in the proceedings of the ESO/EIPC workshop "SN 1987 A and the Other Supernovae", held at Marciana Marina, Isola d'Elba, September 17–22, 1990, and this conference.
59. Sutherland, P.G., 1990, in "Supernovae", Springer–Verlag, ed : A. Petscheck, p. 111.
60. Cassé, M., and Lehoucq, R., 1990, Ecole d'Eté de Physique Théorique des Houches, Session LIV "Supernovae", July 31–September 1st, 1990.
61. Suntzeff et al., 1990, in the proceedings of the ESO/EIPC Woskhop "SN 1987 A and the Other Supernovae", held at Marciana Marina, Isola d'Elba, September 17–22, 1990.
62. Bouchet, P., Danziger, I.J., and Lucy, L.B., 1990, preprint.
63. Woosley, S.E., and Hoffman, R.D., 1990, preprint.
64. Danziger, I.J., Lucy, L.B., Bouchet, P., and Grouiffes, C., 1990, in "Supernovae", ed : S.E. Woosley, in press.
65. Varani, G.F., Meikle, W.P.S., Spyromilio, J., and Allen, D.A., 1990, M.N.R.A.S., 245, 570.
66. Sunyaev, R.A. et al., 1990, Sov. Astron. Lett, 16, 403.
67. Sunyaev, R.A. et al., 1989, Sov. Astron. Lett., 16 171.
68. Lehoucq, R., Cassé, M., and Cesarsky, 1990, in the proceedings of the ESO/EIPC Workshop "SN 1987 A and Other Supernovae", held at Marciana Marina, Isola d'Elba, September 17–22, 1990.

GAMMA-RAY LINES FROM TYPE IB/IC SUPERNOVAE AND SN 1987A

K. NOMOTO, S. KUMAGAI, and T. SHIGEYAMA

Department of Astronomy, Faculty of Science, University of Tokyo, Tokyo

ABSTRACT

Gamma-ray lines expected from helium star models for Type Ib/Ic supernovae (SN Ib/Ic) are presented. Summarized are: (1) presupernova evolution that may lead to the difference between SN Ib and SN Ic, (2) nucleosynthesis and isotopic abundances in the exploding helium stars with masses 3 – 8 $M_{\odot}$ and their mass dependence, (3) 2D hydrodynamical simulations of Rayleigh-Taylor instabilities and mixing, (4) optical light curves of these helium stars and the dependencies on the stellar mass and the degree of mixing, and (5) X-ray/γ-ray spectra and γ-ray line light curves originated from radioactive decays, their dependencies on the stellar mass and mixing. We discuss the possibilities that INTEGRAL and GRO detect γ-ray lines and hard X-rays from nearby SNe Ib/Ic.

Also discussed are the implications of recent change in the bolometric light curve of SN 1987A as well as the updated comparison between the calculated X/γ-ray line light curves and observations.

1. INTRODUCTION

The observations of γ-ray lines and hard X-rays from SN 1987A have provided the solid evidence that ^{56}Ni is synthesized in the supernova explosion and its decay provides energy to power the later phase of the Type II supernova (SN II) light curve. The optical light curves of Type Ia/Ib/Ic supernovae (SNe Ia/Ib/Ic) are also well modeled by radioactive decays of ^{56}Ni and ^{56}Co. Since SNe Ia/Ib/Ic are predicted to produce larger amount of ^{56}Ni than SN 1987A, GRO and INTEGRAL will provide a chance to detect γ-ray lines and hard X-rays from nearby SNe I. Here we examine such possibilities and present some preliminary results. [For detailed discussion, see, e.g., Burrows (1991) and references therein.] We also discuss the implications of recent change in the bolometric light curve of SN 1987A as well as the updated comparison between the calculated X/γ-ray line light curves and observations.

2. TYPE IB AND IC SUPERNOVAE

Supernovae have been spectroscopically classified into Type I (SNe I) and Type II (SNe II) according to the absence and presence of hydrogen in their optical spectra. SNe I are further subclassified into Ia, Ib, and Ic. Unlike SNe Ia, SNe Ib exhibit strong absorption lines of [He I] at early times, and prominent emission lines of [O I], [Ca II], and Ca II at late times. SNe Ic spectroscopically resemble SNe Ib, except that the He I lines are absent at early times (Harkness and Wheeler 1990 for a review). SNe II are subclassified into SNe II-P (*plateau*), SNe II-L (*linear*), and SNe II-BL (*bright linear*) according to the light curve shape (Branch *et al.* 1991).

However, the supernova classification has become more complicated since the discovery of SN 1987K whose spectral classification changed from Type II to Type Ib/Ic as it aged, thereby being called as SN IIb (Filippenko 1988). Hydrogen feature has also been identified in the early time spectrum of SNe Ic 1987M (Fig. 1; Jeffery *et al.* 1991) and 1991A (Filippenko 1991). Interestingly the early time spectra of SN 1987K are very similar to SNe Ic (1983V and 87M) (Fig. 2; Filippenko *et al.* 1990; Wheeler and Harkness 1990).

Wolf-Rayet stars with a wide range of masses have been proposed for the progenitors of SNe Ib and Ic, since most of SNe Ib/Ic are associated with star-forming regions (Wheeler and Harkness 1990 for a review). Recently Shigeyama *et al.* (1990), Hachisu *et al.* (1991), Nomoto *et al.* (1990), and Yamaoka and Nomoto (1991) have calculated the progenitor's evolution, nucleosynthesis, Rayleigh-Taylor instabilities, and optical light curves of exploding helium stars. They have suggested that the helium stars of 3–5 $M_\odot$ (which form from stars with initial masses $M_i \sim$ 12–18 $M_\odot$ in binary systems) are the most likely progenitors of typical SNe Ib/Ic and that SNe Ic progenitors may be slightly less massive than those of SNe Ib. Such low mass helium star models can account for the observations that: (1) the light curves of SNe Ic decline faster than SNe Ib, and (2) the early time spectra of SNe Ic show the presence of hydrogen (Jeffery *et al.* 1991), while hydrogen is absent in SNe Ib. It remains an open question how the presence of hydrogen causes the difference between SNe Ib and Ic in their early time spectra.

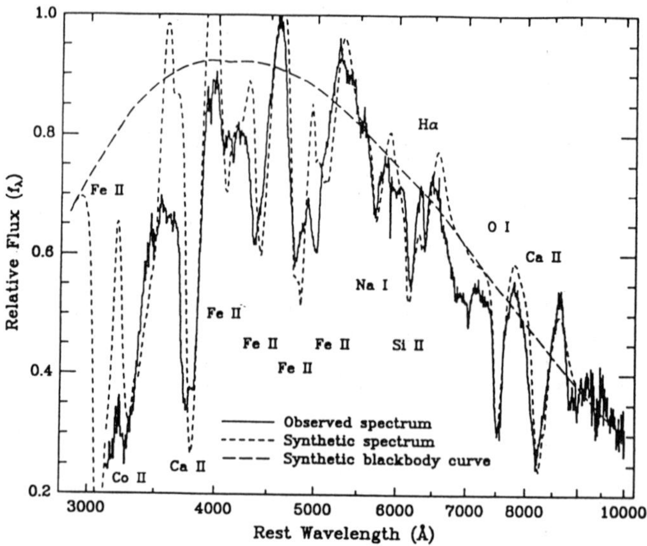

Fig. 1: The observed spectrum of SN Ic 1987M compared with a synthetic spectrum (Jeffery *et al.* 1991).

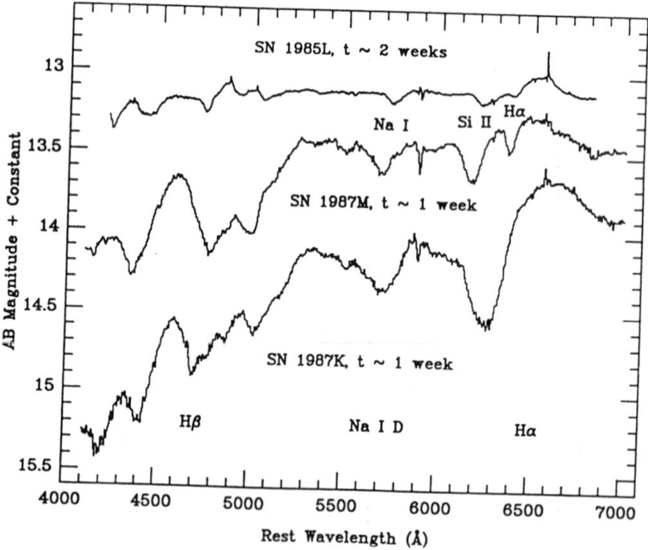

Fig. 2: Comparison of the observed spectra of SN Ic 1987M, SN IIb 1987K, and SN II 1985L (Filippenko *et al.* 1990; Jeffery *et al.* 1991).

2.1. Binary Evolution

The difference in the spectral feature between SNe Ic and Ib may be due to the presence of a thin envelope of hydrogen in SNe Ic immediately prior to the explosion. By evolving massive stars in close binary systems, we examine whether hydrogen can be left on the helium stars after mass exchange and wind-type mass loss. Followings are some preliminary results for two cases 13A and 18A, where the initial masses of the primary stars are $M_{\rm i} = 13\ M_\odot$ (13A) and $18\ M_\odot$ (18A), and their Roche lobe radii are $R_{\rm Roche} = 50\ R_\odot$ (Yamaoka and Nomoto 1991).

After hydrogen exhaustion, the star undergoes Roche lobe overflow forming a helium star of $\sim 3.4\ M_\odot$ (13A) and $\sim 5\ M_\odot$ (18A). Figure 3 (upper) shows the composition structure at the onset of mass transfer during core helium burning (upper) and after the Roche lobe overflow (lower). It is seen that significant amount of hydrogen remains in a relatively thick layer below the surface ($0.5\ M_\odot$ for 13A and $1\ M_\odot$ for 18A).

Whether such a hydrogen layer will be further lost from the helium-rich star depends on the Roche lobe radius and wind mass loss rate. If the helium star is detached from the Roche lobe during helium burning, the star loses its masses in a wind. If the mass loss rate depends on the mass such as $\dot{M} \propto M^{2.5}$ (Langer 1989), it may lead to a difference of the surface abundance between 13A and 18A. For case 13A hydrogen still may remain in the layer down to $\sim 0.2\ M_\odot$ below the surface, whereas for case 18A all hydrogen may be lost in a wind (Yamaoka and Nomoto 1991). If a common envelope forms or the Roche lobe radius has become small enough for the mass and angular momentum to be lost from the system, then all hydrogen will be lost from the star.

Although more parameter study is needed, the present results suggest that more hydrogen may remain on the stellar surface if the helium star had initially smaller main-sequence mass. Also more compact binary systems may lose more hydrogen through common envelope evolution.

2.2. Nucleosynthesis and Radioactivity

Assuming that the helium star progenitors of SNe Ib/Ic are formed in close binary systems as described above, Shigeyama *et al.* (1990) performed hydrodynamical calculations of the explosion of helium stars with masses $M_\alpha = 3.3$, 4, 6, and $8\ M_\odot$. These are presumed to form from the main-sequence stars of masses $M_{\rm i} \sim 13$, 15, 20, and $25\ M_\odot$, respectively. These stars eventually undergo iron core collapse as in SNe II. A shock wave is then formed at the mass cut that divides the neutron star and the ejecta.

Behind the shock wave that propagates outward, materials are processed into nuclear statistical equilibrium (NSE) composition, mostly ^{56}Ni, if the maximum temperature exceeds 5×10^9 K. As derived from the approximate relation $E = 4\pi r^3/3\ aT^4$ with E being the final kinetic energy of explosion (e.g., Woosley 1988), such a high temperature is realized in a sphere of radius $\sim 3700\ (E/10^{51}\ {\rm erg})^{1/3}$ km. With E

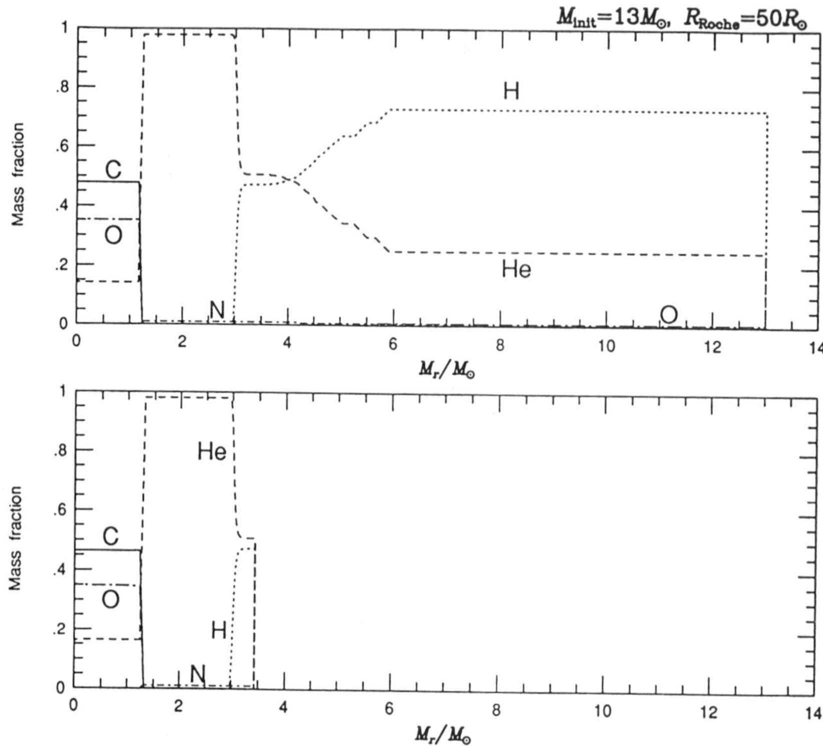

Fig. 3: Change in the composition structure during the binary evolution for the star with $M_i = 13\ M_\odot$ (Yamaoka and Nomoto 1991).

Fig. 4: The expanding supernova matter is excited by the decay of ^{56}Co into ^{56}Fe (H. Nomoto 1989).

= 1×10^{51} erg, this region contains a mass $M_{\rm NSE}$ ($\sim 1.44 - 1.46\ M_\odot$ for $M_\alpha = 3.3$ and $4\ M_\odot$). Then the mass of ^{56}Ni plus neutron-rich iron peak elements is given by $M_{\rm NSE} - M_{\rm NS}$.

The adopted presupernova models (Nomoto and Hashimoto 1988) have the iron core masses as small as 1.18 $M_\odot$ and 1.28 $M_\odot$ for $M_\alpha = 3.3\ M_\odot$ and 4 $M_\odot$, respectively, being significantly smaller than 1.4 $M_\odot$ in the 6 $M_\odot$ star due to the larger effect of Coulomb interactions during the progenitor's evolution. If $M_{\rm NS}$ is approximately equal to the iron core mass, the upper limit to the possible ^{56}Ni masses are obtained as 0.26 and 0.15 $M_\odot$ for $M_\alpha = 3.3$ and 4 $M_\odot$, respectively.

Detailed explosive nucleosynthesis calculations have shown that radioactive ^{57}Ni and ^{44}Ti are also synthesized together with ^{56}Ni by silicon burning and the subsequent α-rich freezeout (Thielemann and Nomoto 1991). The isotopic ratios ^{57}Co/^{56}Co, ^{58}Ni/^{56}Ni, and ^{44}Ti/^{56}Ni are determined from the distribution of neutron excess η near the bottom of the ejecta, which depends on the extent of convective burning shell at various stages, thereby being subject to uncertainties. With these uncertainties taken into account, the masses of several radioactive species are summarized in Table 1.

The masses of oxygen produced in the outer layers are 0.21 and 0.43 $M_\odot$ for M_α = 3.3 and 4 $M_\odot$, respectively. These masses could be consistent with those inferred from the late time spectra of SNe Ib/Ic in view of the strong dependence of the oxygen mass on the temperature of the ejecta (e.g., Uomoto 1986).

Table 1. Radioactivities from SNe Ib/Ic

$M_{\rm i}$ ($M_\odot$)	$\sim$ 13	$\sim$ 15	$\sim$ 20	$\sim$ 25
M_α ($M_\odot$)	3.3	4	6	8
^{56}Ni ($M_\odot$)	0.15 – 0.25	0.1 – 0.15	$\sim$ 0.07	
^{57}Ni ($10^{-3} M_\odot$)	5 – 12	2.5 – 7	1.7 – 3.4	
^{44}Ti ($10^{-4} M_\odot$)	1.3 – 3.6	0.9 – 2.2	0.6 – 1.0	
^{26}Al ($10^{-6} M_\odot$)	1.8	2.7	6.8	9.7
^{22}Na ($10^{-7} M_\odot$)	0.9	0.3	1.3	2.6

2.3. Light Curve Models

Figure 5 shows the observed bolometric light curves of SNe Ia 1972E and 1981B (Graham 1987), SN Ib 1983N (Panagia 1987), and the approximate bolometric light curve of SN Ic 1987M constructed from flux-calibrated spectra (Filippenko *et al.* 1990; Nomoto *et al.* 1990). In each case, the observed light curve has been shifted

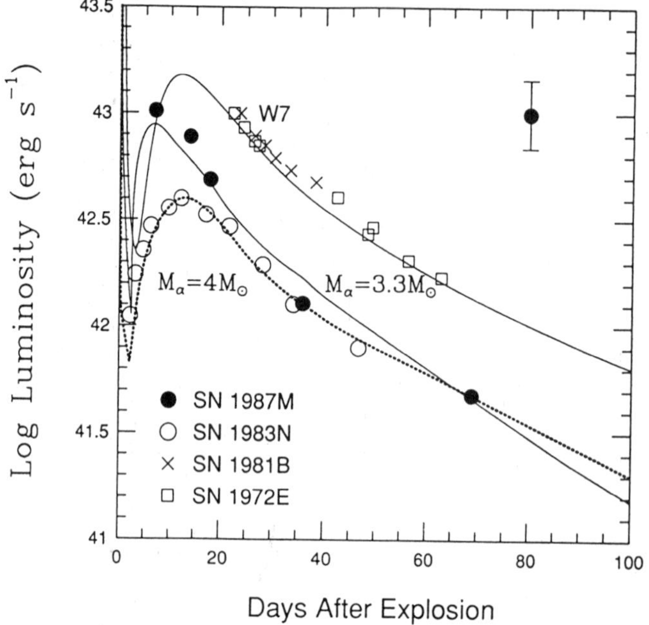

Fig. 5: Approximate bolometric light curve of SN Ic 1987M, and the bolometric light curves of SNe Ia 1972E and 1981B and of SN Ib 1983N. The predicted curves of the 3.3 $M_\odot$ model for SN 1987M, the 4 $M_\odot$ model for SN Ib, and the W7 model for SN Ia are indicated by solid and dotted lines. The error bar illustrates the 2σ photometric uncertainty in the SN 1987M points.

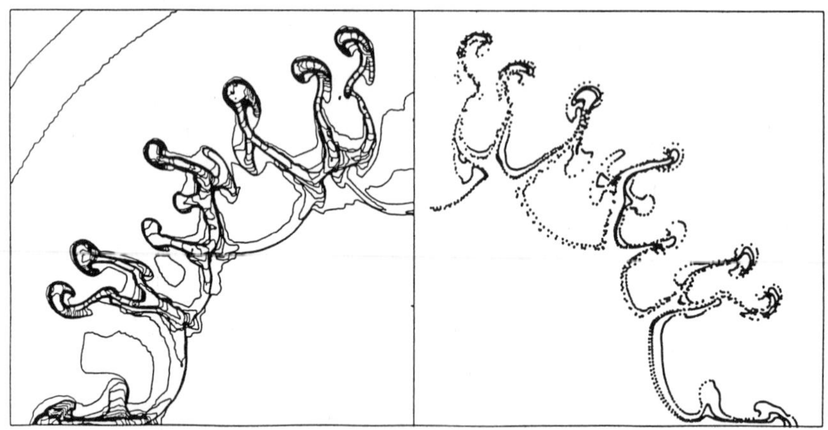

Fig. 6: Rayleigh-Taylor instabilities in the exploding helium star of $M_\alpha = 3.3\ M_\odot$ (t = 180 s). Shown are the density contour map (*left*) and the marker particles at the composition interfaces, He/C+O, O/Si, and Ni/Si from the outerside (*right*) (Hachisu *et al.* 1991).

along the abscissa to match the corresponding theoretical curve. The peak bolometric luminosities assume $H_0 = 60$ km s^{-1} Mpc^{-1}.

The previous Wolf-Rayet star models have some difficulties (1) in reproducing the light curves of typical SNe Ib which decline as fast as SNe Ia (Panagia 1987), and (2) in producing enough ^{56}Ni to attain the maximum luminosities of SNe Ib in relatively low mass helium star models (Ensman and Woosley 1988). In particular, Figure 5 demonstrates two important features of SN Ic 1987M: (1) its brightness fell somewhat more rapidly than that of SNe Ia and SN Ib 1983N (also SN Ic 1983I reported by Tsvetkov 1985), and (2) maximum brightness of 1987M is significantly brighter than in SN Ib 1983N though correction for extinction involves uncertainty in brightness of 1987M (Jeffery *et al.* 1991).

Figure 5 also shows the calculated bolometric light curves of the exploding helium star models with $M_\alpha = 3.3\ M_\odot$ for SN Ic and 4 $M_\odot$ for SN Ib as well as the white dwarf model W7 for SNe Ia (Nomoto *et al.* 1984). The amount of ^{56}Ni is 0.58 $M_\odot$ (W7), 0.26 $M_\odot$ (SN Ic), and 0.15 $M_\odot$ (SN Ib). The helium star models assume uniformly mixed distribution of elements from the center through the layer at 0.2 $M_\odot$ beneath the surface for both cases. Such a mixing may be due to the Rayleigh-Taylor instability during the explosion (Fig. 6; Hachisu *et al.* 1991).

The calculated bolometric light curves of helium stars are powered by the radioactive decays of ^{56}Ni and ^{56}Co (Fig. 4: H. Nomoto 1989). Maximum brightness is higher if the ^{56}Ni mass is larger and the date of maximum earlier. After the peak, the optical light curve declines at a rate that depends on how fast γ-rays from the radioactive decays escape from the star without being thermalized, thereby declining faster if the ejected mass is smaller and if ^{56}Ni is mixed closer to the surface.

The resulted bolometric light curves of $M_\alpha = 3.3\ M_\odot$ and 4 $M_\odot$ are in good agreement with SN Ic 1987M and SN Ib 1983N, respectively. Compared with the 4 $M_\odot$ model, the light curve of the 3.3 $M_\odot$ model is brighter at maximum and declines faster due to the smaller ejected mass, just as observed in SN Ic 1987M.

3. γ-RAY LINES FROM TYPE IB/IC SUPERNOVAE

Table 2 summarize the progenitor's mass $M_{\rm p}$, the ejecta mass $M_{\rm ej}$, the ^{56}Ni mass $M_{\rm Ni}$, the mass included between the surface and the ^{56}Ni layer after mixing $\Delta M_{\rm mix}$, explosion energy $E_{\rm expl}$, and 847 keV line flux at maximum for W7 (SN Ia), helium stars of 3.3 $M_\odot$ (SN Ic), 4 $M_\odot$ (Ib), and 6 $M_\odot$ (Ibw). For flux, distance to the source is assumed to be 1 Mpc for comparison with the previous calculations (e.g., The *et al.* 1991).

Hard X-ray and γ-ray spectra are calculated with Monte-Carlo simulation method as has been made for SN 1987A (Kumagai *et al.* 1988, 1989, 1991b). The calculated spectra as a function of time are shown in Fig. 7. Figure 8 shows the light curves of γ-ray lines (847 keV and 1237 keV) of SNe Ic, Ib, and Ibw. The light curve shapes are basically similar except for the maximum flux which depends on the mass of ^{56}Ni. Mixing of ^{56}Ni to the surface layer makes the maximum earlier ($\sim$ 100 d) than the

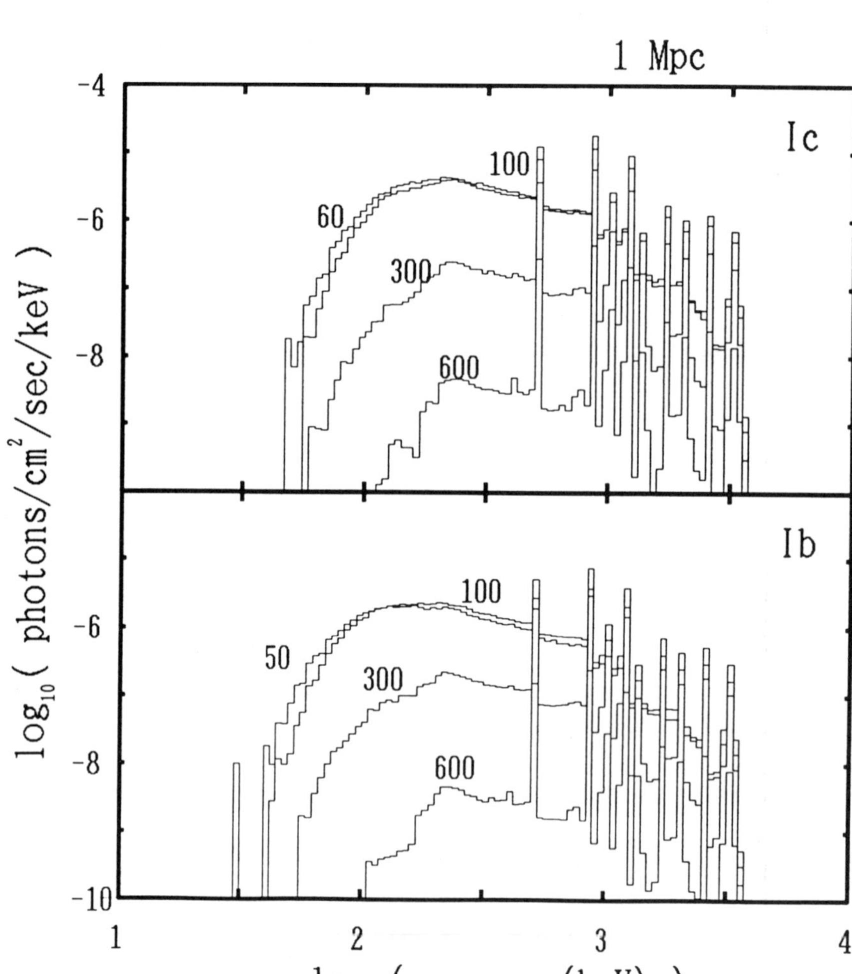

Fig. 7: Calculated hard X-ray and γ-ray spectra due to the decay of ^{56}Co for the exploding helium stars of 3.3 $M_\odot$ (upper: SN Ic) and 4.0 $M_\odot$ (lower: SN Ib) at a distance of 1 Mpc.

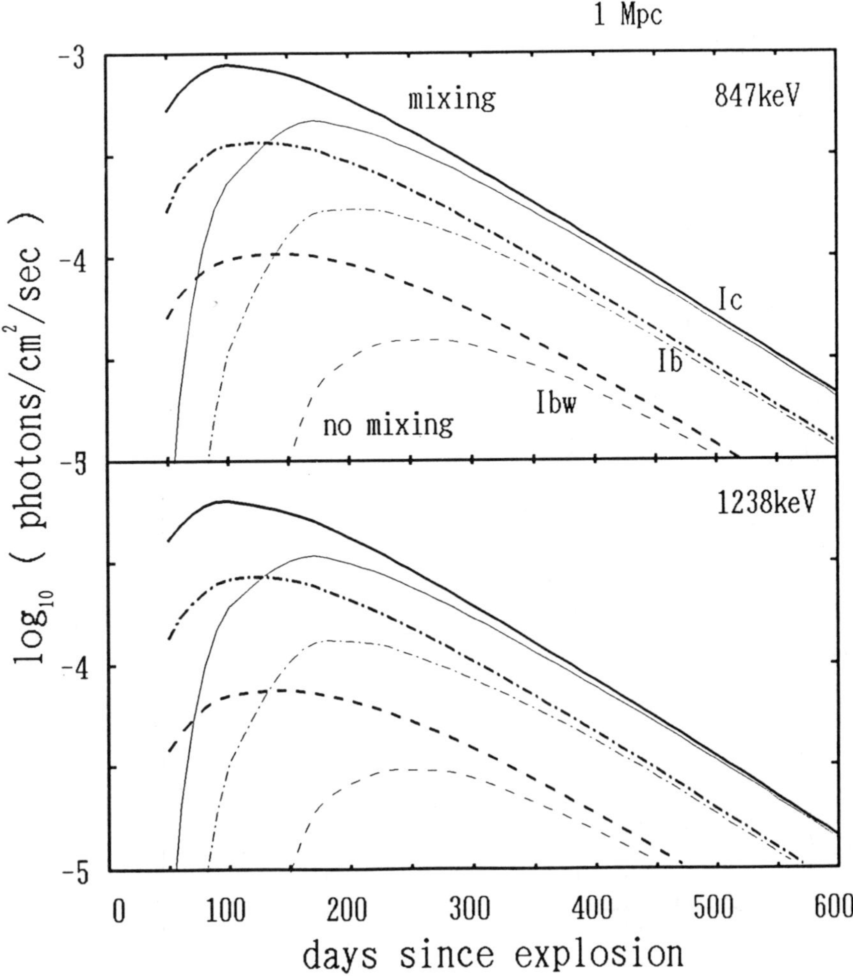

Fig. 8: Calculated light curves for 847 keV and 1238 keV lines for SN Ic, Ib, and Ibw with mixing of ^{56}Ni (thick lines) and without mixing (thin lines) at a distance of 1 Mpc.

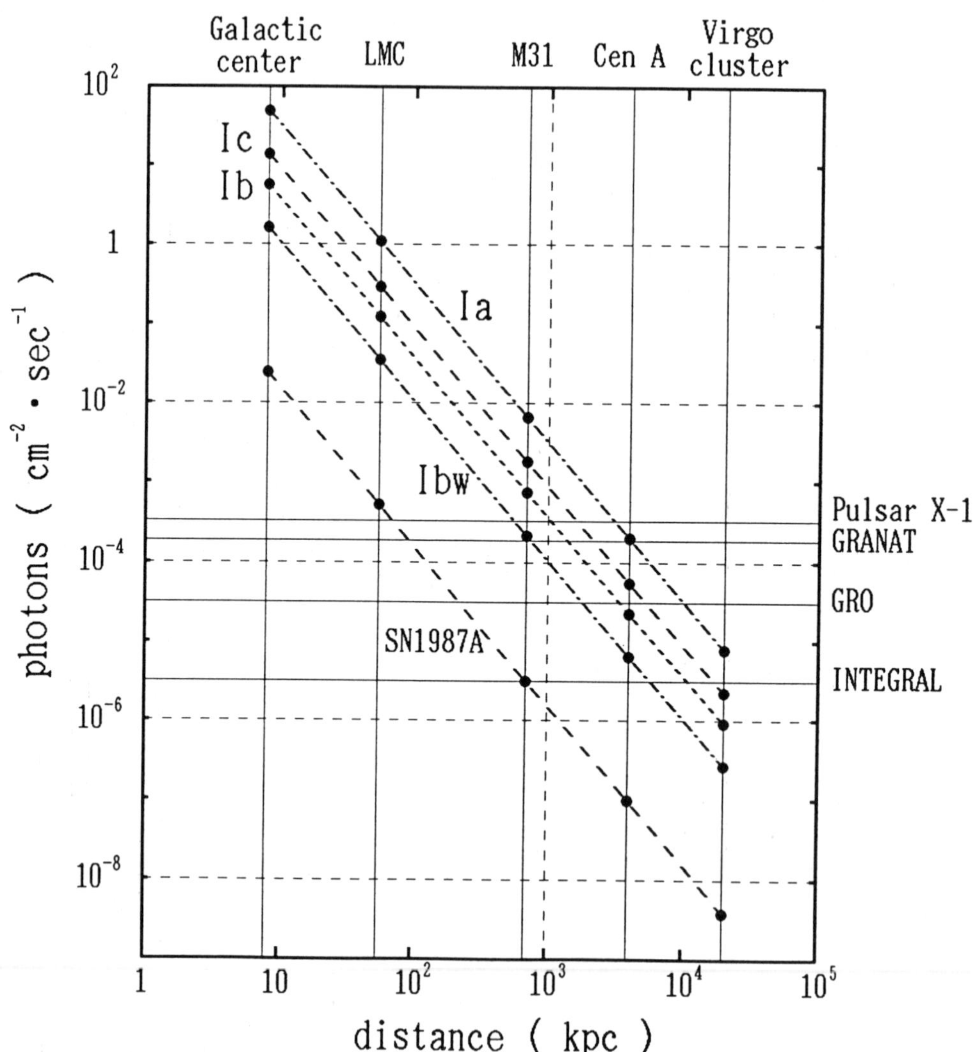

Fig. 9: Fluxes of γ-ray lines from SN Ia, Ib, Ibw, Ic, and SN 1987A at maximum as a function of distance. Their detectability with future missions are summarized.

unmixed case ($\sim$ 170 d), thereby leading to the higher maximum flux by a factor of $\sim 2 - 2.5$.

Figure 9 shows the detectability of these γ-ray lines as a function of distance as compared with SN 1987A (SN II) (adapted from Burrows and The 1989). GRANAT may marginally can detect γ-ray lines from SN Ia and GRO could detect those from SNe Ic at a distance of Cen A. With INTEGRAL, detection of γ-rays from SN Ic in Virgo may be marginal.

It should be noted that the mass of ^{56}Ni from SN Ic in Table 2 must be regarded as an upper limit because the neutron star mass would be significantly larger than the iron core mass. Maximum brightness of SN Ic 1987M involves also uncertainty of extinction.

Table 2. γ-Ray Lines from SNe I

	M_p ($M_\odot$)	M_{ej} ($M_\odot$)	M_{Ni} ($M_\odot$)	ΔM_{mix} ($M_\odot$)	E_{expl} (erg s^{-1})	Flux (mixed) (γ cm^{-2}s^{-1})	Flux (no mix) (γ cm^{-2}s^{-1})
SN Ia ...	1.4	1.4	0.58	0.6	1.3×10^{51}	3.2×10^{-3}	
SN Ic ...	3.3	2.1	0.26	0.2	1.0×10^{51}	8.8×10^{-4}	4.7×10^{-4}
SN Ib ...	4.0	2.7	0.15	0.2	1.0×10^{51}	3.7×10^{-4}	1.7×10^{-4}
SN Ibw ..	6.0	4.4	0.07	0.3	1.0×10^{51}	1.0×10^{-4}	3.9×10^{-5}

4. SN 1987A

4.1. γ-Ray Lines and Hard X-Rays

γ-ray lines predicted to emerge from supernovae have been observed for the first time from SN 1987A. Figure 10 shows the comparison between the calculated γ-ray line light curves and observations with SMM (Leising and Share 1990). The adopted explosion model is 14E1 (Shigeyama and Nomoto 1990) which has the ejecta mass $M = 14.6\ M_\odot$ (4.4 $M_\odot$ core material plus 10.2 $M_\odot$ hydrogen-rich envelope) and explosion energy $E = 1 \times 10^{51}$ erg. The agreement is reasonably good for 3 lines if ^{56}Co is mixed to the outer layers with expansion velocities as high as $v_{exp} \sim 4000$ km s^{-1}, though a slightly flatter distribution of the mixed elements than adopted in Shigeyama and Nomoto (1990) would result in a better agreement than in Fig. 10.

Mixing of ^{56}Co up to $v_{exp} \sim 3000$ km s^{-1} is also suggested in modeling the X-ray light curve at various band (e.g., Sunyaev *et al.* 1990). In particular, the 16 - 28 keV light curve observed with *Ginga* (Inoue *et al.* 1991) can be reproduced if mixing produces *chemically inhomogeneous clumps* (Fig. 11; Kumagai *et al.* 1989).

2D simulations of Rayleigh-Taylor instabilities have shown mixing through $v_{exp} \sim$ 2500 km s^{-1}; further studies of the source of perturbations, 3D effects, and inverse

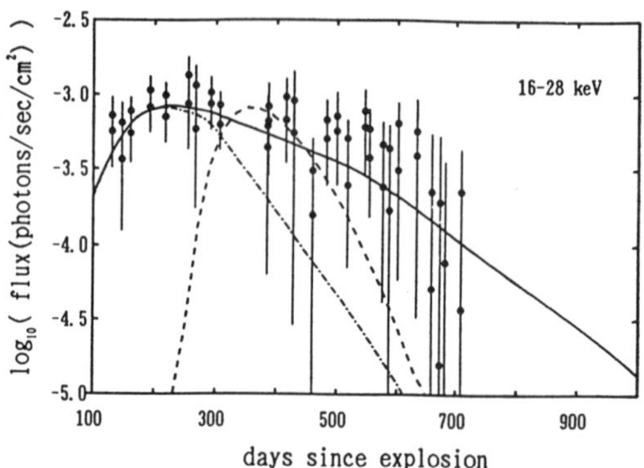

Fig. 10: Calculated X-ray (16 - 28 keV) light curve with 0.076 $M_\odot$ ^{56}Co and 0.0043 $M_\odot$ ^{57}Co. The hydrodynamical model 14E1 with mixed abundance distribution is adopted. Three curves show different degree of clumpiness in the supernova ejecta, i.e., cases with *mixing and clumps* (solid curve), *mixing without clumps* (dash-dotted), and *no mixing* (dashed). The X-ray observations by Ginga is shown

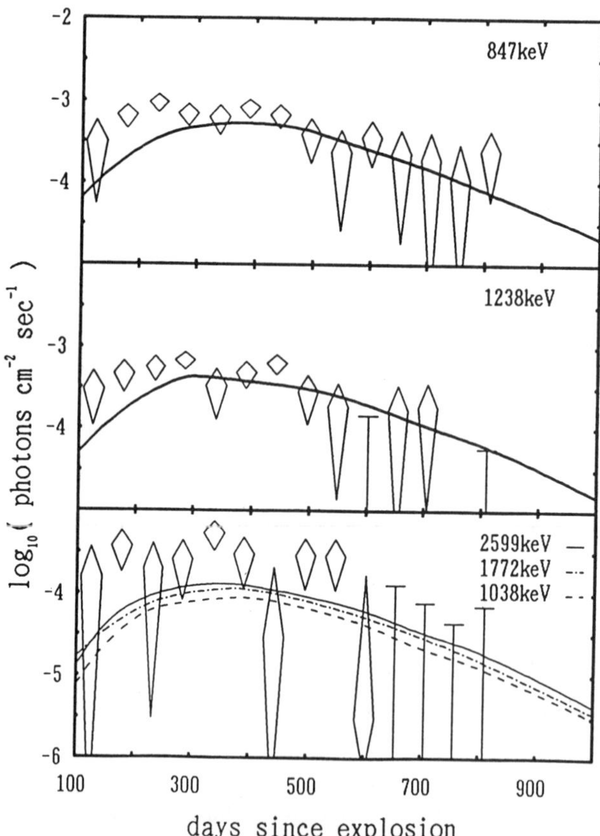

Fig. 11: Calculated γ-ray light curves for 847 keV, 1238 keV, and other lines for the same model as in Fig. 10. For comparison, gamma-ray observations by SMM are shown by the diamonds.

cascading growth are necessary (see, e.g., Arnett 1991; Benz 1991; Hachisu *et al.* 1990).

4.2. Recent Optical Light Curve

The bolometric light curve of SN1987A has recently shown an interesting behavior in its decline (Fig. 12). For the first $\sim$ 800 days, the ESO bolometric light curve (Bouchet *et al.* 1991) is in good agreement with the model 14E1 which includes only the decays of ^{56}Co and ^{57}Co (solid line in Fig. 12). Afterwards the ESO light curve shows the leveling off during $\sim$ day 900 – 1070 with the following decline (Bouchet *et al.* 1991). The bolometric light curve obtained by CTIO (Suntzeff *et al.* 1991) is somewhat below the ESO curve but still its decline rate has been significantly slowed down since $\sim$ day 900.

Here we examine the two possible source of extra heating other than ^{56}Co: (1) radioactive decays of ^{57}Co and ^{44}Ti, and (2) a central magnetized neutron star as either an isolated pulsar or an accreting X-ray pulsar (Kumagai *et al.* 1991a).

4.3. Contributions of ^{57}Co and ^{44}Ti

Explosive nucleosynthesis calculations have shown that radioactive ^{57}Ni and ^{44}Ti are synthesized together with ^{56}Ni by silicon burning and the subsequent α-rich freezeout (e.g., Woosley 1991). The half lives of ^{57}Co and ^{44}Ti are so long as 271 d and 47 yr, respectively, that their decays could dominate the late light curve.

We first calculate the light curve to see how much ^{57}Co is necessary to account for the observed light curve. We then examine how the isotopic ratios depend on the neutron excess near the bottom of the ejecta and to what extent these are constrained from observations. Nucleosynthesis products in the ejecta is taken from Nomoto *et al.* (1991) which contains 0.073 $M_\odot$ ^{56}Ni, 0.0031 $M_\odot$ ^{57}Ni, and 9.2×10^{-5} $M_\odot$ ^{44}Ti. With respect to the solar isotopic ratios, this standard case has $<^{57}\text{Co}/^{56}\text{Co}> \equiv [X(^{57}\text{Ni})/X(^{56}\text{Ni})] \;/\; [X(^{57}\text{Fe})/X(^{56}\text{Fe})]_\odot = 1.7$, $<^{58}\text{Ni}/^{56}\text{Ni}> \equiv [X(^{58}\text{Ni})/X(^{56}\text{Ni})] \;/\; [X(^{58}\text{Ni})/X(^{56}\text{Fe})]_\odot = 1.2$, and $<^{44}\text{Ti}/^{56}\text{Ni}> \equiv [X(^{44}\text{Ti})/X(^{56}\text{Ni})] \;/\; [X(^{44}\text{Ca})/X(^{56}\text{Fe})]_\odot = 1.0$.

Figure 13 shows the calculated UVOIR light curve which is powered by the decays of ^{56}Co (0.073 $M_\odot$) and ^{57}Co with various $<^{57}\text{Co}/^{56}\text{Co}>$ ratios. Here the solid line is the standard case of $<^{57}\text{Co}/^{56}\text{Co}> = 1.7$. The CTIO curve declines consistently with the decay rate of ^{57}Co if $5 \lesssim <^{57}\text{Co}/^{56}\text{Co}> \lesssim 10$ for 14E1 and $8 \lesssim <^{57}\text{Co}/^{56}\text{Co}> \lesssim 16$ for 11E1. For 11E1, which is the same as 14E1 except for the 6.7 $M_\odot$ hydrogen-rich envelope (Shigeyama *et al.* 1988), more γ-rays escape without being thermalized because of larger E/M than in 14E1.

The ^{57}Co/^{56}Co ratio in the ejecta has been estimated from the spectroscopic observations as $<^{57}\text{Co}/^{56}\text{Co}> \lesssim 1.7$ (Danziger *et al.* 1990) and 1 – 2 (Varani *et al.* 1990). Hard X-ray continuum observations with HEXE has given the estimates of $<^{57}\text{Co}/^{56}\text{Co}> \lesssim 1.5$ at $\sim$ day 800 (Sunyaev *et al.* 1990).

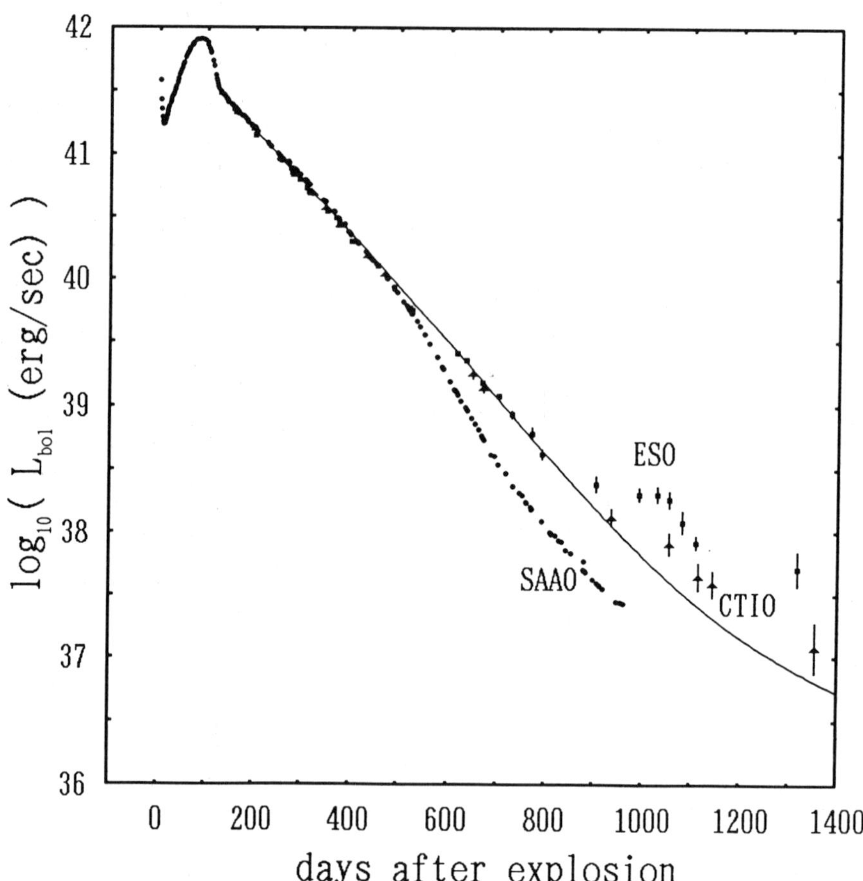

Fig. 12: The calculated UVOIR light curve for the standard abundances of 14E1 (solid line) compared with the U to M SAAO light curve and the UVOIR light curves observed at ESO and CTIO.

If we take the constraint of $<^{57}\mathrm{Co}/^{56}\mathrm{Co}> \lesssim 2$, we may tentatively conclude that the excess brightness of the recent light curve could not be accounted for by the decay of ^{57}Co. In view of possible uncertainties and the model dependence of the observed $<^{57}\mathrm{Co}/^{56}\mathrm{Co}>$, however, it would be useful to use $<^{58}\mathrm{Ni}/^{56}\mathrm{Ni}>$ as a supplementary constraint on $<^{57}\mathrm{Co}/^{56}\mathrm{Co}>$. The mass of ^{58}Ni has been estimated to be $\sim 0.0022 - 0.003\ M_\odot$ (Rank *et al.* 1988; Witterborn *et al.* 1989; Aitken *et al.* 1988; Meikle *et al.* 1989; Danziger *et al.* 1990) which gives $<^{58}\mathrm{Ni}/^{56}\mathrm{Ni}> \sim 0.7 - 1.0$.

Table 3. Isotopic Ratios

$<^{57}\mathrm{Co}/^{56}\mathrm{Co}>$ ...	1.5	1.7	2.0	2.3	4.4
$<^{58}\mathrm{Ni}/^{56}\mathrm{Ni}>$ ...	1.0	1.2	3.0	4.4	14.0
$<^{44}\mathrm{Ti}/^{56}\mathrm{Ni}>$ ...	0.7	1.0	1.1	1.2	2.1

As mentioned in §2.2, the ratios $<^{57}\mathrm{Co}/^{56}\mathrm{Co}>$ and $<^{58}\mathrm{Ni}/^{56}\mathrm{Ni}>$ depend on the distribution of neutron excess η near the bottom of the ejecta which consists of approximately two layers with $\eta = 0.0026$ and $\eta = 0.0117$ in model 14E1 (Hashimoto *et al.* 1989; Thielemann *et al.* 1990). The location of this change in η is subject to uncertainties since it marks the outer edge of the convective oxygen burning shell (Nomoto and Hashimoto 1988; Thielemann *et al.* 1990). By varying this location, we obtain the relations between these ratios in the ejecta as summarized in Table 3 (see also Woosley 1991 for a wider range of parameters). Since $^{58}\mathrm{Ni}/^{56}\mathrm{Ni}$ is much more sensitive to η than other isotopic ratios, the observed upper mass limit to ^{58}Ni provides other upper limits as $<^{57}\mathrm{Co}/^{56}\mathrm{Co}> \lesssim 2.0$ and $<^{44}\mathrm{Ti}/^{56}\mathrm{Ni}> \lesssim 1.1$.

If $<^{57}\mathrm{Co}/^{56}\mathrm{Co}> \gtrsim 4$ is assumed, $<^{58}\mathrm{Ni}/^{56}\mathrm{Ni}> \gtrsim 10$ which would be inconsistent with the observed upper limit. Lehoucq *et al.* (1990) discussed the possible case of $<^{44}\mathrm{Ti}/^{56}\mathrm{Ni}> \gtrsim 10$, but such a large ratio seems unlikely from the above nucleosynthesis constraints.

4.4. Contribution of the pulsar

In modeling the energy supply from the central neutron star, we assume that the neutron star emits X-rays and γ-rays with a luminosity $L_{X\gamma}$ and a spectrum similar to that of an isolated pulsar or an X-ray pulsar (see Kumagai *et al.* 1989 for details; also Woosley *et al.* 1989). The energy deposited in the ejecta is calculated with the Monte Carlo simulation method.

Figure 14 shows the calculated UVOIR bolometric light curves of 14E1 powered by the pulsar with several $L_{X\gamma}$ in addition to the cobalt decays with $<^{57}\mathrm{Co}/^{56}\mathrm{Co}> = 1.7$. The original pulsar radiation is assumed to have a Crab-like power law spectrum.

It is seen that the flat part of the ESO light curve at day 900 - 1070 can be reproduced if $L_{X\gamma} \sim 2 \times 10^{38}$ ergs s^{-1} for 14E1. The subsequent decline corresponds to the decrease in $L_{X\gamma}$ down to $\sim 0.8 - 1 \times 10^{38}$ ergs s^{-1}. Such a rapid decrease

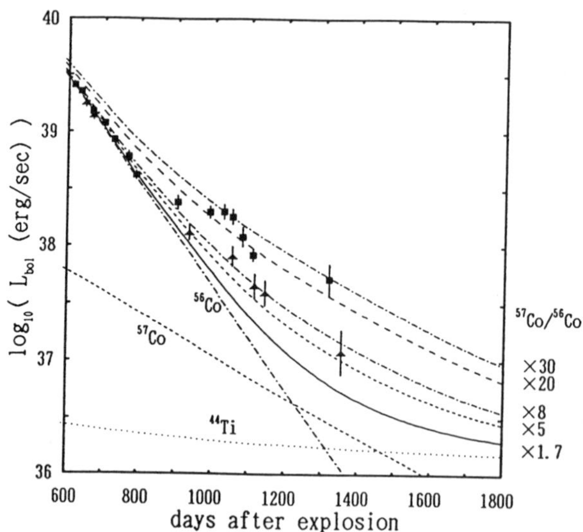

Fig. 13: The calculated UVOIR light curve which is powered by the decays of ^{56}Co (0.073 $M_\odot$), ^{44}Ti (9.2×10^{-5} $M_\odot$), and ^{57}Co with various <^{57}Co/^{56}Co> ratios. The dash-dotted (^{56}Co), dashed (^{57}Co), and dotted (^{44}Ti) lines show the individual contribution for the standard abundances.

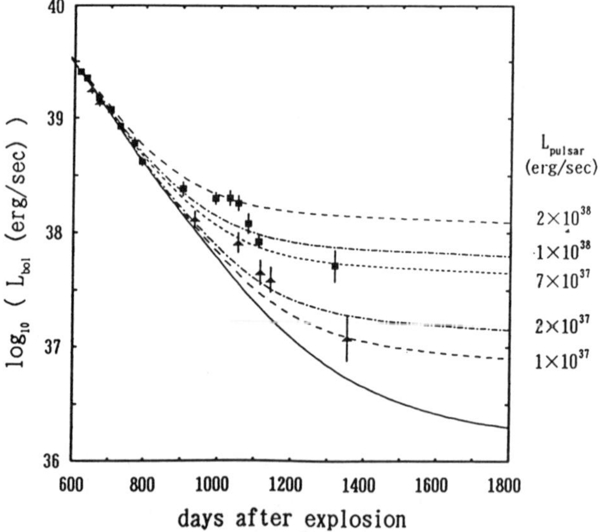

Fig. 14: The calculated UVOIR bolometric light curves of 14E1 powered by the pulsar for several $L_{X\gamma}$ in addition to the cobalt decays with <^{57}Co/^{56}Co> = 1.7. The original pulsar radiation is assumed to have a Crab-like power law spectrum.

has not been predicted by the simple pulsar model and might be related to some instabilities in accreting materials onto the neutron star (Bouchet *et al.* 1991).

The CTIO light curve can be marginally reproduced with the pulsar of $L_{X\gamma} \sim 2 \times 10^{37}$ ergs s^{-1} through day 1350. Afterwards the calculated light curve is predicted to be leveled, so that the next observation is crucial to judge the energy source.

4.5. Predicted γ-ray Lines and Hard X-Rays

Our nucleosynthesis calculations and the spectroscopic constraints on ^{57}Co and ^{58}Ni suggest $<^{57}\mathrm{Co}/^{56}\mathrm{Co}> \lesssim 1.7$ and $<^{44}\mathrm{Ti}/^{56}\mathrm{Ni}> \lesssim 1.1$. Then it is probably less likely that the decay of ^{57}Co alone can supply enough energy to power the recent light curve of SN 1987A. In other words, the buried pulsar would be the more likely energy source, unless the observed luminosity excess is mostly due to infrared echo from circumstellar materials.

However, it is useful to provide a prediction to test the ^{57}Co abundance by the direct observations of line γ-rays with balloon experiments (Durouchoux 1989; Kamae and Takahashi 1991). The predicted light curves of line γ-rays from the decays of ^{57}Co and ^{44}Ti are shown in Fig. 15 (upper) for our standard case. If $<^{57}\mathrm{Co}/^{56}\mathrm{Co}> \gtrsim 5$, the 122 keV line γ-ray flux is predicted to exceed $\sim 2 \times 10^{-4}$ photons cm^{-2} s^{-1} at maximum and 1.5×10^{-4} photons cm^{-2} s^{-1} at $t \sim 1400$ d.

If the observed UVOIR bolometric luminosity is mostly due to the central pulsar, it might be possible to observe X-rays from the pulsar with the current X-ray satellite. The calculated emergent spectrum for the pulsar energy input shows that the soft component below $\sim$ 10 keV is mostly photoabsorbed, so that the most promising energy range to detect pulsar radiation is $\sim$ 20 keV which is covered with the X-ray satellite *Ginga*. Figure 15 (lower) shows the 16 – 28 keV X-ray light curves with the contributions of X-ray pulsar-like (dash-dotted curve: $L_{X\gamma} = 2 \times 10^{37}$ erg s^{-1}) and power-law radiation (dashed curve: $L_{X\gamma} = 2 \times 10^{37}$ and 1×10^{38} erg s^{-1}). It is seen that whether the ejecta will become sufficiently thin for *Ginga* to detect X-rays would be marginal being a competition with the life time of *Ginga*.

We would like to thank H. Yamaoka, T. Tsujimoto, M. Hashimoto, F.-K. Thielemann, A.V. Filippenko, D. Branch, and D. Jeffery for collaborative work on the subjects discussed in this paper. This work has been supported in part by the grant-in-Aid for Scientific Research (01540216, 01790169, 02234202, 02302024) of the Ministry of Education, Science, and Culture in Japan, and by the Japan-U.S. Cooperative Science Program (EPAR-071/88-15999) operated by the JSPS and the NSF.

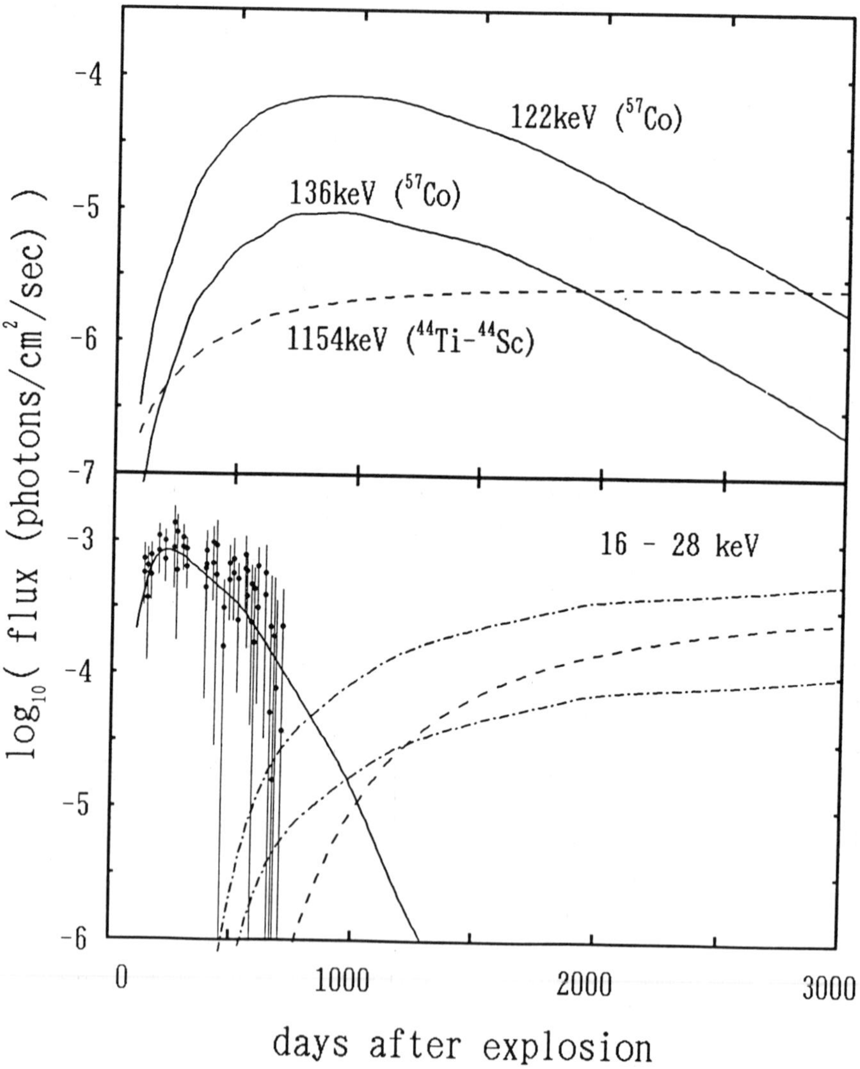

Fig. 15: Light curves of line γ-rays expected from the decays of ^{57}Co and ^{44}Ti for the standard abundances (upper) and the 16 – 28 keV X-rays expected from the Crab-like pulsar (dash-dotted lines for $L_{X\gamma} = 1 \times 10^{38}$ and 2×10^{37} ergs s^{-1}) and the X-ray pulsar (dashed line for $L_{X\gamma} = 2 \times 10^{37}$ ergs s^{-1}). The solid line is due to the cobalt decays and the filled circles are the observed flux with *Ginga*.

REFERENCES

Aitken, D.K., Smith, C.H., James, S.D., Roche, P.F., Hyland, A.R., and McGregor, P.J. 1988, *M. N. R. A. S.*, **235**, 19p.

Arnett, W.D. 1991, in this volume.

Bouchet, P., Danziger, I.J., and Lucy, L. 1991, in *SN 1987A and Other Supernovae*, ed. I.J. Danziger (Garching: ESO), in press.

Branch, D., Nomoto, K., and Filippenko, A.V. 1991, *Comments on Astrophysics*, in press.

Burrows, A. 1991, in this volume.

Durouchoux, P. 1989, private communication.

Ensman, L., and Woosley, S.E. 1988, *Ap. J.*, **333**, 754.

Filippenko, A.V. 1988, *A. J.*, **96**, 1941.

Filippenko, A.V. 1991, *IAU Cir.* No. 5169.

Filippenko, A.V., Porter, A.C., and Sargent, W.L.W. 1990, *A. J.*, **100**, 1575.

Graham, J.R. 1987, *Ap. J.*, **315**, 588.

Hachisu, I., Matsuda, T., Nomoto, K., and Shigeyama, T. 1990, *Ap. J. (Letters)*, **358**, L57.

Hachisu, I., Matsuda, T., Nomoto, K., and Shigeyama, T. 1991, *Ap. J. (Letters)*, **368**, L27.

Harkness, R.P., and Wheeler, J.C. 1990, in *Supernovae*, ed. A. Petschek (Springer-Verlag), p. 1.

Hashimoto, M., Nomoto, K., and Shigeyama, T. 1988, *Astr. Ap.*, **210**, L5.

Inoue, H., Hayashida, K., Itoh, M., Kondo, H., Mitsuda, K., Takeshima, T., Yoshida, K., and Tanaka, Y. 1991, *Pub. Astr. Soc. Japan*, in press.

Jeffery, D., Branch, D., Filippenko, A.V., and Nomoto, K. 1991, *Ap. J.*, submitted.

Kamae, T., and Takahashi, T. 1991, private communication.

Kumagai, S., Itoh, M., Shigeyama, T., Nomoto, K., Nishimura, J. 1988, *Astr. Ap.*, **197**, L7.

Kumagai, S., Shigeyama, T., Hashimoto, M., and Nomoto, K. 1991a, *Astr. Ap.*, in press.

Kumagai, S., Shigeyama, and T., Nomoto 1991b, in *SN 1987A and Other Supernovae*, ed. I.J. Danziger (Garching: ESO), in press.

Kumagai, S., Shigeyama, T., Nomoto, K., Itoh, M., Nishimura, J., and Tsuruta, S. 1989, *Ap. J.*, **345**, 412.

Langer, N. 1989, *Astr. Ap.*, **220**, 135.

Lehoucq, R., Cesarsky, C.J., Cassé, M. 1990, *Astr. Ap.*, submitted.

Leising, M.D., and Share, G.H. 1990, *Ap. J.*, **357**, 638.

Meikle, W.P.S., Allen, D.A., Spyromilio, J., and Varani, G.-F. 1989, *M. N. R. A. S.*, **238**, 193.

Nomoto, H. 1989, *Exploring SN 1987A* (in Japanese), (Tokyo: Kodansha).

Nomoto, K., Filippenko, A.V., and Shigeyama, T. 1990, *Astr. Ap.*, **240**, L1.

Nomoto, K., and Hashimoto, M. 1988, *Physics Reports*, **163**, 13.

Nomoto, K., Shigeyama, T., Kumagai, S., and Yamaoka, H. 1991, in *Supernovae*, ed. S.E. Woosley (Berlin: Springer), p. 176.

Nomoto, K., Thielemann, F.-K., and Yokoi, K. 1984, *Ap. J.*, **286**, 644.

Panagia, N. 1987, in *High Energy Phenomena Around Collapsed Stars*, ed. F. Pacini (D. Reidel), p. 33.

Rank, D.M. *et al.* 1988, *Nature*, **331**, 505

Shigeyama, T., and Nomoto, K. 1990, *Ap. J.*, **360**, 242.

Shigeyama, T., Nomoto, K., and Hashimoto, M. 1988, *Astr. Ap.*, **196**, 141.

Shigeyama, T., Nomoto, K., Tsujimoto, T., and Hashimoto, M. 1990, *Ap. J. (Letters)*, **361**, L23.

Suntzeff, N.B., Phillips, M.M., Depoy, D.L., Elias, J.H., and Walker, A. 1991, *A. J.*, submitted.

Sunyaev, R. *et al.* 1990, *Soviet Astr. Letters*, **16**, 403.

The, L.-S., Clayton, D.D., and Burrows, A. 1991, in *IAU Symposium 143, Wolf-Rayet Stars and Interrelations with Other Massive Stars in Galaxies*, ed. K.A. van der Hucht and B. Hidayat (Kluwer), p. 537.

Thielemann, F.-K., Hashimoto, M., and Nomoto, K. 1990, *Ap. J.*, **349**, 222.

Thielemann, F.-K., and Nomoto, K. 1991, in *Supernovae* (Les Houches, Session LIV), ed. J. Audouze et al. (Elsevier Sci. Publ.), in press.

Thielemann, F.-K., Nomoto, K., and Yokoi, K. 1986, *Astr. Ap.*, **158**, 17.

Tsvetkov, D.Yu. 1985, *Sov. Astr.*, **29**, 211.

Uomoto, A. 1986, *Ap. J. (Letters)*, **310**, L35.

Varani, G.-F., Meikle, W.P.S., Spyromilio, J., and Allen, D.A. 1990, *M. N. R. A. S.*, **245**, 570

Wheeler, J.C., and Harkness, R. 1990, *Rep. Prog. Phys.*, **53**, 1467.

Witteborn, F.C., Bregman, J.D., Wooden, D.H., Pinto, P.A., Rank, D.M., Woosley, S.E., and Cohen, M. 1989, *Ap. J. (Letters)*, **338**, L9.

Woosley, S.E. 1988, *Ap. J.*, **330**, 218.

Woosley, S.E. 1991, in this volume.

Woosley, S.E., Pinto, P.A., and Hartmann, D. 1989, *Ap. J.*, **346**, 395.

Yamaoka, H., and Nomoto, K. 1991, in *SN 1987A and Other Supernovae*, ed. I.J. Danziger (Garching: ESO), in press.

THE NICKEL BUBBLE IN SN1987A

DAVID ARNETT
Steward Observatory and Department of Physics
University of Arizona
and
BRUCE FRYXELL
Steward Observatory and Department of Physics
University of Arizona
and
EWALD MÜLLER
Max Planck Institut für Astrophysik
Garching b. München

ABSTRACT

The status of the study of hydrodynamic instabilities in SN1987A is briefly reviewed, with emphasis upon those aspects relating to the effects of decay of ^{56}Ni and ^{56}Co on the observed velocities, thermal luminosity, and γ and x-ray fluxes.

1. INTRODUCTION

The nature and strength of γ and x-ray radiation for SN1987A and other supernova is sensitive to the distribution of the radioactivities and the structure of the ejected material. To understand these we must understand the instabilities in the supernova explosion itself. Without mixing during the explosion, the new radioactive nuclei would lie beneath the other ejected matter, and so be obscured. The nature of the mixing depends upon the structure of the progenitor star and upon the nature of the explosion itself. Because the newly synthesized ^{56}Ni comes from a deep layer, just outside the "mass cut", its state contains information concerning a major enigma of core collapse supernovae – the explosion mechanism itself.

It might be tempting to regard the mixing process as some sort of "black box", which maps the actual results of a 1D calculation into something that fits some of the observational data. Besides being glib, this approach ignores a wonderful opportunity to learn something entirely new about supernovae.

The challenges are severe. The multi-dimensional modelling of the explosion must be *reliable*. We must be particular about technique, testing and reproducability.

Already we have found subtle problems with method. For example, Fryxell, Müller and Arnett[1] have found that the amount of mixing is a nonmonotonic function of computational resolution. In fact, this implies that lower resolution calculations like those of Hachisu *et al.*[2] do not correctly estimate the degree of mixing. Also, Fryxell, Müller and Arnett[3] find that with further improvement in resolution, a solution results in which the nicely separated "mushrooms" begin to interact and distort into a complex pattern. This is not seen in the SPH results of Herant and Benz[4], which are consistent with with their resolution being comparable to our "moderate" resolution results.

These difficulties should not obscure the significant progress toward a physically well-founded understanding of the hydrodynamics of the SN1987A explosion. The computations of these three groups show that there are robust instabilities with origins in the presupernova structure. The characteristic scale of the instability and its growth from initial ("seed") asymmetries is confirmed. Already this provides interesting confirmation of the basic understanding of the structure of the progenitor star.

2. THE INSTABILITIES

The explosive behavior is complex, and several sorts of instabilities have been found. First is the well known radial instability which causes core collapse. Some other possibile instabilities are mentioned in Arnett, Fryxell and Müller.[5] As the shock moves out past the burning shells, a compressible version of the classical Rayleigh-Taylor (RT) instability develops; this has been the focus of most numerical work.

Low resolution gives rise to numerical errors which act like a dissipation or viscosity, and such errors tend to inhibit growth of an instability. These RT instabilities depend in a relatively sensitive way upon the presupernova structure. Consequently it becomes important to distinguish between these possible causes for differing computational results.

Insofar as we are aware, all calculations beginning from even a vaguely plausible initial model show an RT instability at the H/He interface. Fryxell, Müller and Arnett[1] and Herant and Benz[4] find another RT instability at the CO/He interface for the initial model used by Arnett, Fryxell and Müller.[5] This model was a $15 M_\odot$ star after carbon burning; with no core removed (to be a neutron star) this was thought to be a reasonable representation of the mantle and envelope of a slightly more massive star. Hachisu *et al.*[2], using a different initial model (a $20 M_\odot$ evolved to core instability), find only the instability at the H/He interface. Herant and Benz[4] attribute this to the difference in stellar structure, which is consistent with the results of linear stability analysis (Shigeyama *et al.*[6], Müller, Fryxell and Arnett.[7]

To see if the instability were due to the neglect of the final evolution to core collapse, we have used a $20M_\odot$ initial model (Arnett[8]) after core bounce. This is later than the Hachisu *et al.*[2] initial model, and should be more realistic. Then the inner $1.28M_\odot$ were removed, and an explosion of 2×10^{51} ergs added to the inner $0.1M_\odot$ of the mantle. This produced about $0.09M_\odot$ of ^{56}Ni by explosive burning in the ^{16}O shell, in reasonable agreement with that observed in SN1987A; modifications of the nature of the explosion would alter this yield. The energy was that found appropriate to the early light curve for such a progenitor model (see Arnett[9]; Arnett[10]).

We find two regions of strong instability: (1) the H/He interface, and (2) the He/heavy interface. The He/heavy interface region includes the newly synthesized ^{56}Ni from explosive oxygen burning. We are not yet able to specify the source of this instability more precisely (is it the He/CO region, the Ni region, or ...). Once started, the instability spreads over all all these regions. The two instabilities arise as the reverse shock moves through those regions. Thus the instabilities are at least partially related. Figure 1 shows density contours which illustrate these regions.

3. THE NI INSTABILITY

Because the mean life of ^{56}Ni is short (8.8 days) compared to the diffusion time (months at that epoch) for γ-rays from ^{56}Ni decay, most of the decay energy is thermalized by the local matter. For 2 MeV per decay, this corresponds to $Q \sim 4 \times 10^{16}$ erg g^{-1} of ^{56}Ni , or 7×10^{49} erg$/M_\odot$ of ^{56}Ni . While small compared to supernova energies, this is enough to affect the velocities of slowly moving matter near the mass cut.

To understand the magnitudes involved, let us consider some idealized models. First, assume that an opaque sphere of ^{56}Ni is accelerated by the heat from decay. The root mean square velocity is $\sqrt{2Q} = 2.8 \times 10^3$ km s^{-1}.

This is a bit high because there should be some interaction with matter which is not ^{56}Ni . As a second case, suppose the same opaque sphere, now with an initial outer velocity of 1,000 km s^{-1}, snowplows into homologously expanding matter ($v \propto r$). Then

$$m = m_{Ni}(v/v_o)^3,$$

and

$$\frac{1}{2}m_{Ni}v_0^2 + m_{Ni}Q = \frac{1}{2}mv^2,$$

so that $v_{rms} = 1.5 \times 10^3$ km s^{-1}.

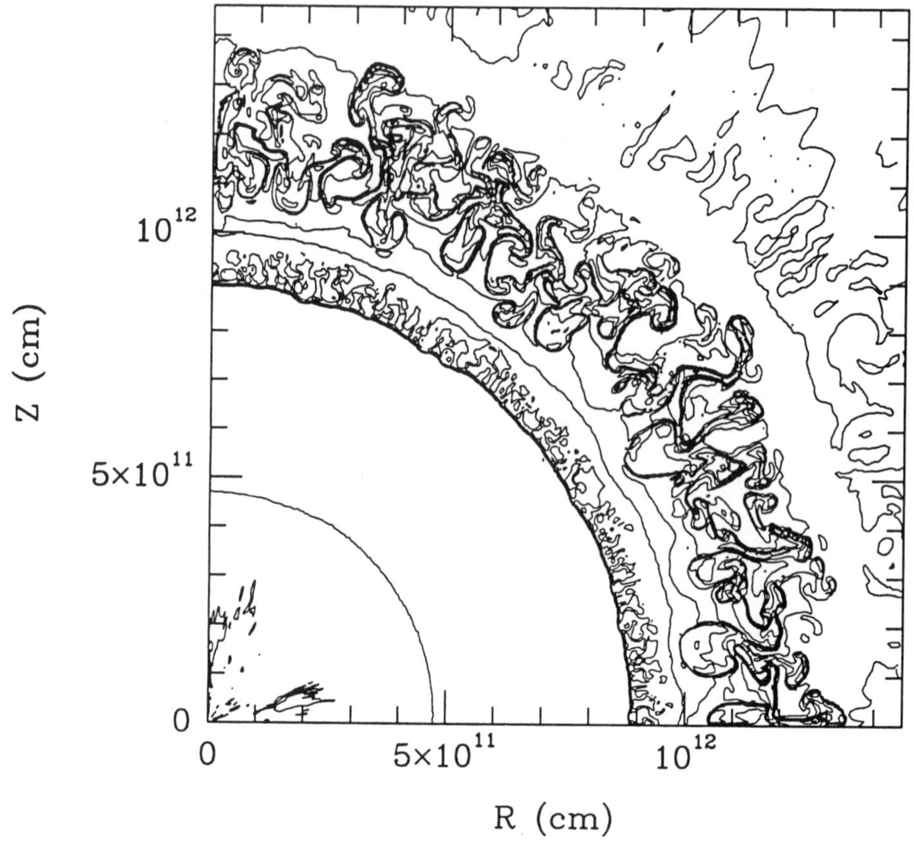

Figure 1. Contours of density at 3.88×10^3 seconds after core collapse. Two regions of instability are evident: at the H/He interface, and at the He/heavy interface. The calculation used a two dimension spherical grid of 500x300 zones.

Observations suggest velocities closer to the first value. To approach case (1) we need:

(a) ^{56}Ni energy to be deposited in the ^{56}Ni itself. This is true, as γ attenuation lengths are short for times $t \leq \tau_{Ni}$.

(b) Holes in the mass distribution of overlying matter for $t \leq \tau_{Ni}$. Something similar may be needed for "early" γ and x-ray luminosity.

(c) A new physical effect which accelerates the ^{56}Ni . An obvious candidate for the source of this effect is the new neutron star, whose rotational energy at least might be available.

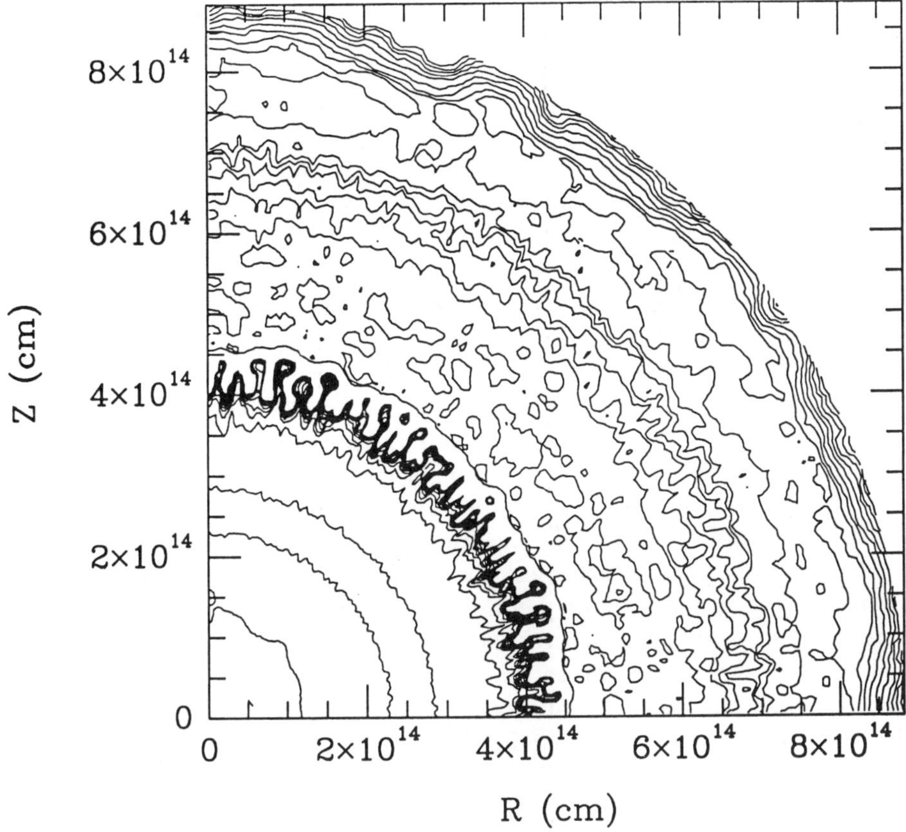

Figure 2. Contours of density for the pure ^{56}Ni model (see text) after 3.54×10^6 seconds. The instability is weak, and this model fails. The calculation used 200x150 zones and a moving grid.

Because of the high degree of interest in the last possibility, it is vital to see if the observational data demand it, by attempting simpler explanations.

The simplest is a pure "Nickel Bubble" model, in which the RT instabilities are deemed irrelevant, and ignored along with possibility (c) above. Figure 2 shows density contours for such a model at a time about 2 months after the core collapse. It was constructed from a 1D shock calculation to 10^5 sec, followed onward in time by a 2D calculation with the PROMETHEUS code (see Fryxell, Müller and Arnett[1]). A low density, hot ^{56}Ni bubble does form. While there is clear evidence for an instability, it is far too weak to break through the snowplow layer, much less give a significantly greater dispersion in ^{56}Ni velocities. Lower velocities result, like the snowplow case above. This model fails.

4. COMBINING RT AND NI INSTABILITIES

Can we reproduce the observations with the combined action of RT and ^{56}Ni instabilities? The RT instability happens on a time scale of hours, not days, after core collapse. It may mix some ^{56}Ni outward to regions of larger velocity. This would give a different "initial" state for the ^{56}Ni decay to work on. Perhaps what we see is the result of a two stage process, with the RT instability acting as a sort of "booster".

This is a more subtle question than that of the nature of the pure ^{56}Ni bubble; in order to begin we must first have the RT instability well in hand. As was mentioned above, that is not yet the case, so that any uncertainties there are propogated into this problem.

For example, low resolution calculations on an Eulerian grid will have numerical diffusion errors which *overestimate* the mixing of ^{56}Ni . However, low resolution calculations of any sort fail to resolve gradients which drive the mixing and breakup of blobs of ^{56}Ni , which *underestimates* the mixing. This is by no means a complete list of the difficulties, but is merely illustrative.

In order to maintain the highest possible resolution, but have sufficient computational efficiency to explore the problem, we have modified PROMETHEUS to use a moving grid. The natural choice is a spherical coordinate system with a radial expansion of the grid so that it moves with the matter. Thus it is, on average, somewhat like a Lagrange grid, but allows for complex vortex motion which a Lagrange grid cannot handle. This gives the "motion with the matter" advantages of Smooth Particle Hydrodynamics (SPH) methods without their loss of resolution.

Preliminary results were shown at the conference; at that time it was not possible to determine whether or not the combined instabilities would give adequate ^{56}Ni velocities (although the numerical results tended to be too low). More complete results are being prepared for publication.

REFERENCES

1. Fryxell, B. A., Müller, E., and Arnett, W. D. 1991a, *Ap. J.* **367**, 619.
2. Hachisu, I., Matsuda, T., Nomoto, K. and Shigeyama, T. 1990, *Ap. J.*, **358**, L57.
3. Fryxell, B. A., Müller, E., and Arnett, W. D. 1991b, submitted.
4. Herant, M. and Benz, W. 1991, *Ap. J. Letters*, in press.
5. Arnett, W. D., Fryxell, B. A. and Müller, E. 1989, *Ap. J.* **341**, L63.
6. Shigeyama, T., Nomoto, K., Tsujimoto, T., and Hashimoto, M., 1990, *Ap. J.*, **361**, L23.
7. Müller, E., Fryxell, B. A., and Arnett, W. D. 1990, in proceedings of the Elba Workshop on *The Chemical and Dynamical Evolution of Galaxies*, ed. F. Ferrini, F. Matteucci and J. Franco, (Casa Editrice Giardini: Pisa).

8. Arnett, W. D. 1991, submitted.
9. Arnett, W. D. 1987, *Ap. J.* **319**, 136.
10. Arnett, W. D. 1988, in *Supernova 1987A in the Large Magellanic Cloud*, ed. M. Kafatos and A. Michalitsianos (Cambridge: Cambridge University Press), p. 301.

HYDRODYNAMICAL INSTABILITIES AND MIXING IN SN1987A: 2D SIMULATIONS OF THE FIRST 3 MONTHS

Willy Benz and Marc Herant
Harvard-Smithsonian Center for Astrophysics
60 Garden Street, Cambridge, MA 02138, USA

Abstract

We present results from numerical simulations of the early stages of the explosion of SN1987A. Using a 2D cylindrical geometry version of a Smooth Particle Hydrodynamics code, we follow the explosion for 3 months to investigate both early hydrodynamical instabilities and the effect of the subsequent radioactive decay of ^{56}Ni and ^{56}Co.

We show that the structure resulting from hydrodynamical instabilities occurring during the first few hours is substantially modified at later time by the radioactive decay of ^{56}Ni and ^{56}Co. The inner cavity of the expanding supernova remnant fills up with nickel and its decay products thus forming a giant "nickel bubble". The peak velocity of the nickel increases by approximately 30% after the decays. While these results adequately model the core of the observed Fe line profiles, they fail to reproduce the high velocity wings of the spectra.

Introduction

Numerical simulations in 2D by Arnett, Fryxell, and Müller [1] and Fryxell, Arnett, and Müller [2] and Hachisu, Matsuda, Nomoto and Shigeyama [3] using realistic models for the progenitor of SN1987A showed the impressive growth of convective fingers during the first few hours following the explosion. These simulations agreed with 1D stability analyses [4,5] showing that the Me/He and He/H interfaces resulting from the evolution of the progenitor are the driving regions of the instabilities. Both groups concluded that these early instabilities are not sufficient to bring a small fraction of nickel to high enough velocities but that the later energy release from radioactive decay of ^{56}Ni and ^{56}Co could possibly achieve the required boost.

We present results from simulations using a 2D cylindrical geometry Smooth Particle Hydrodynamics (SPH) code with 60000 particles. The code will not be described here but is similar to the 3D SPH code described by Benz [7]. We use a realistic equation of state including radiation and gas pressure. Nuclear energy release is treated as a source term in the energy equation since the ejecta are assumed to be optically thick. The internal structure of the progenitor prior to the explosion is taken from [3]. A central explosion (10^{51} ergs) is initiated and evolved using a 1D code to t=300 s. and then transformed into a 2D SPH representation assuming spherical symmetry. We make use of the Lagrangian nature of SPH by increasing the number density of particles in regions of particular interest. We only model a 60 degree wedge centered on the equatorial plane with periodic boundary conditions. The instabilities are seeded by random velocity (± 10%) perturbations introduced behind the shock. Our results show similar dependence on the amplitude of the initial perturbations as in [2] and [3]

Results

To determine the effect of the energy release from the radioactive decay of ^{56}Ni and ^{56}Co we analyzed our results at t=400 min., time at which the instabilities related to the shock propagation are over, and t=90 days when most of the cobalt has decayed. Notice that since we assumed 100% local deposition of radioactive energy our results are really an ***upper limit*** to the effect of the decays, for, in reality, a sizable fraction of this energy may be able to escape.

Figure 1 shows the particle positions at t=400 min. The existence of the two driving regions in the progenitor results in structure inside the main Rayleigh-Taylor "mushrooms". Hence, the densest parts are mainly metals whereas the outer, lower density, regions are essentially helium. Note that hydrogen fills the regions between the fingers and is found with velocities as low as 500 km/s. The peak velocities obtained for the core elements are: 2800 km/s for helium, 2500 km/s for the metals, and 1400 km/s for nickel. All these results are in excellent agreement, both qualitatively and quantitatively with those found by [2] and [3].

Figure 2 shows the particle positions at t=90 days. Not surprisingly, the overall finger-like structure does not change since, away from the nickel, the ejecta are nearly homologously expanding. However, the inner regions are dramatically affected. As radioactive energy is released, the nickel expands, mainly in directions offering the least resistance: inward and sideways. This results in the formation of a large, hot nickel bubble with pieces of crushed fingers forming high density clumps of hydrogen, helium and metals cutoff from the rest of the outer ejecta. Later, the nickel also creeps outward between the convective fingers but due to their low density contrast (factor of a few at most) this process is limited. The peak velocity reached by the nickel is only 1900 km/s. Note also that since the nickel fills the entire inner cavity, the corresponding line profile has a broad core which was not the case at t=400 min. The half width at half maximum of the line obtained from our simulations is $\sim$ 1000 km/s [6], slightly short of the observations (1300 km/s). However, higher velocities could be obtained by increasing the initial explosion energy.

Our simulations fail to produce the small fraction of nickel observed at high velocities by a large amount since only particles beyond the tip of the helium fingers actually travel at $v \sim$ 3500 km/s! The magnitude of this discrepancy makes it unlikely to be the result of inadequate numerical resolution but rather strongly points to the lack of a relevant piece of physics in the models. The most uncertain aspects of the simulations are; 1) the structure of the progenitor, 2) the explosion mechanism and the amount of energy available, 3) differences between 2D and 3D geometry. Further investigations of these points are necessary.

REFERENCES

(1) Arnett, W.D., Fryxell, B., Müller, E., *Ap. J. Lett.* **341**, L63 (1989).
(2) Fryxell, B., Arnett, W.D., Müller, E., *Ap. J.*, in press (1991).
(3) Hachisu, I., Matsuda, T, Nomoto, K., Shigeyama, T., *Ap. J. Lett.*, **358**, L57 (1990).
(4) Ebisuzaki, T., Shigeyama, T., and Nomoto, K.,*Ap. J. Lett.*, **344**, L65 (1989).
(5) Benz, W., and Thielemann, F.-K., *Ap. J. Lett.*,**348**, L17 (1990).

(6) Herant, M., and Benz, W., *Ap. J. Letters*, in press (1991).
(7) Benz, W., 1989, in *Numerical Modeling of Nonlinear Stellar Pulsation*, Ed. J. R. Buchler, (Kluwer Academic Publishers, Dordrecht)

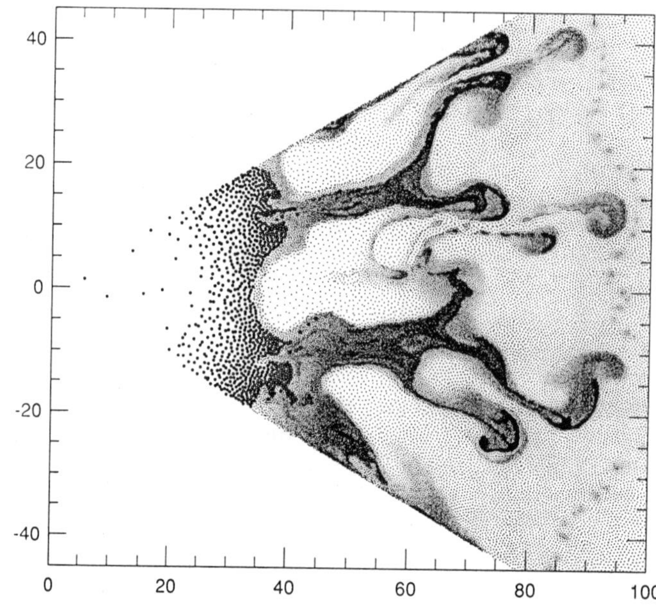

Fig. 1: Particle positions at t=400 min inside 100 $R_\odot$. Nickel particles are plotted as heavy dots.

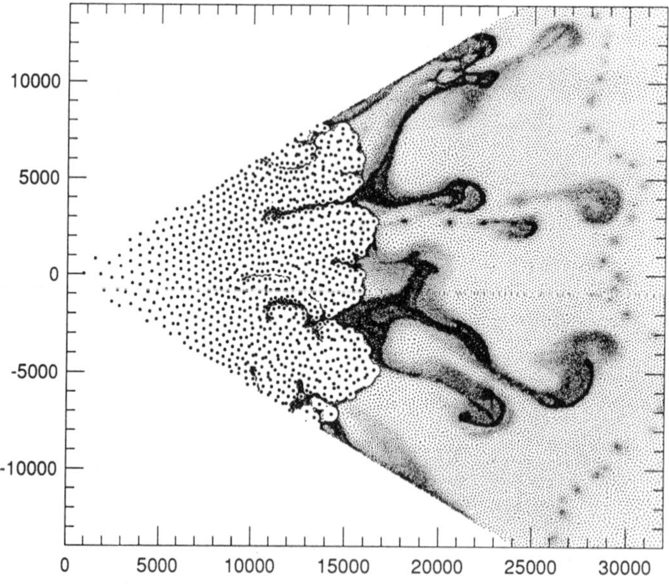

Fig. 2: Particle positions at t=90 days inside 34000 $R_\odot$.

BREMSSTRAHLUNG EMISSION OF THE SN1987A INTERIOR

Lih-Sin The and Donald D. Clayton
Department of Physics and Astronomy, Clemson University, SC 29634, U.S.A.

ABSTRACT

The down-scattering of gamma photons stores a significant fraction of the total energy in recoil electrons. In this paper we present the bremsstrahlung emission of those recoil electrons in SN1987A. If the bolometric luminosity derives from radioactive decay, the bremsstrahlung emissivity must be present, but has not previously been treated. Initially it was our hope to account for the Ginga 16-28 keV measurement of SN1987A or to find a detectable X-ray luminosity near 1 keV for ROSAT. Although this bremsstrahlung emission fails to explain Ginga fluxes, it dominates the X-ray fluxes below 20 keV and can have detectable consequences for the X-ray lines from circumstellar matter. We show our results for day 175, 500, 1000, and 1500.

The gamma-lines observations of SN1987A by the Gamma Ray Spectrometer aboard the SMM satellite and by balloon-borne spectrometer (for a review see Gehrels, Leventhal, and MacCallum[1]) have confirmed the prediction that the light curve of supernovae is powered[2] by the decay chains of $^{56}Ni \rightarrow ^{56}Co \rightarrow ^{56}Fe$ and also could be by $^{57}Ni \rightarrow ^{57}Co \rightarrow ^{57}Fe$ decay chains at later time[3]. The bolometric luminosity, gamma-line light curves, and X-ray luminosity of SN1987A have successfully been modelled by Woosley *et al.*[4], Kumagai *et al.*[5], and The *et al.*[6]. These calculations show that the gamma comptonization does not produce a significant flux below 20 keV. In this paper we present the bremsstrahlung luminosity of W10hmm[4] model of SN1987A. Bremsstrahlung are created as the energetic Compton-recoiled electrons slow down through collisions.

The spectrum and rate of energetic-electron production at each depth within the SN1987A model (W10hmm) is obtained by recording the recoil energy of electrons that collide with gamma photons in the Monte Carlo gamma-ray transport simulation[6]. For the purpose of our approximation, we calculate the time for electron to slow down from its initial kinetic energy E_k to 1 keV and we find it is much smaller than the time evolution of the remnant. We also check the range the electron travels in the ejecta and our analysis shows the electrons move in almost constant density gas, except near the surface where the density drops very steeply. Knowing this, we assume with good approximation that the electrons stop instantly and emit bremsstrahlung emission. Each energetic electron at each depth as it decelerates, emits bremsstrahlung emission and the bremstrahlung is calculated numerically[7] with

$$W^i(\omega_1,\omega_2) = n_e^i(E_k)\int_{\omega_1}^{\omega_2}\int_{\hbar\omega}^{E_k}[dW/d\omega dt]d\omega[dE/(vdE/dx)] \tag{1}$$

and

$$dW/d\omega dt = \sum_Z \hbar\omega(d\sigma/d\omega)vn_Z \tag{2}$$

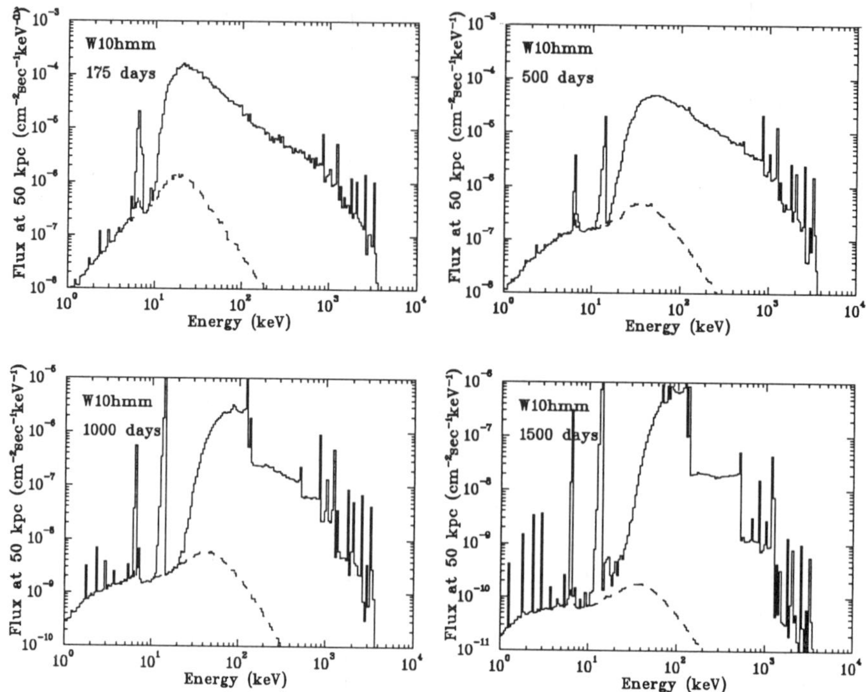

Figure 1. The photon flux (in units of $cm^{-2}s^{-1}keV^{-1}$) at 50 kpc from model 10hmm of SN1987A at day 175, 500, 1000, and 1500. Solid curve is the sum of primary photons from cobalt decays and secondary bremsstrahlung from recoil electrons (dashed curve). Primary photons dominate above 20 keV and bremsstrahlung below 20 keV (except for 14 keV ^{57}Co line and K X-ray lines). Emergent photons are binned logarithmically in 188 bins between 1 keV and 4 MeV with bin boundaries being $E_i(keV)=10^{0.02(i-2)}$ for $i>1$.

where $W(\omega_1, \omega_2)$ is the bremsstrahlung power in the frequency bin (ω_1, ω_2) for each electron of initial kinetic energy E_k and velocity v with electron spectrum $n_e^i(E_k)$ in a certain depth i. This electrons suffer inelastic collisions causing energy loss[8] given by dE/dx and produce bremsstrahlung photons with $d\sigma/d\omega$ being the bremsstrahlung cross section[9] in a gas having number density n_Z of element Z. Summing Eq.(1) over the initial kinetic energy electron spectrum $n_e^i(E_k)$ at each depth, we obtain the bremsstrahlung emissivity of that depth. The bremsstrahlung emissivity varies with energy as $E^{-1.3}$ and is almost depth independent. We compare the bremsstrahlung yield of our calculation for a specific composition at various energies with tabulations of bremsstrahlung yields[8] to ensure the accuracy of our computational prescription, and as it should be, the agreement is very good. The bremsstrahlung yield for W10hmm is of the order 2 x 10^{-3} of the total recoil-electron kinetic energy. This bremsstrahlung production then is injected into a Monte Carlo gamma-ray transport simulation corresponding to the zone where recoil electron was comptonized. We assume the bremsstrahlung to be emitted isotropically and distributed uniformly in that zone.

In Figure 1 we show the emergent bremsstrahlung spectrum (dashed curves) and the total spectrum (solid curves) of model W10hmm of SN1987A for day 150, 500, 1000, and 1500 as flux received at Earth. It shows clearly that accept for K X-ray lines and 14 keV gamma-line of Co^{57} the bremsstrahlung emergence dominates the spectrum below 20 keV. The emergent bremsstrahlung fluxes (not including K X-ray lines) for energy band 1-10 keV are 1.80 x 10^{-6}, 2.50 x 10^{-6}, 1.05 x 10^{-6}, 1.90 x 10^{-7}, 1.29 x 10^{-8}, 1.90 x 10^{-7}, 1.84 x 10^{-9}, and 5.54 x 10^{-10} photons/cm^2/s and for energy band 10-20 keV are 1.03 x 10^{-5}, 6.55 x 10^{-6}, 2.02 x 10^{-6}, 3.23 x 10^{-7}, 2.28 x 10^{-8}, 3.18 x 10^{-9}, 9.18 x 10^{-10} photons/cm^2/s at day 175, 350, 500, 700, 1000, 1250, and 1500 respectively. The bremsstrahlung fluxes in these bands peak around day 250 and their relative ratio to the primary gamma luminosity increases with time. The luminosities in the 0.1 - 2 keV ROSAT sensitive X-ray band are 4.48 x 10^{29}, 7.43 x 10^{28}, 2.64 x 10^{28} ergs/s for day 1000, 1250, and 1500 respectively; they are much less than the ROSAT sensitivity. As faint as our predictions are, they are of unusual interest in that they must be there. Clayton and The[7] show that this low energy X-ray below 20 keV can have a detectable consequence for the K X-ray lines of the circumstellar matter. The bremsstrahlung luminosity we present here is only a lower limit since the electron emitter actually move closer to the surface and radiates in the forward angle rather than isotropically; clumping and ionization can also enhance the low X-ray luminosities. These X-ray luminosities are the smallest SN1987A can have and still be consistent with the theory of the radioactivity-powered light curve.

ACKNOWLEDGMENTS. This research was supported by Naval Research Laboratory Grant N00014-89-J-2034 under the NASA contract for the OSSE Spectrometer DPRS-10987 on Gamma Ray Observatory. We thank Pinto and Woosley for the detailed listing of their model 10hmm. We profitted from several conversations with Adam Burrows.

REFERENCES

1. N. Gehrels, M. Leventhal, and C.J. MacCallum, in AIP Conf. Proc. 170, Nuclear Spectroscopy of Astrophysical Sources, ed. N. Gehrels and G.H. Share (Washington, DC:AIP 1988), p.87.
2. D.D. Clayton, S.A. Colgate, and G. Fishman, Astrophys. J. 155, 75 (1969).
3. D.D. Clayton, Astrophys. J. 188, 155 (1974).
4. P.A. Pinto and S.E. Woosley, Nature 333, 534 (1988).
5. S. Kumagai, T. Shigeyama, K. Nomoto, M. Itoh, J. Nishimura, and S. Tsuruta, Astrophys. J. 345, 412 (1989).
6. L.-S. The, A. Burrows, and R. Bussard, Astrophys. J. 352, 731 (1989).
7. D.D. Clayton and L.-S. The, Astrophys. J. (in press 1991).
8. ICRU Report 37, Stopping Powers for Electrons and Positrons (Bethesda: Int. Commission on Radiation Units and Measurements, 1984).
9. J.M. Jauch and F. Rohrlich, The Theory of Photons and Electrons, 2nd edition (New York/Berlin: Springer Verlag, 1976), p.364.

RADIOACTIVITIES AND GAMMA-RAYS FROM SUPERNOVAE

S. E. Woosley

Board of Studies in Astronomy and Astrophysics
UCO/Lick Observatory, University of California at Santa Cruz
Santa Cruz CA 95064
and
General Studies Group, Lawrence Livermore National Laboratory
Livermore CA 94550

1. PREAMBLE

In the spirit of a workshop, this paper reports recent, mostly unpublished work on several problems relevant to γ-ray astronomy. Certainly the study of γ-ray emission from radioactivities produced by supernovae is a major observational goal of our community. Here I will briefly summarize several calculations, all carried out with collaborators, relevant to estimating the strengths of γ-ray signals from various kinds of explosive events.

The first section (§2) discusses work that graduate student Rob Hoffman and I have done to place limits on the amounts of ^{57}Ni and ^{44}Ti which were produced in SN 1987A and which are expected in other similar Type II explosions. Section 3 discusses recent work with Tom Weaver at Lawrence Livermore National Laboratory on ^{26}Al production in massive stars and supernovae. Section 4 discusses work, also with Weaver, on a "delayed detonation" model for Type Ia supernovae. Section 5 deals with Type Ib supernovae, and in particular a recent model calculated by Norbert Langer of Göttingen, and Weaver, and myself. The last section briefly dwells on a totally different topic, the γ-ray signal (or lack thereof) expected when a bare white dwarf collapses directly to a neutron star. This is work done with Eddie Baron at Oklahoma. I apologize in advance for including so many diverse discussions under one heading, but all seem very relevant to the theme of the meeting.

2. ^{57}Co AND ^{44}Ti IN SN 1987A

As emphasized by others in this volume (Casse, Pinto, and Grebenev), the abundances of ^{57}Co and ^{44}Ti are very important in determining both the late time power budget and γ-ray spectrum of a supernova. The actual abundances of these isotopes depend upon details of both the pre-explosive evolution of the star, which sets the neutron excess in the regions where these species are ultimately produced, and the explosion mechanism, which specifies the temperature-density history and mass of ejected material. However, often simple parameterized nucleosynthesis calculations can be quite useful in placing limits on our expectations based on nuclear systematics. Rob Hoffman and I (*Ap. J. Lettr.*, in press) have recently carried out a series of such calculations to delimit the allowed abundance ratios of ^{57}Co (actually ^{57}Ni in the explosion) and ^{44}Ti to ^{56}Ni. Because this work is published elsewhere, I can summarize the results tersely and dwell more on the reasoning behind and possible loopholes in the argumentation.

We begin with several assumptions. First, the material ejected by SN 1987A in the form of iron-group elements should have ^{56}Ni as its dominant constituent. All models of SN 1987A calculated so far show this to be the case and, moreover, the light curve would have been very different if the amount of ^{56}Ni was much less. Admittedly, one cannot rule out a pathological case in which a great deal of iron was ejected (several tenths of a solar mass or more) with most of it in the form of other isotopes like ^{54}Fe and ^{58}Ni. Such nucleosynthesis would be distinctly non-solar and some form of prompt hydrodynamical mechanism would be required, both of which would fly in the face of current understanding. Perhaps that would be for the best, but for now we assume X(^{56}Ni) $\gtrsim$ 0.5.

Second, we assume that the neutron excess, η, cannot be less than that implied by the initial metallicity of Sk 202-69, about 1/3 solar, and the requirement that the star burn helium before it collapses. In that case the neutron excess cannot be less than that implied by the conversion of all initial CNO isotopes first into ^{14}N, and then into ^{22}Ne by the reaction sequence: $^{14}\mathrm{N}(\alpha,\gamma)^{18}\mathrm{F}(e^{+}\nu)^{18}\mathrm{O}(\alpha,\gamma)^{22}\mathrm{Ne}$. This gives a minimum value of η of about 0.0007. Moreover, the iron producing region must lie in a shell of either silicon or oxygen. If it is silicon, then the material has experienced oxygen burning and its accompanying electron capture. The neutron excess is at least 0.002 regardless of the initial metallicity of the star (Woosley, Arnett, and Clayton 1972; Woosley and Weaver 1982). If it is not silicon, but oxygen, then the material has still experienced carbon and neon burning which usually raise η to about 0.0005 (Arnett and Truran 1969; Woosley *et al.* 1990), even if the initial metallicity is small. All things considered, η = 0.001 seems a reasonable lower bound to the neutron excess in the region where iron is produced. Models for SN 1987A give values more like 0.003 or 0.004.

Armed with these two assumptions, X(^{56}Ni) greater than 50% of the iron group and $\eta \gtrsim$0.001, we can place stringent limits on the ^{57}Ni/^{56}Ni ratio produced by explosive nucleosynthesis and find out some interesting things about ^{44}Ti besides.

Table 1: Alpha-Rich Freeze-Out

T_9	5.5	5.5	5.5	5.5	5.5	5.5	5.5	5.5
ρ	10^7	10^7	10^7	10^7	10^7	10^7	10^7	10^7
η_i	0	0.001	0.002	0.004	0.007	0.01	0.02	0.03
η_f	3.5(-4)	1.0(-3)	2.0(-3)	4.0(-3)	7.0(-3)	1.0(-2)	2.0(-2)	3.0(-2)
X_α	6.3(-2)	6.2(-2)	6.1(-2)	5.9(-2)	5.5(-2)	5.1(-2)	3.9(-2)	2.7(-2)
$X(^{44}Ti)$	1.6(-4)	1.5(-4)	1.4(-4)	1.2(-4)	1.0(-4)	8.3(-5)	3.8(-5)	1.2(-5)
$X(^{56}Ni)$	8.9(-1)	8.7(-1)	8.4(-1)	7.9(-1)	7.0(-1)	6.2(-1)	3.5(-1)	8.5(-2)
$X(^{57}Ni)$	8.3(-3)	2.6(-2)	3.0(-2)	3.6(-2)	4.2(-2)	4.6(-2)	4.8(-2)	2.9(-2)
$X(^{58}Ni)$	7.2(-3)	1.3(-2)	3.8(-2)	8.7(-2)	1.6(-1)	2.4(-1)	5.2(-1)	8.1(-1)
P_{44}^a	0.144	0.138	0.136	0.126	0.117	0.107	0.087	0.117
P_{56}	1	1	1	1	1	1	1	1
P_{57}	0.384	1.24	1.45	1.87	2.46	3.06	5.58	13.8
P_{58}	0.199	0.373	0.910	2.72	5.70	9.53	36.1	234
T_9	5.5	5.5	5.5	5.5	5.5	5.5	5.5	5.5
ρ	10^6	10^6	10^6	10^6	10^6	10^6	10^6	10^6
η_i	0	0.001	0.002	0.004	0.007	0.01	0.02	0.03
η_f	4.2(-4)	1.1(-3)	2.1(-3)	4.0(-3)	7.0(-3)	1.0(-2)	2.0(-2)	3.0(-2)
X_α	3.5(-1)	3.5(-1)	3.5(-1)	3.5(-1)	3.4(-1)	3.4(-1)	3.2(-1)	2.9(-1)
$X(^{44}Ti)$	7.7(-4)	7.5(-4)	7.3(-4)	7.1(-4)	6.7(-4)	6.3(-4)	5.1(-4)	6.9(-5)
$X(^{56}Ni)$	5.9(-1)	5.7(-1)	5.5(-1)	5.0(-1)	4.2(-1)	3.4(-1)	8.3(-2)	1.1(-3)
$X(^{57}Ni)$	1.1(-2)	2.7(-2)	2.9(-2)	3.1(-2)	3.3(-2)	3.3(-2)	2.2(-2)	1.2(-3)
$X(^{58}Ni)$	9.2(-3)	1.1(-2)	3.7(-2)	9.0(-2)	1.7(-1)	2.5(-1)	5.2(-1)	4.7(-1)
P_{44}	1.06	1.07	1.09	1.15	1.29	1.51	4.90	49.0
P_{56}	1	1	1	1	1	1	1	1
P_{57}	0.772	1.96	2.14	2.54	3.21	4.05	10.7	41.7
P_{58}	0.385	0.489	1.68	4.44	9.89	18.2	155	1.0(4)

[a] The production factor here is defined relative to solar (Anders and Grevesse 1989) mass fractions of ^{56}Fe and the final decay product of the given isotope.

Table 2: Ordinary Freeze-Out From NSE

T_9	5.5	5.5	5.5	5.5	5.5	5.5	5.5	5.5
ρ	10^8	10^8	10^8	10^8	10^8	10^8	10^8	10^8
η_i	0	0.001	0.002	0.004	0.007	0.01	0.02	0.03
η_f	1.2(-4)	1.1(-3)	2.1(-3)	4.1(-3)	7.1(-3)	1.0(-2)	2.0(-2)	3.0(-2)
X(^{44}Ti)	2.3(-5)	2.1(-5)	2.0(-5)	1.9(-5)	1.7(-5)	1.3(-5)	9.8(-6)	3.7(-6)
X(^{54}Fe)	9.8(-4)	2.0(-2)	4.0(-2)	8.3(-2)	1.5(-1)	2.1(-1)	4.3(-1)	6.2(-1)
X(^{56}Ni)	8.9(-1)	8.6(-1)	8.3(-1)	7.8(-1)	7.1(-1)	6.3(-1)	3.9(-1)	1.5(-1)
X(^{57}Ni)	3.3(-5)	6.5(-3)	9.3(-3)	1.3(-2)	1.7(-2)	1.9(-2)	2.3(-2)	2.0(-2)
X(^{58}Ni)	5.5(-5)	2.6(-3)	5.8(-3)	1.3(-3)	2.4(-2)	3.7(-2)	8.8(-2)	1.8(-1)
P_{44}	0.021	0.020	0.019	0.019	0.020	0.020	0.020	0.020
P_{54}	0.018	0.379	0.794	1.73	3.42	5.52	18.2	67.6
P_{56}	1	1	1	1	1	1	1	1
P_{57}	0.112	0.313	0.457	0.685	0.977	1.25	2.45	5.50
P_{58}	0.003	0.075	0.169	0.398	0.839	1.43	5.58	19.2

The results summarized in Tables 1 and 2 show that it is virtually impossible for the final iron isotopic ratio, ^{57}Fe/^{56}Fe, to exceed its solar value, 0.024, by more than a factor of 4 unless the dominant iron isotope is not ^{56}Ni, but ^{54}Fe or ^{58}Ni. If one further does not want to overproduce ^{58}Ni by more than a factor of 5, the limit is reduced from 4 to 3. The lower bound on η places a *lower* limit of mass 57 production of about 0.3 and even this value can only be achieved if the α-rich freeze-out was not an important component of nucleosynthesis in SN 1987A. For the most likely value of η, 0.004, the synthesis of mass 57 lies between 0.7 (no α-rich freeze-out) and 2.5 (a strong α-rich freeze-out).

We also note that under no reasonable conditions does the production of ^{44}Ti exceed about twice the value implied by the solar ratio of ^{44}Ca to ^{56}Fe. Even in the case of an extreme α-rich freeze out starting from initial conditions $T_9 = 5.5$, $\rho = 3 \times 10^5$ g cm^{-3}, and $\eta = 0.004$ (not in Tables), a final mass fraction of α-particles of 0.55 still implied a ^{44}Ti production of only 5 times solar. The ratio of ^{57}Ni/^{56}Ni was still only 4 times solar. In this extreme case, ^{56}Ni was only 15% of the ejected material that had achieved nse implying, at a minimum, 0.52 $M_\odot$ of "iron group" material ejected in SN 1987A, far more than any of the models suggest. This is a minimum value since continuity demands some zones of nse without an α-rich freeze out and some zones of incomplete explosive silicon burning also be ejected. Moreover, ^{62}Ni (produced as ^{62}Zn) in this calculation exceeded its solar ratio to ^{56}Fe by a factor of 25.

Thus, though the proof may not be air tight – a very extreme α-rich freeze-out could conceivably have occurred – I stand by my bet with Michel Casse at the meeting. A fine bottle of California white wine says that ^{44}Ti in SN 1987A does not exceed solar by more than a factor of two!

3. NUCLEOSYNTHESIS OF ^{26}Al

The current situation with regard to ^{26}Al synthesis must be regarded as very uncertain, but in my opinion (contrary to that expressed by Prantzos at the meeting), it remains possible, even likely that Type II and Ib supernovae are the major source of the $\sim$2 $M_\odot$ of ^{26}Al inferred for the present abundance in the Milky Way Galaxy. I say 2 $M_\odot$ and not 3 $M_\odot$ because the distance to the center of the Galaxy may be more like 8 kpc than 10 kpc which was assumed in deriving the traditional mass of ^{26}Al implied by the observations. The mass of ^{26}Al might be reduced even more, though not by much, if the ^{26}Al were not concentrated in the center of the Galaxy, but in a ring farther out.

^{26}Al ($\tau_{1/2} = 7.3 \times 10^5$ years) is produced by a number of processes in massive stars: hydrogen burning in the envelope (where the ^{26}Al can be ejected either in the supernova itself or in the wind preceding the formation of a Wolf-Rayet star); explosive neon burning between temperatures of about 2 and 3 $\times 10^9$ K; and by the neutrino process (Woosley *et al.* 1990). All of these except the neutrino process are discussed by Prantzos elsewhere in this volume. In all cases, however, production occurs by the ^{25}Mg(p,γ)^{26}Al reaction and ^{26}Al will be sensitive to, though not necessarily linear in changes in this reaction rate. The recent downward revision of the rate (Iliadis *et al.* 1990) at temperatures appropriate to hydrogen shell burning is therefore highly relevant. At a temperature of 30 to 50 million K, for example, the rate has declined from the value tabulated by Caughlan and Fowler (1988) by a factor of 4.5. On the other hand, at higher temperatures appropriate to neon burning and the ν-process the rate has not decreased. Thus only production in the hydrogen shell is changed.

The numbers given in Table 3 for ^{26}Al synthesis in 15, 20, 25, and 35 $M_\odot$ stars (Woosley and Weaver 1991) include the hydrogen shell component and may now be altered, especially in the lower mass stars, but in cases where substantial ^{26}Al is produced, most of it comes from the heavy element core. The naming of models in the Table follows the convention of giving first the stellar mass in solar masses, then an indication of convective algorithm with "S" meaning a large amount of semiconvection and "N" meaning very little (Woosley and Weaver 1991), and finally a number indicating the factor by which the ^{12}C(α,γ)^{16}O reaction rate (as tabulated by Caughlan and Fowler 1988) was multiplied.

Table 3: ^{26}Al Nucleosynthesis ($M_\odot$) In Massive Stars

Model	15S1	15S2	15S3	15N2	20S2	20N2
Mass	15	15	15	15	20	20
$^{12}C(\alpha,\gamma)^{16}O$	1	2	3	2	2	2
Semiconv.	high	high	high	low	high	low
$10^5 * M(^{26}Al)$	1.2	1.2	1.1	1.0	13.3	2.0

Model	25S1	25S2	25S3	25N1	25N2	25N3
Mass	25	25	25	25	25	25
$^{12}C(\alpha,\gamma)^{16}O$	1	2	3	1	2	3
Semiconv.	high	high	high	low	low	low
$10^5 * M(^{26}Al)$	4.6	4.6	3.8	0.68	2.1	2.6

Model	35S2	35N2	25N2X	25N2Xν	35N2X	35N2Xν
Mass	35	35	25	25	35	35
$^{12}C(\alpha,\gamma)^{16}O$	2	2	2	2	2	2
Semiconv.	high	low	...	...	...	...
$KE_\infty/10^{51}$	...	...	1.5	1.5	1.5	1.5
$10^5 * M(^{26}Al)$	34	17	2.4	4.4	13	18
$M(^{56}Ni)$	...	...	0.08	0.08	0.31	0.31

The first striking thing about the ^{26}Al synthesis reported in Table 3 is that it is highly variable, not even monotonic with stellar mass. Model 20S2 produces a lot more ^{26}Al than either 15S2 or 25S2. The production has a lot to do with the amount of magnesium in the star and this depends on, in addition to the carbon abundance that comes out of helium burning (and thus $^{12}C(\alpha,\gamma)^{16}O$), the location of various convective carbon, neon, and oxygen burning shells (see also Barkat and Marom 1990 for a discussion of non-monotonic behavior in massive star evolution). There are general trends however. More massive stars, especially 35 $M_\odot$ stars make more ^{26}Al. The production does not appear to be too sensitive to the $^{12}C(\alpha,\gamma)^{16}O$ reaction rate itself, but this can be deceiving. Less carbon (from a larger rate for this reaction) means less magnesium and therefore less aluminum synthesis. But a large value for this rate also implies higher entropy in the stellar core and more extensive convective shells – a larger region where ^{26}Al is produced. Extra semiconvection, which means larger carbon-oxygen cores for a given helium core mass, also favors ^{26}Al synthesis.

In two of the models, 25N2 and 35N2, explosion was simulated. A piston was situated at 2500 km in the presupernova model, well outside the iron core, and at a typical distance for which delayed explosions develop (Wilson 1985; Mayle 1985, 1990; Wilson *et al.* 1986; Mayle and Wilson 1990), and given a trajectory such as to produce the ejection of all exterior mass and a total kinetic energy at infinity of 1.5×10^{51} erg. The shock wave was followed through the star and explosive nucleosynthesis determined in each of the stellar zones. An identical calculation was then carried out but including the nucleosynthesis induced by irradiation from neutrinos from the forming neutron star (Woosley *et al.* 1990). A total neutrino energy (all flavors) of 4×10^{53} erg was assumed. The baryon masses interior to the mass cuts in the 25 $M_\odot$ and 35 $M_\odot$ models were 1.72 and 1.86 $M_\odot$. The neutrino burst was assumed to e-fold with a 2 second characteristic time scale, the electron neutrino temperature was 4 MeV and the μ- and τ-neutrino temperatures were 8 MeV. As Burrows emphasized at the meeting, the high energy tails of the neutrino distribution function, to which the neutrino process is sensitive, may not be well represented by a thermal distribution. The parameters adopted here tend to maximize the effect and neutrino-induced nucleosynthesis could, in general, be about a factor of two less efficient than reported. However, the neutrino spectrum tends to get harder at late times and the calculations to which Burrows refers have not yet been carried beyond a few tenths of a second. For now we retain the specified parameterization.

It is interesting that the explosion does not always enhance ^{26}Al synthesis. In Model 35N2X, for example, an explosion without neutrino irradiation, the total ^{26}Al production declined from its presupernova value. The shock heated some portion of the ^{26}Al containing material to such high temperature that it was destroyed. In Model 25N2X on the other hand, the explosion enhanced the ^{26}Al production. In both the 25 and 35 $M_\odot$ models, even those including neutrino irradiation, most of the final abundance of ^{26}Al was created before the explosion in a superheated oxygen-neon shell during the last hour of the star's life. As the iron core begins to contract, even as silicon shell burning is still in progress, this region of the star also contracts to higher temperature and becomes vigorously convective. In a phase that is almost "implosive neon burning" copious ^{26}Al is produced. Obviously the details are sensitive to how time dependent convection is handled and all the other factors affecting the late stages of stellar evolution and the (carbon and) neon abundances. A factor of two error bar on all the numbers in Table 3, either up or down, would not be overly generous.

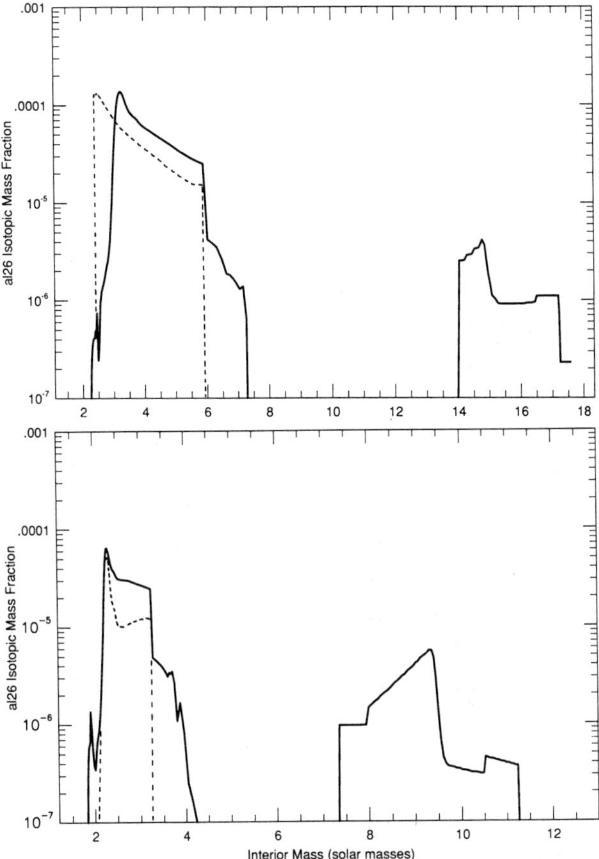

Fig. 1. Distribution of ^{26}Al by mass fraction in a) 25 and b) 35 $M_{\odot}$ model supernovae. The mass of the helium core in the 25 and 35 $M_{\odot}$ models is 9.2 and 14.6 $M_{\odot}$ respectively. The CO-cores were 4.3 and 7.2 $M_{\odot}$. The dashed line shows the abundance before the explosion, the solid line, after.

Figure 1 shows the final ^{26}Al distribution in Models 25N2Xν and 35N2Xν. The production in the hydrogen envelope is obviously negligible compared to that made in the core. With the new rate for ^{25}Mg(p,γ)^{26}Al, it may be even smaller. We expect this to be a general characteristic of very massive stars (M $\gtrsim$ 20 $M_{\odot}$) and, unless such stars fail to explode, the ^{26}Al ejected in the supernova event should greatly predominate over any component ejected in the stellar wind, say while the star was becoming a Wolf-Rayet star. The ^{26}Al distribution in the 25 and 35 $M_{\odot}$ supernova cores largely parallels the distribution of magnesium in these objects. In the 35 $M_{\odot}$ model (Figure 1), the sharp peak at 3.2 $M_{\odot}$ is a result of explosive neon burning and the tail from 4 to 6 $M_{\odot}$ is a mixture of ^{26}Al made before the explosion and by the neutrino process in comparable amounts.

Many more models remain to be done, including stars of 20, 30, 40, and 60 $M_{\odot}$, but the following preliminary conclusions are offered. A reasonable prescription for ^{26}Al synthesis in Type II supernovae might be, in units of 10^{-5} $M_{\odot}$: 0.5, 1, 4, 5, and 15 for stars of 12, 15, 20, 25, and 35 $M_{\odot}$ respectively (though again note the non-monotonic behavior in Table 3). For stars bigger than 35 $M_{\odot}$, including the Ib progenitors that may result from such stars, also use 1.5×10^{-4} $M_{\odot}$. Note that the Ib's, if they come from massive stellar cores (§5), will be very important and might contribute, per event, substantially more ^{26}Al than the Type II's. As remarked previously, all these numbers have an error bar of at least a factor of two.

Next the rates for Type II and Ib supernovae are critical. Tammann and Van den Bergh (1990) have discussed the Galactic supernova rate concluding that a likely value for core collapse events is two per century. However, as they point out, this is very much less than inferred from the 6 historical supernovae in the last 1000 years (all within 3 or 4 kpc of the Earth and within a 50 degree pie slice of the Galactic disk) from which they infer the rate *could* be as frequent as one every 10 years (this will shortly be limited by neutrino and γ-ray detectors, the 7 year lifetime of SMM already placing an interesting upper bound). Perhaps our Galaxy, or our region of the Galaxy, is undergoing a burst of star formation. Let us adopt a rate of 4 core collapse supernovae per century as a reasonable compromise between the arguments of Tammann and Van den Bergh and the observed local Galactic rate. One Type Ib (derived from a massive Wolf-Rayet star) once a century (or any other 35+ $M_{\odot}$ supernova) plus two or three explosions per century around 15 – 25 $M_{\odot}$ could provide all the ^{26}Al necessary to explain the observations. On the other hand, the present error bars are large enough that some other source could dominate.

If the ^{26}Al does come from supernovae (rather than novae or AGB stars), then these calculations suggest that the most massive stars will predominate. One then expects the ^{26}Al to have a distribution in the Galaxy that resembles not only Population I, but extreme Population I, *i.e.* like the O-stars in H II regions and molecular cloud complexes.

4. THE DELAYED DETONATION MODEL FOR TYPE Ia SUPERNOVAE AND ^{60}Fe SYNTHESIS

Considerable debate has recently surrounded our understanding of the physics of degenerate carbon burning. There has been, perhaps prematurely, a group consensus that Type Ia supernovae come from the explosions of carbon-oxygen white dwarf stars pushed nearly to the Chandrasekhar mass by accretion from a binary companion. Further, the story continues, a flame is born somewhere near the middle and this flame propagates *subsonically*, though precariously close to the sound speed, through the rest of the white dwarf burning a large fraction to ^{56}Ni plus some silicon, calcium, and other elements closer to the surface. The success of this "carbon deflagration" model in reproducing many observational data, especially highly constraining spectroscopic data, is well known (*e.g.* review by Woosley and Weaver 1986a). Some of its shortcomings are also known – chiefly nucleosynthetic (too much ^{54}Fe and ^{58}Ni), and possible problems with light curve and late time spectrum. Here is not the place to review all that.

What is important to the γ-ray astronomer, however, is that some of the most recent and realistic models (see especially Khoklov 1990ab; Woosley and Weaver 1991), have the carbon burning flame initially propagating for an extended period of time, about 1 second, at a very slow speed (at least compared to the sound speed). One consequence of this is that a small amount of material experiences a great deal of electron capture. As we shall see, this may imply interesting nucleosynthesis of ^{60}Fe, a well know γ-line candidate (Clayton 1971). A second interesting aspect of these new solutions is that the flame ultimately accelerates to supersonic speeds, that is the explosion, born a deflagration, ends a detonation. The star has been "pre-expanded" by the slow burning to the point where, in the outer layers at least, the heat capacity of the radiation field is sufficient to keep the material from burning to nuclear statistical equilibrium. Intermediate mass elements can be formed, a new twist for detonation models. However, considerably more material is ejected as ^{56}Ni and less as carbon and oxygen. The explosion energy is also larger, all this being very good news for the γ-ray astronomer. The ^{56}Ni and ^{56}Co γ-line signatures of these models are much brighter (see Pinto this volume).

Why should the flame have this peculiar behavior of starting slow and accelerating? It is a manifestation of the Rayleigh-Taylor instability without which there would be no Type Ia supernova. When propagated by simple electron conduction, the flame moves very slowly, only about 50 km s^{-1} (Woosley 1986; 1990), far too slow to make a supernova. But as it becomes wrinkled owing to crossed gravitational and density gradients across the flame surface, its area increases and for the same reason a corrugated radiator dispels heat quicker than a flat sheet of metal, the rate of mass consumption rises. Were there nothing to set the smallest scale of deformation, the rate of mass consumption could become very large (like the coastline of Britain measured with a ruler of ever decreasing scale), but there is a minimum scale to the problem set by the smallest deformation that can grow before the (conductive) flame washes over it. This is

$$\lambda_{min} = \frac{4\pi v_{cond}^2}{g_{eff}}$$

where g_{eff} is the effective acceleration due to gravity because of the density inversion, $g_{eff} = (\Delta\rho/\rho)(GM(r)/r^2)$. There is also a *maximum* scale of deformation that can exist given by the size of the region that has already burned. The effective speed of the flame, measured in terms of how fast it consumes mass, is $v_{cond}(\lambda_{max}/\lambda_{min})^{D-2}$ where $D(t)$ is the fractal dimension of the flame front (Woosley 1990). Finally there is the time dependence of D itself. If the small scale instabilities are very efficient at digesting the material enveloped by the large ones, or if there is not much range between λ_{min} and λ_{max}, then D will be close to 2, that is the surface will not be very wrinkled. If, on the other hand, intermediate wavelengths have time to grow, become non-linear, and float away seeding new regions of burning, D will increase. Such a phenomenon is also favored by the fact that the deflagration is preceded by a stage of convective carbon burning. As the burning time becomes short compared to the convective cycle time, blobs of burning carbon may float away so that the star is ignited at multiple locations.

The evolution of $D(t)$ may ultimately be studied using a multi-dimensional hydrodynamics code. For now, at least in Weaver's and my formulation, it is simply a way of parameterizing the flame propagation, but in a fashion which we feel has some physical motivation lacking in previous studies that have characterized the flame propagation as "convection." It is clear that D begins close to 2 because, for a time, λ_{min} is not so much less than λ_{max} that there is time for non-linear growth of the Rayleigh-Taylor instability. It also seems reasonable that when there is more than an order of magnitude difference between these two scales and when the flame has moved a pressure scale height, which might characterize the path length of individual burning blobs, that D should increase. For what it is worth, in nature turbulent structures seem to frequently be characterized by $D \approx 2.6$ to 2.7 (Mandelbrot 1983). We are thus led to naturally consider a flame that starts very slowly (D $\approx$ 2, or maybe 2.2 to 2.3) and then accelerates at late times (D $\approx$ 2.7). Of course, $D = 3$, a foamy flame filling the entire white dwarf at all scales down to λ_{min} is an extreme upper bound.

Table 4: Delayed Detonation Model Summary

Model	DD1	DD2	DD3	DD4
$KE_\infty/10^{51}$ erg	1.56	1.44	1.37	1.22
$M^a(^{12}C)$	1.1(-3)	2.8(-3)	1.9(-3)	3.3(-3)
$M(^{16}O)$	1.9(-2)	3.9(-2)	3.6(-2)	1.0(-1)
$M(^{20}Ne)$	3.0(-4)	6.8(-4)	7.0(-4)	1.2(-3)
$M(^{24}Mg)$	1.4(-3)	2.2(-3)	2.2(-3)	9.0(-3)
$M(^{28}Si)$	6.6(-2)	1.2(-1)	1.4(-1)	2.6(-1)
$M(^{32}S)$	3.9(-2)	6.8(-2)	8.3(-2)	1.6(-1)
$M(^{36}Ar)$	5.7(-3)	9.5(-3)	1.8(-2)	3.5(-2)
$M(^{40}Ca)$	7.9(-3)	1.3(-2)	1.9(-2)	3.7(-2)
$M(^{54}Fe)^b$	2.0(-1)	1.9(-1)	7.1(-2)	7.0(-2)
$M(^{56}Ni)$	9.7(-1)	8.7(-1)	9.3(-1)	6.3(-1)
M("Fe")^c	7.2(-2)	7.1(-2)	7.1(-2)	8.9(-2)

[a] Solar masses of ejecta.

[b] Includes some component of ^{58}Ni produced in the α-rich freeze out.

[c] All other iron group isotopes; mostly ^{56}Fe.

Table 5: Some Details of Model DD3

Time (sec)	Mass ($M_\odot$)	r_b (10^8 cm)	D	g_{eff} (10^8 cm s^{-2})	λ_{min} (10^3 cm)	v_{eff} (10^8 cm s^{-1})	c_s (10^8 cm s^{-1})	ρ_c (10^7 g cm^{-3})
0.201	0.0162	0.142	2.3	31.8	165	0.246	9.74	303
0.530	0.0665	0.262	2.3	35.7	78.2	0.269	9.32	228
0.698	0.106	0.354	2.3	33.5	40.4	0.251	8.86	157
0.964	0.161	0.572	2.3	19.6	14.6	0.181	7.55	63.8
1.350	0.232	1.15	2.4	6.99	3.24	0.281	5.67	14.0
1.478	0.269	1.46	2.5	5.05	2.66	0.765	5.62	8.70
1.499	0.319	1.58	2.6	5.07	4.02	2.30	6.10	8.12
1.505^a	0.385	1.66	2.7	5.54	2.36	7.99	5.66	7.95

[a] Shortly after this, conduction was terminated. A detonation wave had formed.

Figures 2 and 3 and Tables 4 through 6 give some details of the delayed detonation model. See also Khoklov (1990ab) for an alternate description of the physics involved in the delayed detonation model which leads to similar results. Table 4 gives the nucleosynthesis from several models that used varying prescriptions for $D(t)$ (see Woosley and Weaver 1991). Tables 5 and 6 and the figures give some particulars of one of these models.

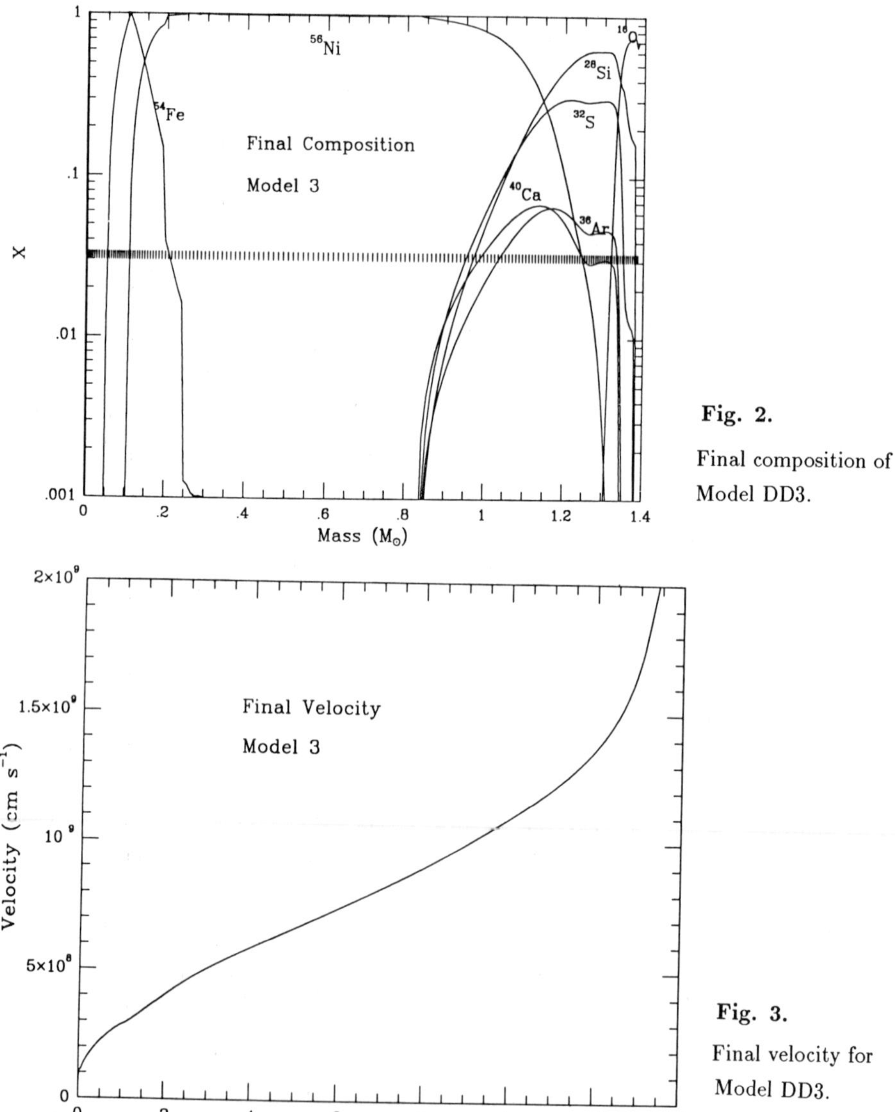

Fig. 2. Final composition of Model DD3.

Fig. 3. Final velocity for Model DD3.

Table 6: Electron Mole Number in Delayed Detonation Model

Y_e	η	Mass ($M_\odot$)	ΔM ($M_\odot$)	nuclei
0.43 – 0.44	0.12 – 0.14	0.0022	0.0022	$^{64,66}Ni$, ^{60}Fe, ^{48}Ca
0.44 – 0.45	0.10 – 0.12	0.0095	0.0073	^{54}Cr, ^{58}Fe
0.45 – 0.46	0.08 – 0.10	0.026	0.016	$^{56,58}Fe$
0.46 – 0.47	0.06 – 0.08	0.054	0.029	^{56}Fe
0.47 – 0.48	0.04 – 0.06	0.085	0.031	$^{54,56}Fe$
0.48 – 0.49	0.02 – 0.04	0.122	0.037	^{54}Fe
0.49 – 0.50	0.00 – 0.02	1.38	1.26	^{56}Ni, ^{28}Si, etc.

For $Y_e = 0.4311$, the mass fraction of ^{60}Fe in the final frozen out abundances is about 9%; for $Y_e = 0.4375$, it is 11%. The exact amount produced in the supernova depends sensitively upon the density at which the carbon runs away, but for our standard model ($\rho_i = 3.7 \times 10^9$ g cm^{-3}) we estimate an ^{60}Fe yield of roughly 10^{-4} $M_\odot$. If one Type Ia supernova occurred in our Galaxy every 200 years (a compromise between the longer interval suggested by by Tammann and shorter values elsewhere), a steady concentration of roughly one $M_\odot$ would accumulate in the 2×10^6 year mean life of ^{60}Fe. This would give a signal about an order of magnitude weaker than that of ^{26}Al and should definitely be sought by the Gamma-Ray Observatory and future missions. Because the ^{60}Fe is produced in the innermost layers, its velocity is low, $\lesssim 10^8$ cm s^{-1}. The strong emission lines are at 59 keV and 1.17 and 1.33 MeV (equal strengths). If detected, these lines would teach us a lot about Type Ia supernovae, especially their Galactic distribution, frequency, and, indirectly, the density at which carbon ignites. It would also be powerful positive evidence for the slow initial flame propagation favored in the current models. Interestingly the number suggested here is smaller than the limit set by SMM (Share elsewhere in this volume), but not by a large factor.

5. TYPE Ib SUPERNOVAE

Examples of this class of supernova based upon the explosion of 3, 4, and 6 $M_\odot$ helium cores of massive stars that lost their hydrogen envelopes (but not much else) have been given by Ensman and Woosley (1988) and by Shigeyama *et al.* (1990; see also Nomoto, this volume). Here I present a model of another type. This one is based upon a 60 $M_\odot$ star studied by Norbert Langer, Tom Weaver, and myself. The star experienced extensive mass loss, especially after becoming a WR star, and ended up as a 4.25 $M_\odot$ WO star (Fig. 4). It is to be emphasized that this residual is quite different both in composition and structure from the 4 $M_\odot$ core of a 15 $M_\odot$ star that lost only its hydrogen envelope. Among other things, the entropy is higher as the core remembers its past history as a massive star. Thus the star is less centrally condensed and the density gradients are shallower. The CO-core is also much larger. As a result, explosion (simulated by placing a piston near the edge of the iron core), produces a large amount of ^{56}Ni. Because the density gradient is shallower, a larger mass of ejected material experiences temperatures in excess of 5×10^9 K. For an explosion of 1.7×10^{51} ergs, a ^{56}Ni mass of 0.3 $M_\odot$ is made. A preliminary light curve for the model has been given in Woosley and Weaver (1991), but a correct calculation, still to be carried out, will require an examination of the time dependent spectrum to see the role of Doppler broadened lines in sculpting the "bolometric" light curve. The γ-ray light curve for this model is given by Pinto elsewhere in this volume.

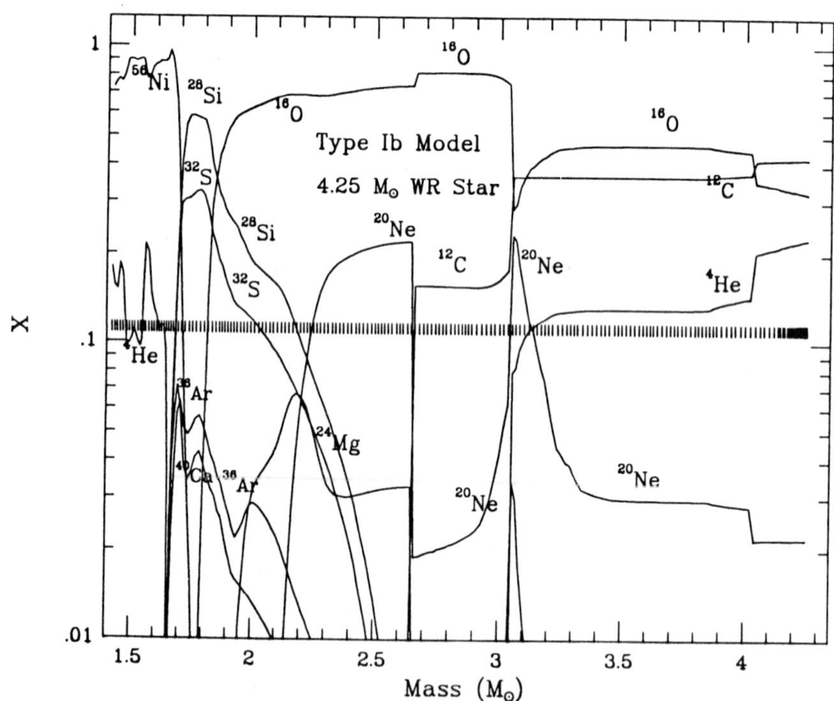

Fig. 4. Final composition of an exploded Wolf-Rayet star model for a Type Ib supernova (Woosley, Langer, and Weaver 1991).

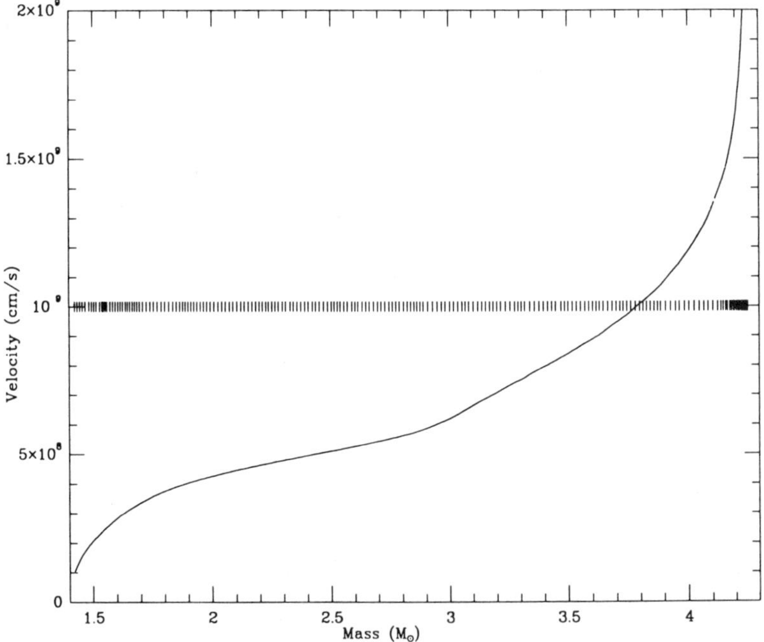

Fig. 5. Final velocity for the model in the previous figure.

Legitimate concern may be expressed both regarding the frequency of death for such massive stars and the assumed large efficiency for mass loss. The key to the latter may be a mass-dependent mass loss rate for WR-stars. Langer (1989) presents arguments that mass loss in WR-stars should be large and may proceed at a rate proportional to the 2.5 power of the (remaining) mass. This results in convergence on lower masses for the final presupernova star. To address the frequency issue, one may speculate that mass exchange in a binary system leads to stars less than the canonical 40 $M_\odot$ value already being stripped of their hydrogen envelopes and that the highly efficient mass loss rate of the bare WR-star takes the mass down from there.

The alternate solution, discussed by Nomoto elsewhere in these Proceedings, invokes a narrow mass range for a common phenomenon. Moreover, it makes some very optimistic assumptions about the explosion in order to get out as much ^{56}Ni as possible. The light curve requires that at least 0.15 $M_\odot$ and probably several tenths $M_\odot$ of ^{56}Ni be ejected. In the currently favored delayed explosion models for core collapse supernovae, however, a portion of the mantle accretes onto the protoneutron star before sufficient energy is deposited by the neutrinos to revitalize an outwardly propagating shock wave. Most likely the material that would have been ^{56}Ni in Shigeyama and Nomoto's model falls into the neutron star. The model presented here circumvents this problem, at least partly, by having a shallow density gradient around the core. This allows high temperatures to develop for a substantial amount of mass even if the mass separation occurs farther out.

Any way you look at it though, we still have a way to go before we understand Type Ib supernovae.

6. ACCRETION INDUCED COLLAPSE

To conclude this *potpourri* of γ-ray emitting explosions, a very different kind of supernova is briefly discussed, one which, it has been hypothesized (Paczynski 1986; Goodman 1986; Goodman, Dar, and Nussinov 1987; Ramaty *et al.* 1991), might produce γ-ray bursts at cosmological distances. Under certain circumstances an accreting white dwarf, be it carbon and oxygen or neon and oxygen, can forestall ignition to a central density of about 10^{10} g cm^{-3} (Canal, Isern, and Labay 1990; Isern, Canal, and Labay 1990; Nomoto and Kondo 1990). Ignition at this density will lead, not to a Type Ia supernova, but to direct collapse to a neutron star. A similar phenomenon may occur at high accretion rates where carbon ignites off center in a mild flash and a carbon-oxygen white dwarf is turned into one composed of neon and oxygen (Saio and Nomoto 1985; Woosley and Weaver 1986b; Nomoto and Iben 1986).

Leaving aside many interesting considerations regarding the pre-collapse evolution, the suggestion is that neutrino annihilation (ν_e annihilating with $\bar{\nu}_e$, ν_μ, with $\bar{\nu}_\mu$, and ν_τ, with $\bar{\nu}_\tau$ to produce $e^+ + e^-$) in the region above the photosphere of the collapsing dwarf will give a pair-fireball (Goodman, Dar, and Nussinov 1987; Cavallo and Rees 1978). Annihilation and thermal radiation from this high temperature cloud would then produce a γ-ray burst and perhaps provide positrons relevant to the Galactic diffuse 511 keV emission as well (Ramaty *et al.* 1991).

Paczynski (1990), though he speculates that such events are γ-ray bursts, has shown that the visible electromagnetic signal from such an event can easily be obscured by even a modest wind. He finds

$$L^{\gamma}_{out} \approx 3 \times 10^{48} \left(\frac{\dot{Q}}{10^{50}}\right)^3 \left(\frac{\dot{M}}{10^{27}}\right)^{-8/3} \quad \text{erg s}^{-1},$$

where $\dot{Q}$ is the energy input at the base of the envelope in erg s^{-1} and $\dot{M}$ is the mass loss rate in g s^{-1}. Further, the temperature of the radiation escaping from the photosphere, for mass loss rates that are not too high, is

$$T_{out} \approx 4 \times 10^8 \left(\frac{\dot{Q}}{10^{50}}\right)^{9/4} \left(\frac{\dot{M}}{10^{27}}\right)^{-8/3} \quad \mathrm{K}.$$

Eddie Baron and I have recently constructed a numerical model of the collapse of a white dwarf (initial model provided by Nomoto, 1984) to a neutron star. No prompt hydrodynamical mass ejection was found, but the deposition of neutrino energy in the outer envelope of the collapsing star did lead to a powerful wind. At a time roughly 0.3 s after core bounce, for example, the neutrino luminosity was 8×10^{52} erg s^{-1}, shared roughly equally among the six varieties of neutrinos. The neutrinosphere was situated at a radius of 30 km. Chiefly by neutrino capture on nucleons, an energy of about 3×10^{50} erg s^{-1} was deposited external to the neutrinosphere, but in a region where the matter is still very optically thick ($\rho \approx 10^8$ to 10^9 g cm^{-3}; T $\approx$ 2 to 2.5 MeV). We found that it is very important that the region external to this energy deposition be finely zoned. Earlier studies of collapse in this same model (Baron *et al.* 1987) did not observe a wind owing to coarse zoning, about 0.01 $M_\odot$, employed in the calculation. Here we found zoning of about 10^{-7} $M_\odot$ was necessary to accurately treat the problem. Powered by the neutrino energy deposition, a wind of about 0.002 $M_\odot$ s^{-1} was found to blow from the surface of the collapsing dwarf. Its terminal velocity, at least at the time studied, a few tenths of a second, was about 0.1 c. It is important that this wind, though very rapidly moving, is *not* extremely relativistic. That is $\dot{Q}$ is less than $\dot{M}c^2$. Paczynski has pointed out that one prerequisite for hard emission is a relativistic wind.

Instead we find that the wind occludes the γ-ray burst and prevents pair ejection. For typical values given above, the electromagnetic radiation would only have a luminosity of about 10^{40} to 10^{41} erg s^{-1} and would not be in the form of γ-rays or even x-rays, but something much softer. Certainly one cannot make γ-ray bursts at cosmological distances this way. A brief transient of this sort, lasting only a few seconds, would even be very difficult to detect in our own Galaxy.

Still such collapses may not be without observable consequence. The composition of this wind is extremely neutron-rich. Moreover it is ejected at such speeds that the freeze-out from nuclear statistical equilibrium leads to the production of rare isotopes heavier than the iron group. Rob Hoffman and I are studying this nucleosynthesis, both in the context of the collapsing dwarf and in the similar case of an ordinary Type II supernovae exploding by the delayed mechanism. We find that the wind that blows off of the protoneutron star in both cases is an excellent site for production of the classical r-process.

Even in those situations where a full r-process does not operate, neutron-rich nuclei between the iron group and about A = 100 are abundantly produced. The rarity of such species in the sun suggests that the collapse of a bare white dwarf to a neutron star probably cannot occur more frequently in our Galaxy than about once every 10,000 years.

It is well to keep in mind, however, that only 10^{-7} $M_\odot$ of ^{56}Ni (or similar short lived radioactivity from the r-process – ^{89}Sr, ^{91}Y, ^{95}Zr, ^{95}Nb, ^{99}Mo, and ^{103}Ru, for example, all have half-lives in the range 2 to 65 days) would give a signal of 10^{-4} photons cm^{-2} s^{-1} at a distance of 10 kpc. It would be difficult for a supernova of any type, even the collapse of a bare white dwarf, to occur in our Galaxy without its γ-ray emission being discovered. Indeed continuous surveillance by a γ-ray spectrometer of modest sensitivity may be the best way to make sure that no explosion is overlooked. All core and white dwarf collapses would give neutrino bursts but exploding white dwarfs (Type Ia?) would not.

This research has been supported by NASA (NAGW 1273) and the NSF (88-13649).

REFERENCES

Anders, E., and Grevesse, N. 1989, *Geochim. Cosmochim. Acta.*, **53**, 197.

Arnett, W. D., and Truran, J. W. 1969, *Ap. J.*, **157**, 339.

Barkat, Z., and Marom, A. 1990, in *Supernovae*, Proceedings of Jerusalem School for Theoretical Physics, Volume 6, eds. J. C. Wheeler, T. Piran, and S. Weinberg, (World Scientific: Singapore), p. 95.

Baron, E., and Cooperstein, J., Kahana, S., and Nomoto, K. 1987, *Ap. J.*, **320**, 304.

Canal, R., Isern, and Labay, J. 1990, *Ann. Rev. Astron. and Ap.*, **28**, 183.

Caughlan, G. A., and Fowler, W. A. 1988, *Atomic Data and Nuclear Data Tables*, **40**, 238.

Cavallo, G., and Rees, M. J. 1978, *MNRAS*, **183**, 357.

Clayton, D. D. 1971, *Nature*, **234**, 291.

Ensman, L., and Woosley, S. E. 1988, *Ap. J.*, **333**, 754.

Goodman, J. 1986, *Ap. J. Lettr.*, **308**, L47.

Goodman, J., Dar, A., and Nussinov, S. 1987, *Ap. J. Lettr.*, **314**, L7.

Illiadis, Ch., Schange, Th., Rolfs, C., Schröder, U., Somorjai, E., and 7 others 1990, preprint.

Isern, J., Canal, R., and Labay, J. 1990, preprint submitted to *Ap. J.*

Khoklov, A. M. 1990a, Max Planck Inst. für Astrophysik preprint 513, *Astron. and Ap.*, in press.

Khoklov, A. M. 1990b, Max Planck Inst. für Astrophysik preprint 528, *Astron. and Ap.*, in press.

Langer, N. 1989, *Astron. and Ap.*, **220**, 135.

Mandelbrot, B. B. 1983, *The Fractal Geometry of Nature*, (New York: W. H. Freeman).

Mayle, R. W. 1985, Ph. D. Thesis, Lawrence Livermore National Laboratory reprint, UCRL–53713

Mayle, R. W. 1990, in *Supernovae*, ed. A. G. Petsheck, (Springer Verlag: New York), p. 267.

Mayle, R. W., and Wilson, J. R. 1990, in *Supernovae*, Proceedings of the Tenth Santa Cruz Summer Workshop, ed. S. E. Woosley, (Springer Verlag: New York), p. 333.

Nomoto, K. 1984, *Ap. J.*, **277**, 791.

Nomoto, K., and Kondo, Y. 1990, preprint submitted to *Ap. J.*

Nomoto, K., and Iben, I. 1985, *Ap. J.*, **297**, 531.

Paczynski, B. 1986, *Ap. J. Lettr.*, **308**, L43.

Paczynski, B. 1990, *Ap. J.*, **363**, 218.

Ramaty, R., Dar, A., Nussinov, S., and Kozlovsky, B. 1991, in *Proceedings of the Los Alamos Gamma-Ray Burst Workshop*, eds. C. Ho, R. I. Epstein, and E. E. Fenimore, (Cambridge Univ. Press: Cambridge), in press.

Saio, H., and Nomoto, K. 1985, *Astron. and Ap.*, **150**, L21.

Shigeyama, T., Nomoto, K., Tsujimoto, T., and Hashimoto, M. 1990, *Ap. J. Lettr.*, **361**, L23.

Tammann, G., and Van den Bergh, S. 1990, preprint to appear in *Ann. Rev. of Astron. and Ap.*

Wilson, J. R. 1985, in *Numerical Astrophysics*, ed. J. M. Centrella, J. M. LeBlanc, and R. L. Bowers, (Jones and Bartlett: Boston), p. 422.

Wilson, J. R., Mayle, R., Woosley, S. E., and Weaver, T. A., 1986, *Ann. N.Y. Acad. Sci.* **470**, 267.

Woosley, S. E. 1986, in *Nucleosynthesis and Chemical Evolution*, 16th Advanced Course of the Swiss Soc. of Astron. and Ap., "Saas-Fee Lecture Notes," ed. B. Hauck, A. Maeder, and G. Meynet, (Geneva Observatory: Geneva), p. 1.

Woosley, S. E. 1990, in *Supernovae*, ed. A. Petschek, (Springer Verlag: New York), p. 182.

Woosley, S. E., Arnett, W. D., and Clayton, D. D. 1972, *Ap. J.*, **175**, 731.

Woosley, Langer, N., and Weaver, T. A. 1991, in preparation for *Ap. J.*

Woosley, S. E., and Weaver, T. A. 1982, in *Essays in Nuclear Astrophysics*, ed. C. A. Barnes, D. D. Clayton, and D. N. Schramm, (Cambridge Univ. Press: Cambridge), p. 377.

Woosley, S. E., and Weaver, T. A. 1986a, *Ann. Rev. Astron. and Ap.*, **24**, 205.

Woosley, S. E., and Weaver, T. A. 1986b, in *Nucleosynthesis and Its Implications on Nuclear and Particle Physics*, ed. J. Audouze and N. Mathieu, (D. Reidel: Dordrecht), p. 145.

Woosley, S. E., Hartmann, D., Haxton, W., and Hoffman, R. 1990, *Ap. J.*, **356**, 272.

Woosley, S. E., and Weaver, T. A. 1991, in *Les Houches Lecture Session LVI 1990*, ed. J. Audouze, R. Mochkovitch and J. Zinn-Justin, (Elsevier Science Publishers) and in preparation for *Ap. J.*.

HEAO 3 LIMITS ON THE ^{44}Ti YIELD IN TYPE I SUPERNOVAE

W. A. Mahoney, J. C. Ling, and W. A. Wheaton
Jet Propulsion Laboratory 169-327, Pasadena, CA 91109

and

J. C. Higdon
Joint Science Department, The Claremont Colleges, Claremont, CA 91711

ABSTRACT

Data from the high-resolution gamma-ray spectroscopy experiment on HEAO 3 have been searched for line emission from the decay of ^{44}Ti created in recent galactic supernova explosions. Because the 78-year mean life of ^{44}Ti is comparable to the average time between galactic supernovae, the gamma-ray line emission from its decay should appear as a galactic point source. No evidence was found for such emission from a point source anywhere in the galactic plane, with a 1 σ limit of 8.3×10^{-5} photons cm^{-2} s^{-1}. A detailed model was developed to simulate the galactic gamma-ray emission from the decay of ^{44}Ti produced in type I supernovae. This model was used with the measured gamma-ray line limits to constrain the supernova yields and recurrence periods. For example, if the supernovae recurrence period is 36 years, the average ^{44}Ti yield per event must be $< 4 \times 10^{-4}$ $M_{\odot}$ at the 90% level of confidence.

INTRODUCTION

Stellar nucleosynthesis produces a number of radioisotopes which are potentially observable through their gamma-ray line emission. Among these is ^{44}Ti, produced largely in type I supernovae[1]. Type I models involving carbon deflagration of accreting carbon/oxygen white dwarfs[2,3] give ^{44}Ti yields in the range of $(2 - 8) \times 10^{-5}$ $M_{\odot}$. Helium detonation models of accreting carbon/oxygen white dwarfs[4] give yields of $(0.2 \text{ to } 4.5) \times 10^{-4}$ $M_{\odot}$ per supernova. Finally, rare detonations of helium white dwarfs[4] could produce up to 0.01 $M_{\odot}$ of ^{44}Ti in a single explosion. To span this wide range of estimated ^{44}Ti yields, the calculations which follow consider the range of 10^{-6} to 10^{-2} $M_{\odot}$.

Simple arguments involving isotopic abundance ratios have been used by Woosley[5] to show that the average galactic production rate of ^{44}Ti, in the models discussed above may significantly *underestimate* the true galactic production rate. According to Woosley, if type I supernovae are the prime sources of both ^{44}Ca (produced entirely as ^{44}Ti), and ^{56}Fe (produced primarily as ^{56}Ni), as implied by most supernova models[1,4,6], then the ^{44}Ti/^{56}Ni isotopic ratio in the supernova yield should be about the same as the corresponding ^{44}Ca/^{56}Fe ratio measured in solar material[7], approximately 1.2×10^{-3}. Since each type I supernova is believed to create[2,8] about 0.5 $M_{\odot}$ of ^{56}Ni, approximately 6×10^{-4} $M_{\odot}$ of ^{44}Ti should also be produced. This corresponds to about 2×10^{-5} $M_{\odot}$ per year in the Galaxy, about an order of magnitude higher than estimated above.

The higher ^{44}Ti production could also help account[5] for the strong diffuse galactic 511 keV line emission[9] resulting from electron/positron annihilation. About 15% of this 511 keV emission is explained by the annihilation of positrons from the decay of ^{26}Al in the interstellar medium[10-12]. Positrons from the decay of ^{44}Ti produced at the rate of 6×10^{-4} $M_{\odot}$ per supernova could account for the remainder. This high ^{44}Ti production rate should result in point sources of gamma-ray line emission above the the HEAO threshold. Thus HEAO 3 would be expected to either measure a source of ^{44}Ti or place meaningful limits on the galactic production rate.

The average galactic production rate of ^{44}Ti depends directly on the galactic supernova frequency which is not well known. Tammann[13] has estimated that type I supernovae occur about once every 36 years in the Galaxy based on both the observed supernova rates in other galaxies and on the frequency of historical galactic supernovae. Other investigations[14,15] indicate a recurrence period up to 3 times longer. Unfortunately, the estimates are subject to large systematic uncertainties. Thus the modeling calculations cover recurrence periods of 10 to 1000 years, a range which includes the rare detonation of helium white dwarfs mentioned above.

DATA ANALYSIS AND RESULTS

The 78-year mean life of ^{44}Ti is comparable to the expected recurrence time of galactic supernovae. For such sources, it has been shown[16] that the observed gamma-ray flux will be dominated by a few recent events. However, there has not been an observed galactic supernova recent enough to have presently detectable line emission. Thus the approach in the present investigation involved the search for young, unknown, visually obscurred galactic supernovae. For the analysis of HEAO 3 gamma-ray data, a technique known as the scan-by-scan approach[17], has been developed which is believed relatively free of systematic errors but which currently requires the definition of a specific multi-parameter model that is then compared to the data. For the present analysis, we have studied a series of models, each consisting of background plus a single point source in the galactic plane. Although the instrument field-of-view was 30° FWHM, we chose a 10° spacing for the assumed point sources to assure the maximum practical instrumental response to any real source in the galactic plane. This gave a total of 36 candidate sources uniformly spaced around the Galaxy.

The decay of ^{44}Ti to ^{44}Sc (τ = 78.2 years) is accompanied by the emission of gamma-ray lines at 68.85 keV and 78.38 keV. The ^{44}Sc subsequently decays to ^{44}Ca (τ = 5.67 hours) with the emission of a 1157 keV gamma ray. These lines will be broadened by the rms velocity dispersion of the material at the time of decay which we have taken as 6000 km s^{-1} for a typical supernova. For such a velocity dispersion, the two low energy lines would be broadened to about 4.4 keV FWHM, including the instrumental energy resolution of 2.6 keV FWHM, making them the most easily seen by HEAO 3. Furthermore, the 78.38 keV line falls in a region with no background features[18], making it the more sensitive of the two. On the other hand, the intrinsic broadening of the 1157 keV line would be over 50 keV, rendering the HEAO 3 instrument less sensitive to its detection. Thus we have chosen to search for the two lower-energy lines by investigating the 32-keV energy band from 58 to 90 keV, broken into 16 2-keV channels.

HEAO 3 was a scanning mission with a spin period of about 20 minutes and with the spin axis generally pointed at the Sun. The high-resolution gamma-ray spectrometer consisted of a cluster of four high-purity germanium detectors in an active anticoincidence shield[19]. Data analysis involved calculating the net cosmic flux for each scan of the assumed point source. In the present application, the measured counts as a function of scan angle were fitted to a model consisting of three free parameters: a cosmic point source at a given location whose flux is modulated by the known instrument transmission function, a constant background, and a term proportional to the instrumental background rate as a function of geomagnetic latitude. The fitting was

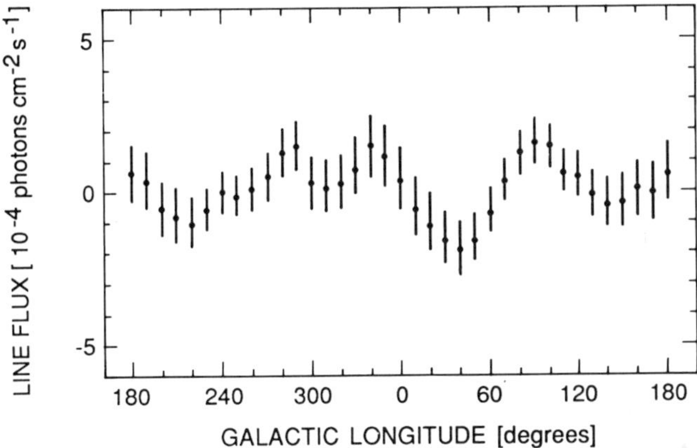

Figure 1. The measured net line flux from the decay of ^{44}Ti for each of the 36 assumed point sources uniformly distributed around the galactic plane.

performed independently for each of the four germanium detectors and for each of the 16 energy bands. The calculated fluxes were weighted together over the four detectors and over hundreds of scans to give the final result for each source model.

The 36 net cosmic spectra from 58 to 90 keV were then fitted to a quadratic background plus a Gaussian function with an energy of 78.38 keV and a width of 4.4 keV FWHM. Similar fits were performed independently for a line at 68.85 keV. For each of the 36 assumed point sources, the weighted average of the net 68.85 keV and 78.38 keV line intensities was obtained. These are plotted in Figure 1. Clearly there is no evidence for emission from a point source anywhere in the galactic plane. The correlation between adjacent points results because the data were not independent. The scatter of the data around zero is consistent with that expected for the full 36 points as well as for the subset of 12 points spaced every 30°. Given the 30° field-of-view of the instrument, the latter represent nearly independent observations.

The errors vary from point to point in Figure 1 owing to differences in exposure time on the galactic plane. The geometry of the scan plane and the galactic plane gave the largest exposure at galactic longitudes of approximately 90° and 270°. We have therefore defined the average error as

$$\sigma_{rms} = (\Sigma\, \sigma_i^2 / N)^{1/2}$$

where σ_i is the error for a given source location and N is the number of sources considered. For the 36 points shown in Figure 1, the rms error is 8.3 x 10^{-5} photons cm^{-2} s^{-1}. The corresponding rms error for the subset of 12 points spaced every 30° is almost identical as is the straight arithmetic average.

MODELING

Following Higdon and Fowler[16], a Monte Carlo simulation was employed to model the ages and locations of young type I supernovae in our Galaxy. It is based on the premise that the frequency of type I supernovae scales with the visual emissivity.

The distribution of visual emissivity is significantly different in the galactic disk and spheroid[20]. To determine whether a particular supernova occurred in the disk or spheroid, a random number, distributed uniformly between 0 and 1, was determined and compared to 1/6, the ratio of visual luminosity in the spheroid to that in the disk.

Supernovae were assumed to occur randomly in a cylindrically-symmetric galactic disk with a stellar density, $n(\rho, z)$, proportional to

$$\exp[-|z|/\sigma_z-(\rho-R_o)/\rho_h]$$

where (ρ,ϕ,z) indicates a position in a galactocentric, cylindrical coordinate system. The scale height of the supernovae perpendicular to the galactic plane, σ_z, was set equal to 0.325 kpc, a value indicative of an old disk population[21]. The radial scale height of the disk, ρ_h, of 3.5 kpc, and distance of the Sun from the galactic center, R_o, of 8 kpc, were taken from de Vaucouleurs and Pence[22]. Supernovae were assumed to occur randomly in a galactic disk with integral probability distributions $P(<\rho)$, $P(<z)$, and $P(<\phi)$ based on this stellar density distribution:

$$P(<\rho)=1-[1+(\rho/\rho_h)\exp(-\rho/\rho_h)]$$

$$P(<\phi)=\phi/2\pi$$

$$P(<z)=1-\exp(-|z|/\sigma_z)$$

To locate a random disk supernova at (ρ,ϕ,z) three random numbers, distributed uniformly over 0 to 1, were determined from a pseudorandom number generator[23] and set equal to the above probability distributions. The location of each supernova was determined by inverting these probability distributions. The distance of each supernova from the Earth was derived, $d=[R_o^2+\rho^2+z^2-2R_o\rho\cos\phi]^{1/2}$.

Supernovae were assumed to occur randomly in a spherically-symmetric galactic spheroid with a stellar density, $n(R)$, proportional to

$$(R/R_o)^{-6/8}\exp[-10.09(R/R_o)^{1/4}+10.09], \quad R\leq 0.03R_o$$

$$(R/R_o)^{-7/8}\{\exp[-10.09(R/R_o)^{1/4}+10.09]\}[1 - 0.087/(R/R_o)^{1/4}], \quad R\geq 0.03R_o,$$

where (R, θ, ψ) indicate a location in a galactocentric spherical coordinate system[24,25]. To locate a spheroid supernova and to determine its distance from the Earth, random numbers were chosen and set equal to the corresponding integral probability distributions in analogy to the disk calculation.

To determine supernova ages we generated random numbers, t, distributed uniformly between 0 and $10\tau_{44}$, 782 yr, where τ_{44} is the mean life of ^{44}Ti. For each supernova we determined the γ-ray line flux at the Earth,

$$F=[M_{ej}/(\tau_{44}m_{44})]\exp(-t/\tau_{44})/(4\pi d^2)$$

where M_{ej} is mass of ^{44}Ti ejected in the supernova and m_{44} is the mass of a ^{44}Ti atom.

We investigated broad ranges of both supernova recurrence times, τ_{rec}, and ^{44}Ti ejected mass, M_{ej}. τ_{rec} was sampled every 0.1 in $\log(\tau_{rec})$ from 10 to 10^3 yr. In addition the number of supernovae per test galaxy was set equal to a random number drawn from a Poisson distribution[23] based on the mean number of expected supernovae, $(10\tau_{44})/\tau_{rec}$. M_{ej} was sampled every 0.1 in $\log(M_{ej})$ from 10^{-2} to 10^{-6} $M_\odot$. For each pair of τ_{rec} and M_{ej} an ensemble of 250,000 test galaxies was generated, and the probability distribution of the brightest candidate supernova was determined.

For each ensemble, the probability that the HEAO 3 gamma-ray spectrometer would have observed line emissions from ^{44}Ti decay was derived in the following manner. The probability that the HEAO 3 experiment could have resolved the ^{44}Ti decay lines is clearly dependent on line intensity. Given the 1 σ limit of 8.3×10^{-5} photons cm^{-2} s^{-1} derived in the previous section, HEAO 3 would have detected a source with a line strength above 1.93×10^{-4} photons cm^{-2} s^{-1} 99% of the time. Thus a

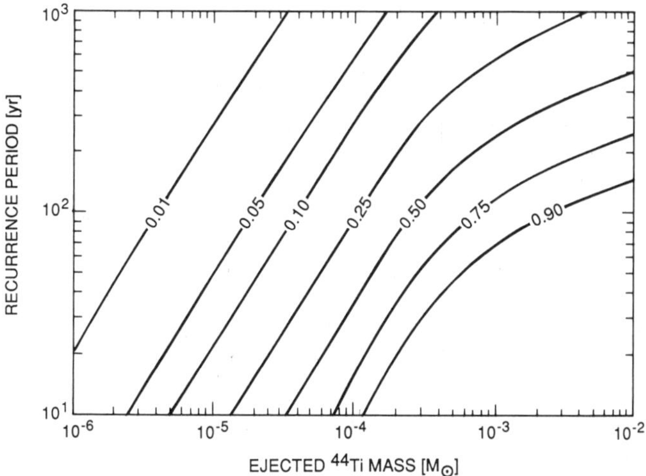

Figure 2. The contours represent the probability that HEAO 3 would have observed a point source from the decay of ^{44}Ti somewhere in the Galaxy for a given supernova recurrence period and average yield of ^{44}Ti.

Gaussian detection probability with a mean flux of 1.93 x 10^{-4} photons cm^{-2} s^{-1} and a dispersion of 8.3 x 10^{-5} photons cm^{-2} s^{-1} was used. The probability of detecting the ^{44}Ti decay lines for each τ_{rec} and M_{ej} was determined by integrating the probability distribution of the brightest candidate supernova over the Gaussian detection probability. The results of these calculations are displayed in Figure 2 where the detection probability is shown by a contour plot as a function of τ_{rec} and M_{ej}.

SUMMARY

Data obtained by the high-resolution gamma-ray spectrometer on HEAO 3 have been searched for evidence of line emission from the decay of ^{44}Ti produced in galactic type I supernovae. While no line emission was detected at 68 or 78 keV, a 1 σ limit of 8.3 x 10^{-5} photons cm^{-2} s^{-1} was obtained for emission in the 68 or 78 keV lines from a point source anywhere in the Galaxy. In order to translate this gamma-ray flux limit into a limit on the average ^{44}Ti yield in a type I supernova, a detailed Monte Carlo model was created. From Figure 2 it can be seen that at the 90% level of confidence, the average type I supernova yield of ^{44}Ti must be less than 4 x 10^{-4} $M_\odot$ if the recurrence period is 36 years as suggested by Tammann[13].

Based on the assumption that the bulk of the galactic ^{44}Ti and ^{56}Fe arose from type I supernovae, it has been estimated[8] that the average yield of ^{44}Ti from each supernova must be 6 x 10^{-4} $M_\odot$. It is interesting that our limit is somewhat below this value and suggests that not all the galactic ^{44}Ca was created in typical type I supernovae. This conclusion is consistent with the suggestion by Woosley[8] that the bulk of the galactic ^{44}Ca may have been produced in rare detonations of helium white dwarfs. Figure 2 indicates that these events would be consistent with the HEAO 3 observations. If type II supernovae contribute to the galactic production of ^{44}Ti, then the limits on the type I yields derived above become even more stringent.

ACKNOWLEDGEMENT

The authors wish to thank A. Dunklee and R. Radocinski for their outstanding support in developing the analysis software and in data reduction. The research described in this paper was carried out by the Jet Propulsion Laboratory, California Institute of Technology, under contract with the National Aeronautics and Space Administration.

REFERENCES

1. S. E. Woosley, T. S. Axelrod, and T. A. Weaver, *Comments on Nucl. Part. Phys.*, **9**, 185 (1981).
2. K. Nomoto, F. -K. Thielemann, and K. Yokoi, *Ap. J.*, **286**, 644 (1984).
3. F.-K. Thielemann, K. Nomoto, and K. Yokoi, *Astron. Astrophys.*, **158**, 17 (1986).
4. S. E. Woosley, R. E. Taam, and T. A. Weaver, *Ap. J.*, **301**, 601 (1986).
5. S. E. Woosley, *Ap. J. (Letters)*, submitted (1988).
6. K. Nomoto, F. -K. Thielemann, and J. C. Wheeler, *Ap. J. (Letters)*, **279**, L23 (1984).
7. A. G. W. Cameron, in *Essays in Nuclear Astrophysics*, eds. C. A. Barnes, D. D. Clayton, and D. N. Schramm, (Cambridge: Cambridge University Press), 23 (1982).
8. S. E. Woosley, in *Supernovae*, ed. A. G. Petschek (Berlin: Springer Verlag), (1989).
9. G. H. Share, et al., *Ap. J.*, **326**, 717 (1988).
10. J. C. Higdon, in *Nuclear Spectroscopy of Astrophysical Sources,* eds. N. Gehrels and G. Share (New York: AIP), 167 (1988).
11. W. A. Mahoney, J. C. Ling, W. A. Wheaton, and A. S. Jacobson, *Ap. J.*, **286**, 578 (1984).
12. G. H. Share, el al., *Ap. J. (Letters)*, **292**, L61 (1985).
13. G. A. Tammann, in *Supernovae: A Survey of Current Research*, eds. M. J. Rees and R. J. Stoneham (Dordrecht: Reidel), 371 (1982).
14. S. van den Bergh, R. D. McClure, and R. Evans, *Ap. J.*, **323**, 44 (1987).
15. E. Cappellaro and M. Turatto, *Astron. Astrophys.*, **190**, 10 (1988).
16. J. C. Higdon and W. A. Fowler, *Ap. J.*, **317**, 710 (1987).
17. W. A. Wheaton, et al., *Ap. J.*, submitted (1990).
18. W. A. Wheaton, et al., in *High Energy Radiation Background in Space*, eds. A. C. Rester, Jr. and J. I. Trombka, (New York: AIP), 304 (1989).
19. W. A. Mahoney, J. C. Ling, A. S. Jacobson, and R. M. Tapphorn, *Nucl. Instr. Meth.*, **178**, 363 (1980).
20. J. N. Bahcall and R. M. Soneira, *Ap. J. (Suppl.)*, **44**, 73 (1980).
21. D. Mihalas and J. Binney, *Galactic Astronomy: Structure and Kinematics*, 2nd Ed., (San Francisco: W. H. Freeman), 251 (1981).
22. G. de Vaucouleurs and W. D. Pence, *A. J.*, **83**, 1163 (1978).
23. W. H. Press, B. P. Flannery, S. A. Teukolsky, and W. T. Vetterling, *Numerical Recipes: The Art of Scientific Computing*, (Cambridge, UK: Cambridge University Press), 196 and 207 (1986).
24. P. J. Young, *A. J.*, **81**, 807 (1976).
25. J. N. Bahcall, M. Schmidt, and R. M. Soneira, *Ap. J. (Letters)*, **258**, L23 (1982).

THE THEORY AND DIAGNOSTIC PROMISE OF SUPERNOVA GAMMA-RAYS

Adam Burrows
Departments of Physics and Astronomy
University of Arizona, USA

1. INTRODUCTION

The oft-predicted[1,2,3] radioactivity of supernovae has been directly and stunningly verified by the recent detections from the Type II supernova, SN1987A, of ^{56}Co gamma-ray lines and their hard X-ray Comptonization continuum.[4–14] The production of the parent ^{56}Ni by explosive nucleosynthesis in supernova shock waves has long been a staple of supernova lore and has been central to the modeling of Type Ia optical light curves for decades.[15–19] Type Ia's are thought to contain between 0.4 and 0.8 $M_\odot$ of ^{56}Ni and to involve the explosive disassembly of an entire C/O white dwarf at the Chandrasekhar mass.

A comparison shown in Figure 1 of the blue light curve for Type Ia supernovae compiled from Doggett and Branch[20] with the theoretical total and 847 keV line emissions calculated by Burrows and The[21] for model W7 of Nomoto, Thielemann, and Yokoi[22] serves to demonstrate the energetic importance of radioactivity in Type Ia's. The observations of the Type II SN1987A tell us that between 0.07 and 0.08 $M_\odot$ of ^{56}Ni was created and ejected in its explosion. Mayle and Wilson[23] find that the collapse of an ONeMg white dwarf might yield $\sim$0.002 $M_\odot$ of ^{56}Ni. Therefore, both theory and observation imply that the ^{56}Ni burden of a supernova explosion might span three orders of magnitude from 0.001 to 1.0 $M_\odot$ and that supernovae as a class provide a rich and varied range of gamma-ray line and continuum signatures.

In addition to isotopes of the decay chain, $^{56}\mathrm{Ni} \rightarrow {}^{56}\mathrm{Co} \rightarrow {}^{56}\mathrm{Fe}$, it is thought that ^{57}Co, ^{44}Ti, ^{26}Al, ^{22}Na, and ^{60}Co pepper the supernova debris in varying amounts. They too could provide characteristic gamma-ray and positron signals from a suitably close event. The gamma spectroscopy of supernovae has been the subject of several studies over the years.[21,24–32] In principle, gamma-rays can reveal a supernova's debris mass, progenitor, explosion energy, velocities and velocity profiles, hydrodynamic instabilities, heavy element abundances, and explosion mechanism, among other particulars. Conversely, the explosions of C/O white dwarfs, helium cores, Wolf-Rayet stars, red supergiants, blue supergiants, and the accretion-induced collapse/explosion of an ONeMg white dwarf all have characteristic hard photon signatures. Each type of supernova (Ia, Ib, Ic, IIp, IIL, IIpec, and SN1987A-like) must have a unique gamma print that profitably can be used to distinguish it. With this paper, I hope to give a preliminary and broad-brush glimpse of the diagnostic power of gamma- and X-ray observations of supernovae of all types. I outline the basic features of supernova gamma-ray emissions and the beginnings of a program to characterize the entire family of supernovae in gamma light.

2. GAMMA-LINE LIGHT CURVES

There are more than fifty distinct gamma lines emitted during the various radioactive decay sequences in supernovae, each with its own characteristic branching ratio and energy. The decay of ^{56}Ni ($\tau \sim 8.8$ days) provides important gammas at 158 keV (98.8%), 269 keV (35.6%), 480 keV (35.6%), *750 keV (49.9%), *812 keV (87.4%), and 1562 keV (14%). The decay of ^{56}Co ($\tau \sim 111.3$

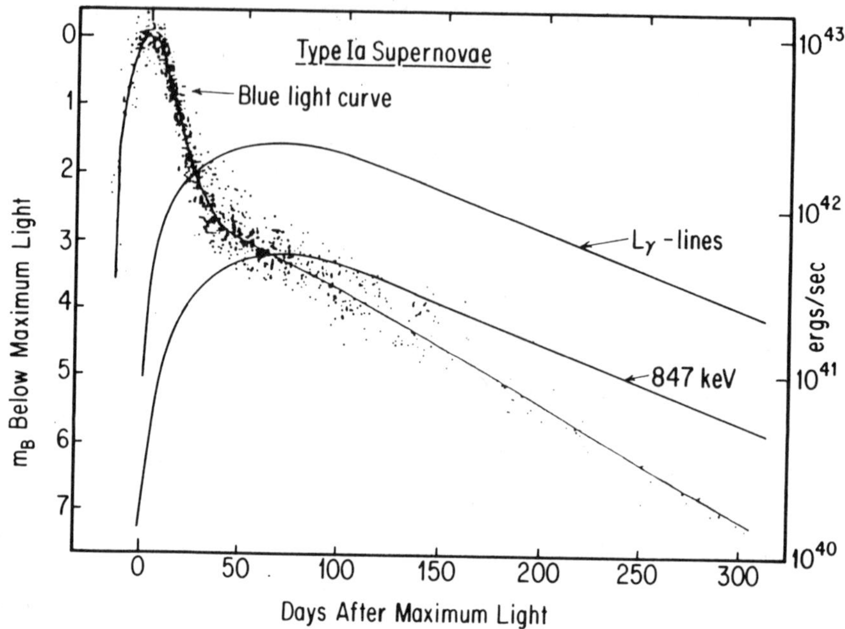

Figure 1: A comparison of the collective blue magnitude light curves of Type Ia supernovae compiled by Doggett and Branch[20] with the corresponding total and 847 keV gamma-line luminosities, as calculated by Burrows and The[21], for Nomoto, Thielemann, and Yokoi's model W7.[22] The right scale refers to the gamma luminosities alone. Time on the abscissa is in days after maximum blue light.

day) yields important gammas at *847 keV (99.9%), 1038 keV (14%), *1238 keV (67.6%), 1772 keV (15.7%) and 2598 keV (16.9%), along with a positron 19% of the time. Small abundances of the interesting isotopes ^{57}Co ($\tau \sim 391$ days) and ^{44}Ti ($\tau \sim 68$ years) are produced with lines at 14.4 keV (86%), *122 keV (86%), and 136 keV (11%), and 68 keV (100%), 78 keV (100%), and *1157 keV (100%), respectively. ^{44}Ti decay also provides a positron 94% of the time. (Those lines that are most important for diagnostic purposes are identified with an asterisk.)

The light curves of the various gamma-ray lines result from a battle between decay and the progressive transparency of the expanding and thinning debris. Supernovae whose progenitors have low mass and large expansion speeds (e.g. Type Ia's) become transparent and reveal their radioactive burdens before they have decayed away and thereby are very bright at peak in the dominant gamma lines. Conversely, supernovae with lower expansion speeds (e.g. Type IIp) from more massive progenitors (red supergiants?) become transparent only after a large fraction of the dominant ^{56}Ni and ^{56}Co have decayed away and yield only modest gamma line emission. Since the latter are also thought to produce less ^{56}Ni ($\sim$0.1 $M_\odot$ versus $\sim$0.6 $M_\odot$ for a Ia), their peak 847 keV emissions are only one-thousandth those of a Type Ia. Intermediate mass progenitors (for Type Ib's?) should provide intermediate displays. Therefore, a gamma-line light curve, with its rise and peak times, its flux at peak, etc., can be a powerful diagnostic and discriminant of the global progenitor and explosion characteristics, as well as of mixing and clumping in the debris.[33] A range of theoretical 847 keV light curves for different supernova types at 1 Megaparsec is shown in Figure 2. Characteristically, a small-mass ^{56}Ni-rich supernova (Type Ia) peaks early (~80 days) and bright, while a more massive supernova (e.g. SN1987A, ~16 $M_\odot$) peaks later (~350–400 days) and much dimmer. The systematics can be reproduced crudely by the standard analytic theory[25] that predicts that the time to peak, t_p, is proportional to $(\kappa_\gamma \frac{M^2\tau}{E})^{1/3}$ and $F(\text{peak}) \propto \frac{1}{\tau} e^{-\frac{3t_p}{2\tau}}$, where τ is the mean-life, M is the supernova mass, E is the explosion kientic energy, and κ_γ is the Compton opacity, which at 1 MeV is ~0.06 cm^2/gm ($\sim\frac{1}{3}\,\kappa_{\text{Thomson}}$).

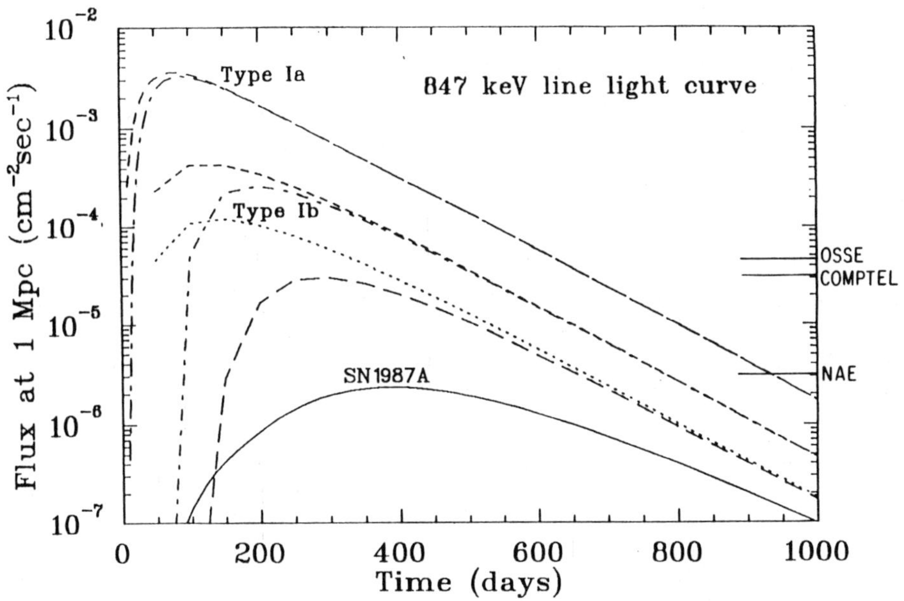

Figure 2: The 847 keV line flux (in photons/[cm^2·s]) at 1 Mpc versus the time (in days) since the explosion of various model supernovae. These curves were obtained for the Type Ia's from Burrows and The[21], for the Type Ib's from The, Clayton, and Burrows[37], and for SN1987A from The, Burrows, and Bussard.[38] The 10^6 second sensitivities at 847 keV in OSSE and COMPTEL on the GRO and in the NAE are indicated on the right-hand-side.

Since the Compton scattering cross section is a decreasing function of energy, even lines with moderate to low branching ratios can emerge with respectable luminosities if they are hard. In this way, the 1238 keV line ($\sigma_{\text{KN}} = 0.285\sigma_T$) can outshine the higher-yield 847 keV line ($\sigma_{\text{KN}} = 0.344\sigma_T$) in the early phases and the weak line blends of ^{56}Co decay near 2.02 MeV ($\sigma_{\text{KN}} \sim$

$0.219\sigma_T$) and 3.2 MeV ($\sigma_{\rm KN} \sim 0.166\sigma_T$), with collective branching ratios of only 10%, can compete. For some supernovae, the latter could be more visible in COMPTEL on GRO, due to its significant sensitivity above 2 MeV, than the former would be in OSSE[30,34] (see Figure 8 below). Figure 3 depicts the theoretical gamma-line light curves for the brightest ten lines of ^{56}Co decay for Type Ia model W7 of Nomoto *et al.*[22] The complicated line crossings during the first 50 days demonstrate the effect of different Compton depths. The line ratios change dramatically during the early phase. The different Compton opacities, mean lives, and branching ratios of the dominant lines should make gamma line ratios and their evolution useful diagnostics of supernova type. Furthermore, due to the short ^{56}Ni decay time, the ^{56}Ni line emissions at 812 keV and 750 keV and the 812/847 keV ratio are very sensitive to explosion mass and, for Type Ia's, to the explosion mechanism (e.g. detonation/deflagration). They can vary orders of magnitude from model to model even when the 847 keV fluxes vary only slightly.[21,30] This makes the ^{56}Ni lines sensitive model probes, if detected. In addition, the velocity distribution of an isotope maps cleanly onto the line profiles, centroids, and FWHM's that may be discernable with the proposed high resolution ($\lesssim$2 keV) Nuclear Astrophysics Explorer[35] (NAE) or Ge detectors flown on balloons (if the supernova is nearby)[14,30,36]. Thus, we expect that detailed gamma-line studies could provide unprecedented glimpses of the supernova interior and structure.

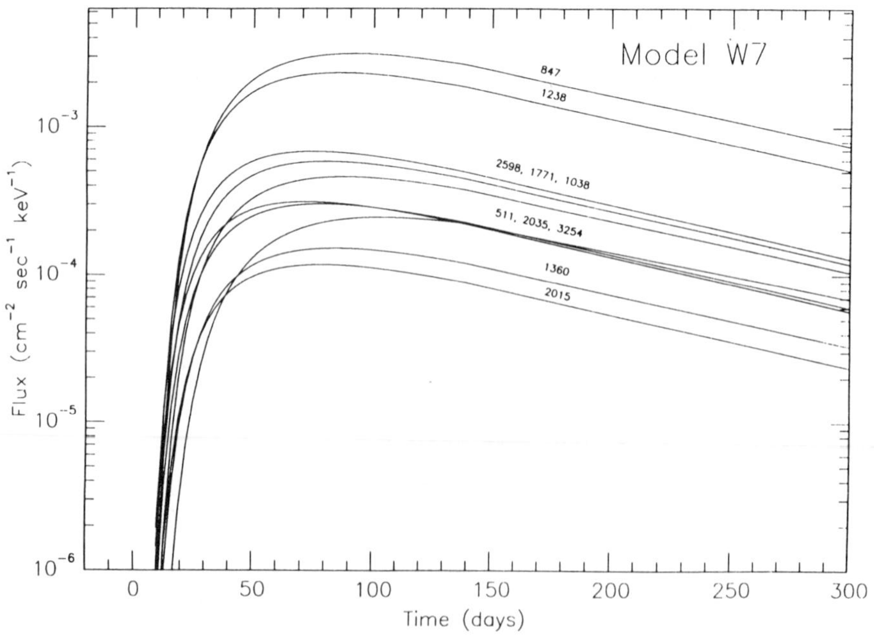

Figure 3: The line flux (in photons/[cm^2·keV·s])versus time (in days) for the ten brightest ^{56}Co lines from Type Ia model W7 of Nomoto, Thielemann, and Yokoi.[22] The line energies in keV mark the corresponding light curves.

3. THE COMPTONIZATION CONTINUUM

Those gamma-line photons that don't escape unimpeded from the supernova will Compton-scatter off its electrons. With every scattering, a photon loses energy. Either the photon eventually emerges at lower energies from the debris after a series of scatterings or it is photoelectrically absorbed. In this way, not only are many of the sharp gamma lines converted into a hard X-ray continuum that can extend down to as low as 10 keV, but simultaneously a significant portion of the radioactive energy is deposited as heat. As SN1987A and Type Ia supernovae demonstrate, this heat can be a crucial power source for a supernova's optical, IR, and UV emissions. Importantly, and for the first time, the associated emergent Compton continuum was observed from SN1987A, starting near day 150, and was accurately modeled by many.[4,5,8,10–12,38–46] Qualitatively, Compton scattering in an electron cloud is diffusion in wavelength space, with every scattering shifting the photon's wavelength down by the electron Compton wavelength ($\frac{h}{m_e c}$). Therefore, the first few scatterings entail the largest energy transfers. The number of scatterings required to achieve an energy ϵ_γ is, after a few scatterings, independent of the initial photon energy and approximately equal to $\frac{m_e c^2}{\epsilon_\gamma}$. Hence, to reach 20 keV requires $\sim$25 scatterings[46] and a debris Compton depth to 1 MeV gammas larger than 3. This is possible during the mid-stages of the supernova's expansion, when it is not so opaque that no hard X-rays can emerge before being photoelectrically absorbed, but when it is not so transparent that the gammas can escape before scattering down to lower energies. The continuum spectrum is the result of an evolving competition between down-scattering, escape, and photoelectric absorption and is usefully model-specific.

In all supernova models, the X-ray continuum emerges before the gamma lines themselves. For instance, as calculated by BT, during the first 3 weeks of a Type Ia explosion in both model W7 of Nomoto, Thielemann, and Yokoi[22] and model cdtg7 of Woosley and Weaver[47], the total continuum luminosity exceeds the total gamma line luminosity by as much as a factor of 3. Despite this similarity between models W7 and cdtg7 (called WW2 in BT), there are interesting differences between them in these Comptonization continuum emissions that exemplify the potential diagnostic usefulness of hard photon observations. Both models involve a white dwarf progenitor at roughly the same Chandrasekhar mass and with the same C/O composition. The only crucial difference between them is that in WW2 the deflagration wave is slower and fizzles earlier than in W7. This one difference translates into a smaller ^{56}Ni yield distributed more deeply, a lower explosion energy, and, hence lower expansion speeds, and fewer intermediate mass elements (e.g. Ca, Ar, Si, S). The consequences for the gamma-line and continuum signatures are significant. WW2 has lower peak ^{56}Ni and ^{56}Co line fluxes and narrower line profiles. Importantly, the continuum hard X-ray flux from WW2 is as much as an order of magnitude lower at 10 days, its X-ray/line ratio stays high longer, and, due to its lower heavy element burden, it has a lower hard X-ray photoelectric energy cutoff.

Figure 4 depicts the flux (in photons/[cm^2·s·keV]) at a distance of 1 megaparsec from various Type Ia models at 20 days. Models W7A and W7E are mixed versions of W7. Figure 3 shows the typical low-energy cutoff, peak, and high energy power-law structure of a supernova hard-photon spectrum before significant transparency. The differences between models WW2 and W7 in Figure 3 are quite striking and suggest an important diagnostic role for emissions below 100 keV. The differences between models W7A and W7E and model

W7 serve to illustrate the effects of various levels of hydrodynamic instability and mixing. The order-of-magnitude larger flux from WW2 below 50 keV may spell the difference between a detection and an upper limit in some satellite instruments (e.g., MIR/HEXE[5]; GINGA[10,11]).

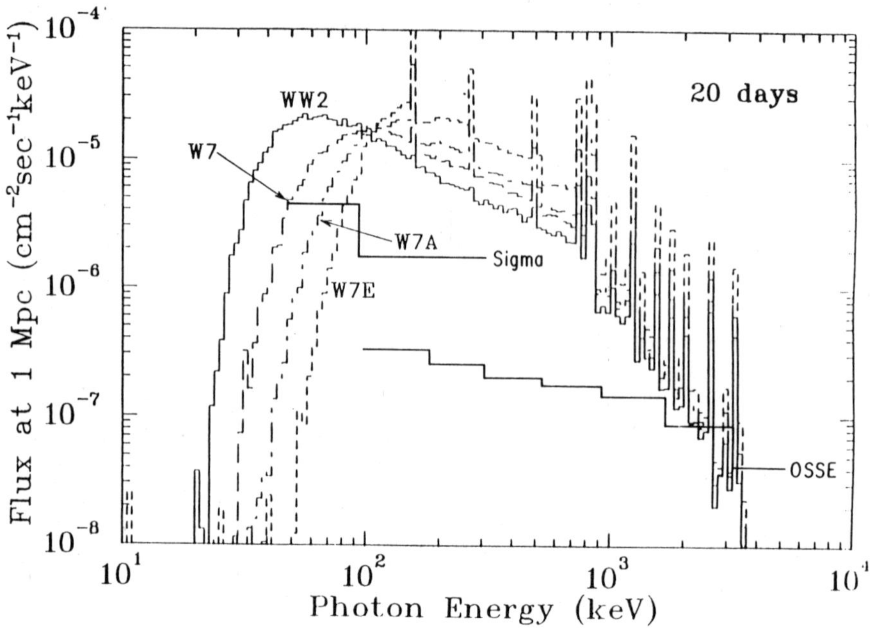

Figure 4: The continuum flux (in photons/[cm^2·s·keV]) at 1 Mpc versus photon energy (in keV) for Type Ia models WW2 (cdtg7 from Woosley and Weaver[47]), W7 (from Nomoto, Thielemann, and Yokoi[22]), W7A, and W7E, as calculated by Burrows and The.[21] Superposed are the approximate continuum sensitivities of the Sigma/GRANAT and OSSE gamma-ray detectors. The significant differences between the models below 100 keV are noteworthy. After day 60, these spectra merge and become almost indistinguishable.

Some gross distinguishing characteristics of the hard X-ray continuum from the variety of supernova types are the peak energy (ϵ_p), the peak total flux, the time at peak, the low-energy cutoff, the heights of the spectral steps across the strong gamma-ray lines, and the X-ray band ratios or "colors".[37] The latter may provide crude, but robust, signatures of various models. Another important feature of the continuum comes from the 3γ (75%) annihilation in positronium of the ^{56}Co positron (19%). At late times ("late" being determined for a specific mass and explosion energy), much of the continuum flux below 511 keV comes from this 3γ decay. BT have shown how much more important the <511 keV step is as a diagnostic feature in Type Ia's than in Type II's from supergiants (e.g., SN1987A). Similarly, much of the continuum flux below 122 keV at late times comes from the Comptonization of the 122 keV and

136 keV lines of ^{57}Co. Despite the low ^{57}Co abundance expected in supernovae $((\frac{Fe^{57}}{Fe^{56}})_{\odot} \sim \frac{1}{43})$, the low energies and, hence, high Compton cross sections, of its dominant lines ensure a Compton tail, even when the 847 keV line is more prominent. A shelf of continuum just below 122 keV was confidently predicted for SN1987A, but, unfortunately, it was too dim to be unambiguously detected by either Sigma/GRANAT[48] or MIR/HEXE.[49]

As with the ^{56}Ni lines, the continuum emissions speak most eloquently early during the supernova explosion. For Type Ia's, different models show radically different continua only during the first 50 days. Afterwards, they converge and lose their ability to reflect details that can distinguish different supernova models for the same supernova type. However, even at late times, the continua can be used to distinguish gross supernova and progenitor types. Figures 5 and 6 directly demonstrate the quick diagnostic power of the hard photon spectrum. They depict theoretical spectra for, respectively, a Type Ia (W7[22]) and a Type Ib (WR4A[38]) supernova model at the same time (200 days) and distance (1 Mpc). The stark differences between the spectral slopes, the low-energy cutoffs, the ^{57}Co and 511 keV line fluxes, the 3γ continua, and the line widths are readily apparent. (Note the high line resolution of these figures, in contrast to that in Figure 4.)

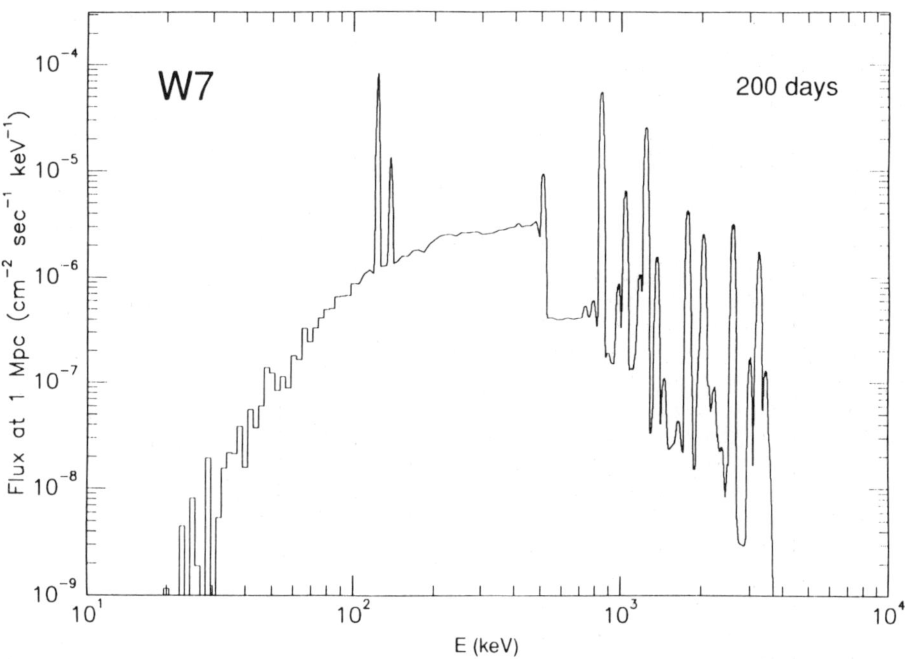

Figure 5: A high-resolution (for ϵ_γ >100 keV) theoretical spectrum at 200 days and 1 Mpc for Type Ia model W7.[22] The flux is in photons/[cm^2·s·keV] and the photon energy is in keV. This figure is to be compared with Figure 6.

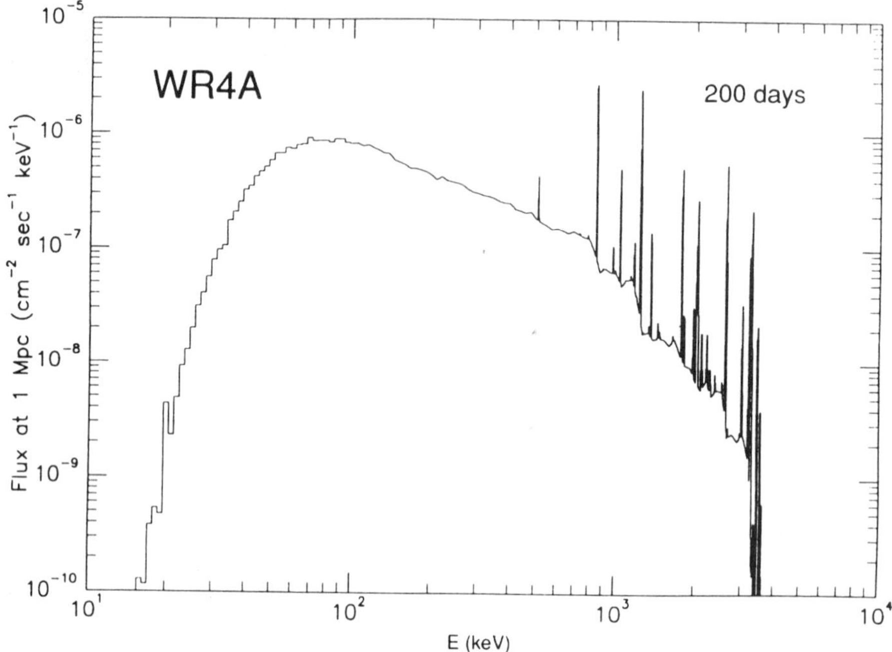

Figure 6: A high-resolution (for ϵ_γ >100 keV) theoretical spectrum at 200 days and 1 Mpc for Type Ib model WR4A of Ensman and Woosley as analysed by The, Clayton, and Burrows.[38] The flux is in photons/[cm^2·s·keV] and the photon energy is in keV. [Compare to Figure 5].

At late times, Compton photons desert the low-energy region of the spectra. The gammas are escaping after only a few scatterings. Indeed, for Type Ia's, ϵ_p increases quickly with time from tens of keV at 10–20 days to 100's of keV at 150 days. For supergiants, there is a more gradual, but no less systematic, hardening of the spectrum with time.

The time to peak for the continuum is very, very roughly proportional to the progenitor mass. Therefore, all else being equal, massive stars rise to maximum more slowly. In SN1987A, the continuum peaked near day 350, while for the much lighter Type Ia's it is predicted to peak near day 35 (BT). Furthermore, the ratio of t_p (continuum) to t_p (847 keV) is an increasing function of M^2/E. In this way, while for Type Ia's this ratio is predicted to be 35 days/80 days, for SN1987A it was almost 1:1. Such a simple ratio is but one of the progenitor-distinguishing features of the hard photon emissions of supernovae. However, because the continuum requires some Compton opacity to form, the X-ray flux peak should never follow that for the 847 keV gamma-ray line.

4. HYDRODYNAMIC INSTABILITIES

The UVOIR, hard X-ray, and gamma-ray fluxes from SN1987A could not be fitted without the assumption that some of the ^{56}Co source was "mixed" up from its production site at <500 km/s to ~3000 km/s (see Arnett *et al.* [50] and references therein). This implies that bullets or blobs of the isotope penetrated through both the heavy element core and the helium shell to the hydrogen envelope and that the spherical onion-skin structure of the progenitor was violated during explosion. Furthermore, the observations of line and continuum polarization[51], very-high-resolution ripples in the OI line profiles[52], clumpy, far IR-emitting/optical-absorbing dust clouds[53] and the asymmetric and corregated line profiles of FeII at 1.257 μm[54], all point to asphericities and hydrodynamic instabilities in the debris of SN1987A. Interestingly, the redward shift of 500–800 km/s in the IR FeII line observed after day 700 by Spyromilio, Meikle, and Allen[54] was used to infer a receding clump of ^{56}Fe that also explains the redward shift of the 847 and 1238 keV lines of ^{56}Co observed by Tueller *et al.* [14] on days 433 and 613. Such mutually confirming data in such disparate spectral bands hints at the potential power of complementary gamma-ray line data for studying future supernovae.

That the explosion of Sanduleak –69°202 was not spherically symmetric implies a frightening richness for the supernova family: all supernova explosions may involve hydrodynamic instabilities and ^{56}Ni mixing at some level. Theorists are just now trying to come to grips with the multi-dimensional nature of these explosions[33] and even the gross systematics have not yet been worked out. Which supernova types mix, where, and by what mechanisms? On the answers to these basic questions might hang our future understanding of much of the supernova display.

As they were for SN1987A, the hard photon emissions from supernovae of all types are sensitive indicators of the degree of ^{56}Ni mixing. Mixing and clumping lead to earlier hard X-ray and gamma-ray emergence, higher peak luminosities, higher continuum energy cutoffs, higher continuum peak energies, and broader gamma lines. The shifting of the continuum peak and cutoff due to partial (W7A) and complete (W7E) mixing in a Type Ia model (W7) can be seen in Figure 4 (see BT for a more complete discussion). Mixing to lower gamma depths truncates the Comptonization cascade to lower energies and enhances the effective photoelectric depths of the debris. The net effect is a pronounced shifting of the hard X-ray spectrum out of various standard hard X-ray bands. Some of the wide variation in Type Ib 847 keV light curves in Figure 2 can be traced to ambiguities concerning mixing. "Mixing" affects so many features of the gamma-line and continuum signature of a supernova that an inversion of future hard photon data from a supernova could not fail to reveal crucial aspects of the entire event. A collaboration of myself, K. Van Riper, D. Arnett, B. Fryxell, and A. Shankar will soon perform Monte Carlo calculations on the multi-dimensional hydrodynamic models of Arnett, Fryxell, and Müller[33] in an attempt to illuminate the detailed gamma-ray consequences of these problematic instabilities.

5. THE LOWEST-ENERGY X-RAYS

Lest the reader be lulled into believing that only very hard X-rays and gamma rays issue from supernovae, mention should be made of the variety of softer X-ray signatures expected. Supernovae from red supergiant-, blue supergiant-, or helium core-progenitors with their larger fraction of light elements will indeed have Comptonization flux below 20 keV. In addition, depending on the production of ^{57}Ni, the 14.4 keV gamma line of ^{57}Co might be discernable. However, the collision of the ejecta with the progenitor's fossil wind (predicted to occur within $\sim$10 years for SN1987A) will begin an X-ray renaissance that should gradually lead into the supernova remnant stage for which both AXAF (<10 keV) and ROSAT (<2 keV) are well suited. Furthermore, as pointed out by Woosley, Pinto, and Hartmann[55] and Chevalier[56], an X-ray pulsar powered either by fall-back accretion or the standard pulsar mechanism might progressively leak soft X-rays, particularly if the debris is clumpy. This core emission has yet to be seen from either SN1987A or any other supernova, but, in principle, is not excluded.

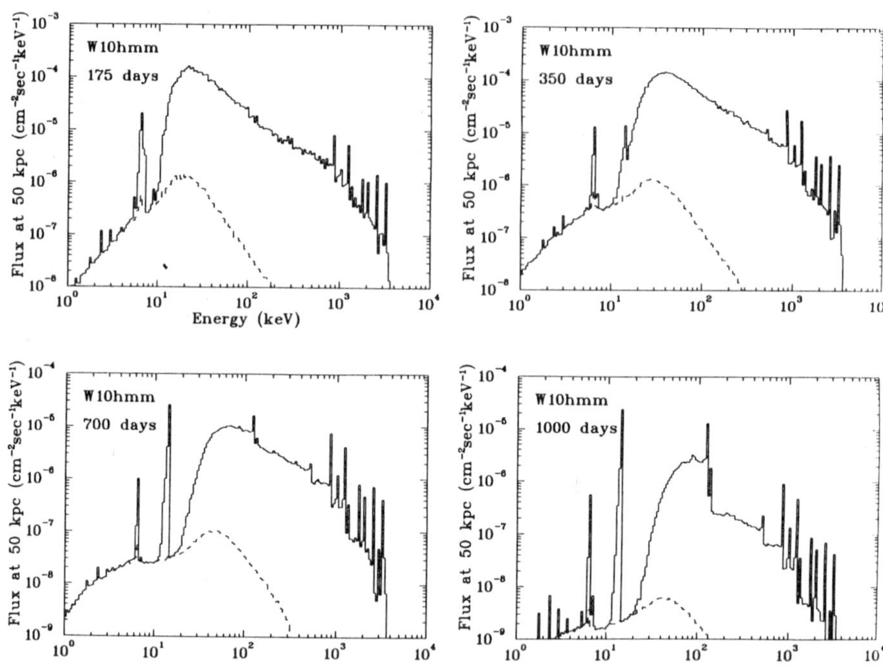

Figure 7: The flux (in $cm^{-2}sec^{-1}keV^{-1}$) at 50 kpc from the SN1987A model W10HMM of Woosley and Pinto[31] versus energy (in keV) from 1 to 10^4 keV at 175, 350, 700, and 1000 days. The bremsstrahlung contribution to the flux above 10 keV is identified with a dashed line. Notice that fluorescence and bremstraulung components fill in the spectrum between 1 and 10 keV, despite the significant photoelectric absorption and the absence of a Comptonization contribution. (Figure from Clayton and The[57].)

Three other potential classes of supernova X-rays below 10 keV are worth mentioning. The first class is comprised of the K and L X-ray fluorescence lines from the heavy element-rich debris. Pumped by Comptonization electrons and the ambient hard radiation field, not only the standard iron lines at 6.4 keV, but those at lower energies from Si, S, and Ca, etc. might be expected. Photoelectric absorption would be the most severe inhibiter of the latter, but proximity and clumpiness might reveal them. The second is bremsstrahlung of the Compton electrons. Figure 7 shows the results of a recent calculation by Clayton and The[57] that includes the Compton bremsstrahlung and fluorescence lines at various epochs during the evolution of SN1987A. As much as 0.1% of all the supernova flux might be in this component, some of which is between 1 and 10 keV. However, the third class is the most intriguing: the EUV/soft X-ray break-out burst. When the supernova shock wave first emerges from the stellar surface, the shock radiation floods out of the star in an ultraviolet/soft X-ray burst that for SN1987A had a temperature between 5×10^5 and 10^6 K. The actual duration, temperature, and luminosity depend predominantly on the progenitor's radius[58], so that if this flash is detected, it will provide an independent and direct indicator of the progenitor's geometry. Each supernova type will have a distinctive flash history and spectrum. That from a Type Ia should last only several seconds, but be in the hard X-ray. That from a red supergiant should be in the near UV and last days. The flash from SN1987A was missed by only a day, as it lasted only hours, but it caused the swept-up ring of material from its progenitor's fossil red giant wind to fluoresce. This fluorescence was observed by the IUE.[59]

6. A SYNOPTIC GAMMA SIGNATURE LOOK-UP TABLE

One can get a summary sense of the correspondence between the gamma-line and hard X-ray continuum emissions and the progenitor and explosion parameters by studying Table 1. Vertically are listed various hard photon observables and horizontally are listed a representative sample of supernova characteristics one would like to derive about a specific event. The entries should be self-explanatory. While at first glance this table might seem daunting, it contains useful information concerning supernova gamma-ray spectroscopy and its promise. The black dots at the intersection of row and column identify those observable model pairs that are most intimately associated. Since almost every model feature is coupled to almost every observable, black dots could, in principle, fill the entire table and render it useless. To avoid this problem, I have highlighted only the most important correspondences and the best observables for a given progenitor feature and have been a tad idiosyncratic. For instance, "^{44}Ti lines" and "EXPL. NUCL." (explosive nucleosynthesis) are matched because the identification of ^{44}Ti would reveal the nature of the explosive nucleosynthetic environment in the progenitor. "Mass," "^{56}Ni Yield," "Exp. Energy," and "Mixing" are coupled to the peak ^{56}Co line and "Comptonization continuum" fluxes because it is these model parameters that most directly influence them. Furthermore, "Break-Out Burst" is tied to "Initial Radius" because it is on this radius that the duration, energy, luminosity, and color temperature of the burst most depend. In addition, the "511 keV" line is tagged with "Ionization" because the ratio of the 2γ to the 3γ positron annihilation emissions depends on the ionization fraction of the debris[60], while "FLUX $<$ 20 keV" is coupled with "Mixing" and "Heavy Abund." because of the latter's important effect on the photoelectric opacity at low X-ray energies. It is recommended that the reader stop to peruse this table and puzzle out, where the reason for the black dot is not obvious, the why of its presence. It is in the full supernova gamma-ray literature cited in this text that the variety of full explanations can be found, though the positioning of most of the dots is self-evident.

7. DETECTORS AND RANGES

Hard X-rays and gamma-ray lines were observed from SN1987A by a series of NASA balloons, the SMM satellite, HEXE and PULSAR X–1 on the Soviet Mir space station, and the Japanese satellite, Ginga. The future promises much more. The American Gamma-Ray Observatory (GRO) soon will be launched with its instrument complement of OSSE, COMPTEL, and BATSE.[34,61] The

	MASS	^{56}Ni YIELD	EXP. ENERGY	VELOCITIES	VEL. PROFILE	"MIXING"	ASYMMETRY	^{56}Ni PROFILE	INITIAL RADIUS	HEAVY ABUND.	EXPL. NUCL.	IONIZATION	DEFL./DETON.			Y_e OF EJECTA	REMNANT AGE	PULSAR	WIND INTER.			TYPE Ia	BLUE SG	RED SG	ONeMg A.I.C.	TYPE Ib	~EXPL. TIME
^{56}Co LINES:																											
PEAK FLUXES	●	●	●	●		●							●									●	●	●	●	●	
TIMES AT PEAK	●		●	●		●							●									●	●	●	●	●	
RISE TIMES	●		●	●		●																					
LINE RATIOS				●		●																					
LATE FLUXES		●																							●		
FWHM				●		●							●									●	●	●	●		
LINE CENTROIDS	●			●		●	●						●									●	●	●	●		
LINE PROFILES			●	●	●	●	●	●					●														
511 keV	●											●										●			●	●	
^{56}Ni LINES:																											
PEAK FLUXES	●			●		●							●									●			●	●	
812/847 RATIO						●							●									●			●	●	
158 keV						●																●			●	●	
^{57}Co LINES:																											
122/136 keV											●					●							●	●			
14.4 keV										●	●					●							●	●			
^{44}Ti LINES:											●						●										
^{26}Al LINE:											●																
^{22}Na LINE:											●																
COMPTON CONT:																											
PEAK TOTAL FLUX	●	●	●	●		●							●									●	●	●	●	●	
TIME AT PEAK	●			●		●																	●	●		●	
t_p^x/t_p^{847}	●			●																					●		
SPECTRAL PEAK E						●				●												●			●	●	
SPECTRAL SHAPE	●			●						●																	
SPECTRAL STEPS	●			●												●											
LOW-E_γ CUTOFF						●				●			●														
BAND RATIOS	●			●												●											
F_x/F_γ						●				●																	
RISE TIME	●			●		●																					
BREAK-OUT BURST:																											
EUV/X-RAY									●													●	●	●	●	●	●
FLUX <20 keV						●				●								●	●								
Fe FLUOR:		●								●		●															

TABLE 1

Soviet/French Sigma/GRANAT is now in orbit[62], as is the German ROSAT. The American AXAF is on a trajectory for a mid- to late-90's launch[63], while the European HIREX, the Soviet Spectrum-Xγ, and the American NAE[35] are at encouraging stages in their development. The nineties should indeed see an international efflorescence of gamma- and X-ray astronomy that bodes well for the high-energy study of nearby supernovae.

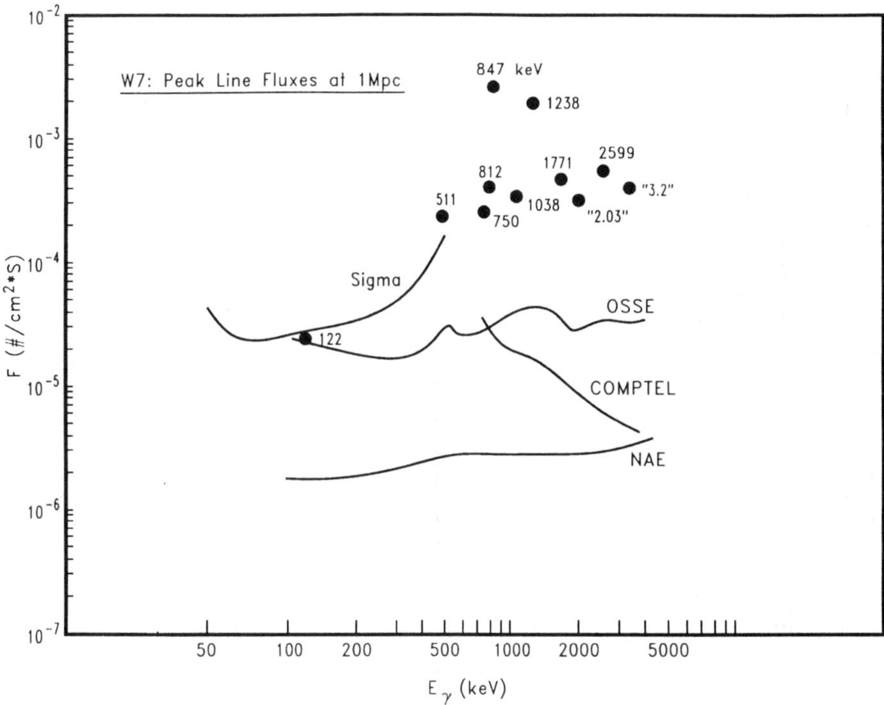

Figure 8: The peak line fluxes (in photons/[$cm^2 \cdot s$]) versus photon energy (in keV) at 1 Mpc for various dominant gamma-ray lines from Type Ia model W7 of Nomoto, Thielemann, and Yokoi.[22] The numbers that accompany the dots are in keV, except for the "2.03" and "3.2", which are in MeV. The approximate line sensitivities of the Sigma/GRANAT, OSSE, COMPTEL, and NAE detectors are included for comparison.

Figure 8 depicts various peak gamma-line fluxes at 1 Mpc from Type Ia model W7. The approximate line sensitivities in Sigma, OSSE, COMPTEL, and the proposed NAE are shown for comparison. One megaparsec is a fiducial distance and with its use we do not mean to imply that the average supernova is expected to be that close. It is not. However, as the comparison between the theoretical fluxes and the sensitivities implies, a simple scaling yields a Type Ia detection range of ~10 Mpc for both the 847 keV line by OSSE and the "3.2" MeV line blend by COMPTEL. These lines would be visible in the NAE out to distances of 15–30 Mpc. The sensitivities in the OSSE, COMPTEL, and NAE at 847 keV are also indicated in Figure 2 and the approximate continuum sensitivities in OSSE and Sigma/GRANAT are indicated in Figure 4. Importantly, OSSE has approximately the same range for continuum detection as line detection (10 Mpc), while Sigma can now see out to ~3 Mpc in the continuum above ~100 keV and to 0.75 Mpc near 50 keV, where OSSE has no sensitivity. Given the Type Ia supernova event rates inferred from the historical and modern record[64], at least one such supernova should be caught by GRO during its active life. It is encouraging to note that the recent Type Ia's, 1986G and 1989B, were within GRO's range. In general, a Type Ia whose blue magnitude is brighter than 11.5 will be visible by the GRO and whose blue magnitude is brighter than ~14 will be visible in the proposed NAE.[32,65]

Other supernova types are expected to be dimmer in high-energy photons. Hence, the various satellite instruments will have correspondingly shorter ranges for their detection. However, the 847 keV, 1238 keV, and hard continuum fluxes from various helium core models of Type Ib's might be visible in OSSE out to ~4 Mpc (see Figure 2 and The, Clayton, and Burrows[37]). Both of the recent Type Ib's, 1983N and 1985F, skirted this distance.

The characteristics of a good gamma line and continuum supernova detector are: a resolution ($\frac{E}{\Delta E}$) of 500 or greater, sensitivity from 5 kev to 4 MeV, an effective area of 10^3 cm^2, and a time resolution of at most days (except to observe the break-out flash!). Sensitivity is far more important than angular resolution, which need be no better than 30′. These specifications are all within reach today and should be achieved this decade, if only collectively through the overlap of the various national and international programs. Soon, all we will need will be a few obliging nearby supernovae.

ACKNOWLEDGMENTS:

The author is indebted to Anurag Shankar, Rich Lingenfelter, Peter Sutherland, Don Clayton, Dave Arnett, Phil Pinto, Stan Woosley, Bruce Fryxell, Roger Bussard, and Lisa Ensman. This work was supported by the U.S. NSF and NASA's Long-Term Space Astrophysics Research Program through grants no. AST89–14346 and NAGW–2145, respectively.

REFERENCES

1. Woosley, S. E. and Weaver, T. A., 1986, *Ann. Rev. Astro. Ap.*, 24, 205.
2. Nomoto, K., 1983, in *Proc. of the Workshop on Challenges and New Developments in Nucleosynthesis*, (University of Chicago Press).
3. Thielemann, F.-K., Nomoto, K. and Yokoi, K., 1986, *Astron. Astrophys.*, 158, 17.
4. Cook, W. R., Palmer, D. M., Prince, T. A., Schindler, S., Starr, C. H., and Stone, E. C., 1988, *Ap. J. (Letters)*, 334, L87.
5. Trumper, J., Reppin, C., Pietsch, W., Englhauser, J., Voges, W., Kendziorra, E., Bezler, M., Staubert, R., Sunyaev, R., Kaniovski, A., Efremov, V., Grebenev, M., Kuznetsov, A., Melioranskiy, A., Stepanov, D., Chulkov, I., 1988, in *Proc. of the George Mason Workshop*, "Supernova 1987A in the Large Magellanic Cloud," ed. M. Kafatos and A. Michalitsianos (Cambridge U. Press), p. 355.

6. Matz, S. M., Share, G. H., Leising, M. D., Chupp, E. L., Vestrand, W. T., Purcell, W. R., Strickman, M. S., and Reppin, C., 1988, *Nature*, 331, 416.
7. Mahoney, W. A., Varnell, L. S., Jacobson, A. S., Ling, J. C., Radocinski, R. G., and Wheaton, WA., 1988, *Ap. J. (Letters)*, 334, L81.
8. Sandie, W. G., Nakano, G. H., Chase, Jr. L. F., Fishman, G. J., Meegan, C. A., Wilson, R. B., Paciesas, W. S., and Lasche, G. P., 1988, *Ap. J. (Letters)*, 334, L91.
9. Sunyaev, R., et al., l988, *Sov. Astron. Lett.*, 14, 579.
10. Tanaka, Y., l988, in IAU Colloquium 108, Atmospheric Diagnostics of Stellar Evolution, ed. K. Nomoto (Springer-Verlag), p. 499.
11. Tanaka, Y., l988, in *Proc. of the George Mason Workshop*, "Supernova l987A in the Large Magellanic Cloud," ed. M. Kafatos and A. Michalitsianos (Cambridge U. Press) p. 349.
12. Wilson, R. B., Fishman, G. J., Meegan, C. A., Paciesas, W. S., and Pendleton, G. N., l988, to appear in *Proc. of the Workshop on Nuclear Spectroscopy of Astrophysical Sources*, AIP Conf. Proc. 170, ed. N. Gehrels and G. H. Share (Washington, DC:AIP).
13. Teegarden, B. J., Barthelmy, S. D., Gehrels, N., Tueller, J., Leventhal, M., and MacCallum, C. J., l989, *Nature*, 339, 122.
14. Tueller, J., Barthelmy, S., Gehrels, N., Teegarden, B. J., Leventhal, M. and MacCallum, C. J., 1990, *Ap. J. (Letters)*, 351, L41.
15. Pankey, T., 1962, Ph.D. Thesis, Howard University.
16. Clayton, D. D., Colgate, S. A., and Fishman, G. l969, *Ap. J.*, 155, 75.
17. Colgate, S. A. and McKee, C., 1969, *Ap. J.*, 157, 623.
18. Colgate, S. A., Petschek, A. G., and Kriese, J. T., 1980, *Ap. J. (Letters)*, 237, L81.
19. Arnett, W. D., l982, *Ap. J.*, 253, 785.
20. Doggett, J. B. and Branch, D., 1985, *Astron. J.*, 90, 2303.
21. Burrows, A. and The, L.-S., 1990, *Ap. J.*, 360, 626 (BT).
22. Nomoto, K., Thielemann, F.-K., and Yokoi, K., 1984, *Ap. J.*, 286, 664.
23. Mayle, R. and Wilson, J. R. 1988, *Ap. J.*, 334, 909.
24. Ambwani, K. and Sutherland, P., 1987, *Ap. J.*, 325, 820.
25. Sutherland, P., 1990, in *Supernovae*, ed. A. Petschek (Berlin: Springer-Verlag), p. 111.
26. Clayton, D. D., l974, *Ap. J.*, 188, 155.
27. Fishman, G. J., 1985, in *Proc. Workshop on Cosmic Ray and High-Energy Gamma-Ray Experiments for the Space Station Era*, ed. W. V. Jones and J. P. Wefel (Baton Rouge: LSU Press), p. 400.
28. Chan, K. W. and Lingenfelter, R. E., l987, *Ap. J. (Letters)*, 318, L51.
29. Chan, K. W. and Lingenfelter, R. E., 1988, in *Nuclear Spectroscopy of Astrophysical Sources*, ed. N. Gehrels and G. Share (New York:AIP), pp. 110-115.
30. Chan, K. W. and Lingenfelter, R. E., 1990, *Ap. J.*, in press.
31. Woosley, S. E. and Pinto, P. A., l988, in *AIP Conf. Proc. 170, Nuclear Spectroscopy of Astrophysical Sources*, eds. N. Gehrels and G. H. Share (Washington, DC:AIP).
32. Gehrels, N., Leventhal, M., MacCallum, C. J., l987, *Ap. J.*, 322, 215.
33. Arnett, W. D., Fryxell, B., and Müller, E., 1989, *Ap. J. (Letters)*, 341, L63.
34. Kurfess, J. D., l988, in *AIP Proc. 170, the Workshop on Nuclear Spectroscopy of Astrophysical Sources*, ed. N. Gehrels and G. H. Share (Washington, DC).

35. Matteson, J., 1989, in *The Proceedings of the Xth Santa Cruz Summer Workshop, Supernovae*, ed. S. E. Woosley, Santa Cruz, CA, July 10-21.
36. Bussard, R. W., Burrows, A., and The, L.-S., 1989, *Ap. J.*, 341, 401.
37. The, L.-S., Burrows, A., and Bussard, R. W., 1990, *Ap. J.*, 352, 731.,
38. The, L.-S., Clayton, D. D., and Burrows, A., 1990, "X-Ray and Gamma-Ray Signatures of Wolf-Rayet Supernova Explosions," in the proceedings of the IAU Symposium No. 143: "Wolf-Rayet Stars and Interrelations with Other Massive Stars in Galaxies," Denpasar, Indonesia, June 18–22, 1990.
39. Ebisuzaki, T. and Shibazaki, N., 1988, *Ap. J.*, 328, 699.
40. Grebenev, S. A. and Sunyaev, R. A., 1988, *Sov. Astron. Lett.*, 14, 675.
41. Kumagai, S., Itoh, M., Shigeyama, T., Nomoto, K., and Nishimura, J. 1988, *Astron. and Astrophys. (Letters)*, 197, L7.
42. McCray, R., Shull, J. M., and Sutherland, P., 1987, *Ap. J.*, (Letters), 317, L69.
43. Pinto, P. A. and Woosley, S. E., 1988, *Ap. J.*, 329, 820.
44. Sunyaev, R., et al., 1987, *Nature*, 330, 227.
45. Ubertini, P., Bazzano, A., Sood, R., Staubert, R., Sumner, T. J., and Frye, G., 1989, *Ap. J.*, 337, L19.
46. Xu, Y., Sutherland, P., McCray, R., and Ross, R. P., 1988, *Ap. J.*, 327, 197.
47. Woosley, S. E. and Weaver, T. A. 1986, in *Radiation Hydrodynamics in Stars and Compact Objects*, eds. D. Mihalas and K.-H. A. Winkler, (Springer-Verlag), p.91.
48. Cassé, M., 1990, in the proceedings of the ESO/EIPC Workshop *SN1987A and Other Supernovae*, held at Marciana Marina, Isola d'Elba 17–22 September.
49. Efremov, V. V., 1990, in the proceedings of the ESO/EIPC Workshop *SN1987A and Other Supernovae*, held at Marciana Marina, Isola d'Elba 17–22 September.
50. Arnett, W. D., Bahcall, J. N. Kirshner, R. P., and Woosley, S. E., 1989, *Ann. Rev. Astron. and Astrophy.*, 27, 629.,
51. Jeffrey, D. 1990, in the proceedings of the ESO/EIPC Workshop *SN1987A and Other Supernovae*, held at Marciana Marina, Isola d'Elba 17–22 September.
52. Stathakis, R. and Cannon, R. D., 1988, Anglo-Australian Observatory Newsletter No. 45.
53. Lucy, L. and Danziger, I. J., 1990, in the proceedings of the ESO/EIPC Workshop *SN1987A and Other Supernovae*, held at Marciana Marina, Isola d'Elba 17–22 September.
54. Spyromilio, J., Meikle, W. P. S. and Allen, D. A., 1990, *Mon. Not. R. astr. Soc.*, 242, 669.
55. Woosley, S. E., Pinto, P. A. and Hartmann, D., 1990, *Ap. J.*, in press.
56. Chevalier, R. 1990, in the proceedings of the ESO/EIPC Workshop *SN1987A and Other Supernovae*, held at Marciana Marina, Isola d'Elba 17–22 September.
57. Clayton, D. D. and The, L.-S., 1990, submitted to *Ap. J. (Letters)*.
58. Woosley, S. E., 1990, preprint.
59. Panagia, N. 1990, in the proceedings of the ESO/EIPC Workshop *SN1987A and Other Supernovae*, held at Marciana Marina, Isola d'Elba 17–22 September.
60. Bussard, R., Ramaty, R., and Drachman, R., 1979, *Ap. J.*, 228, 928.
61. Neal, V., Fishman, G. and Kniffen, D., 1990, *Mercury*, July/August issue, p. 98.

62. Durouchoux, P., 1989, in the Proceedings of the Xth Santa Cruz Workshop, "Supernovae," ed. S. E. Woosley (Springer-Verlag).
63. Gorenstein, P. 1990, in the proceedings of the ESO/EIPC Workshop *SN1987A and Other Supernovae*, held at Marciana Marina, Isola d'Elba 17–22 September.
64. Evans, R., van den Bergh, S., McClure, R. D., 1989, *Ap. J.*, 345, 752.
65. Arnett, W. D., Branch, D., and Wheeler, J. C., 1985, *Nature*, 314, 337.

COMPACT OBJECTS

GAMMA RAY BURSTS AND SPECTRAL FEATURES

G. Vedrenne and E. Jourdain

Centre d'Etude Spatiale des Rayonnements - CNRS/UPS
B.P. 4346 - 31029 TOULOUSE CEDEX

ABSTRACT

The situation concerning the presence of features in gamma ray burst (GRB) spectra is analyzed taking into account recent results obtained especially with the PHOBOS probes and GINGA satellite. If the existence of cyclotron features seems now to be generally accepted the situation for the other features reported around 400-500 keV is not completely clarified. The presence of such features is discussed. Moreover some aspects of the high and low energy variations in the GRB and on the characteristics of their total spectrum are also reviewed. Finally future missions which might have a great impact in the GRB spectral analysis will be shortly considered.

INTRODUCTION

One important way to understand the gamma ray burst (GRB) origin rests on the determination of their spectral characteristics. Here we will concentrate on the more recent results obtained with two missions : GINGA and PHOBOS and mainly on the analysis of possible features in the GRB spectra. For GINGA which is still operating two detectors are used : a proportional counter and a scintillation counter covering respectively the energy ranges 1.5 - 28 KeV and 14 - 375 KeV with effective area of about 60 cm^2 each. On the PHOBOS probe two types of detectors were also used : one CsI scintillator, diameter : 10 cm , height : 10 cm (APEX experiment) and two small cleaved NaI scintillators, diameter : 5 cm , height : 3 cm situated at each extremity of the solar pannels (LILAS experiment). These detectors were operating during the whole mission duration : between July 1988 and March 1989. The search for features at low energy (tens of keV) and high energy (400 - 1000 keV) was made easiest by the large number of energy channels available for the spectral analysis.

In this presentation we will review successively the observed features below few tens of keV and in the 500 keV - 1 MeV region. Then we will present very shortly recent observations of the GRB continuum and its rapid variations. Finally we will consider the future missions which might contribute decisively to the understanding of GRB spectral characteristics.

1) Features in GRB spectra :

The first reports of features in GRB spectra were due to the KONUS experiments on Venera 11 and 12 probes [1]. Two kinds of features were observed : one at low energy, below ~ 40 keV and one at higher energy around 400 - 500 keV (figures 1-2). The two next tables I and II give the main characteristics of the features summarized by K. HURLEY, 1987 [2].

Table I. GRB Line Feature Detections

		NUMBER OF EVENTS		
Spacecraft	Instru-mentation	A. Total	B. With L.E. Features	C. With H.E. Features
ISEE-3	Ge	≥24	(not observable)	1
HEAO-A	NaI	21	2	1
Venera 11,12	NaI(KONUS)	143	~ 30(~ 21%)	15(10%)
	NaI(SIGNE)	39	1	1
Venera 13,14	NaI(KONUS)	130	≥ 20(>15%)	29(22%)

Table II. Properties of GRB Line Features

Energy, keV	Type	Instan-taneous line to continuum flux ratio	Line width keV	Duration, sec
~ 27-70	Absorption	1-18%	~ 3->28	≤4
350-500	Emission	3-30%	200-990	≤0.25->20

a) Low energy features :

At low energies the features were in general absorption features, or very often breaks in the continuum spectra fitted by an optically thin thermal bremsstrahlung (OTTB) continuum. Much contreversies surrounded these results [3] , because the break at low energy naturally depends strongly on the choice of the continuum. Without reviewing this debate, it is clear that if for instance one choose OTTB continuum as opposed to a simple exponential fit, the absorption feature may be present or it may disappear [4]. This simply illustrates one difficulty with this energy domain. Nevertheless, Mazets and his group reported features which were not only breaks at low energy but also real absorption features compatible with the resolution of the detectors. Of course, due to the small area of these detectors, the significance level was generally low. Nevertheless the HEAO A 4 experiment [5] confirmed this low energy behavior of GRB spectra. But the discussion of the reality of this very

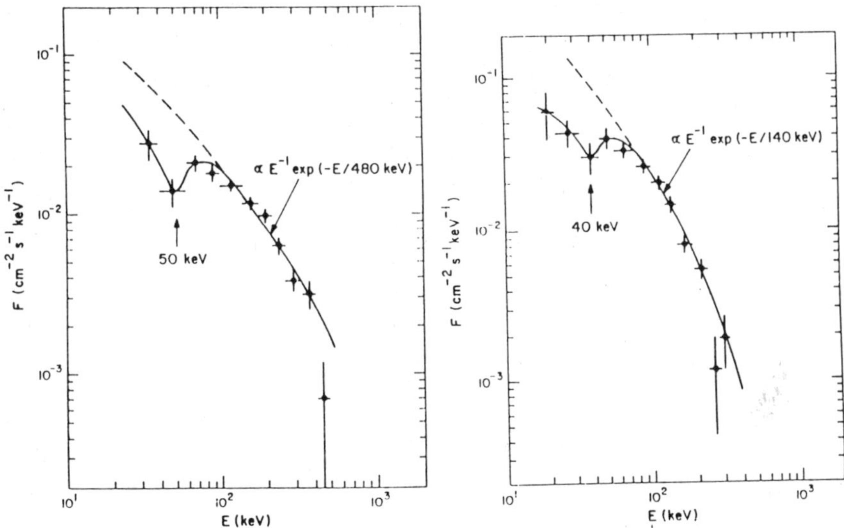

Fig. 1 (Left) KONUS spectrum of γ-ray burst GB790412, showing a low-energy dip at ≈ 50 keV. (Right) KONUS spectrum of γ-ray burst GB790612 showing a low-energy dip at ≈ 40 keV. The continuum in both cases is assumed to be ∞ E^{-1} exp (-E/kT) (from Mazets et al., 1981).[1]

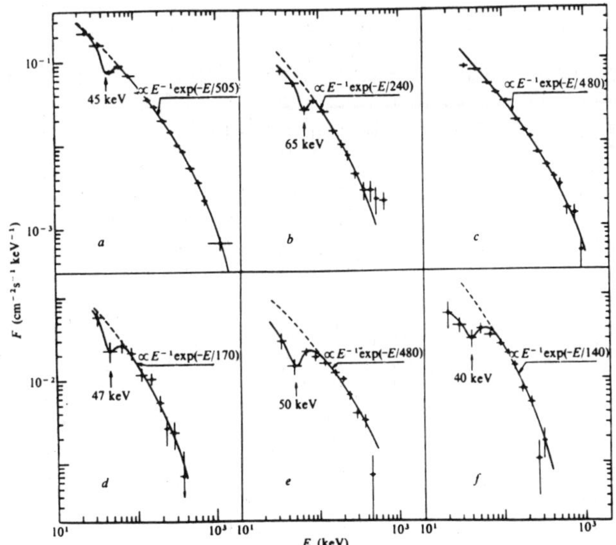

Fig. 2 GRB spectra with "annihilation" lines a) GB790305, b) GB780918, c) GB790418, d) GB790116, e) GB781104, f) GB781119 (initial stage) (from Mazets et al., 1981).[1]

important characteristic of GRB spectra was not closed. GINGA results have to be waited to clearly demonstrate the evidence for absorption features . The great advantage of the GINGA detectors is their ability to measure the spectra below 20 keV (the limit of the KONUS experiments) to a few keV. The results reported by Murakami et al. [6] (figure 3) and analysed in detail by Fenimore et al. [7] conclusively demonstrated the presence of features at two energies at least, which can be considered to be the first and second cyclotron absorption harmonics in the presence of strong magnetic field : (B $\approx 2.10^{12}$ Gauss) which can exist at the surface of neutron stars. This is of primary importance because, as we have not identified any counterpart to GRB sources this is an evidence that GRB (although perhaps not all) are produced at the surface of neutron stars. Recently Murakami reported the observation of a third GRB presenting low energy absorption features [8] . The characteristics of the 3 GRB with significant low energy absorption features are given in the table III. (Murakami, 1990)[8] (figure 3).

Table III. Characteristics of the three GRB events observed by Ginga[8]

Event ID	E_1	W_1	E_2	W_2	E_2/E_1
GB870303	20.4±0.7	3.5±2.7	40.6±2.6	12.3±6.3	1.99±0.3
GB880205	19.3±0.7	4.1±.2.2	38.6±1.6	14.4±4.6	2.0±0.15
GB890929	26.3±1.5	4.2	46.6±1.7	7.7	1.8±0.2

The energies of the features detected from three GRBs in unit of keV.
The E_1 and E_2 are the center energies of the 1st and 2nd harmonic and the W_1 and W_2 are the widths of each line in unit of keV (FWHM) respectively.

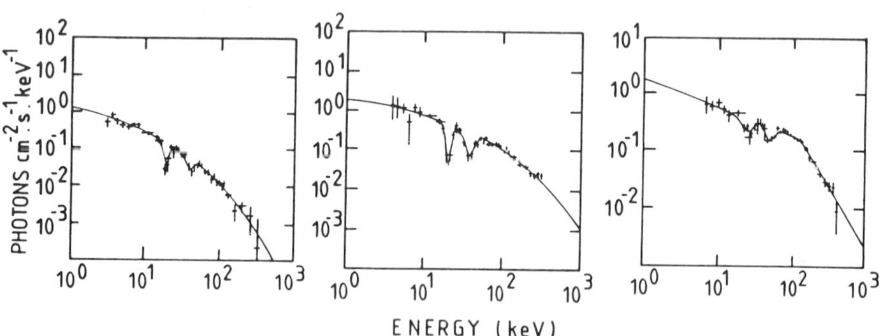

Fig. 3 Three cases of possible cyclotron features in GRB spectra. The third GRB, GB 88 09 29 has a center energy of 26.3 keV ± 1.5 keV in the first harmonic (from Murakami, 1990).[8]

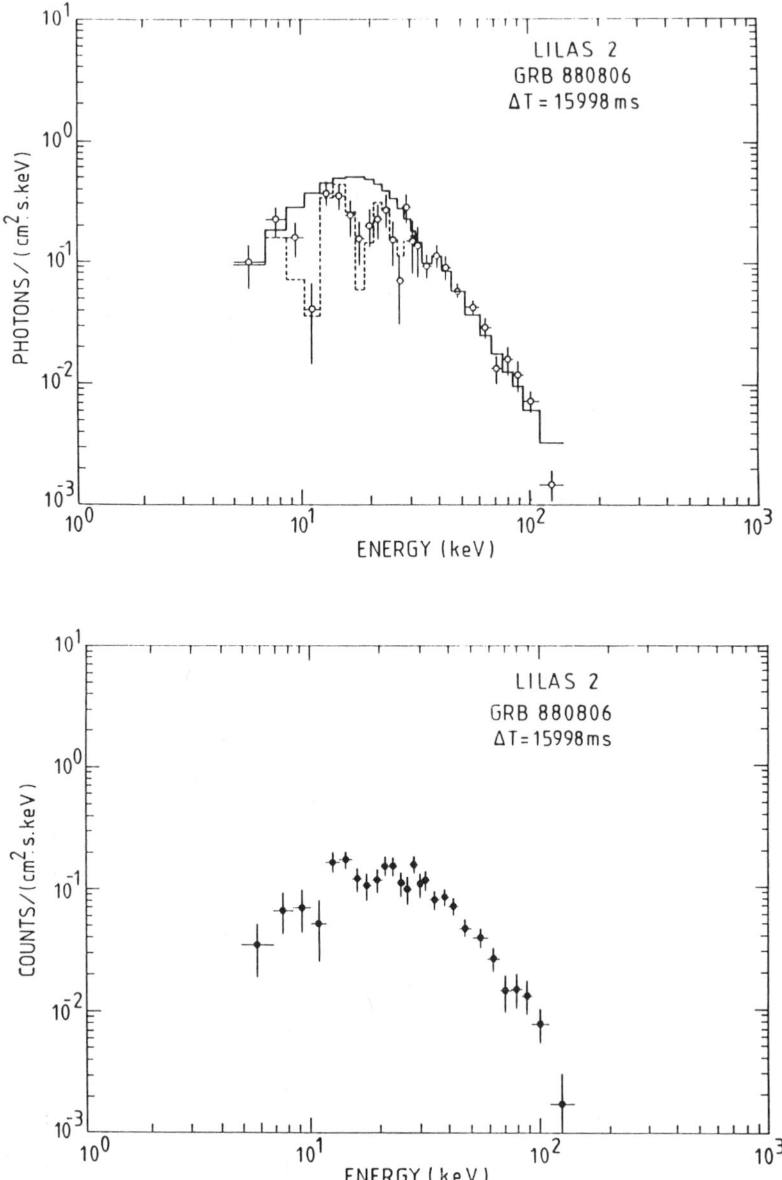

Fig. 4 a) Count rate obtained with the LILAS experiment for GRB GB880806 - b) Photon spectrum after deconvolution. The solid line indicates the best fit continuum while the dashed line gives the best fit taking into account the features (from Barat et al. ,1990).[9]

On the PHOBOS mission, the LILAS experiment also observed absorption features. One example has been analysed recently by C. Barat et al. [9] ; the counting rate and resulting photon spectra are given in figure 4.

All of these recent observations confirm, at least for some bursts, the presence of cyclotron absorption features with two harmonics and perhaps even three. However, the detailed analysis and particularly the choice of the continuum spectrum have a decisive influence on the strengths of the reported lines and their significance levels. Besides if the count spectra are considered, some features quite visible in one observation may be completely absent for the same GRB observed by another detector. Figure 5 illustrates this point for one event detected by two LILAS experiments on PHOBOS 1 and 2. Unfortunately due to the very short lifetime of PHOBOS 1 such a comparison between 2 identical detectors on 2 different probes looking the same direction was not possible except during one month.

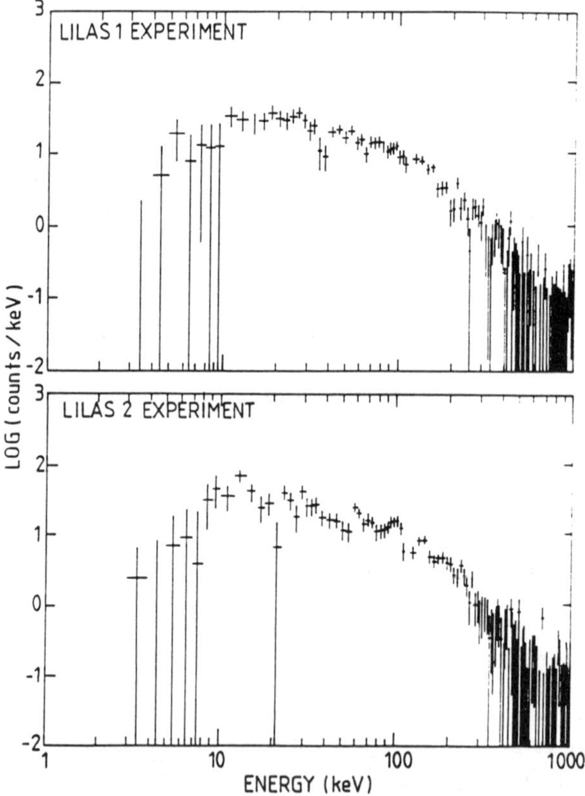

Fig. 5 LILAS spectra for the GRB GB880825 seen by the two probes PHOBOS 1 and 2 and corresponding to the same part of the burst. The two detectors are oriented in the same direction.

b) Annihilation features :

These were first reported from the KONUS experiments for more than 10 % of the observed GRB [1]. Very often these features are bumps with full width half maxima of 200 - 250 keV. These were subject to controversy [10, 11], the statistical significance level very often being low ; but in some cases features were reported by two different detectors.

The Germanium spectrometer onboard the ISEE 3 satellite observed two lines for the GRB : GB 781119 . These lines at 420 ± 20 keV and 738 ± 10 keV might be a redshifted positron annihilation line and the first nuclear deexcitation line from ^{56} Fe at 847 keV. But here again the statistical significance of this observation (~ 3.5 σ) is rather low [12].

With APEX experiment aboard the PHOBOS probe, seventy bursts were detected by the CsI detector (Ø = 10 cm, h = 10 cm). The large dimensions and the spectral analysis (108 channels, and a time to spill method for accumulating the spectra) allowed a search for features. Eight of the strongest bursts have been examined in detail with a time resolution for each spectrum varying between tens of milliseconds for the strongest event and half a second for the weakest one. A systematic analysis of the excesses at the 3σ level has been done after fitting the continuum by a four parameter polynomial function. Looking at the excesses which were observed, (some of them of statistical origin), it appears that most cluster in the regions around 400 keV, 500 keV and 800 keV [13] although none has a high confidence level. Figure 6 shows an example of a feature at 475 keV with a >3σ confidence level observed for 2.4 sec. (corresponding to the first peak of the GRB : GB890223).

Another example is particularly interesting because it corresponds to a very strong burst observed on 1988, October 24. The main peak of this strong event was analyzed with a time resolution of about 30 ms. Several features are quite visible : a strong absorption at low energy in spectrum b which disappears after 30 ms and two lines in spectra b and a respectively at 530 ± 30 keV and 820 ± 60 keV (3.2 σ and 3.9 σ excesses [14,15]). These lines are also quite transient, their durations are around 30 ms (figure 7). This example demonstrates the difficulties of the spectral analysis when such features have durations as short as a few tens of ms. It means that only if the detector is very large or the burst very intense, which is quite rare, can the search for features be positive. And if this is the case, it also means that many spectra within a particular burst have to be treated before a feature appears ; under these conditions the statistical significance of the feature will be strongly decreased due to the large number of trials needed to reveal something.

As for low energy features the difficulty to be sure about the reality of possible features observed by only one experiment can be illustrated by one example. It concerns the GB881024 observed by SMM [16] and by APEX instrument. A broadened line feature near 2.2 MeV at 3.3 σ confidence level is quite visible in the SMM spectrum but absent in APEX data. This again confirms that a lot of care must be taken in the identification of lines when several spectra with many channels are inspected.

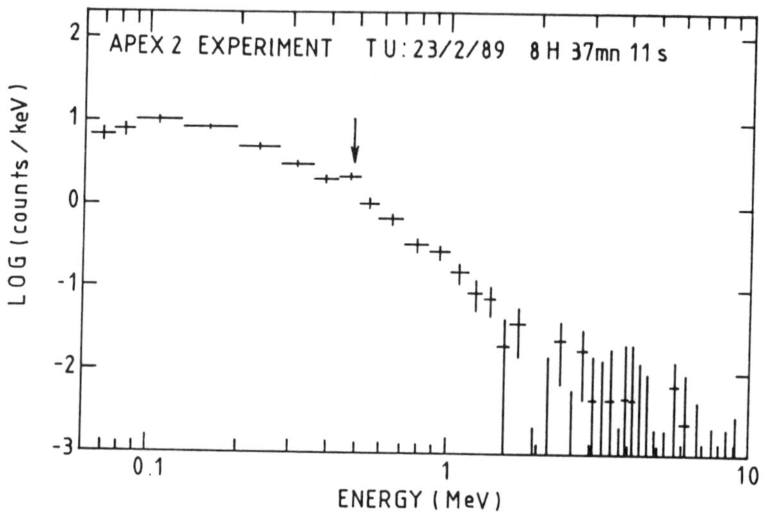

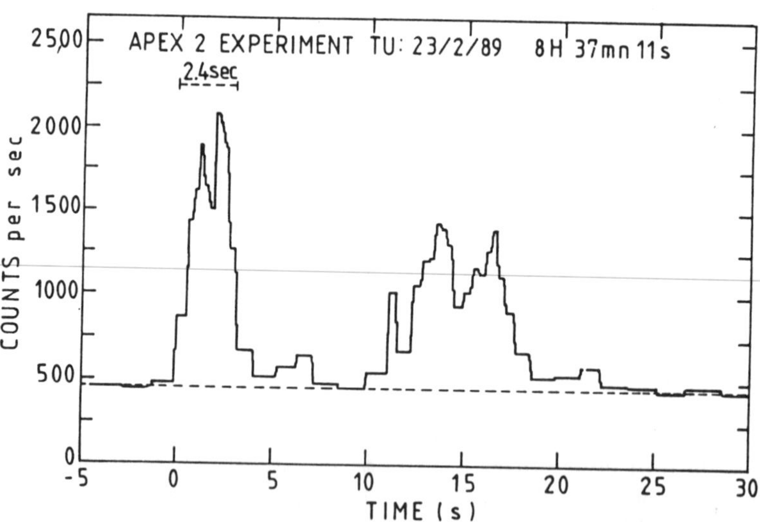

Fig. 6 A feature observed at 475 keV for the GRB : GB890223 at the time of the first peak of the burst (2.4 sec.).

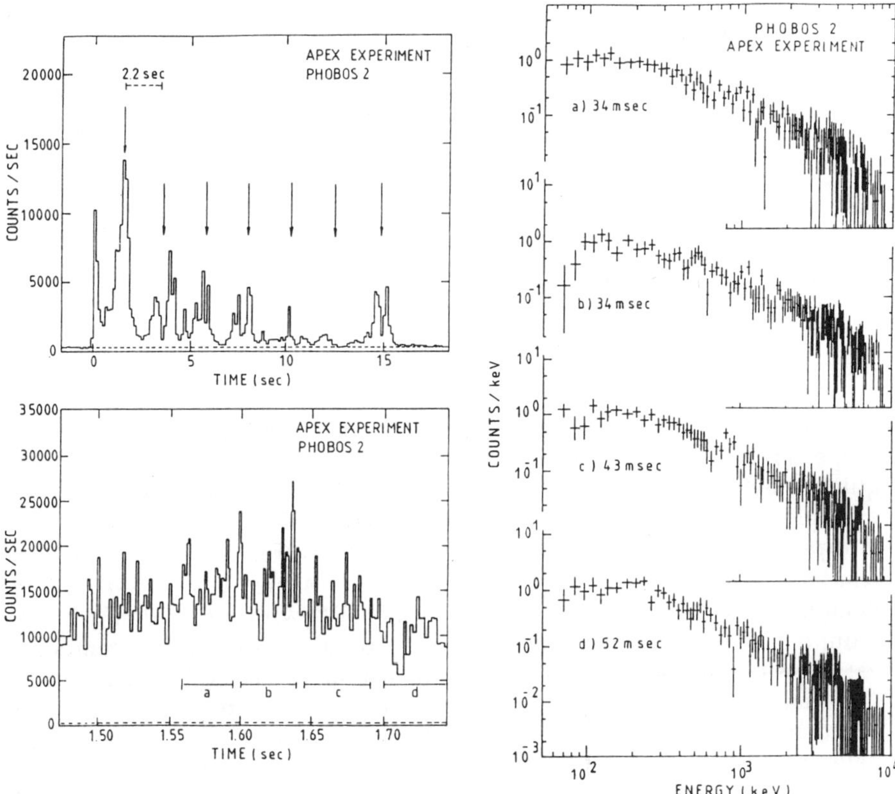

Fig. 7 The strong GRB : GB 881024 - Time profile and time to spill analysis during the main peak of the burst. Spectra are taken in the four time intervals a, b, c, d showing the fast variability of GRB spectra at the level of few tens of ms.

For APEX the spectral analysis has been carried out to look for features whose energy widths are compatible with the detector resolution, (40 - 50 keV). A more difficult task has to be performed to search broader features ; in this case to reach a definitive conclusion the photon spectrum must be obtained, which requires the knowledge of the direction of the burst with respect to the detector axis. Moreover a continuum fit must be chosen, and this is not easy. Depending on this choice, the presence or absence of a broad feature can be debated ; see for instance the discussion on GB781104 [10] . Such a discussion can be pursued for other examples: for instance the strong GB781119 detected by the KONUS and SIGNE experiments . The continuum chosen by Mazets et al. is the traditional optically thin thermal bremsstrahlung and the figure 8 indicates the presence of a broad excess above the continuum, which has been attributed to a broad annihilation feature. If we look at the SIGNE results for the same period (figure 9) it appears that during the few seconds of the first peak the spectrum was changing very rapidly ; in the first time interval there was a very flat spectrum below 500 keV and after that a rapid evolution towards a power law spectrum. In this case, to explain the spectra it is not necessary to introduce a broad annihilation feature. The shape can be explained by the addition of different spectra changing very quickly over <0.6 sec. ; this is really what is observed. Many other examples in the SIGNE data, particularly the GB781104 which has been interpreted by us as a broad annihilation feature, confirming Mazets results [18], might also be considered in the same way as GB791119, demonstrating only the fast variability of GRB with very often a break around 500 keV which appears as a pivot point in the fast temporal variation of the spectrum. Thus, the reality of the large bump in many GRB can be debated. Due to the difficulty in demonstrating the reality of a broad line at, or around 511 keV, at present it is really impossible to determine the percentage of GRB which present a possible annihilation feature in their spectrum.

The preceding analysis finds another kind of confirmation if we consider the GRB spectra integrated for the total duration of the burst. In this case no feature are seen either by SMM with an integration time of 16 sec for the GRB spectra or on the PHOBOS mission with APEX.

Therefore GRB spectra have to be studied on short time scales to look for features. Only large detectors, such as the BGO detectors on GRANAT (PHEBUS experiment) large NaI detectors on GRO (BATSE experiment) , will be able to define the fast variability of GRB spectra with high confidence level and provide possible evidence for transient annihilation features.

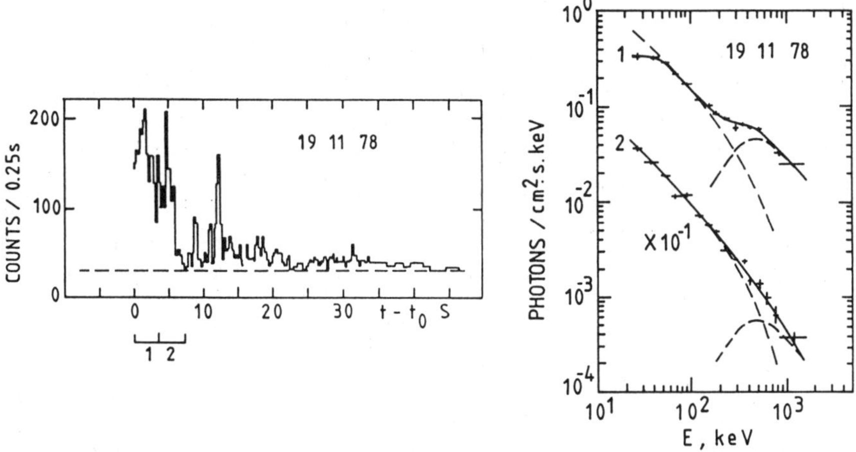

Fig. 8 KONUS observations of the same event : time profile and spectra taken in the two time intervals 1 and 2 (from Mazets et al., 1981)[19]

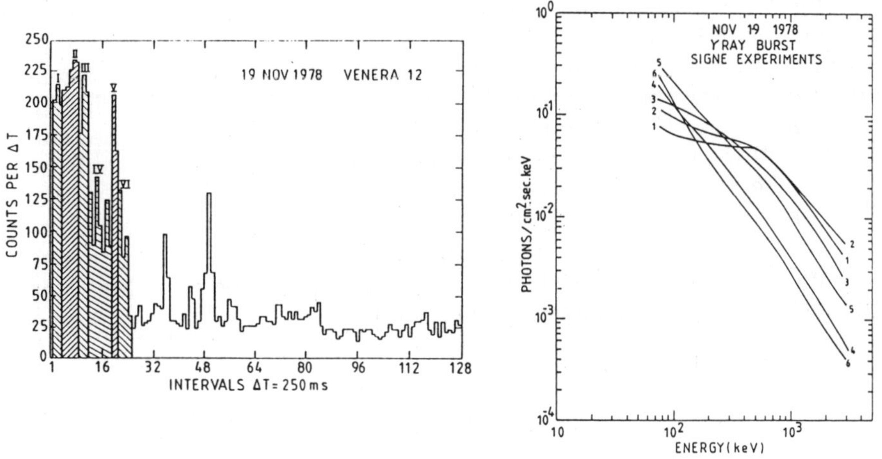

Fig. 9 GRB : GB781119 - Time profile and evolution of the spectrum taken in the time intervals noted from 1 to 6 (from Barat et al., 1981).[17]

2) Time variation of GRB spectra at low and high energies :

Here, high energy means energies higher than 1 MeV. With the APEX CsI detector, it has been possible to detect the high energy portion of GRB spectra and to show that in many cases the time variations of the high and low energy parts of these spectra are quite independent. Figure 10 gives two examples of this behavior. Moreover, there seems to be an anticorrelation between high and low energy variations (figure 10), [13,20]. This might be an indication that there are two different regions for the production of high and low energy photons. But, as is very often the case for GRB, it seems that the GRB properties which are illustrated by some examples are not so easily generalized to the whole GRB population. This is not surprising if in fact GRB have different origins and possibly different energy production mechanisms with no a single model explaining all of them. This point can also be illustrated by the study of the hardness ratio (HR). This was introduced by Mazets [21] to show a clear correlation between the HR and the time profiles of GRB. Later SMM results[22] demonstrated for other bursts and in a different energy range, a different behavior, with an HR decreasing systematically from the beginning of a GRB peak to its end.

With APEX, the strong GB891024 was studied, taking for the hardness ratio the ratio of the 64 keV - 200 keV and 1030 keV - 9218 keV counting rates . For this burst a generally good correlation is observed between the HR evolution and the GRB time profile. This behavior doesn't confirm the SMM results but it again might prove that different bursts do not obey to the same rules. Here it has to be noted that the HR for SMM was chosen differently : 300 - 350 keV for the high energies and 52 - 182 keV for the low energies. A comment on this point : it would be better, if the hardness ratio has some physical meaning, to agree on the energy range which should be considered. It makes no much sense to use energies below 1 MeV for the high and low energy parts of the ratio in one case and in other case (APEX) to use the flux above 1 MeV, as representative of the high energy domain.

3) Total GRB spectra :

A large variety of models have been proposed to represent the continuum of the GRB spectra : The last results of SMM indicate a power law spectrum above 300 keV is in general a good fit to the data with the exponent α between 2 and 3 [16] . With the APEX experiment on PHOBOS the use of a large CsI detector allows the detection of high energy photons with a good efficiency as for SMM. More recently the PHEBUS experiment which is part of the GRANAT mission has been designed to explore the high energy part of GRB spectra (100 keV - 100 MeV). The large BGO detectors which are used (Ø = 7.8 cm ; h = 12 cm) are very efficient even for photons with energy above 10 MeV. Figure 11 gives an example of a

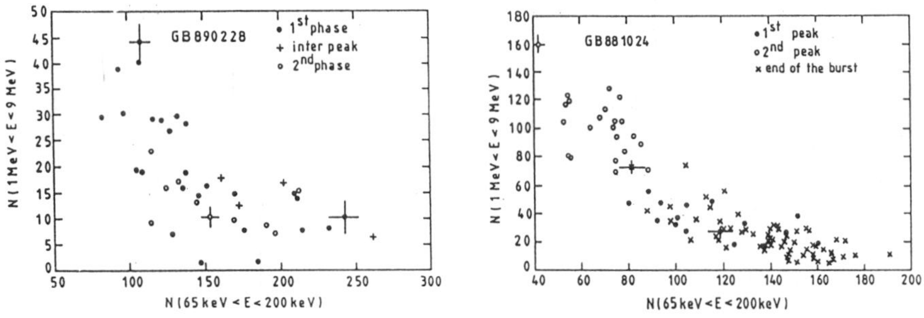

Fig. 10 Example of the anticorrelation between hight and low energy count rates during two GRB seen by the APEX experiment (from E. Jourdain, 1990).[13]

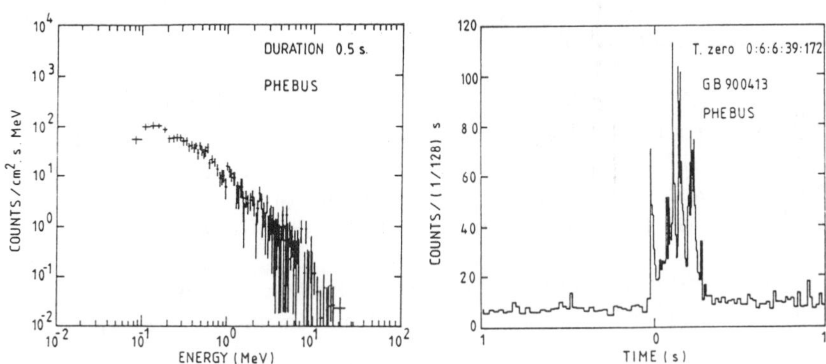

Fig. 11 Short GRB seen by the PHEBUS experiment on the GRANAT mission : time profile and an example of the BGO spectrum (integrated during 0.5 sec) which extends beyond 10 MeV.

GRB spectrum observed with one of the 6 BGO scintillators. The extension of the spectrum above few MeV is quite visible as it is often the case with the SMM and APEX detectors.

The APEX results show that for the total spectrum of a GRB (the spectrum cumulated for the duration of the burst) very often a break is present around 500 keV with two power laws above and below this point which appears quite often as a pivot point in the spectrum (figure 12) and as we have already shown, this point may have the same function when the GRB spectra are observed on a very short time scale (≤ 0.5 sec) .

Before affirming such a characteristic for the total spectrum of many GRB the photon deconvolution has to be done with the difficulty to know for each burst its direction with respect to the detector axis.

This behavior has been also searched for GRB spectra observed by SMM. For instance it is quite evident for the GRB GB80805 [23] (Figure 13). Such spectra with a break in the region of 500 keV (broken power law) are well explained by the model proposed by Zdiarski and Lamb [24,25] .

4) Future missions :

The recent results demonstrate the difficulties for revealing features in GRB spectra due to many reasons :

- the source being transient without any knowledge of the recurring time only one observation is possible.

- the SMM and APEX results show no evident narrow feature when the total spectrum of the burst is considered [26] . Thus the features has to be searched on a short time scale at any place in the GRB. Many trials have to be done, decreasing considerably the significance of the feature which is in general low when the spectra is integrated on a short time scale.

- the continuum being a fortiori unknown and not fitted by a unique law, the choice of this law has a strong implication on the reality or (and) the significance of the feature.

- the spectra may present a strong variability within a GRB and in this case broad features may result from the addition of rapidly changing spectra with time scale shorter than the integration time. We have illustrated this possibility in paragraph 1.

For these many reasons to search for GRB features, large detectors with high sensitivity are needed and to increase the confidence level of the observations it is better to have two detectors with similar characteristics observing the same bursts. Here again we have shown

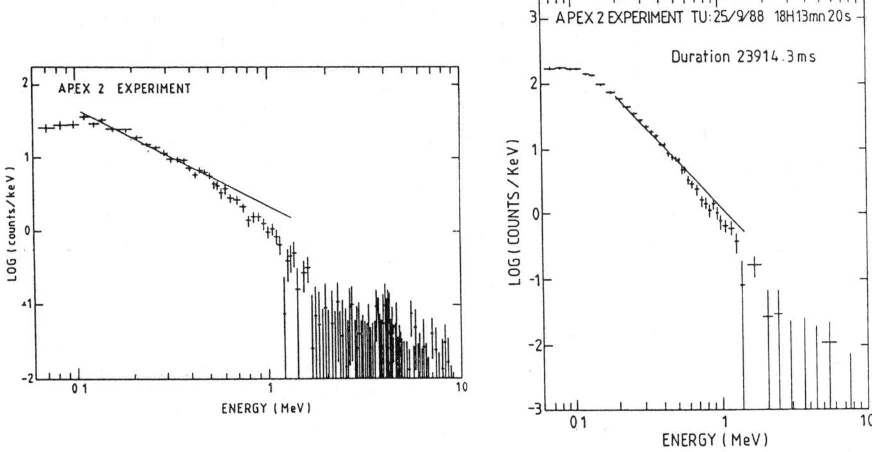

Fig. 12 Examples of the total spectra obtained for two GRB seen by the APEX experiment. A "eye ball" fit shows a break in the spectra around 500 keV.

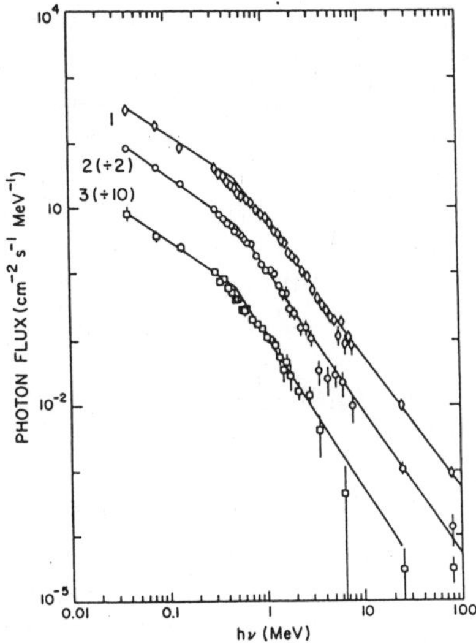

Fig. 13 SMM photon spectra in 3 consecutive 16 second time intervals for the GRB : GB840805 (from Share et al., 1986)[25]. fitted by two power law by Zdziarski and Lamb, 1986[25].

(figure 5) on one example that features may be visible in one instrument and completely absent in the other.

In the next future two kinds of missions will be very useful to answer with the highest confidence to the question of the existence and especially the percentage of features, bumps or lines which are present in GRB spectra.

-BATSE on GRO with eight NaI detectors (Ø = 50.8 cm ; h = 1.27 cm) optimized to detect gamma-rays between 20 and 600 keV and three NaI detectors (Ø = 12.7 cm ; h = 7.62 cm) with enhanced energy resolution for spectroscopy.

At the same time if GRANAT satellite is still working it will be possible for the same GRB to compare the spectral characteristics and possible features above few hundred keV using BATSE and PHEBUS detectors;

- Fine spectroscopy devices with Germanium detectors onboard : MARS OBSERVER MISSION, WIND and SPECTRUM X. The fine spectroscopy which can be achieved with these detectors will be particularly important to explore the low energy part of the GRB spectra. The region from 10 keV to hundred keV will be studied by the same detector with an energy resolution at least a factor of 10 better than the previous ones. Besides if these satellites are working at the same time , at least few months , it will be possible to compare the spectra obtained by the same type of detectors looking at the same bursts. This is particularly important because the area of these detectors are small : few tens of cm^2.

Being perhaps optimistic we think such missions will likely have a decisive impact for the understanding of the spectral characteristics of the GRB especially the presence of annihilation features and nuclear lines.

REFERENCES

1. MAZETS, E.P., S.V. GOLENETSKII, R.L. APTEKAR, Yu. A. GUR'YAN & V.N. IL'INSKII, 1981. Nature, 290, 378.

2. HURLEY, K., 1988. AIP Conf. Proc. 170, 258 Editors : N. GEHRELS & G. SHARE.

3. FENIMORE, E.E, J.G. LAROS, R.W. KLEBESADEL, R.E. STOCKDALE & S. KANE, 1981. Proceedings of the Workshop on Gamma Ray Transients and Related Astrophys. Phenomena at University of California San Diego, La Jolla, CALIFORNIA.

4. FENIMORE, E.E., R.W. KLEBESADEL & J.G. LAROS, May 1982. Proceedings of XXIV COSPAR Meeting, Ottawa, CANADA.

5. HUETER, G.J. & D.E. GRUBER, 1982. Accreting Neutron Stars MPE Report N° 177, Editors W. BRINKMANN & J. TRUMPER.

6. MURAKAMI, T., M. FUJII, K. HAYASHIDA, M. ITOH, J. NISHIMURA, T. YAMAGAMI, J.P. CONNER, W.D. EVANS, E.E. FENIMORE, R.W. KLEBESADEL, A. YOSHIDA, I. KONDO & N. KAWAI, 1988. Nature, 335, 234.

7. FENIMORE, E.E., J.P. CONNER, R.I. EPSTEIN, R.W. KLEBESADEL, J.G. LAROS, A. YOSHIDA, M. FUJII, K. HAYASHIDA, M. ITOH, T. MURAKAMI, J. NISHIMURA, T. YAMAGAMI, I. KONDO & N. KAWAI, 1988. Astrophys. J. Lett. 335, L 71.

8. MURAKAMI, T., 1990. Proceedings of XXVIII Cospar Meeting, The Hague, NETHERLANDS.

9. BARAT, C., R. TALON, R. SUNYAEV, N. BLINOV, A. KUZNETSOV & O. TEREKHOV, August 1990. Workshop on Cosmic Gamma-Ray Bursts, TAOS NEW-MEXICO.

10. FENIMORE, E.E., R.W. KLEBESADEL, J.G. LAROS, R.E. STOCKDALE & S.R. KANE, 1982. Nature, 297, 665.

11. NOLAN, P.L., G.H. SHARE, E.L. CHUPP, D.J. FORREST & S.M. MATZ, 1984. Nature, 311, 360.

12. TEEGARDEN, B.J. & T.L. CLINE, 1980. Astrophys. J. 236, L 67.

13. JOURDAIN, E., 1990. Thèse Doctorat TOULOUSE (France).

14. ATTEIA, J.L., C. BARAT, A. CHERNENKO, V. DOLIDZE, A. DYATCHKOV, E. JOURDAIN, N. KHAVENSON, A. KOZLENKOV, RR. KUCHEROVA, I. MITROFANOV, L. MOSKALEVA, M. NIEL, A. POZANENKO, O. SCHEGLOV, Yu. SURKOV, G. VEDRENNE & A. VILCHINSKAYA, 1991. Astron. Astrophys. (in press).

15. MITROFANOV, I., N. KHAVENSON, A. KOSLENKOV, R. KUCHEROVA, A. POZANENKO, A. CHERNENKO, V. DOLIDZE, A. DYATCHKOV, N. POZANENKO, A. VILCHINSKAYA, J.L. ATTEIA, C. BARAT, E. JOURDAIN, M. NIEL, G. VEDRENNE, S. KHARIUKOVA, L. MOSKALEVA, O. SCHEGLOV & Yu. SURKOV, 1991. Planet Space Sci, (in press).

16. SHARE, G.H., D.C. MESSINA, A. JADICCO, S.M. MATZ, E. RIEGER & D.J. FORREST, August 1990. Workshop on Cosmic Gamma-Ray Bursts, TAOS, NEW MEXICO.

17. BARAT, C., G. CHAMBON, K. HURLEY, M. NIEL, G. VEDRENNE, I.V. ESTULIN, A.V. KUZNETSOV & V.M. ZENCHENKO. 1981 Astrophysics and Space Science, 75, 83.

18. BARAT, C., K. HURLEY, M. NIEL, G. VEDRENNE, I.G. MITROFANOV, I.V. ESTULIN, V.M. ZENCHENKO & V. SH. DOLIDZE, 1984. Astrophys. J. 286, L11.

19. GOLENETSKII, S.V., E.P. MAZETS, R.L. APTEKAR, Y.A. GURYAN & V.N. ILYNSKII, 1986. Astrophysics and Space Science 243, 278.

20. JOURDAIN, E., J.L. ATTEIA, M. NIEL, G. VEDRENNE, I. MITROFANOV, A. CHERNENKO, V. DOLIDZE, A. DYATCHKOV, N. KHAVENSON, A. KOSLENKOV, R. KUCHEROVA, A. POZANENKO, A. VILCHINSKAYA, S. KHARIUKOVA, L. MOSKALEVA, O. SCHEGLOV & Yu. SURKOV, August 1990. Workshop on Cosmic Gamma-Ray Bursts, TAOS NEW-MEXICO.

21. GOLONETSKI, S., V. ILYINSKII & E. MAZETS, 1984. Nature, 307, 41.

22. NORRIS, J.P., G.H. SHARE, D.C. MESSINA, B.R. DENNIS, U.D. DESAI, T.L. CLINE S.M. MATZ & E.L. CHUPP, 1986. Astrophys. J. 301, 213.

23. SHARE, G.H., S.M. MATZ, D.C. MESSINA, P.L. NOLAN, E.L. CHUPP, D.J. FORREST & J.F. COOPER, 1986. Adv. Space Res., 6, N° 4, 15.

24. ZDIARSKI, A.A. & D.Q. LAMB, 1986. Astrophys. J. 309, L 79.

25. ZDIARSKI, A.A. & D.Q. LAMB, 1986. Adv. Space Research, 6, 85

26. NOLAN, P.L., G.H. SHARE, D.J. FORREST, E.L. CHUPP, S. MATZ & E. RIEGER, 1983. AIP Conf. Proceedings, 83, 59. Editors : M.L. BURNS, A.K. HARDING & R. RAMATY.

FEATURE IN THE CRAB PULSAR GAMMA-RAY SPECTRUM

B.Parlier, B.Agrinier, E.Barouch, R.Comte
Service d'Astrophysique, CEA Saclay, France

G.Cussumano, B.Sacco, L.Scarsi
Istituto di Fisica Cosmica, CNR, Palermo, Italy

G.Gerardi,
Università di Palermo, CNR, Palermo, Italy

P.Mandrou, M.Niel, J.F.Olive
CESR, Université P.Sabatier,Toulouse, France

E.Massaro, G.Matt
Istituto Astronomico,Università "La Sapienza", Roma,Italy

J.L.Masnou
Observatoire de Paris-Meudon,Meudon, France

E.Costa
Istituto di Astrofisica Spaziale, Frascati, Italy

M.Salvati
Osservatorio Astrofisico di Arcetri,Firenze, Italy

ABSTRACT

During the third flight of the FIGARO II experiment on July 9th 1990 the Crab pulsar PSR0531+21 has been kept in the axis of the telescope during 7.5 hours.
On line analysis showed the pulsar light curve with only 5 minutes of data acquisition time.
The complete analysis of the observation will probably help to search for gamma ray line in the pulsed part of the spectrum.
Particulary a line around 450 Kev has been previously detected by some experiments but at low statistical level and attributed ambiguously to the Nebula.

INTRODUCTION

The FIGARO II (French-Italian GAmma Ray Observatory) experiment is a large area, actively shielded, non collimatd detector for photons in the energy range 0.150-6 Mev, to be operated at high altitude by means of stratospheric balloons. It is especially designed to observe cosmic sources with a well established time signature as for instance pulsars.

In this contribution we present preliminary results of a 7.25 hours observation of the Crab pulsar performed on July 9th 1990 in the course of the third flight of the experiment between Sicily and Spain. In particular we present the detection of an emission feature in the spectrum of the second peak in the pulse profile.

EXPERIMENT AND OBSERVATIONAL PARAMETERS

The experiment is described extensively elsewhere (Ref.1). It consists of an actively shielded, large area (3600 cm2), set of 9 NaI crystals 5 cm thick. The angular aperture is large (77 degrees FWHM). The emphasis is placed on the timing accuracy wich is better than 10 microseconds. The detector can be oriented in the direction of the observed source by means of a two-axis pointing system.

Each detected gamma ray is analysed in energy in 256 channels. The energy word is transmitted to the ground receiving station where the time is marked with an atomic clock with 10 microseconds accuracy. The absolute accuracy, in UT scale, is better than 50 microseconds. In parallel, the arrival time of the event is marked with an onboard quartz clock with 100 microseconds accuracy and is registered on board with the energy information on a magnetic tape recorder. For the third flight we reduced the energy range to .150-4.2 Mev. During the Crab observation the residual atmospheric column changed very little between 4.35 and 4.50 gr/cm2.

DATA ANALYSIS

The phase histogram of the Crab pulsar for the whole energy range is presented in figure 1. It was obtained by means of the folding technique: The arrival times of the accepted photons were converted to the solar system barycenter using the JPL DE200 ephemeris. The ephemeris of the pulsar itself at the observation date have been derived from a nearly contemporaneous radio measurement (1990 July 15) (Ref.2)

The pulse shape is typical of this energy range (Ref.3) the second peak (p2) is higher than the first one (p1) and separated by 0.40 unit of phase. The zero phase is the same given by radio astronomers. For the phase resolved spectroscopy study we have divided the period in several intervals corresponding to the various features: 0.95-0.05 for p1, 0.05-0.27 for the interpulse region, 0.27-0.47 for p2 and 0.47-0.77 for the off-pulse interval.

RESULTS AND DISCUSSION

In order to confirm a marginal detection obtained during the 1986 flight (Ref.3), we have looked for spectral features in the second pulse. For this purpose we have substracted the spectrum of the events corresponding to the off-pulse component from the spectrum with phase between 0.27 and 0.47, normalizing each spectrum to the corresponding phase interval lenght. The result is shown in figure 2. We observe a feature centered at about 0.450 Mev, consistent with the 1986 observation both in position and intensity.
A systematic effect is ruled out by performing the same analysis on different parts of the light curve having similar spectral indexes: nothing was found in the spectra of the first pulse and the interpulse. The width of the observed feature is 45+/-10 kev comparable with the instrumental resolution within the limited statisics.
The excess observed in 3 channels around 450 Kev above an power law fit to the continuum is larger than 3 sigmas (1.8 sigma for the 1986 observation, which was only 2.5 hours long), and implies a flux of $(1.6+/-0.5)10^{-4}$ photons/cm2sec.
We stress that our result was made possible only due to the unprecedented sensitivity of FIGARO and that previous observations of lines have provided positive results (Ref.4,5,6) around 400 Kev for Crab Nebula but not for the pulsar alone. Conversely if, at the time of our observation, the pulsed spectrum contained a line as intense as those detected previously at lower energies, it should appear in the data at an unquestionable level of significance.
An appealing explanation of the detected feature is a redshifted electron-positron annihilation line: it would then be the first direct evidence for positrons in the vicinity of a pulsar, as predicted by the acceleration models of the "gap" type (Ref.7). With this interpretation, the redshift would be 12%, which implies $M_o/r_6 = 0.77$, where M_o is the pulsar mass in units of one solar mass and r_6 is the distance of the annihilation region in units of 10^{+6} cm. A narrow line and a steady position derive from a scenario where the annihilation occur at the pulsar surface, in which case the measured redshift has important applications for the neutron star structure.

References

1) G. Agnetta et al. 1989, Nucl.Instr.Methods A,281,197.

2) A.G. Lyne and R.S.Pritchard , Private communication

3) B. Agrinier et al. 1990, Ap.J., 355, 645.

4) M. Leventhal, C.J. MacCallum and A.C. Watts 1977, Nature, 266, 696.

5) C.A. Ayre, P.N. Bhat, Y.Q. Ma, R.M. Myers and M.G. Thompson 1983, Mon.Not.R.Astr.Soc. 205, 285.

6) G.V. Jung 1989, Ap.J.,338,972.

7) M.A. Ruderman and P.G. Sutherland 1975, Ap.J.,196, 51.

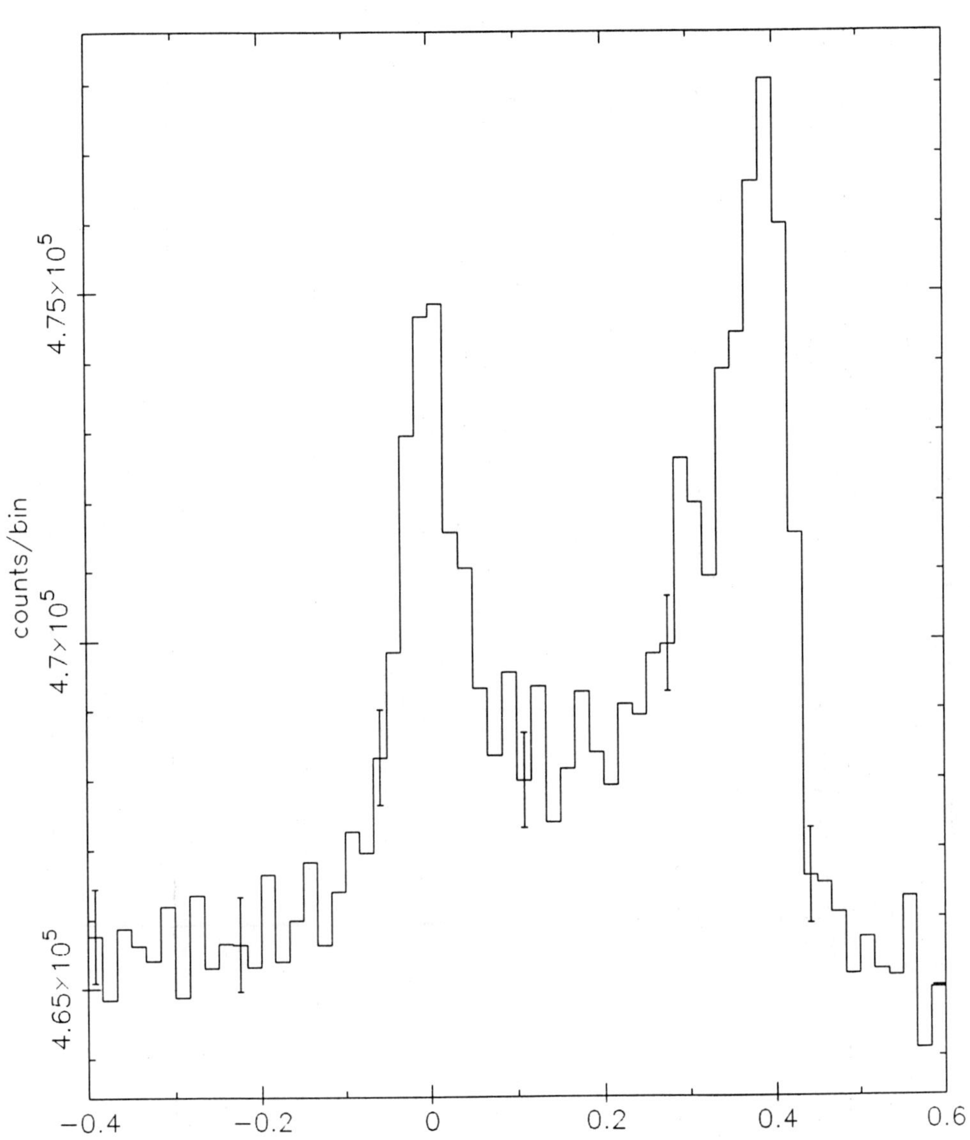

Fig.1: Phase Histogram, Energy Range 0.150−4.2 Mev

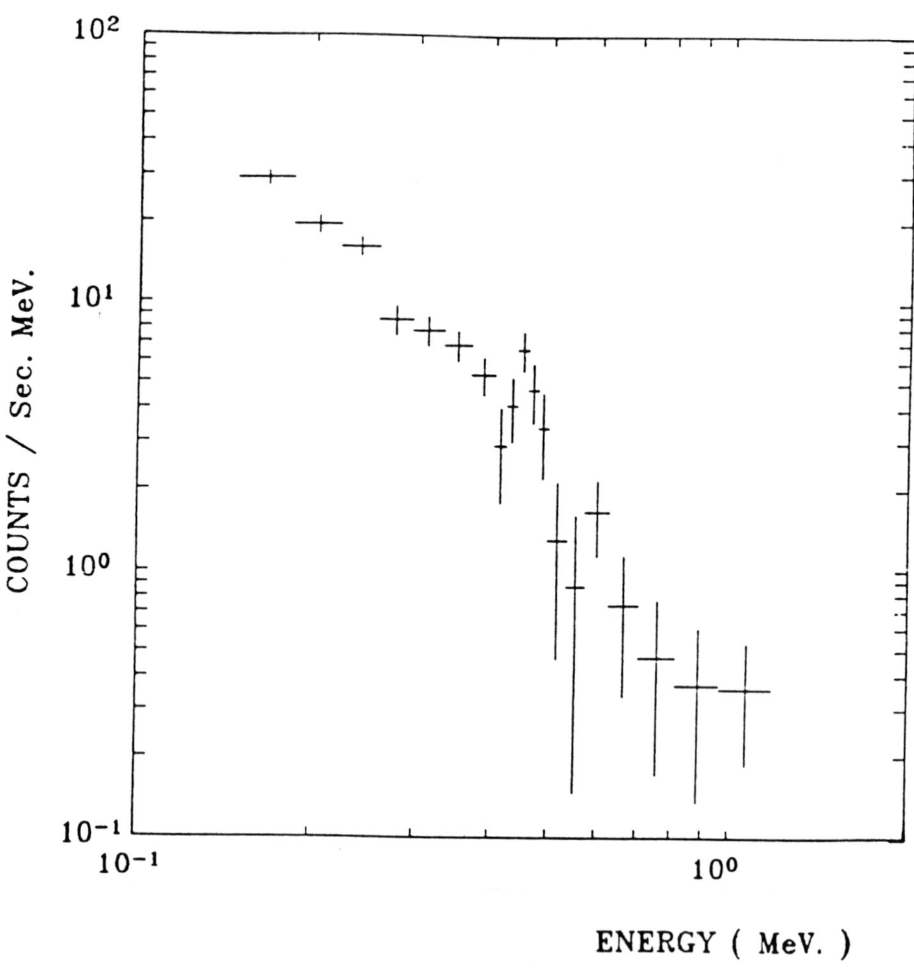

Fig. 2 : Second pulse (Phase 0.27-0.47) Energy Spectrum

GAMMA-RAY LINE AND ANOMALOUS CONTINUUM EMISSION FROM THE CRAB

Alan Owens[1]
Laboratory for High Energy Astrophysics, NASA/GSFC, Greenbelt MD 20771

ABSTRACT

We review the observations of transient γ-ray line and sporadic continuum emissions from the Crab over the last two decades and use these measurements to explore the emission mechanisms and regions. When examined in detail, the measurements require an emission region(s) considerably more complex than predicted by simple synchrotron models, suggesting behavior reminiscent of gamma-ray bursts.

INTRODUCTION

The Crab has long been thought of as the canonical example of a strong, stable isolated point source. It is the only astrophysical object to be detected at all wavelengths from radio through PeV γ-rays. At the highest energies, the bulk of the emission emanates from the pulsar itself. Its phased-averaged spectrum, from infra-red to GeV energies, can be described by a minimum of three power law components resulting from the synchrotron process occurring within the pulsar magnetosphere. In addition to its apparently well behaved continuum spectrum, there have been a number of reports of transient γ-ray line emission[1-10], as well as a sporadic hard spectral component extending into the MeV region of the spectrum[11-14] and bursts of PeV γ-rays[15]. The observed line features are generally narrow and occur in the low energy region of the γ-ray spectrum at energies around 75 keV and 400 keV. The first line is interpreted as cyclotron emission in the tera-Gauss magnetic field of the pulsar and the second as redshifted annihilation radiation produced at, or near, its surface.

LINE MEASUREMENTS

The observations to date are summarized in Table 1 and, while none are statistically compelling in themselves, taken as a whole they are persuasive by the shear weight of 3 σ effects. From the Table, we see that features in the range 73 to 79 keV have been observed on five occasions. There is some evidence that these emissions are pulsed, implying a pulsar origin. For instance, the feature reported by Strickman *et al.*[5] was detected only in the pulsed spectrum and specifically in the phase region encompassing the main and inter- pulses. Lines near 400 keV have also been observed on a number of occasions. As with the cyclotron features, there is evidence that they are pulsed. For example, the report of a photon excess at ~ 400 keV by the FIGARO experiment[9], was discovered as secondary structure in the pulsar light curve. Interestingly, the structure appears at phases 0.4 and 0.8; the same separation as the main pulse and interpulse, but shifted in phase by 0.2 of a period. We note that the observation of additional structure in the light curve is not new. For example, Kurfess[16], working in the energy range 100 to 400 keV, explicitly points out the presence of interpulse emission and a later comprehensive analysis of HEAO-1 data by Knight[17] showed that two pulsed components are required in order to explain both the pulse peaks and the "light-bridge". Most recently, Parlier[10] reported the detection of a 4.4 σ feature during a second balloon flight of the FIGARO instrument on July 9,

---

[1]NRC/NAS Resident Research Associate

TABLE 1 Reported spectral line observations from the Crab. Here E is the line center energy. Upper limits are at the 3σ level.

OBSERVER	DATE	(73 - 79) keV E(keV)	(73 - 79) keV photons $cm^{-2}s^{-1}$	(400 - 455) keV E(keV)	(400 - 455) keV photons $cm^{-2}s^{-1}$
Ling *et al.*[2]	10 June 1974	73.3 ± 1	$(3.8 \pm 0.9) \times 10^{-3}$	-	$< 1.7 \times 10^{-3}$
Leventhal *et al.*[1]	10 May 1976	‡	-	400 ± 1	$(2.24 \pm 0.65) \times 10^{-3}$
Strickman *et al.*[5]	11 May 1976	76.6 ± 2.5	$(3.0 \pm 0.9) \times 10^{-3}$	‡	-
Yoshimori *et al.*[3]	30 Sept. 1977	‡	-	400 ± 2	$(7.4 \pm 5.4) \times 10^{-3}$
Jung[8]	Sept. 1977 March 1978 Sept. 1978	-	-	~ 400	$(7.5 \pm 3.0) \times 10^{-4}$
Knight[17]	29 March 1978 29 Sept. 1978	-	$< 1.0 \times 10^{-3}$	-	-
Manchanda *et al.*[4]	26 Aug. 1979	73 ± 4	$(5.0 \pm 1.5) \times 10^{-3}$	‡	-
Mahoney *et al.*[25]	23 Sept. to 10 Oct. 1979 and 24 Feb. to 4 April 1980	-	$< 2.8 \times 10^{-4}$**	-	$< 1.8 \times 10^{-4}$**
Schwartz *et al.*[26]	17 Oct. 1979	-	$< 2.0 \times 10^{-3}$	-	-
Hameury *et al.*[27]	25 Sept. 1980	-	$< 6.2 \times 10^{-4}$	-	$< 1.7 \times 10^{-3}$
Ayre *et al.*[6]	6 June 1981	†78.8 ± 1	$(1.1 \pm 0.3) \times 10^{-2}$†	†405 ± 1	$(7.2 \pm 2.1) \times 10^{-3}$†
Hasinger *et al.*[28]	28 Sept. 1981	-	$< 2.9 \times 10^{-3}$	‡	-
Watanabe[7]	29 Sept. 1982	78 ± 2	$(3.5 \pm 1.9) \times 10^{-3}$	‡	-
McConnell *et al.*[29]	2 Oct. 1984	‡	-	-	$< 3.2 \times 10^{-3}$
Ubertini *et al.*[30]	5 August 1985	-	$< 2.3 \times 10^{-3}$	‡	-
Agrinier *et al.*[9]	11 July 1986	‡	-	~ 400	$(1.0 \pm 0.55) \times 10^{-4}$
Perotti *et al.*[31]	29 July 1986	-	$< 3.8 \times 10^{-3}$	‡	-
Perotti *et al.*[32]	15 July 1987	-	$< 2.7 \times 10^{-3}$	‡	-
Parlier *et al.*[10]	9 July 1990	‡	-	455 ± 10	$(1.05 \pm 0.25) \times 10^{-4}$

‡Line energy outside the energy range of the detector. **50 day integrations. †values taken from ref. 33

1990. This result is particularly interesting since the emission was determined to be entirely pulsed, with the bulk of the events arriving in the interpulse. At this time, it is not clear whether the flux distribution in phase is significantly different from the earlier FIGARO measurement of Agrinier *et al.*[9].

From Table, 1 we see that the center energies of both the cyclotron and annihilation lines have systematically increased over the last two decades. The corresponding change in the derived redshift of the annihilation line, from 0.28 in 1976 to 0.26 in 1981, may be significant since both measurements[1,6] were carried out using Ge detectors; consequently systematic errors of more than a few tenths of a keV in the energy calibrations can be ruled out. The shift has been attributed to movement of the emission region due to magnetic instabilities near the surface[6]. However, as we attempt to show, this may not be entirely correct. Noting that Strickman *et al.*[5] observed the 77 keV line within one day of Bell/Sandia's observation (Leventhal, McCallum and Watts[1]) of the 400 keV line, and correcting the unshifted line energy derived from the measurements of Ayre *et al.*[6] for the redshift derived from the Bell/Sandia result, yields a shifted energy of (77.9 ± 2) keV, in agreement with the value measured by Strickman *et al.*[5]. The implication is that the apparent energy variability is not due to a basic change in field configuration but solely to a variable surface redshift. If this is true, then z has changed from 0.36 in 1974 to 0.12 in 1990. Since there is no plausible reason why the pulsar mass should change by $> 0.05\ M_{\odot}$, we conclude that the emission regions of both the annihilation and cyclotron lines have moved radially outwards. Again, using the measurements of Leventhal *et al.*[1]/ Strickman *et al.*[5] and Ayre *et al.*[6], we find that between 1977 and 1981 this distance is ~ 600 m. This suggests a common emission region which in turn suggests production from the same population of particles. Unfortunately, the question of common emission cannot be addressed directly from the observations, since only two of the experiments that detected lines had sufficient energy range to detect both features: one detected both lines[6] and the other did not[2].

ANOMALOUS CONTINUUM MEASUREMENTS

As well as line features there have also been a few reports of a sporadic hard spectral component extending into the MeV region of the spectrum. For example, Hillier *et al.*[11] observed an excess at MeV energies during a balloon flight in 1970 and later, Baker *et al.*[12] reported a 2.2 σ excess during balloon-borne observations in September 1972. A similar effect was reported by Gruber and Ling[13]. The excess was detected at the 3 σ level in each of 2 broad energy bands above an MeV in the unpulsed spectrum and was equivalent to the total luminosity. Most recently, in an extended analysis of HEAO-3 data, Ling *et al.*[14] report a 10 σ effect which is qualitatively and quantatively similar to the measurements of Gruber and Ling[13]. The emission lasted at least 50 days and was entirely unpulsed. The spectrum may be approximated by a power law or, alternately, a broadened Gaussian in analogy to the observed MeV excesses from Cygnus X-1 (see [18] and references therein), implying that such emissions may be a common property of compact objects in general. At PeV energies there have been numerous (and in some instances contradictory) reports of burst-like activity from the Crab. However, the case for such emission has been greatly strengthened by the simultaneous observation[15] of transient emission by the Kolar Gold Field and Baksan extensive air shower arrays on February 23, 1989. The emission was found to be pulsed with all the events arriving in the first half of the phasogram, which includes the main and inter- pulses. As with the observations of Cygnus X-3 at these energies, the observed showers are muon rich, implying that either photon and hadron interactions are similar at high energies, or the primaries are neutrinos (interacting with a substantially increased cross section at these energies) or a new strongly interacting stable particle. There were no reports of period discontinuities around this date. It had been previously suggested that TeV emission is correlated with pulsar timing glitches observed to occur approximately every two years. For example, Fazio *et al.*[19] reported observable levels of 10^{11} eV flux, following (within a 60 day delay) 'glitches' on September 29, 1969, August 1, 1971 and October 26, 1971. (Other reported glitches[20–22] occurred on February 4, 1975, August 22, 1986 and August 29, 1989). The same is probably not true for the reported lines or the MeV emission, since the phase analysis of both the Durham[6] and UCSD[13] groups exclude this possibility around the time of their respective measurements.

Finally, the period analysis of Grindlay, Helmken and Weeks[23] is particularly interesting and may tie in with the later FIGARO line measurement of Agrinier *et al.*[9]. Based on observations during December 1973, they report measuring TeV fluxes with ground-based Cherenkov counters whose light curve consisted of a single profile shifted from the main pulse by phase 0.2.

DISCUSSION

The phase-averaged continuum spectrum from the Crab can be understood as synchrotron emission produced in the magnetosphere of the pulsar, the system being powered entirely by the rotational energy-loss of the pulsar. From the above menagerie of observations, we summarize the pertinent recurring trends and common peculiarities over the canonical Crab electron synchrotron model.

1) There are transient emission lines near 75 and 400 keV.
2) The line fluxes range from 10^{-4} to 10^{-2} $\gamma\,cm^{-2}s^{-1}$.
3) Both lines appear unbroadened.
4) In both cases, the lines have shifted by ~ 6 keV in 10 years and possibly by 50 keV in 20 years.
5) The 75 keV line is pulsed, with the bulk of the emission emitted in the phase region 0 to 0.4.
6) There is evidence that the 400 keV feature is also pulsed, with the emission shifted in phase by 0.2 of a period from the main and interpulses, or entirely in the interpulse.
7) A related phenomena may be the sporadic TeV gamma-rays which also appear to be produced at a phase shifted 0.2 of a period from the bulk of the pulsed emission.
8) The Crab occasionally emits a hard, unpulsed component in the MeV region of the spectrum which persists on the order of months.
9) There are short bursts of pulsed PeV emission, emitted in the phase region 0 to 0.4.
10) At this time, there is no evidence that the observed lines, MeV, TeV and PeV emissions are related.

From the above summary we can form the following tentative picture. The cyclotron emission is generated in a region near the neutron star surface. From the measured line widths we can place a constraint on the radial extent of the emission region of < 100 m. The annihilation line is probably produced by the same population of particles that gave rise to the cyclotron γ-rays. However, the results of Agrinier *et al.*[9] and Parlier[10] suggest that the positrons must annihilate in a different region, possibly at the light cylinder or the opposite pole. The TeV γ-rays are probably produced in the same region because of the similarities in phase. Unfortunately, there aren't enough PeV events to glean much information from the light curve. Taking both the FIGARO results at face value, suggests that there may be steady state emission of the ~ 400 keV feature at the 10^{-4} $\gamma\,cm^{-2}s^{-1}$ level, occasionally flaring to several times 10^{-3} $\gamma\,cm^{-2}s^{-1}$.

Finally, we note that cyclotron and annihilation features have also been detected in GRB along with ~ 100 MeV emission. Therefore, in order to construct the micro and macro physics behind the emission from the Crab, it would beneficial to bear in mind that mature theories of multi-component transient emission exist for GRB (for a review, see ref. 24).

AO acknowledges a NAS Resident Research Associateship.

REFERENCES

[1] Leventhal, M., McCallum, C.J., and Watts, A.C., Ap. J., **216**, 491 (1977).
[2] Ling, J.C., Mahoney, W.A., Willett, J.B., and Jacobson, A.S., Ap. J., **231**. 896 (1979).
[3] Yoshimori, M., Watanabe, H., Okudaira, K., Hirasima, Y. and Murakami, H., Aust. J. Phys., **32**, 375, (1979).
[4] Manchanda, R.K., Bazzano, A. La Padula, C.D., Polacaro, V.F., and Ubertini, P., Ap. J., **252**, 172 (1982).
[5] Strickman, M.S., Kurfess, J.D., and Johnson W.N., Ap. J., **253**, L23 (1982).
[6] Ayre, C.A., Bhat, P.N., Ma, Y.Q., Myers, R.M., and Thompson M.G., M.N.R.A.S., **205**, 285 (1983).
[7] Watanabe, H., Ap. Space Sci., **111**, 157 (1985).
[8] Jung, G.V., Ap. J., **335**, 972 (1989).
[9] Agrinier, B., *et al*, Ap. J., **355**, 645 (1990).
[10] Parlier, B. *et al*., this conference.
[11] Hillier, R.R., Jackson, W.R., Murray, A., Redfern, R.M., and Sale, R.G., Ap. J., **162,** L177 (1970).
[12] Baker, R.E., Lovett, R.R., Orford, K.J., and D. Ramsden, Nature Phys. Sci., **245**, 18 (1973).
[13] Gruber, D.E., and Ling, J.C., Ap. J., **213**, 802 (1977).
[14] Ling, J.C., and Dermer, C.D., this conf.
[15] Acharya, B.S., Rao, M.S.V. Sivaprasad, K., Sreekantan, B.V., and Vishwanath, P.R., Nature **447**, 364 (1990).
[16] Kurfess, J.D., Ap. J., **168**, L39 (1971).
[17] Knight, F.K., Ap. J., **260**, 538 (1982).
[18] Owens, A., and McConnell, M.L., submitted to Comments on Astrophysics.
[19] Fazio, G.G., Helmken, H.F., O'Mongain, E., and Weeks, T.C., Ap. J., **175**, L117 (1972).
[20] Lohsen, E., Nature, **258**, 688 (1975).
[21] Lyle, A.G., Pritchard, R.S. and Smith, F.G., M.N.R.A.S., **233**, 667 (1988).
[22] Lyle, A.G., and Pritchard, R.S., IAU Circular No. 4845 (1989).
[23] Grindlay, J.E., Helmken, H.F., and Weekes, T.C., Ap. J., **209**, 592 (1976).
[24] Harding, A.K., Physics Reports, in press.
[25] Mahoney, W.A., Ling, J.C., and Jacobson, A.S., Ap. J., **278**, 784 (1984).
[26] Schwartz, R., Lin, R.P., Pelling, R., and Hurley, K., B.A.A.S., **12**, 542 (1980).
[27] Hameury, J. M., Boclet, D., Durouchoux, Ph., Cline, T.L., Paciesas, W.S., Teegarden, B.J., Tueller, J., and Haymes, R.C., Ap. J., **270**, 144 (1983).
[28] Hasinger, G., Pietsch, W., Reppin, C., Trumper, J., Voges, W., Kendziorra, P., and Staubert, R., Adv. Space Res., **3**, no. 10-12, 63 (1983).
[29] McConnell, M.L., Dunphy, P.P., Forrest, D.J., Chupp, E.L., and Owens, A., Ap. J., **321**, 543 (1987).
[30] Ubertini et al., Ap. J., in press.
[31] Perotti, F., *et al*., Ap. J., in press.
[32] Perotti, F., *et al*., Ap. J., **356**, 467 (1990).
[33] Owens, A., Myers, R.M., and Thompson, M.G., Proc. 19th ICRC, **1**, 145 (1985).

OBSERVATIONS OF 4U1700-377 WITH THE SIGMA TELESCOPE

P. Laurent, A. Goldwurm, F. Lebrun and J. Paul
Service d'Astrophysique
Centre d'Etudes Nucléaires de Saclay
91191 Gif-sur-Yvette Cedex, France

S. Mereghetti, D. Barret, L. Bouchet, J.P. Roques
Centre d'Etude Spatiale des Rayonnements
9, Avenue du Colonel Roche
BP 4346, 31029 Toulouse Cedex, France

E. Churazov, M. Gilfanov, A. Kuznetzov, R. Sunyaev,
I. Chulkov, A. Dyachkov, N. Khavenson and B. Novikov
Institute for Cosmic Research (IKI)
USSR Academy of Sciences
Profsoyouznaya, 84/32, Moscow 117296, USSR

ABSTRACT

The massive binary X-ray source 4U 1700-377 was observed three times in 1990 by the SIGMA telescope in the energy range 35 KeV to 1.3 MeV. The source, which was detected up to 170 KeV, exhibits a strong flaring activity, reaching the luminosity of the Crab Nebula in the 40-77 keV energy band during nearly 20 min on September 12. Its spectrum has been found to be of thermal Bremsstrahlung type with a temperature kT ≈ 50 KeV.

INTRODUCTION

The binary system 4U 1700-377/HD 153919 was discovered by Uhuru in 1970 [1]. The estimate of its distance, derived from the optical study of the O6.5f primary, ranges from 1.2 kpc [2] to 2.2 kpc [3]. The X-ray emission is powered by accretion from the dense stellar wind from the supergiant onto a compact object with a mass of ≈ 1.3 solar mass [4], most likely a neutron star, although X-ray pulsations have not been conclusively identified (see e.g. Murakami et al. [5] and references therein). The X-ray light curve is dominated by X-ray flaring activity that typically involves intensity variations by a factor of 10 to 100 on all time scales from hours to seconds, and by regular eclipses at the orbital period of 3.412 days. The observed variability is probably the result of variation in the accretion rate due to inhomogeneities in the stellar wind of the primary [6].

The relatively strong flux from this source has allowed the study of its properties in the hard X-ray regime with several balloon and satellite instruments (see e.g. Pietsch et al. [7] and references therein). However, none of these experiments was able to localize the source of the hard X-rays, although indirect evidence, mainly based on temporal and spectral behavior, was given for an hard X-ray emission from 4U 1700-377. Spectral analysis of the 20-180 KeV radiation from this region of the sky [7] have produced a high-energy spectrum which can be well fitted by a thermal Bremsstrahlung continuum with a temperature kT ≈ 28 KeV, independent of the source intensity.

Although rather precise, these measurements may be affected by the presence of other sources and/or diffuse emission in the field of view and suffer the problem inherent to the on/off source chopping technique : the reference field may also be contaminated by diffuse emission and/or other sources. Here we report new results, obtained with the SIGMA coded aperture telescope, on the first arcmin imaging observations in the 35 keV to 1.3 MeV energy band of the region around 4U 1700-377, which avoid all previously mentioned limitations.

OBSERVATIONS

The SIGMA telescope is an hard X-ray and soft gamma-ray instrument designed to obtain arcmin resolution images of the sky in the 35 KeV to 1.3 MeV energy range. The detector consists of a NaI(Tl) crystal optically coupled to an array of 61 hexagonal photomultipliers tubes. A tungsten coded mask aperture, based on an URA pattern, is placed 2.5 m in front of the detector plane. The angular resolution is of about 15 arcmin over the entire field of view ($4.7^\circ \times 4.3^\circ$ at full sensitivity and about $10^\circ \times 10^\circ$ at half sensitivity). A detailed description of SIGMA could be found in Paul et al. [8].

On March 24, 1990, the SIGMA telescope was pointed for ≈ 24 hours on the pulsar OAO 1657-415 which is 4° away from 4U 1700-377, which was then in the partially coded field of view with a sensitivity reduction of 30 %. This observation was carried out in the "spectral-imaging" mode, the standard mode of SIGMA, which yields two different sets of images : the "fine images" (pixel size 1.6 arcmin) accumulated in 4 consecutive energy bands to provide the highest source-location accuracy, and the "spectral images" which give sky pictures in 95 channels covering the whole SIGMA energy range. The pixel size and integration time of the spectral images are twice that of the fine images.

The second observation of 4U 1700-377 was carried out in September 12, 1990, also in the spectral-imaging mode. The telescope was pointing on GX 349+2, a soft X-ray source located only 1.5° away from 4U 1700-377, so that both sources were fully coded. The third observation of the source was in September 27, 1990, with the instrument pointing on 4U 1700-377 in the "variability-imaging" mode, where the spectral images are replaced by a series of consecutive, short exposures taken in a single energy band.

The SIGMA telescope, when operating in one of both imaging modes, does not record the arrival time of each detected photons. The only data available for time analysis are the total number of counts detected in each of the fine image energy bands, integrated over consecutive 4 s intervals. Although these counts rates are housekeeping data, they have proven to be very useful for the study of strong variable sources like 4U 1700-377.

RESULTS

The March 24 observation. As the source was observed in the partially coded field of view, absolute flux estimate require a more elaborated processing which is going on by now. Only intensity variations by at least a factor of two on timescale of hours could be inferred so far on the basis of the fine images. The source intensity is too weak to allow the use of the 4 s counting rate to trace the erratic source variations at shorter timescales. In addition, a search for periodic variations from 10 to 6000 seconds was unsuccessful. Specifically, nothing significant has been found at the previously reported frequencies of 67.4 s [5], 24 min [9] and 97 min [10].

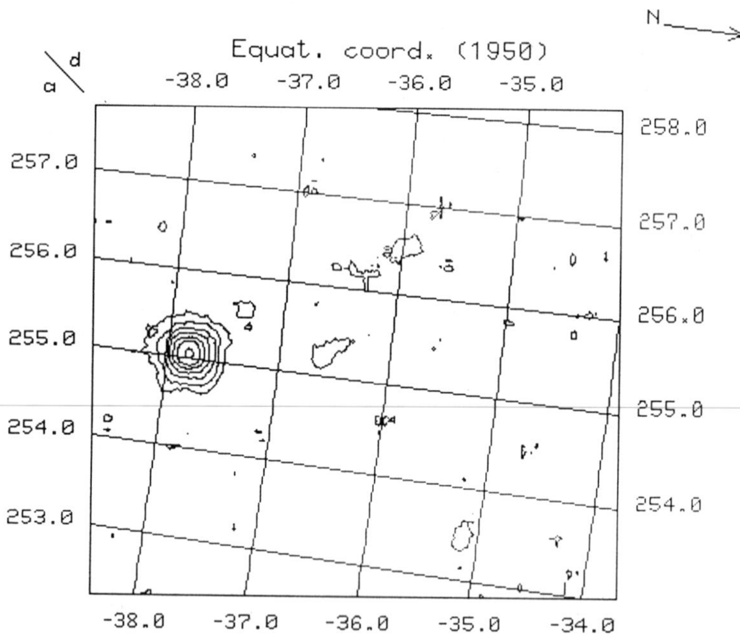

Figure 1 Skymap of the region around 4U 1700-377 as derived in the 40-77 keV energy band from the SIGMA data recorded on September 12, 1990. The contours are in statistical significance beginning at the 3 σ level and spaced by 3 σ.

<u>The September 12 observation</u>. The image of the sky region around 4U 1700-377, as derived from the <u>SIGMA</u> data recorded in the 40-77 keV energy band is shown in Figure 1. To obtain this picture, the fine images have been first corrected for spatial non-uniformities intrinsic to the detector or due to the background and then deconvolved using standard techniques [11]. The best fit position of the source is at right ascension $\alpha_{1950} = 255.145^{\circ}$ and declination $\delta_{1950} = -37.773^{\circ}$, consistent with the position of the star HD 153919 at $\alpha_{1950} = 255.136^{\circ}$ and $\delta_{1950} = -37.775^{\circ}$.

The analysis of the fine images shows that 4U 1700-377 was by far the strongest source in the total SIGMA field of view. We can therefore study its variability on shorter timescales using the 4 s counting-rate data, rebinned in 2 min intervals (Figure 2). The time corresponding to the end of the X-ray source eclipse is at orbital phase 0.11. The light curve is dominated by the strong flaring activity of 4U 1700-377 : particularly strong flares, the highest hard X-ray flares ever observed from that source, are visible at phases 0.24 and 0.26. At that moment, the source has reached the luminosity of the Crab Nebula in the 40-77 KeV energy range. The 4 s counting-rate data were also used to search for periodicities using phase folding and Fourier transforms methods. Here again, no significant results were found for periods between 10 and 6000 s although this analysis is strongly affected by the erratic source variations.

Figure 3 shows the energy spectrum derived from the third spectral pose and deconvolved using the energy response matrix [13]. It could be fitted with a thermal Brems-

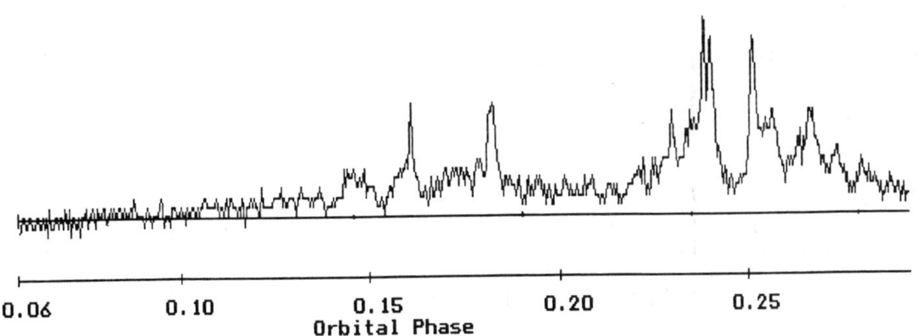

Figure 2 September 12, 1990 counting rate histogram of 4U 1700-377 in the 40-77 keV band, integrated over 120 s, as a function of the source orbital phase computed using the ephemeris given by Haberl <u>et al</u>. [12].

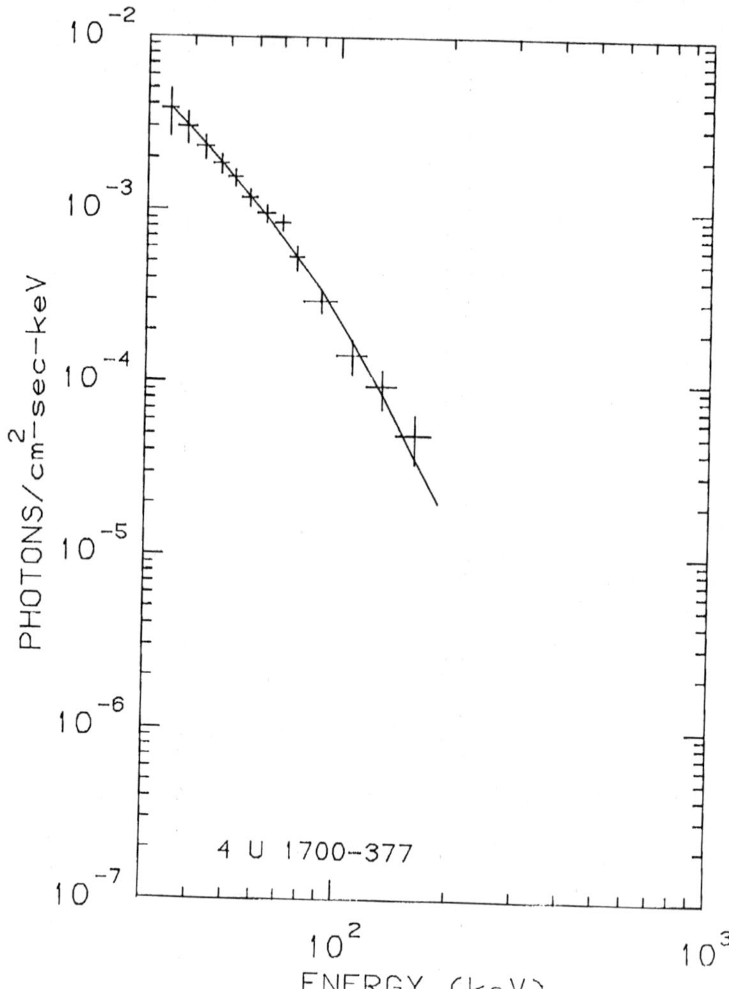

Figure 3 Photon spectrum of 4U 1700-377 derived from the SIGMA data collected during the strongest flares, i.e. from orbital phase 0.23 to 0.29. The solid line is a thermal Bremsstrahlung continuum with kT = 50 keV.

strahlung continuum with kT = 50 ± 10 KeV. Possible spectral shape variations with intensity are currently under investigation.

The September 27 observation. The strong flaring activity of the source is still clearly seen even if the average source intensity was a factor 2 lower than that observed on September 12.

CONCLUSIONS

We have observed the flaring activity of the source on timescales of minutes to hours. In a preliminary analysis, no periodicities have been found during our observations. When the source intensity was the highest, its spectrum can be well fitted with a thermal Bremsstrahlung law with kT ≈ 50 KeV. Although slightly higher than the temperature of ≈ 28 keV found earlier [7], our result is not yet precise enough for the difference to be significant.

The luminosity of the source at its maximum was found to be 4 x 10^{36} erg s^{-1}, i.e. comparable to the Crab Nebula luminosity. Pietsch et al. [7] have advocated that a white dwarf model was quite unlikely, mainly on the basis of their high observed luminosity. The peak luminosity we observed, two times that of Pietsch et al., ruled out definitively such a model. Complete timing analysis, as well as a more detailed study of the spectral shape dependence on the source intensity, has to be done to yield the valuable parameters required to improve our understanding of high-luminosity accretion in wind-driven binary X-ray source as 4U 1700-377 (see e.g. Ho and Arons [14]).

REFERENCES

1. Jones, C. et al. Ap. J. (Letters) **181**, L43 (1973).
2. Hutchings, J.B. in X-ray Binaries, NASA SP-389, 531 (1976).
3. Cruz-González et al. Rev. Mex. Astron. Astrof. **1**, 211 (1974).
4. Hutchings, J.B. Ap. J. **192**, 677 (1974).
5. Murakami, T. et al. Pub. Astr. Soc. Japan **36**, 691 (1984).
6. White, N.E. et al. Ap. J. **269**, 264 (1983).
7. Pietsch, W. et al. Ap. J. **237**, 964 (1980).
8. Paul, J. et al. in Advances in Space Research, proc. of XVIII COSPAR Symp. ME.4 (Pergamon Press, Oxford, in the press).
9. Branduardi, G. et al. M.N.R.A.S. **185**, 137 (1978).
10. Matilsky, T. et al. Ap. J. (Letters) **224**, L119 (1978).
11. Laudet, P. & Roques, J.P. Nucl. Instr. and Meth. **A267**, 212 (1988).
12. Haberl, F. et al. Ap. J. **343**, 409 (1989).
13. Barret, D. & Laurent, P. in preparation.
14. Ho, C. & Arons, J. Ap. J. **316**, 283 (1987).

SIGMA OBSERVATION OF THE SKY REGION CONTAINING THE COS-B SOURCE 2CG135+01

L.Natalucci, L.Bouchet, Y.Gaigé, M.Niel
Centre d'Etude Spatiale des Rayonnements, Toulouse, France

L.Salotti, J.Ballet, F.Lebrun, J.Paul
Service d'Astrophysique, CEA Saclay, France

E.Churazov, M.Gilfanov, A.Kuznetsov, R.Sunyaev,
I.Chulkov, A.Diachkov, N.Khavenson, B.Novikov
IKI, Moscow, U.S.S.R.

ABSTRACT

The region of the sky containing the high energy gamma-ray source 2CG 135+01 was observed by SIGMA on 1990, July 28-30th. The images obtained were searched for positive emission from the known candidate counterparts. No significant excess was detected. Upper limits in the hard X-ray range are presented and discussed in the light of the present observational status.

INTRODUCTION

The gamma-ray source discovered by COS-B near the position l=135°.0, b=1°.5[1] belongs to the set of 8 sources of the 2CG catalog which have been confirmed as "pointlike" following a recent analysis of the COS-B data[2]. A number of searches for counterparts at other wavelenghts have been performed in the past, yielding positive results from radio to the soft gamma-ray range. Nevertheless, the problem of the identification of the gamma-ray source is still open, due to the presence of at least two peculiar objects within the COS-B error box ($\sim$ 4 square degrees) and to the poor angular resolution of the already flown hard X-ray/soft gamma-ray experiments.

The possible identification with a faint X-ray source located at the limits of the COS-B error box was first proposed by Apparao et al.[3] due to its identification with a luminous, low redshift quasar (QSO 0241+622). Hard X-ray emission from this object was detected by OSO-8 and HEAO-1 A2[4] in the energy range 2-50 keV, with a weak indication of variability in the X-ray flux below 10 keV. Good evidence of variability on a time scale of months was provided by the Einstein IPC measurements between 0.5 and 4.5 keV[5]. A similarity with the well-known gamma-ray emitting quasar 3C 273 has been proposed by Worrall et al.[4] on the basis of the IR to gamma-ray spectral shapes and luminosities. Another possible counterpart is represented by the flat spectrum, periodically variable radio source GT 0236+61 discovered by Gregory & Taylor[6], located at about 1°.5 from the quasar. This source, identified with a B0 star at a distance of 2.3 kpc (LSI+61°303), emits typical 26.5 day periodic radio outbursts[7]. The outburst flux is found to be modulated over a time scale of 4 years[8]. X-ray emission from this star has been detected by the Einstein

IPC & HRI instruments[9] and from the HEAO A1 detector[10]. The high variability of radio emission and the high X-ray/optical flux ratio ($\sim 10^{-2}$) make LSI+61 303 a very peculiar object.

In addition to the above observations, hard X-ray emission above 10 keV from the general direction of 2CG135+01 was detected by a number of experiments[11,12,13]. Among them, ARIEL-V[12] and MISO[13] report positive detections up to about 1 MeV. The large error boxes associated with these instruments (typically a few degrees) do not allow to discriminate between the two main candidate objects. Clearly, a sensitive search in this energy range capable of separating the emission from each individual source could provide valuable informations. The SIGMA telescope[14], imaging the sky in the energy range 35-1300 keV with an angular resolution of 13 arcminutes and a 4°.7x4°.3 full sensitivity field of view (FSFOV), seems particularly suited for this type of search. In this paper we describe the results of observations performed during a three days period, and briefly compare them to the already existing data.

OBSERVATIONS AND RESULTS

Three separate observations of the same sky region centered at α=39°.67, δ=62°.8 have been carried out on 28,29,30 July 1990, yielding more than 45 hours of useful data. The images obtained in the range 35-1300 keV have been searched for positive emission in different energy bands. No positive signal was detected in any of the images despite the long observation time. The 3 sigma upper limits derived for the sum of the observations are reported in Fig.1, along with data from other experiments. These limits apply to a localized source in the FSFOV. As no excesses greater than $\sim$ 4 sigma were found in the reconstructed sky images, we estimate that the upper limits corresponding to a serendipitous source in the FSFOV are about 1.5 the reported values.

Inspection of Fig.1 shows that, while at low energies the SIGMA upper limits are consistent with the extrapolation above 50 keV of the OSO-8 spectrum measured from the quasar, in the range 100-400 keV they are at least a factor 2-3 lower than the reported MISO fluxes. The FSFOV of SIGMA crosses more than 70% of the $\sim$ 2°x6° MISO error box, while the remaining fraction is partially coded at a sensitivity $\gtrsim$ 80% the full sensitivity. In order to explain this difference one is led to consider possible variability of the soft gamma-ray source. If the latter has to be associated to the radio source, then the hard X-ray flux is likely to be correlated with the periodic radio light curve. Several binary models have been proposed which could explain emission from radio to gamma-rays[7,8,9] from LSI+61°303. For example, it is possible for hard X-rays to be created by scattering with stellar photons of energetic particles which can be produced in the vicinity of the compact object during the outburst phase. On the basis of the published radio ephemerides[8] the SIGMA observations have been performed at an epoch close to the outburst maximum. If confirmed with more recent radio data, this will possibly put further constraints on the emission scenario of this binary source. Anyway, further high resolution observations are needed in order to help investigate the origin of the gamma-ray emission and better assess the significance of presently available results.

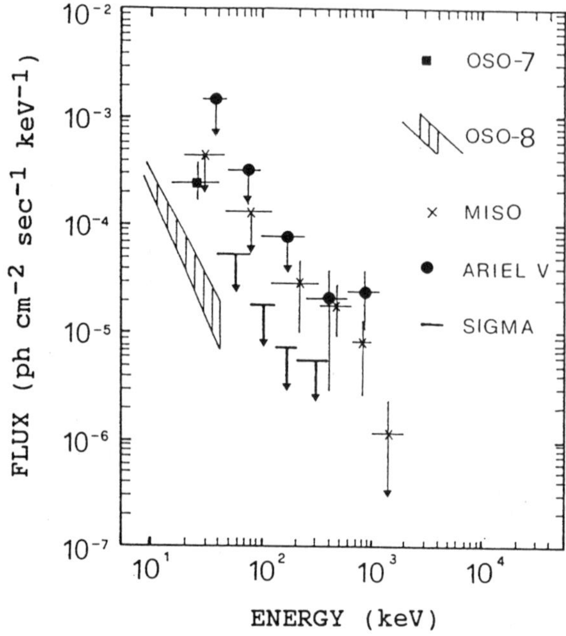

Fig. 1. Energy spectrum from the region of 2CG 135+01 in the hard X-ray/soft gamma-ray range, with our 3 sigma upper limits. With the exception of the OSO-8 data,which have to be associated with QSO 0241+622, the positive results reported are affected from low angular resolution, with typical error boxes of several degrees.

REFERENCES

1. W. Hermsen et al., Nature 269, 494 (1977)
2. H.A. Mayer-Hasselwander and G. Simpson, Proc. 27th COSPAR Meeting, Spoo, Finland (1988)
3. K.M.V. Apparao et al., Nature 273, 450 (1978)
4. D.M. Worrall et al., Ap.J. 240, 421 (1980)
5. S.Mereghetti, G.F. Bignami and P.A. Caraveo, Astron. Astrophys. 142, 37 (1985)
6. P.C. Gregory and A.R. Taylor, Nature 272, 704 (1978)
7. A.R. Taylor and P.C. Gregory, Ap.J. 255, 210 (1982)
8. P.C. Gregory et al., Ap.J. 339, 1054 (1989)
9 G.F. Bignami et al., Ap.J.(Lett) 247, L85 (1981)
10. G.H. Share et al., Proc. 21st COSPAR Meeting, Innsbruck (1978)
11. L. Maraschi et al., Nature 272, 679 (1978)
12. M.J. Coe, J.J. Quenby and A.R. Engel, Nature 274, 343 (1978)
13. F. Perotti et al., Ap.J.(Lett.) 239, L49 (1980)
14. J. Paul et al., Proc. 28th COSPAR Meeting, The Hague (in press)

SIGMA RESULTS ON THE FAST VARIABILITY OF CYG X-1 IN THE HARD X-RAY/GAMMA-RAY ENERGY BAND

S.Mereghetti*, S.Bonazzola, P.Mandrou, J.P.Roques
Centre d'Etude Spatiale des Rayonnements, Toulouse, France.

L.Salotti, B.Cordier, A.Goldwurm, A.Lambert
Service d'Astrophysique, CEA Saclay, France.

E.Churazov, M.Gilfanov, A.Kuznetzov, R.Sunyaev,
I.Chulkov, A.Dyachkov, N.Khavenson, B.Novikov
IKI, Moscow, USSR.

ABSTRACT

The black-hole candidate Cyg X-1 has been one of the first targets of the SIGMA γ-ray telescope on board the Soviet GRANAT spacecraft. SIGMA, an imaging telescope based on the coded mask principle, operates in the 35 keV-1.3 MeV energy range and, when used in "fast variability" mode, has a time resolution of one millisecond. We present here the preliminary results of a "fast variability" observation of Cyg X-1 performed in March 1990.

INTRODUCTION

X-ray observations of black-hole candidates can give more insight on the physics of the accretion disks than the observations of x-ray pulsars. In the latter we are dazzled by the emission from the accretion column, which can be much stronger than that from the inner disk. On the contrary, we can observe the x-ray emission of the whole accretion disk of a black-hole. In addition, the absence of a strong and organized magnetic field simplifies the physics and, we hope, the task of understanding the microphysics of an accretion disk.

Cyg X-1 is one of the binary systems that is though to contain a black-hole. The matter supplied by the O-type supergiant companion feeds the accretion disk and its gravitational energy is radiated via not steady state x-ray emission. The time variations and the energy spectrum are the two tools we have to understand the origin of the emitted radiation.

DATA ANALYSIS AND RESULTS

We have observed Cyg X-1 on March 23, 1990, using the coded mask SIGMA telescope[1] in fast variability mode. In this operating mode, the arrival time (1 ms accuracy), and the energy of each count are recorded. The observation lasted about 5000 seconds, with an average total counting rate of about 440 counts s^{-1} over the 40 keV - 1.3 MeV energy range.

---

* on leave from Istituto di Fisica Cosmica del CNR, Milano, Italy

We have computed the partially unbiased[2] autocorrelation functions (ACF) for Cyg X-1 in different energy intervals. The curves shown in Fig.1 have been obtained by averaging about 450 ACF relative to data intervals of 10 s each (bin size=0.1 s), and then applying a running average filter. While the curve for E>115 keV is compatible with that expected for an uncorrelated poissonian noise, typical Cyg X-1 behavior is visible in the lower energy band. lFigure 1.b, comparing the ACF for the energy intervals 40-80 keV and 80-115 keV, shows that correlated variability on time scales of ~1 second is present also at energies greater than 80 keV. In order to measure the shortest variations, we have also computed ACF with high time resolution. These have been fitted, for small values of the lag τ between 0 and $\tau_{max} < \tau_o$, with a straight line of equation $y(\tau)=a-b\tau$ (i.e. the first two terms of the exponential law $y(\tau) = \alpha \exp(-\tau/\tau_o)$). We claim the existence of fluctuations on timescales as short as τ_{min}, where τ_{min} is the smallest value of τ_{max} for which we obtain a value of b statistically different from 0. A τ_{min} of 80 ms has been obtained for the energy range 40-115 keV (at a confidence level of about 4σ).

To assess the presence of variability on different timescales, we have used the following method, which allows to establish the presence of variability on short timescales without being influenced by longer trends in the counting rate[3]. The whole observation is first divided into N segments of length T. The n_i counts in each segment are then binned into 10 bins of equal duration, and the corresponding χ^2 value, with respect to a uniform distribution with average $n_i/10$, is computed. The resulting distribution of χ^2 values is then compared to the expected theoretical distribution of χ^2 with 9 d.o.f.. The above procedure has been repeated for various values of T, and in different energy bands. The results are summarized in Table I, where the probabilities of obtaining the observed χ^2 distributions are reported. These have been computed with a Kolmogorov-Smirnov test, or, for the cases in which n_i was small, by performing numerical simulations.

CONCLUSIONS

The variability properties of Cyg X-1 have been traditionally interpreted in terms of shot noise models[2,4], which, although not specifically linked to any physical emission process, provide a convenient mathematical frame to describe the observed phenomena. The most detailed analysis have been made for energies <~30 keV, thanks to the data with high signal to noise ratio collected from Cyg X-1 since the seventies. These studies have demonstrated the existence of shots with a distribution of durations, ranging from a few milliseconds to a few seconds[5], and with overall characteristics compatible with models invoking magnetic flares or instabilities located at different radii of the accretion disk. Some constraints on the models could come from the study of the energy dependance of the shot noise properties, particularly in the hard x-ray/soft γ-ray energy range. Our preliminary results in the range 40-115 keV are compatible with those derived by Nolan et al.[6] for similar energies. In addition, although a precise quantitative analysis is limited by the low signal to noise ratio in our high energy data, they indicate that shot noise variability is present even for E>80 keV.

TABLE I

keV	T=10 s	T=5 s	T=2.5 s	T=1.25 s	T=0.625 s
40-115	2×10^{-11}	2×10^{-7}	2×10^{-5}	3×10^{-5}	
40-80	2×10^{-4}	4×10^{-3}	6×10^{-3}	4×10^{-6}	$<10^{-3}$
80-115	7×10^{-3}	0.1	0.3	5×10^{-4}	$\sim2\times10^{-2}$

REFERENCES

1. Paul, J. et al., Advances in Space Research, in press (1991).
2. Sutherland, P.G. et al., Ap.J. 219, 1029 (1978).
3. Meekins, J.F. et al., Ap.J. 278, 288, (1981).
4. Terrel, N.J., Ap.J. 174, L35 (1972).
5. Lochner,J.C. et al., Ap.J. in press (1991).
6. Nolan,P.L. et al., Ap.J. 246, 494 (1981).

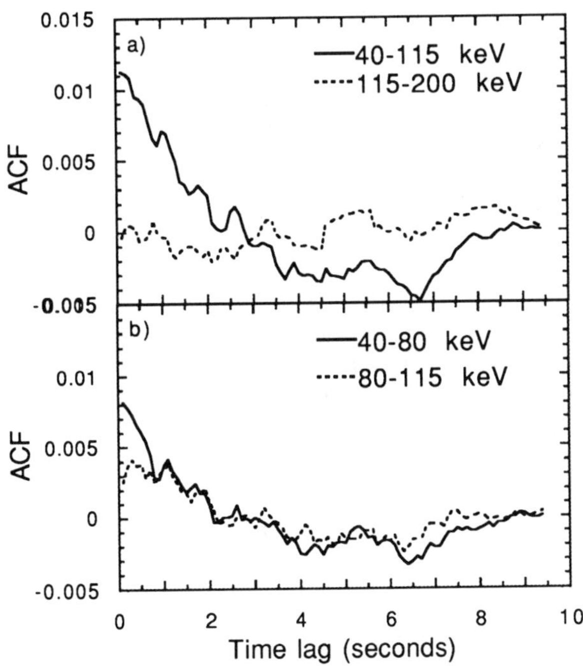

Fig. 1 - Cyg X-1 autocorrelation functions

THE POSITION OF GRS1758−258

G. K. Skinner
School of Physics and Space Research
University of Birmingham, B15 2TT, UK

ABSTRACT

Observations with the *Spacelab-2* and TTM coded mask telescope instruments identify GRS1758−258 with a weak source detected in the *Exosat* LE and hence lead to a very well defined position.

INTRODUCTION & OBSERVATIONS

GRS1758−258 is a hard source lying only about 40′ from GX5−1. It was first reported by Mandrou *et al.*[1] who observed it with the ART-P and SIGMA instruments on GRANAT and suggested that it is the origin of the hard flux which has been reported from GX5−1.

We have examined the data from the *Spacelab-2* coded mask telescope (XRT) and from TTM, a similar but smaller telescope on the *Kvant* module of MIR. Both instruments operate in the 3–30 keV band. In each case only observations with limited sensitivity are available (the SL2 observations are short and the region is close to the edge of the available fields of view of both instruments, particularly in the case of TTM). Nevertheless in the SL2 data, and marginally in that from TTM, we found a hard source close to, but outside, the error circle

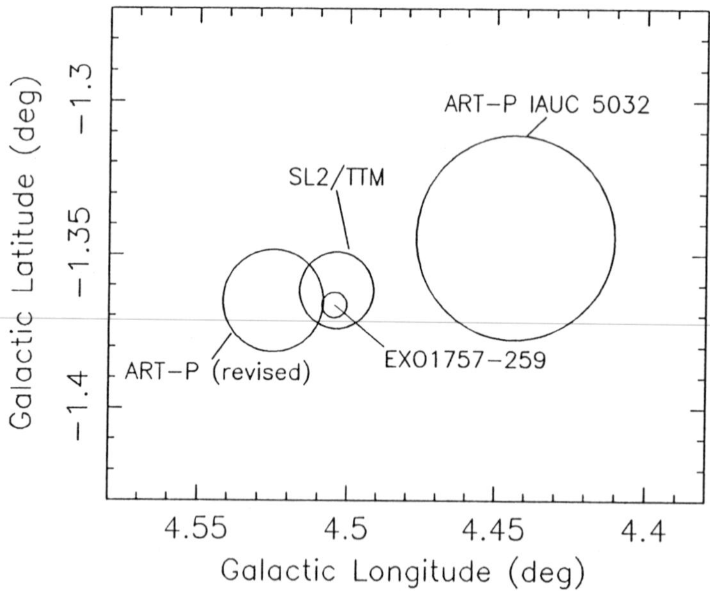

Fig. 1. 90% confidence error circles for GRS1758−258 and EXO1757−259 (galactic coordinates).

Fig. 2. Solid line and crosses: *Exosat* ME spectrum (photon index 1.70 and $N_H = 1.9 \times 10^{22}\mathrm{cm}^{-2}$(an approximate correction for collimator response has been applied on the assumption that the flux seen comes from EXO1757−258). Box: *Exosat* LE flux for EXO1757−258 (converted to flux assuming ME spectral shape). Dotted line and square symbols : *Spacelab-2* XRT spectrum (photon index 1.75 and $N_H = 12 \times 10^{22}\mathrm{cm}^{-2}$). Diamonds: *GRANAT* ART-P data[1]. Dashed line: TTM flux (data are poorly defined due to proximity to edge of the field of viewand are not shown for clarity; ME spectral shape, which is consistent with data, assumed; error bar shows uncertainty in fitted normalisation).

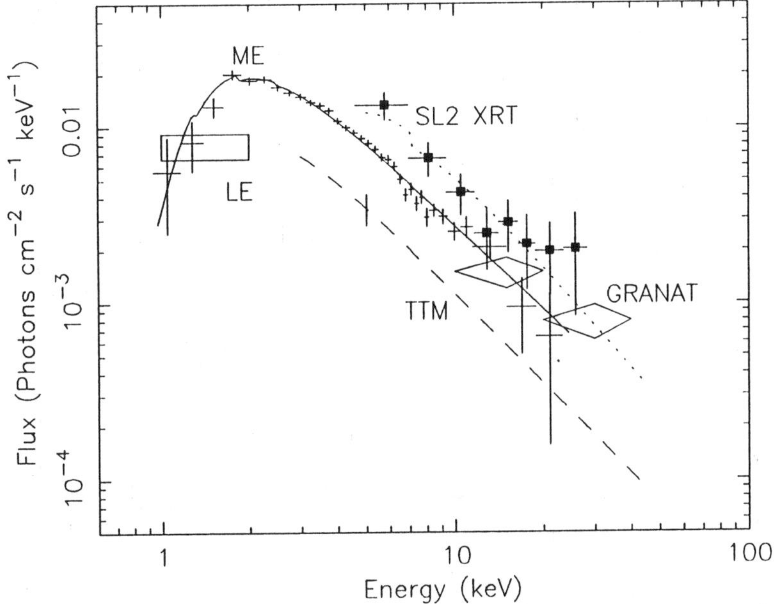

given by Mandrou *et al.* (Fig. 1). This lead to a re-examination of the GRANAT data and a corrected and improved ART-P error circle (R. Sunyaev, private communication) overlaps our 45″ radius (90% confidence) circle.

A search of the *Exosat* database revealed a source, EXO1757−259, close to the the centre of our error circle and only marginally outside the revised ART-P one. The field was observed because there had been indications of a new source when a slew carried the 45′ (FWHM) *Exosat* ME field of view through the region. However, because of the proximity of GX5−1 the ME data obtained were of limited value and as only a weak source was detectable in the LE image little more could be leant at that time (A. Parmar, private communication).

The *Spacelab-2* and TTM instruments provide spectrally resolved images and an accurate location for a hard source, present in 1985 August and in 1989 March–September, respectively, which is presumably the same as the source seen in 1990 March–April with GRANAT. The *Exosat* observations were made in 1985 September, less than 2 months after the *Spacelab-2* ones. The ME spectrum is very similar to that found for the *Spacelab-2* source (Fig. 2) and so is presumably also flux from GRS1758−258 and not leakage from GX5−1. The source seen in the LE image obtained at the same time not only lies within the *Spacelab-2* error circle but has a count rate which is consistent with that expected given the ME spectrum (which is well fitted with a power law with a photon index of 1.70 and $N_H = 1.9 \times 10^{22}\mathrm{cm}^{-2}$). Thus we may conclude that although the LE position lies slightly outside the revised ART-P error circle for GRS1758−258 these objects are in fact the same.

CONCLUSIONS

GRS1758−258 appears to be a persistent source and not simply a transient. Comparison of the preliminary intensity figures suggests that it is somewhat variable. If the X-ray column density is due to interstellar absorption it implies an extinction of $A_V \sim 8.5$ and hence $E_{B-V} \sim 3$. The *Exosat* LE error circle is 15″ radius (90% confidence) and centred on Right Ascension $17^h\ 58^m\ 7^s.2$, Declination −25° 44′ 54″ (1950). The ESO R Survey plate reveals several objects within the circle and comparison the UK Schmidt J survey plate shows that two of these are heavily reddened (Fig. 3). It is quite possible, however, that the optical counterpart of GRS1758−258 is not detectable at these wavelengths down to the plate limit ($m \sim 22.5$).

ACKNOWLEDGEMENTS

I wish to thank members of the *Spacelab-2* XRT and TTM teams for assistance, R. Sunyaev for providing information about the ART-P results and A. Parmar, P. Barr and J. Osborne for advice about the *Exosat* data.

REFERENCES

1. P. Mandrou, I.A.U. Circular 5032 (1990).

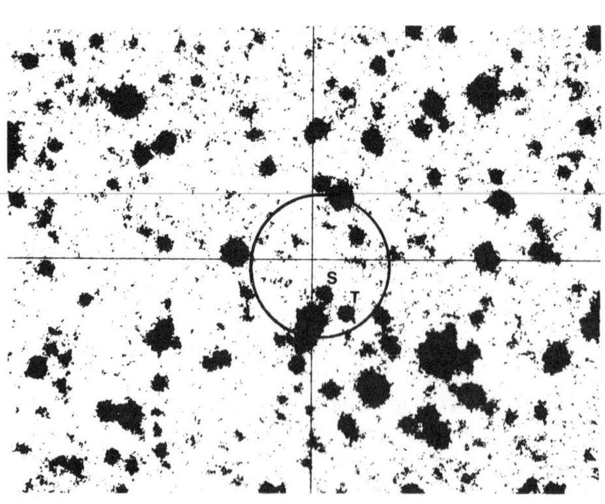

Fig. 3. The error circle for EXO1757−259 shown on the ESO R survey plate image. North is to the top, East to the left. Stars marked S and T are heavily reddened.

HEAO 3 OBSERVATIONS OF STRONG VARIABLE 0.5-3 MEV EMISSION FROM THE TAURUS REGION

James C. Ling
Jet Propulsion Laboratory, California Institute of Technology, Pasadena, CA 91109 &
NASA Headquarters, Washington, D.C. 20546

Charles D. Dermer
Space Sciences Laboratory, University of California, Berkeley, CA 94720

ABSTRACT

We briefly report preliminary results of observations of strong (>10σ) variable MeV emission from the Taurus region, possibly the Crab nebula, in 1980 by the HEAO 3 Gamma-Ray Spectroscopy experiment. The ~50-day time-averaged spectrum of the Crab shows a new component, which consists of a hard spectral "knee" in the 0.4-1 MeV range followed by a higher energy component extending to > 3 MeV, superposed on the canonical Crab power law spectrum with a spectral index of 2.2. This new spectral component was not observed 4 months earlier in the fall of 1979, when the spectrum was consistent with a single power law with the index of 2.2, the same as that observed in 1980. There is no indication that the γ-ray emission was pulsed. Such variable emission could be interpreted as Comptonization of soft photons by relativistic electrons in the magnetic field surrounding the neutron star.

HEAO 3 OBSERVATIONS

The HEAO 3 Gamma-Ray Spectroscopy experiment[1] scanned the Taurus region for ~20 days in 1979 between September 24 and October 13, and for ~50 days in 1980 between February 21 and April 11. In both of these periods, the data were collected from two alternate modes of observations: (1) the normal-scan (NS) mode when the scan plane was perpendicular to the earth-sun line, and (2) the galactic-plane scan (GPS) mode when the spin axis was pointed toward a galactic pole, allowing the instrument to scan in the equatorial plane of the galaxy. In 1979, the observations consisted of ~4 days of NS, followed by two weeks of GPS and 3 days of NS, for a total of ~20 days. In 1980, the observations started with ~2 weeks of NS followed by 5 cycles alternating between GPS (~3-4 days) and NS(~3-4 days), for a total of ~50 days.

The 1979-1980 light curves[2] of the 45-140 keV and 466-1914 keV fluxes (1-day average), respectively, of the Crab showed contrasting behaviors. First, the hard x-ray (45-140 keV) flux remained fairly stable from 1979 to 1980. The gamma-ray (466-1914 keV) flux, however, increased from a mean value of $(-0.06 \pm 0.26) \times 10^{-5}$ to $(1.43 \pm 0.14) \times 10^{-5}$ photons cm^{-2} s^{-1} keV^{-1} (>10σ significance). The 1980 gamma-ray flux was also quite stable over the entire 50-day period. The reduced χ^2 of a fit to the 50 daily-averaged fluxes, assuming a constant γ-ray flux, is 0.8. Figure 1 shows the time-averaged spectra of the Crab for these two periods. The solid lines are the best-fits to a single power law whose parameters are consistent with those of the canonical Crab spectrum. The new gamma-ray component shown in the 1980 spectrum consists of a spectral "knee" in the 0.4-1 MeV region plus a higher energy component extending to >3 MeV. The observed fluxes in the 1-3 MeV range are about a factor of 4-6 stronger than the fluxes extrapolated from the X-ray spectrum. There is no evidence that the MeV emission is pulsed with the 33 ms pulsed period[3], although our study shows that the preferred source location for emitting gamma rays is from the direction of the Crab rather than from other near-by objects such as Geminga, or possible unknown sources located within 5° of the Crab.

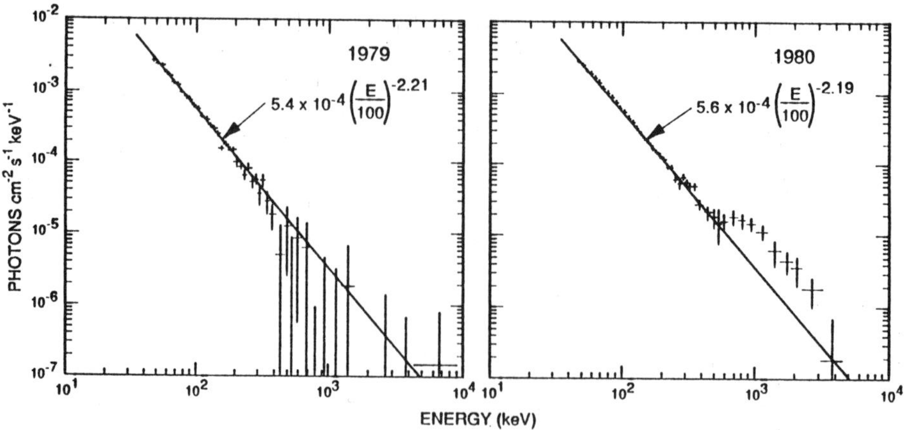

Figure 1: The time-averaged spectrum of the Crab observed in the Spring of 1980 consists of a new γ-ray component superposed on the canonical power-law spectrum with index of 2.2. This is in contrast to the spectrum observed 4 months earlier in the fall of 1979 which showed a single power-law spectrum with the same spectral index.

Gamma-ray flux >0.5 MeV possibly associated with the Crab nebula was also observed by several balloon experiments in the 1970's[4-5], although at lower statistical significance. These spectra showed a hard component extending to ~10 MeV. An explanation of such MeV emission should enhance our understanding of the physics of the neutron star system, especially interactions between the neutron star and its environment. Finally, we have also searched for a possible narrow annihilation feature associated with the 1980 MeV emission, and obtained a 3σ upper limit of 5 x 10^{-4} photons cm^{-2} s^{-1}. Evidence for such a correlation between the line and MeV fluxes has been reported in two black hole systems[6-7].

DISCUSSION

The nonpulsed emission observed from the direction of the Crab between radio and gamma-ray frequencies is generally interpreted as synchrotron radiation produced by relativistic electrons spiralling in a nebular magnetic field with strength ~10^{-4} G. Models for the nonpulsed spectrum of the Crab generally involve injection of electrons and positrons from the pulsar into the surrounding nebula[8-10]. Comparison of the synchrotron energy-loss time scale for spatially-resolved emission implies that acceleration of electrons beyond the pulsar is required. This may occur through the strong, low-frequency 30 Hz wave radiated by the neutron star[11-12], through resonant acceleration of electrons by MHD turbulence in the magnetic field of the Crab supernova remnant[13], or through shock acceleration at the MHD shock front[9-10].

The magnitude of the nonpulsed gamma-ray component observed in the Crab spectrum in the Spring of 1980 is nearly a factor of ten greater than the pulsed component[3]. It is therefore likely that this new component is formed outside the light cylinder of the Crab pulsar. We propose a scenario in which the secondary acceleration of electrons and positrons is modified by a source of soft photons emitted for a duration of at least several weeks in the Spring of 1980. These soft photons could be emitted by the Crab pulsar itself, and represent a source of energy loss of the electrons and positrons through Compton scattering. According to the theory of Schlickeiser[14-15], the competition between first-and second-order Fermi acceleration and Compton scattering can lead to the formation of a pileup distribution of relativistic electrons and positrons. If the escape time of the particles from the acceleration/energy-loss region is large, the monoenergetic portion of the pileup distribution dominates the formation of the spectrum.

We have derived the spectrum of Compton-scattered thermal photons under the assumption that the pileup distribution can be approximated by a monoenergetic source of electrons and positrons. We also assumed that the direction of the radiating electrons was random, so that the photons are isotropically scattered. We are able to obtain a good fit[2] to the new gamma-ray spectral component of the Crab for a thermal photon source with temperature 0.4 keV, and for a particle Lorentz factor γ~20. An exact calculation of the Comptonized spectrum requires the solution to the electron and positron distribution function subject to both acceleration and energy losses, but the preliminary results from this simple model are encouraging.

This scenario for the emergence of a gamma-ray component in the Crab spectrum makes a number of straightforward predictions. Foremost is the existence of a separate soft photon source that appears coincident with the gamma rays. There should also be correlated variability at other wavelengths; for example, we expect a synchrotron feature from the monoenergetic part of the pileup electron/positron distribution, and Compton and synchrotron spectral features should be formed by the power-law tail of the pileup distribution[14]. Multiwavelength studies of the Crab correlated with γ-ray observations are important in determining the processes taking place in this system.

ACKNOWLEDGEMENTS

We want to thank W. Mahoney and Wm. Wheaton for their valuable comments of this manuscript. The research described in this paper was carried out by the Jet Propulsion Laboratory, California Institute of Technology, under contract with the National Aeronautics and Space Administration. The research by CD was partially supported by grant 90-13 from the LLNL Institute for Geophysics and Planetary Physics.

REFERENCES

1. Mahoney et al., Nuclear Instrument and Method, 178, 363, (1980).
2. Ling et. al., in preparation to be submitted to Ap. J., (1991).
3. Mahoney, Ling and Jacobson, Ap. J., 278, 784, (1984).
4. Baker et al., Nature, 245, 18, (1973).
5. Gruber and Ling, Ap. J., 213, 802, (1977).
6. Riegler et al., Ap. J. (Letters), 294, L13, (1985).
7. Ling and Wheaton, Ap. J. (Letters), 343, L57, (1989).
8. Kundt and Krotscheck, Astron. Ap., 83, 1, (1980).
9. Kennel and Coroniti, Ap. J., 283, 694, (1984).
10. Kennel and Coroniti, Ap. J., 283, 710, (1984).
11. Pacini, Nature, 219, 145, (1968).
12. Ostriker and Gunn, Ap. J., 157, 1395, (1969).
13. Barnes and Scargle, Ap.J., 184, 251, (1973).
14. Schlickeiser, Astron. Ap., 136, 227, (1984).
15. Schlickeiser, Astron. Ap., 143, 431, (1985).

HARD X-RAY OBSERVATION OF SCO X-1 IN THE RANGE 15-160 KEV

P. Ubertini, A. Bazzano, M. Cocchi, C. La Padula, and R.K. Sood*

Istituto di Astrofisica Spaziale (CNR), Frascati, Italy

*University of N.S.W., ADFA, Canberra, Australia

ABSTRACT

The region containing the galactic source Sco X-1 has been observed on May 17, 1989 in the energy range 15 ÷ 160 keV. The measurement was performed with the balloon borne experiment "POKER" based on an array of three high pressure Multiwire Proportional Counters (MWPC) with energy resolution of 10% at 120 keV. The total sensitive area was 8,000 cm^2 and the experiment was passively collimated with copper hexagonal collimator with 1.9 degree FWHM field of view. The preliminary data analysis confirmed the detection of the source at energies up to 60 keV.

I. INTRODUCTION

Sco X-1, apart from the sun, was the first X-ray cosmic source discovered[1] and is the brightest non-transient X-ray object in the sky at energies lower than 20 keV[2]. From the optical point of view the source shows a binary period of 0.787 day[3] characterized by the presence of an old Nova. The high luminosity and the X-ray behaviour of the source, not showing any fast periodicity, indicate the presence of a collapsed non magnetised companion like a neutron star[4], a black hole[5] or a white dwarf[6]. At low energy ($E \leq 20$ keV) the source shows evidence of a high temperature thermal emission in at least three different states: quiescent, active and flaring[7,8]. Long and repeated observations and monitoring have been performed on this source at low energy, while only a few high energy observations are available above 20 keV, apart from many early observations that do not provide a strong constrain on the emission mechanisms. The most intriguing result characterizing the high energy behaviour of Sco X-1 was the reported presence of a "hard " tail, showing up above ≃ 50 keV[9,10,11,12,13,14], where thermal emission mechanisms cannot be present any more. This tail was not detected on different occasions[15,16,17,18,19,20]. The "existence" or not of this hard tail is of a crucial importance to understanding the nature of the collapsed object, its presence having a strong impact on the energy balance of the system.

II. THE INSTRUMENT

The "POKER" experiment, described in detail elsewhere[21,22], was designed to perform high sensitivity observations of cosmic sources in the hard X-Ray range (10÷180 keV). The telescope consists of an array of three high pressure Xenon Multiwire Proportional Counters (hereinafter MWPC) with a sensitive area of 2,500 cm^2 each. The detectors named SPC1, SPC3 and SPC4 were filled at a pressure of 2.6, 3.6 and 3.8 bar respectively with a mixture of Xe/CO_2 (SPC1) and Xe/Isobutane (SPC3 and SPC4) to provide an efficiency greater than 20% respectively for 13 keV and 120 keV photons. The MWPCs spectral resolution was 14% at 60 keV. The fields of view of the three MWPCs were coaligned and limited by means of hexagonal copper collimators with an aperture of 1.9 degree FWHM. The X-ray telescope is mounted on the central cage of the steering platform which includes the elevation drive system and azimuthal control. The azimuthal reference is obtained with the use of a three axes flux gate magnetometer measuring the earths magnetic field. The elevation angle is controlled by a shaft encoder relying on the local vertical, measured with a set of three independent inclinometers, on each detector, to measure the elevation and cross-elevation angle. Changes of the payload azimuth beneath the balloon are achieved by the combined use of a reaction wheel mounted on the payload and a torsion relief drive in the payload bearing. The signals from the magnetometers are on-board processed to compute the azimuth angle of the payload. This angle is compared with the desired target azimuth either telecommanded or calculated by the on-board microprocessor. The actual and target azimuth data are used to generate an error signal to drive the motors. The reaction wheel is directly driven by the signal error while the torsion relief system is driven with the use of velocity feedback from a tacho generator integrally fitted to the reaction wheel drive. Two CCD cameras, coaligned with the detector axis, have been used to check the absolute telescope aspect in real time. The two cameras have been equipped with different objectives and filters to be used differently, the first one as a Sun Sensor, over a very narrow f.o.v. of 1.9 deg, and the second one as a Star Sensor with a 4 deg f.o.v. The video data are transmitted via a 1 Mhz bandwidth high power transmitter. During the first part of the flight aspect check were performed on the moon. The in-flight positioning accuracy and stability was better than 0.1 deg in elevation and azimuth and the orientation of the telescope axis was known with an accuracy of ± 0.1 degree.

III. OBSERVATION AND DATA ANALYSIS

The telescope was launched from the Balloon Launching Station in Alice Springs (Australia) at 09.30 U.T. on May 17, 1989. The Scorpius X-1 region was studied between 13.03 and 14.20 U.T. using both tracking and scanning observations. During this period the energy response of the instrumentation was found to be stable within 2%. The data from each MWPC are separately analyzed and the results statistically combined. In this paper we present the preliminary data analysis of about one third of the whole data.

In order to minimize possible systematic background variations related to elevation and/or azimuthal changes during this observation, the background was evaluated before and after the source observation. Correlation between the background counting rate and potential sources of systematic variation such as altitude, longitude, latitude and steering coordinates were carefully investigated to allow the appropriate correction to the background noise subtraction.

During the flight the only significant source of systematic background variation was found to be related to the float altitude changes. Sco X-1 and the associated background data were collected at a float altitude corresponding to an atmospheric pressure ranging between 3.35 and 3.40 mbar. This variation resulted in a negligible effect if compared with the errors due to the statistics. The total integration time of useful data on the source was 1710 seconds, while 4500 seconds were spent in background evaluation.

The raw detected excess was interpreted as originating from Sco X-1 and folded through the matrix describing the detector with a standard reiterative χ^2 procedure assuming a power law, a thermal bremsstrahlung spectrum, with and without Gaunt correction[23], and a Wien law[24,25].

The detector matrices were determined by taking into account the detector parameters (gas mixture, geometry of the absorption cell, escape effects etc.), the electronics thresholds, the collimator effect, and the atmospheric absorption. The precision of this deconvolution method has been confirmed by observation of the Crab Nebula during other flights[26,27,28]. The resulting photon spectra and the relative best fits are listed in Table I. The errors are at the 68% confidence level for joint parameters estimation[29]. In figure 1 the spectral points are shown together with the best fit obtained with a thermal bremsstrahlung spectrum with Gaunt factor correction. In the figure 1 the Sco X-1 POKER spectrum is also compared with the high energy HEAO-1[15] and MPI/AIT[20] data. As can be seen our data are not consistent with a power law and we found no evidence of the early reported hard X-ray tail. In fact the "POKER" upper limits are about one order of magnitude below the reported hard tail flux. A preliminary timing analysis on the same data set shows no presence of the reported high frequency pulsation at 2.93 ms[30] neither the presence of QPO[31] at energies above 15 keV.

Acknowledgements

We thank Prof. R. Sunyaev for very useful scientific discussions on the emission mechanisms of this source.

We also are grateful to the NSBF/NASA crew, led by R. Kubara, for the flight operations conducted with high professionalism during a long and difficult campaign.

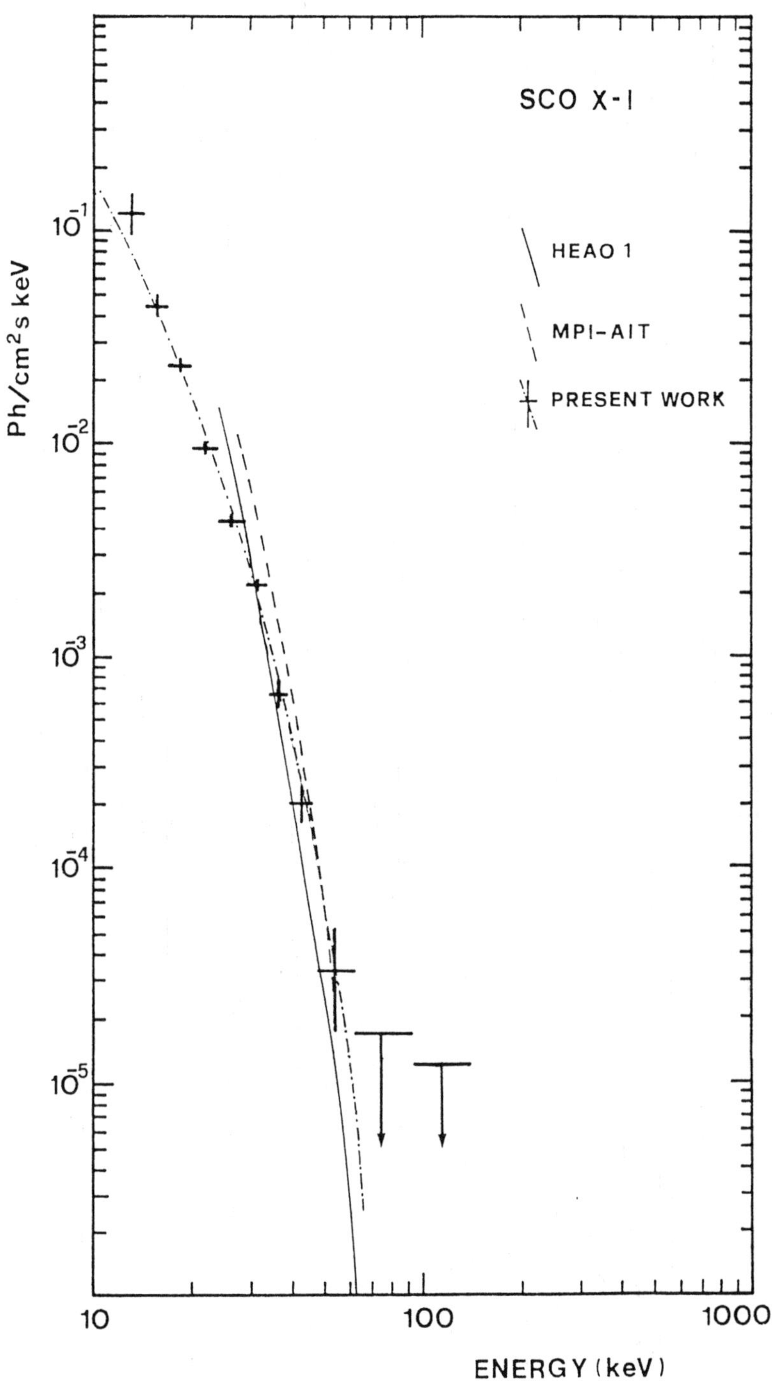

Figure 1. The SCO X-1 Spectrum reduced at the top of the atmosphere

TABLE I

Sco X-1 Hard X-Ray Spectra: Model fit parameters
POKER Experiment 1989 May 17

Model	Parameter	Value
Single power law: $A(E/30keV)^{-\alpha}$		
	Intensity: $A(10^{-3})$	$1.82^{+0.10}_{-0.12}$
	Slope: α	$5.15^{+0.15}_{-0.14}$
	$\chi^2_{d.o.f}$ (d.o.f.)	5.78 (9)
Wien law: $A*(E/30keV)^2*e^{-(E-30)/kT}$		
	Temperature: kT(keV)	$3.90^{+0.13}_{-0.17}$
	Intensity: $A(10^{-3})$	$2.45^{+0.16}_{-0.20}$
	$\chi^2_{d.o.f}$ (d.o.f.)	3.64 (9)
Thermal bremsstrahlung: $A*(E/30keV)^{-1}*e^{-(E-30)/kT}$		
	Temperature: kT(keV)	$6.78^{+0.37}_{-0.36}$
	Intensity: $A(10^{-3})$	2.26 ± 0.14
	$\chi^2_{d.o.f}$ (d.o.f.)	2.16 (9)
Thermal bremsstrahlung: $A*(E/30keV)^{-1-\alpha}*e^{-(E-30)/kT}$ with $\alpha=0.37(30/kT)^{0.15}$ (with Gaunt factor)		
	Temperature: kT(keV)	$7.54^{+0.42}_{-0.4}$
	Intensity: $A(10^{-3})$	$2.24^{+1.34}_{-1.55}$
	$\chi^2_{d.o.f}$ (d.o.f.)	2.12 (9)

Energy range: 15 ÷ 160 keV; Errors indicated are joint parameters variation at 68% confidence level calculated according to Lampton et al., 1976.

REFERENCES

1.Giacconi, R., *et al.*, 1962, Phys. Rev. Letters, **9**, 439
2.Forman, W., *et al.*, 1978, Ap. J. Supp., **38**, 357
3.Gottlieb, E.W., *et al.*, 1975, Ap. J. Letters, **195**, L33
4.Zel'dovich, Ya. B., and Sakura, N. I., 1969, Astr. Zh., **46**, 225
5.Basko, M. M., and Sunyaev, R., 1973, Ap. Space Sci., **23**, 117
6.Kylafis, N. D., and Lamb, D. Q., 1979, Ap. J. Letters, **228**, L105
7.White, N. E., *et al.*, 1976, M.N.R.A.S., **176**, 91
8.Miyamoto, S., and Matzuoka, M., 1977, Space Sci. Rev., **20**, 687
9.Peterson, L. E., and Jacobson, A. S., 1966, Ap. J., **145**, 962
10.Riegler, G. R., *et al.*, 1970, Nature, **226**, 1041
11.Haymes, R.C., *et al.*, 1972, Ap.J. (Letters), **172**, L47
12.Matsuoka, M., *et al.*, 1972, Astrophys. Space Sci., **18**, 472
13.Greenhill, J.G., *et al.*, 1979, M.N.R.A.S., **189**, 563
14.Dudlig, M., L., *et al.*, 1983, Astrophys. Space Sci., **95**, 137
15.Soong, Y., and Rothschild, R.E., 1983, Ap.J., **274**, 327
16.Rothschild, R.E., *et al*, 1980, Nature, **286**, 786
17.Johnson, W. N., *et al.*, 1980, Ap. J., **238**, 982
18.Coe, M. J., *et al.*, 1980, Ap. J., **237**, 148
19.Lewin, W. H. G., *et al.*, 1970, Ap. J. Letters, **162**, L109
20.Jain, A., *et al.*, 1984, Astron. & Astrophys., **140**, 179
21.Bazzano, A. *et al.*, 1983, Nucl. Instr. & Meth. **214**, 481.
22.Ubertini, P. *et al.*, 1985, SPIE **579**, 74.
23.Mätzler, C., *et al.*, 1978, Ap. J., **223**, 1058
24.Sunyaev R. A., and Zel'dovich, Ya. B., 1980, Ann. Rev. Astron. Astrophys., **18**, 537
25.Sunyaev, R. A., and Titarchuk, L. G., 1980, Astron. Astrophys., **86**, 138
26.Ubertini, P., Bazzano., A., Sood, R., Staubert, R., Sumner, T.J. and Fry, G., 1989, Ap.J. (Letters), **337**, L19
27.Bazzano, A. *et al.*, 1989, SPIE, **1159**, 126
28.Ubertini, P. *et al.*, 1990, Submitted to Ap.J.
29.Lampton, M.,Margon,B.,and Bowyer,S., 1976, Ap. J.(Letters), **208**, 177
30.Damle, S.V. *et al.*, 1989, Proc 23rd ESLAB Symp, EAS **SP-269**, 377
31.Hasinger, G., and van der Klis, M., 1989, Astron. & Astrophys., **225**, 79

LOW STATE HARD X-RAY OBSERVATION OF CYG X-1

A. Bazzano, C. La Padula, R.K. Manchanda*, V.F. Polcaro, P. Ubertini
Istituto di Astrofisica Spaziale (CNR), Frascati, Italy

R. Staubert, E. Kendziorra
Astronomisches Institut der Universitat Tubingen, Tubingen, Germany

ABSTRACT

We report a "super-low" state observation of the black hole candidate Cygnus X-1 in the energy range 15 - 120 keV. The data were obtained with the "POKER" experiment, designed to perform high sensitivity observations of cosmic sources in the hard X-Ray range (15 - 120 keV). The telescope consisted of an array of three high pressure Xenon Multiwire Proportional Counters (MWPC) with a total sensitive area of 7,500 cm^2. The detectors were filled at a pressure of 2.6 bar with a mixture of Xe/Argon/Isobutane to provide an efficiency greater than 20% for photon energies from 15 keV to 110 keV. The MWPC spectral resolution was 13% at 60 keV. The fields of view of the three MWPCs were coaligned and limited by means of hexagonal copper collimators with an aperture of 5.0 degree FWHM. In order to provide imaging capability to the telescope one of the MWPC was equipped with two co-rotating Rotation Modulation Collimators, developed at AIT, and modulating a geometrical area of 1,600 cm^2 of the detector. The telescope was launched from the Milo Base, Sicily (Italy), on 1985 August 5, and scanning and pointed observations were carried out on the Crab Nebula, A0535+26, MCG 8-11-11, NGC 4151, Cygnus X-1 and Her X-1. The Cygnus X-1 and Crab photon spectra are well described by single power laws, with photon indeces of $\alpha = 1.87$ and $\alpha = 2.17$ and intensities of 2.57×10^{-3} and 2.32×10^{-3} ph cm^{-2} s^{-1} keV^{-1} at 50 keV, respectively. The low hard X-ray emission from Cyg X-1 confirms that during the POKER observation the source was in a "low" state as also supported by EXOSAT data collected at lower energies on August 12.

THE INSTRUMENT

The "POKER" experiment, described in detail elsewhere[1,2], basically consists of an array of three high pressure Xenon Multiwire Proportional Counters (MWPCs) named SPC1, SPC3 and SPC4, with a sensitive area of 2,500 cm^2 each. The fields of view of the detectors were coaligned and limited by means of hexagonal copper collimators with an aperture of 5.0 degree FWHM. One of the MWPCs (SPC1) was equipped with two co-rotating Rotation Modulation Collimators [3] (hereafter RMC), developed at AIT, and modulating a geometrical area of 1,600 cm^2 of detector SPC1. The orientation of the telescope axis was known with an accuracy of ± 0.1 degree.

---

*Permanent Address: Tata Institute of Fund. Research, Bombay, India

---

OBSERVATION AND DATA ANALYSIS

The telescope was launched from the Milo Base, Sicily (Italy), at 05.00 U.T. on 1985 August 5. The Crab Nebula was studied between 9.22 and 12.00 U.T. by both tracking and drift-scan observations. The total integration time of useful data on the source was 1740 seconds, while 2600 seconds were spent on the background.

A single drift scan was performed on the sky region containing the black-hole candidate Cygnus X-1 between 21.50 and 23.10 U.T. The total integration time of useful data was 1350 s on-source and 3100 s off-source.

Crab Nebula and Pulsar

The excess count rate spectra, interpreted as originating from the Crab, were detected over the whole operative energy range of the three MWPCs (SPC1: 10-84 keV, SPC3: 10-120 keV, and SPC4: 10-100 keV) at a confidence level corresponding to 22 σ, 58 σ and 46 σ respectively. The lower statistical significance of SPC4 is due to 300 s of bad on-source data not used in the spectral analysis. The three raw excess spectra were folded through the matrices describing the detectors with a standard reiterative χ^2 procedure assuming a power law emission from the source. The best fit for the combined data is given by:

$$dN/dE = (7.03\pm 0.11) \times 10^{-3} \, (E/30keV)^{-2.17\pm0.04} \quad (ph\ cm^{-2}s^{-1}keV^{-1})$$

with a reduced χ^2 of 0.82 for 41 degrees of freedom. The errors are at the 68% confidence level for joint parameters estimation[4].

The spectra obtained by the individual detectors differ by less than 2% over the entire energy range.

Cygnus X-1

The same procedure as described for the Crab was used to evaluate the excess count rate from Cygnus X-1. The set of on-source/off-source differences from the three MWPCs were binned in 7, 15 and 15 energy channels respectively. The source was detected at a confidence level corresponding to 5.6 σ, 18.7 σ and 18.1 σ for SPC1, SPC3 and SPC4, respectively.

The Cygnus X-1 data were fitted with power law spectra as well as with a single temperature thermal law. The best fit results were obtained with a power law as follows:

$$dN/dE = (6.69\pm 0.33) \times 10^{-3} \, (E/30keV)^{-1.87+0.11} \quad (ph\ cm^{-2}s^{-1}keV^{-1})$$

with a reduced χ^2 of 0.56 for 31 degrees of freedom. The uncertainty on the estimation of the source intensity due to systematic errors associated with background and instrumental gain variation and detector efficiency is less than 3%. The Cygnus X-1 and Crab spectra are shown in figure 1.

DISCUSSION

Cygnus X-1 is considered to be one of the most variable high energy sources (for a comprehensive review see Liang and Nolan 1984 and Ubertini *et al.* 1991 and references therein)[5,6]. Recent observations show a more "normal" scenario at high energies. Our suggestion is that the POKER/EXOSAT-1985 and the MIFRASO/GINGA 1987[6,7] observations represent the normal behaviour of Cygnus X-1, the high energy emission due to accretion processes onto the central object surrounded by a two temperature accretion disk, following the Sunyaev and Titarchuk Comptonization model[8], accounting for the X-ray emission from a few keV up to at least 300 keV.

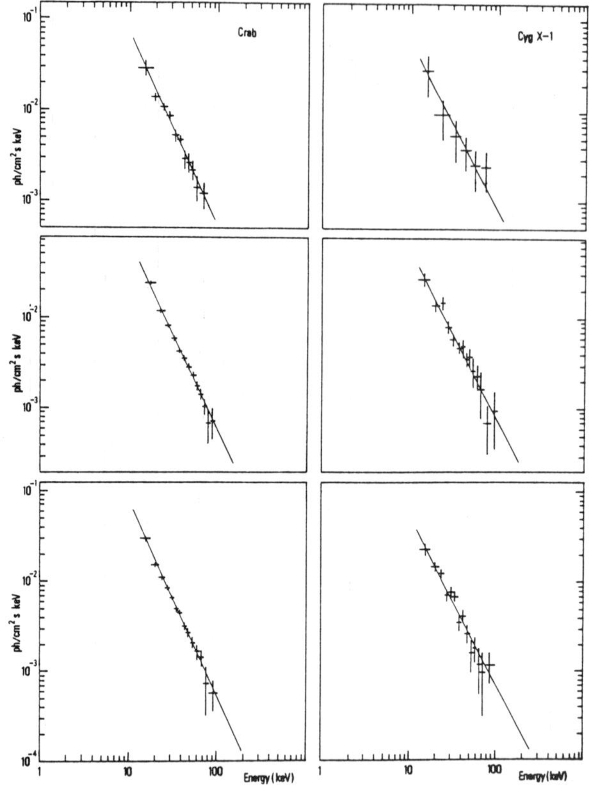

Figure 1. Deconvolved spectra and best fit of Crab and Cygnus X-1 reduced to the top of the atmosphere.
The data are independently obtained from the three MWPCs:
Top: SPC1,
Middle: SPC3,
Bottom: SPC4.

Acknowledgement

We are grateful to the technical staff of the IAS for the decennal support to this project. We acknowledge the ASI-Milo Balloon Base, CNES and INTA personnel for the flight operations and recovery. R.St. and E.Ke. acknowledge the support of the Deutsche Forschungsgemeinschaft (Sta173/8).

REFERENCES

1. Bazzano A. *et al.*, 1983, Nucl. Instr. & Meth. **214**, 481.
2. Ubertini P. *et al.*, 1985, SPIE **579**, 74.
3. Theinhardt, J., *et al*, 1984, Nucl. Instr. and Meth., **221**, 288.
4. Lampton M.,Margon, B., and Bowyer,S., 1976, Ap.J.(Letters),208,177
5. Liang, E.P., and Nolan, P.L., Space Sci. Rev., 1984, **38**, 353
6. Ubertini P. *et al.*, 1991, Ap.J., Jan. 10 Issue, in press
7. Miyamoto,S., and Kitamoto, S., 1989, Nature, **342**, 773
8. Sunyaev, R.A., and Titarchuk, L.G., 1980, Astron. Astrophys. **86**, 121

GALACTIC X-RAY SOURCES WITH EXTREMELY HARD SPECTRA. "KVANT" MODULE OBSERVATIONS

R.A.Sunyaev, V.A.Aref'ev, K.N.Borozdin, E.M.Churazov, V.V.Efremov, M.R.Gilfanov, A.S.Kaniovskiy, A.V.Kuznetsov, A.S.Melioranskiy, N.S.Yamburenko
Space Research Institute, Academy of Sciences, Moscow, USSR

W.Pietsch, S.Doebereiner, J.Englhauser, C.Reppin, J.Truemper
Max Planck Institut für Physik and Astrophysik, Institut für Extraterrestrische Physik, 8064, Garching, FRG

E.Kendziorra, M.Maisack, B.Mony, R.Staubert
Astronomisches Institut der Universität Tübingen, 7400, FRG

G.K.Skinner, T.G.Patterson, A.P.Willmore, O.Al-Emam, H.Pan, M.R.Nottingham
Department of Space Research, University of Birmingham, Edgbaston, Birmingham B15 2TT, U.K.

A.C.Brinkman, J.Heise, J.J.M.In't Zand, R.Jager
Space Research Laboratory, Utrecht, the Netherlands

A number of galactic X-ray sources with extremely hard spectra were observed by *Kvant* X-ray detectors, including Crab nebula, Cyg X-1, and X-ray transients GS2000+25 and GS2023+338 detected by *Ginga* satellite[1,2,3]. X-ray instrumentation allows to fulfil spectral and timing investigations in 2-1000 keV energy range and imaging in 2-30 keV band (15x15 deg FOV, 2 arc min. angular resolution[4]). More detailed description of results obtained is given in Soviet Astronomy Letters[5,6] and by Doeberainer et al[7].

On Fig.1 we present the spectra of these sources. The X-ray novae characteristics as measured by *Ginga* and *Kvant*, and also optical measurements show that these novae are accreting objects with low mass companions, and are obvious black hole candidates of the type of A0620-00.

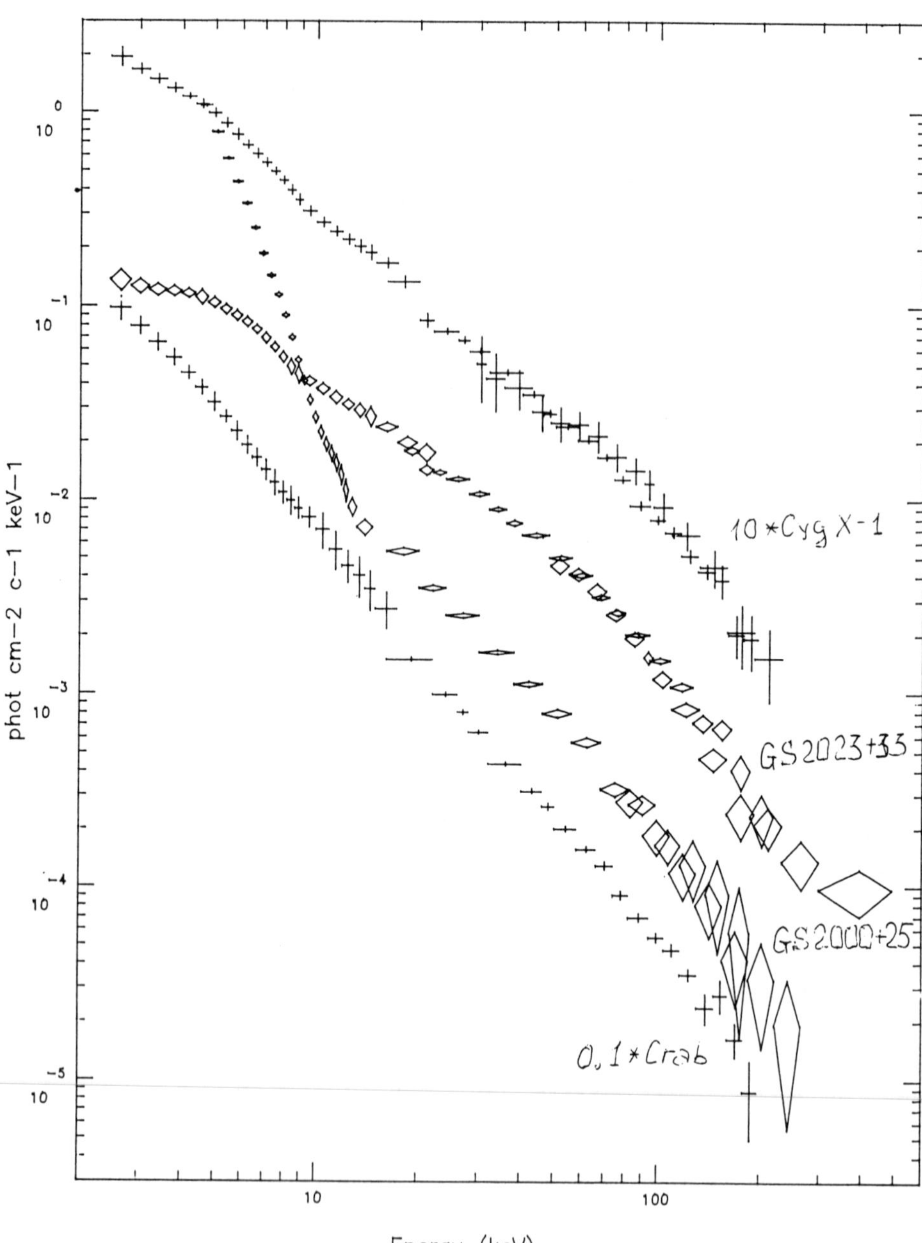

Fig.1. Spectra of Cyg X-1 (multiplied by 10), obtained on April, 25 1989 by *TTM*, *HEXE* and *Pulsar X-1* telescopes, GS2023+338 (June, 10, 1989, *TTM*, *HEXE*, *Pulsar X-1*), GS2000+25 (June, 1988, *GSPC*, *HEXE*, *Pulsar X-1*), and 10% of Crab nebula.

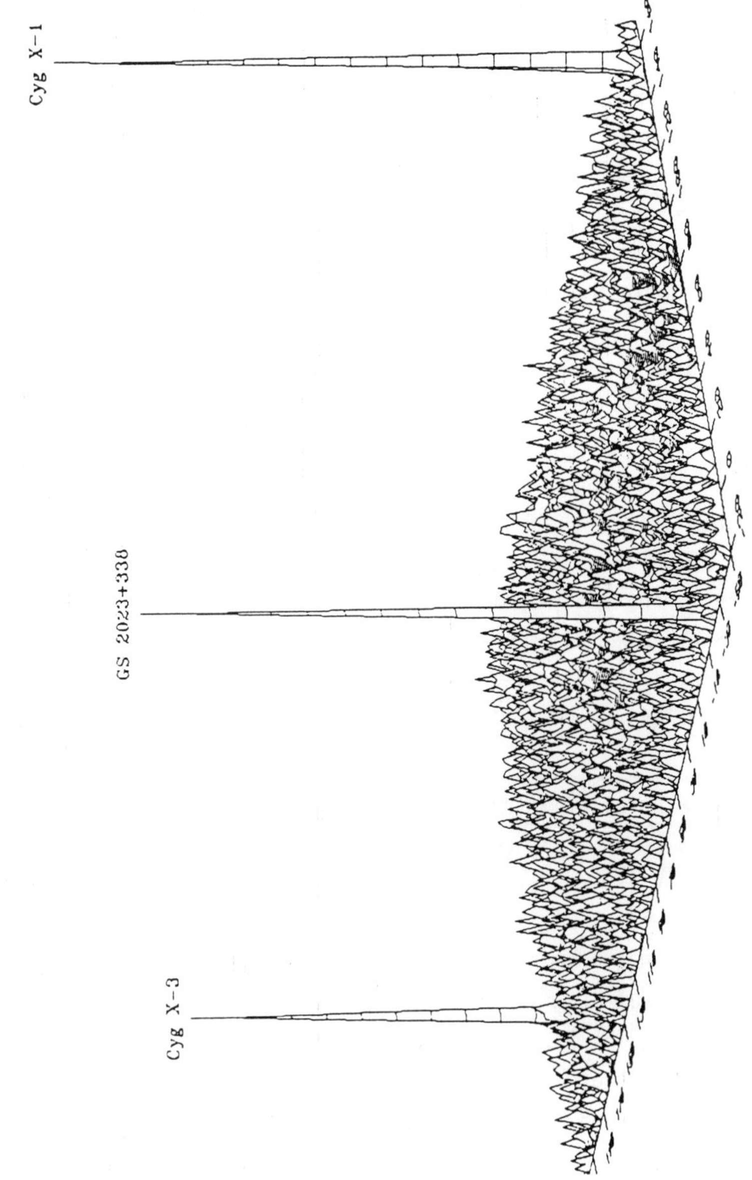

Fig.2. The narrow slice (0°.9 by 6°.2) of *TTM* image in 2-27 kev energy band during observations on June, 9 1989. Peak height is proportional to the significance of the sources. Coordinates shown correspond to pixels (1 pix= 1.86 within the *TTM* field of view. GS2023+338 peak position correspond to *R.A.* = $20^h 22^m 06^s.5$, *Dec.* = $\delta 33^{\circ} 42' 14''$ (1950 equinox) with 25 arc min accuracy. *V404 Cyg* is within error box.

On Fig.3 three spectra of GS2023+338 are presented, which were obtained in June, July and August of 1989. Hard component seems to be more complex than of Cyg X-1 and GS2000+25 and can not be fitted neither by comptonization model[8] nor by power law or exponential laws. No high energy cutoff was detected. One can see soft excess in the spectrum obtained in June. Strong and highly variable soft absorption was detected in early June observations, while the spectra above 20 keV were quite stable (see also Tanaka et al., 1989). Double power law plus soft absorption fit over June, 8, 1989 data is presented on Fig.4.

Timing analysis on the *HEXE* data revealed similar flickering both for Cyg X-1 (Fig.5a) and GS2023+338 (Fig.5b) with power density spectrum of f^{-1} type in 10^{-2}- 5 Hz band. For GS2023+338 power density spectra with equal intensity (with respect to 10% accuracy) were obtained both for 20-35 and 65-180 keV bands, e.g. at energies lower and higher than the change of the energy spectra slope. For this source timing analysis was possible only for July set of data.

Much lower amplitude of low frequency noise in the case of GS2000+25 (Fig.5c) together with bright soft component strongly differ this sources from GS2023+338 and Cyg X-1.

1. Makino et al., IAU Circ. No.4587 (1988)
2. Makino et al., IAU Circ. No.4882 (1989)
3. Tanaka,Y., Proc. of 23rd ESLAB Symp, 3 (1989)
4. Sunyaev R. et al., Sov. Astron. Lett. 13, 1027 (1987)
5. Sunyaev R. et al., Sov. Astron. Lett. 14, p.771 (translated No. 5, 327) (1988)
6. Sunyaev R. et al., Sov. Astron. Lett. 17, in press (1991)
7. Doeberainer et al., Proc. of 23rd ESLAB Symp, 387 (1989)
8. Sunyaev R. and Titarchuk L., Astron.Astrophys. 86, 121 (1980)

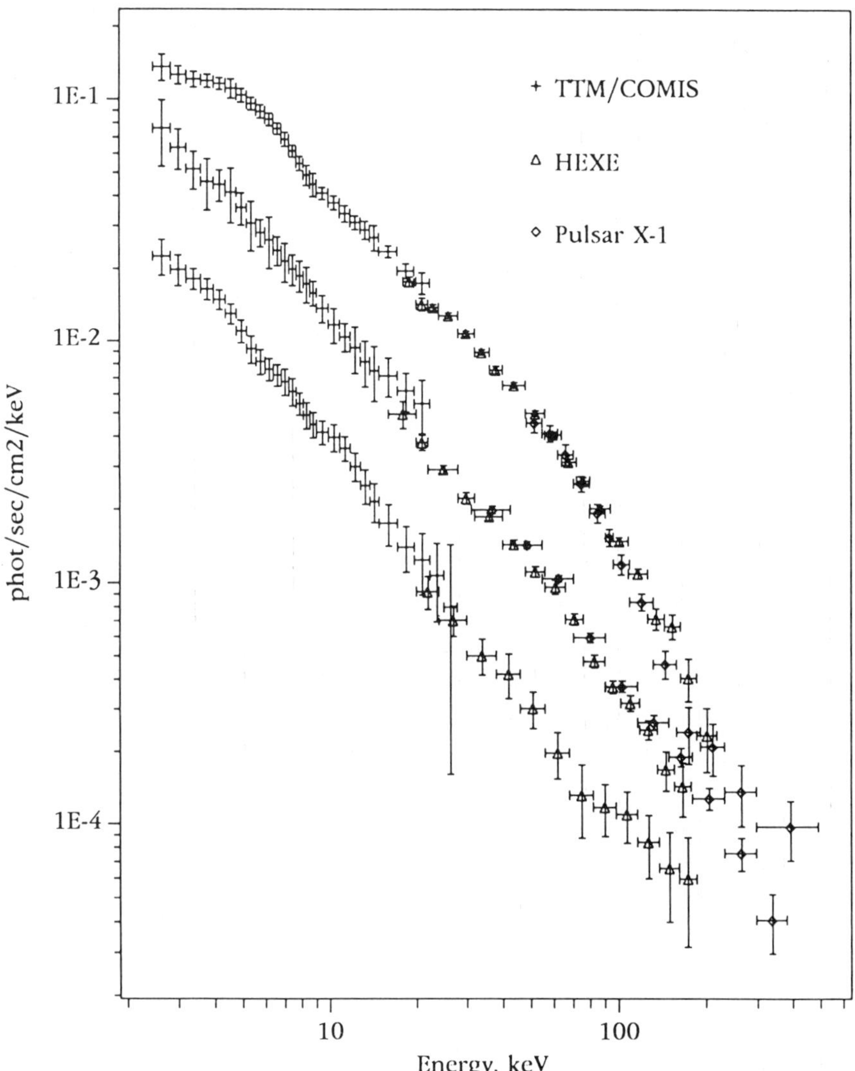

Fig.3. The X-ray spectra of GS2023+338 obtained on June,10 (tc spectrum), July,8-9 (middle) and August, 19, 1989 (bottom) by *TTN* *HEXE* and *Pulsar X-1*.

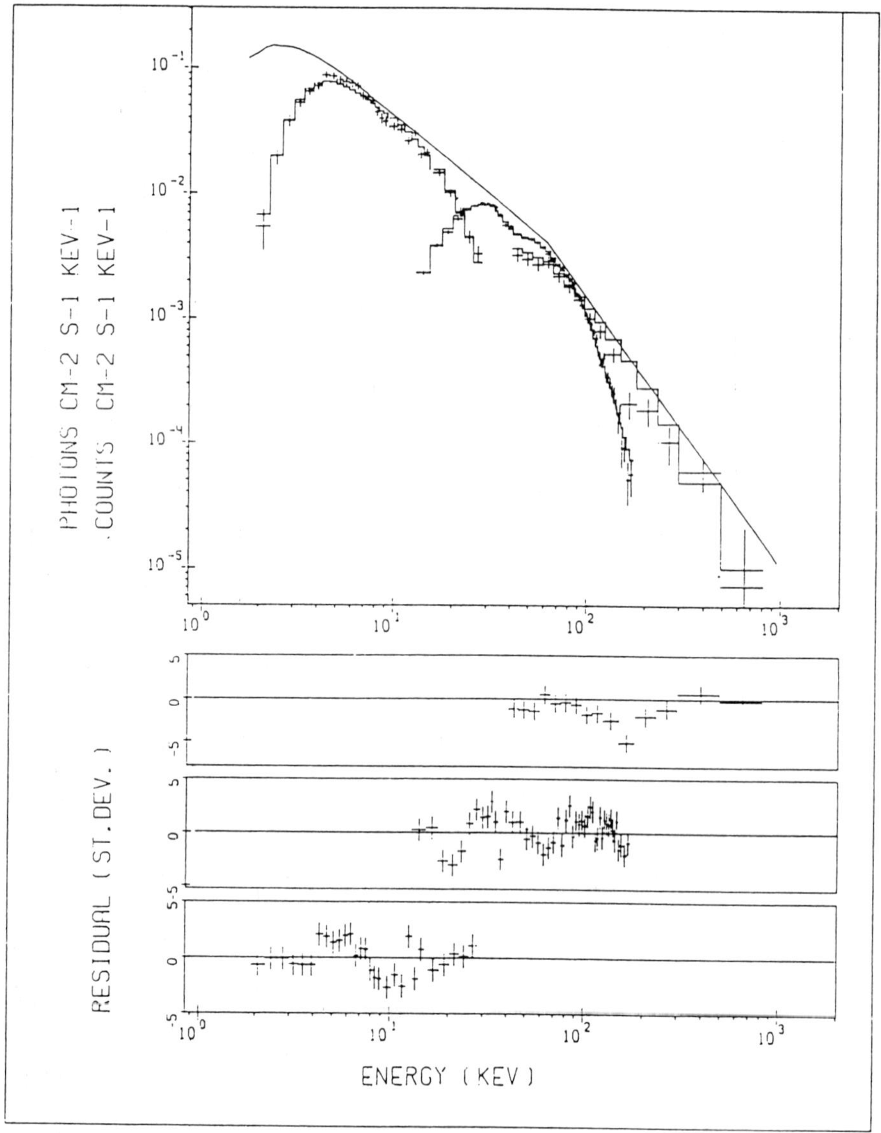

Fig.4. Double power law plus low energy absorption fit over GS2023+338 spectrum measured in June,8 1989. Fit parameters: $\alpha(E<E_{break})=1.28\pm.13$, $\alpha(E>E_{break})=2.14\pm.21$, $E_{break}=61.0\pm1.6$, $N_H=(1.97\pm.50)\cdot10^{22}$, χ^2/d.o.f=2.4.

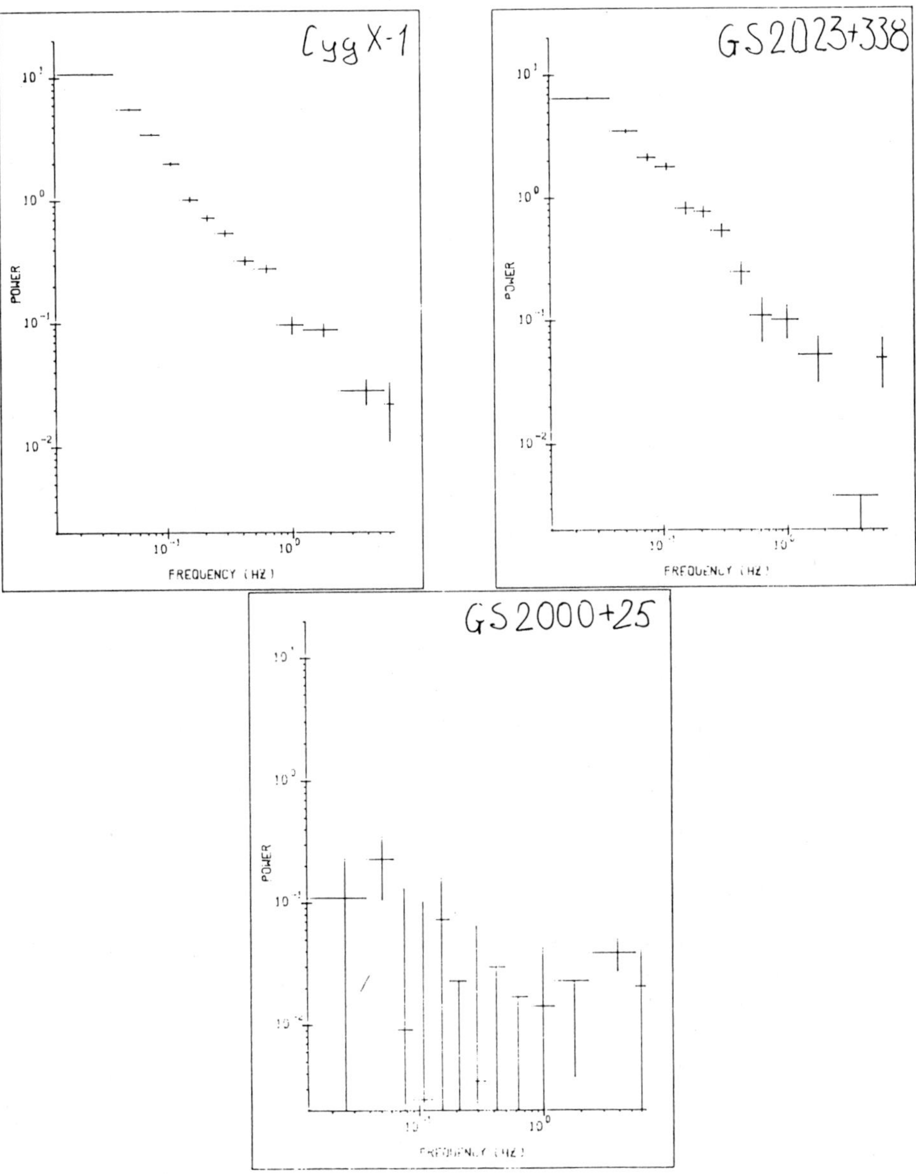

Fig.5. Power density spectra of Cyg X-1, GS2023+38 and GS2000+25 (energy range 30-100 keV).

LINE AND CONTINUUM SPECTROSCOPY AS DIAGNOSTIC TOOLS FOR GAMMA RAY BURSTS

Edison P. Liang
LLNL, University of California
Livermore, CA 94550

ABSTRACT

We review the theoretical framework of both line and continuum spectra formation in gamma ray bursts. These include the cyclotron features at 10's of keV, redshifted annihilation features at ~ 400 keV, as well as other potentially detectable nuclear transition lines, atomic x-ray lines, proton cyclotron lines and plasma oscillation lines. By combining the parameters derived from line and continuum modeling we can try to reconstruct the location, geometry and physical conditions of the burst emission region, thereby constraining and discriminating the astrophysical models. Hence spectroscopy with current and future generations of detectors should provide powerful diagnostic tools for gamma ray bursters.

INTRODUCTION

Study of spectral lines has been the backbone of astrophysics. It provides us with critical informations about the physical conditions of remote environments and exotic phenomena. Line spectroscopy is the standard diagnostic tool of objects from stars to galaxies. Though this is a symposium on gamma ray lines, in the case of gamma ray bursts (GRBs), it is not possible to discuss the study of spectral lines without first addressing the issues related to the formation of the continuum, hence the title of this review talk. Since the bulk of the observational results is already reviewed in Vedrenne's talk (this volume), I will concentrate on the theoretical issues, reviewing the observational facts only briefly when necessary.

Adopting a somewhat biased approach, I will consider the highest near-term priority in the study of gamma ray bursts to be the verification of the strong field ($B \sim 10^{12}$G) Galactic isolated neutron star hypothesis. If this TGNS (teragauss Galactic neutron star) paradigm can indeed be proven beyond doubt, then the next goal of gamma ray burst spectroscopy would be to diagnose the detailed geometry and physical conditions of the emission and line forming regions. This will in turn constrain the various astrophysical and energy source models, ranging from star quakes to cometary impacts. Beyond that the study of GRB as the most populous species of Galactic neutron stars will tell us much about neutron star origin and evolution (e.g. Hartmann, this volume), and perhaps even the structure and evolution of the Galaxy as a whole. But first we must consolidate the evidences for the TGNS picture and resolve all of its remaining puzzles and difficulties. Hence in this talk I will concentrate on the bearing of line and continuum spectroscopy on the TGNS picture. Recent reviews of GRB spectra can be found in Harding (1991), Hurley and Lamb (1991), Higdon and Lingenfelter (1990), Liang (1989) and the monographs edited by Ho and Epstein (1991) and Liang and Petrosian (1986).

TIME-AVERAGED SPECTRA

Much of our interpretations of GRBs come from their time-averaged spectra, with integration times ranging from ~fraction of a second in SIGNE and PROGNOZ experiments (Atteia et al. 1987) to 16 seconds for SMM (Nolan et al. 1984, Share et al. 1986). But the true spectral evolution time scale is likely much shorter, say typically < 10 ms, especially for narrow spectral features (e.g. GB790305). Hence when we discuss the generic GRB spectra using time-averaged samples, we should keep in mind that these may not be truly representative of the physical condition on short time scales.

Fig. 1 reproduces the generic GRB spectrum. It consists of an exponential shape from 10's of keV to ~ MeV followed by a power law tail above an MeV of photon index 2 - 3. At least for a few events this tail extends out to ~100 MeV and GRO should be able to see if it goes much higher. The continuum typically breaks to a slope of flatter than index 2 at a few hundred keV, continuing to flatten towards lower energies. In cases where x-rays are detected the x-ray spectrum below 10 keV seems to be consistent with a simple extrapolation of the gamma ray spectrum, leading to a x/gamma ray luminosity ratio of < 2%. However, the x-rays seem to persist long after the gamma rays die off, and in a few instances there is also an x-ray precursor (Murakami 1991). The spectrum of the x-ray tail and the precursor is much softer, consistent with a blackbody of $\lesssim 2$ keV. Hence it likely represents a different emission component. In addition to x-rays, archival optical transients have been detected in several GRB error boxes (Schaefer 1981) which may correspond to ~ 10^{-3} of the flux of the gamma rays. However, recent works on plate defects cast doubts on the meaning of the optical images and we can only conservatively put upper limits on the optical flux. Similarly, there are only up limits on the emission of other low frequency radiation from radio to infrared (Fig. 2).

In 10-25% of the events the spectra show evidence of time-dependent line features. Recent GINGA confirmation of multi-harmonic absorption dips at ~20 - 40 keV (Murakami et al. 1988) gave credence to the interpretation of these features as cyclotron harmonics in a teragauss magnetic field, first observed in KONUS data (Mazets et al. 1980). Less certain are the alleged redshifted annihilation features at ~ 400 keV reported by KONUS, which was not confirmed by SMM observations. The reality of these features must await confirmation by GRANAT or GRO, but it is at least worth discussing here their implications if they are real. In at least one event (GB781119) there was evidence of additional line features at 740 keV and higher energies, which may be interpreted as nuclear lines. So far there has been no observation of atomic lines in the x-ray range, either in the main part of the burst or in the x-ray tail or precursor. But future x-ray observations with HETE and other experiments may change that.

SPECTRAL EVOLUTION

Almost all observations to date suggest that the spectral features, whether it is the cyclotron features at 10's of keV or annihilation features at ~400 keV, vary with time during a single burst. Fig. 3 illustrates the spectral evolution of GB880205 observed by GINGA showing that the double cyclotron features appear most prominent only during the brief middle part of the spike. More recent studies (Murakami 1991) suggest that the line center position also seem to evolve downward even within that time interval. As for the annihilation line, the most celebrated example is that of GB790305, which may be present only during the first 24 milliseconds of the burst (Barat et al. 1983).

The continuum itself shows evidence of rapid evolution down to 10's of ms (Mitrofanov 1991) . Study by Norris et al. (1986) and other groups suggest that the spectral hardness peaks before total count rate and the overall evolution of each spike is from hard to soft (Fig.4). Golenetskii et al. (1984) on the other hand claimed that there is a close correlation between luminosity and spectral hardness. Despite its controversy this result has recently been confirmed by the study of SIGNE data (Liang et al. 1991, Fig.5). It is found that the luminosity is tightly correlated with the best-fit bremsstrahlung temperature according to $L \sim T^{1.65 \pm .28}$ down to 0.5 sec. Note that this result need not conflict with that of Norris et al. (1986) since luminosity is proportional to count rate times temperature T. So even though the rising phase of the burst has fewer counts its luminosity may still be high due to the hardness of the photons. If confirmed this result has interesting implications for the burst mechanism.

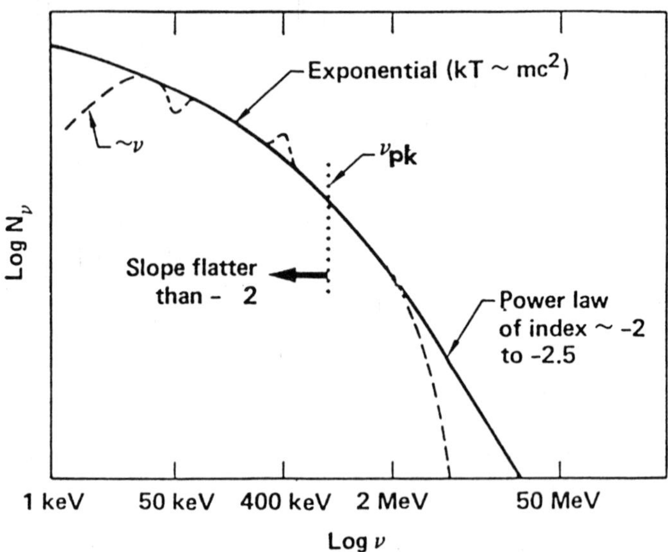

Fig. 1 Generic GRB spectrum. Dashed features appear only in a fraction of observed spectra (from Liang 1987)

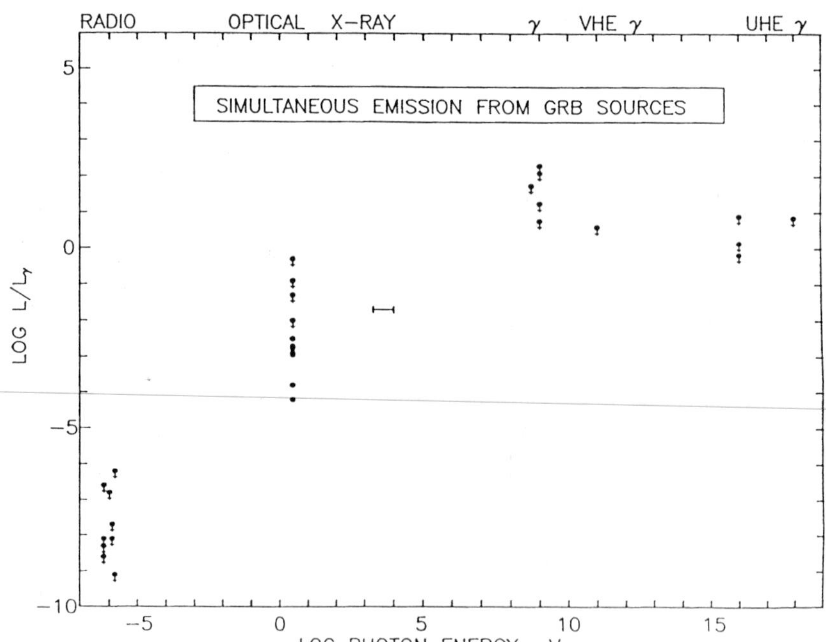

Fig. 2 Limits on GRB continuum emissions from radio to >100 MeV (from Hurley, 1991)

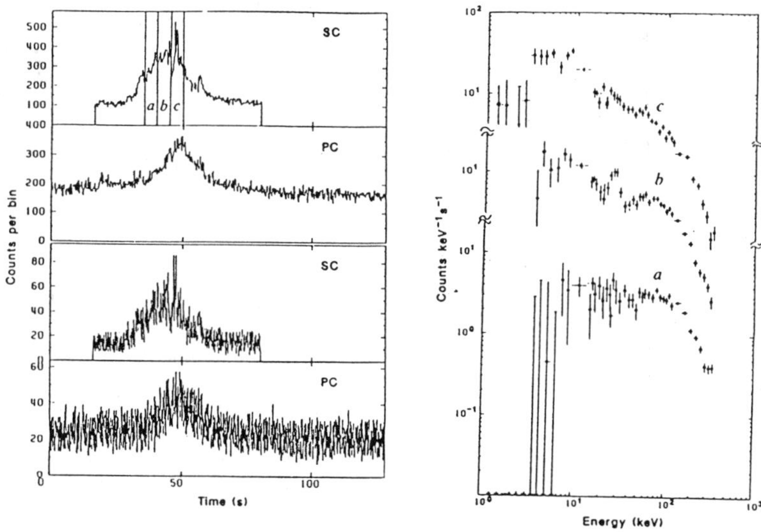

Fig. 3 Time profile and spectra of GB880205 observed by GINGA (from Murakami et al 1988)

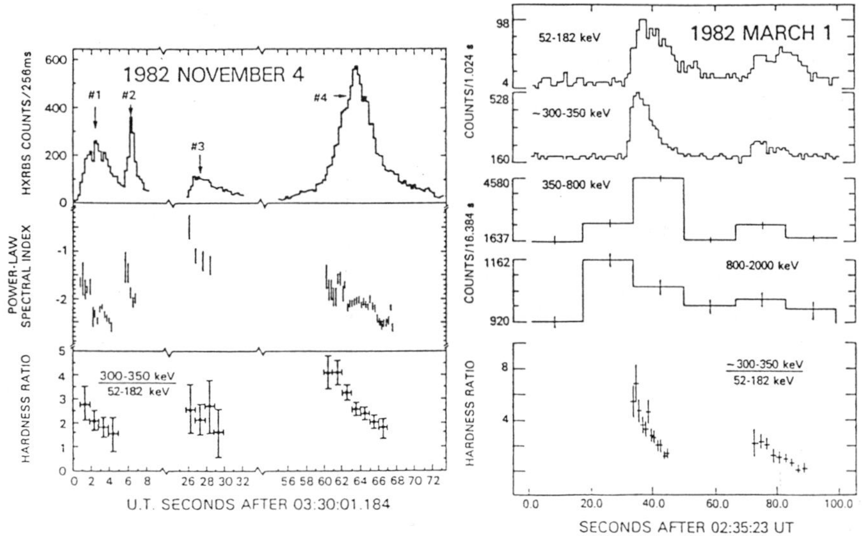

Fig. 4 Evolution of hardness ratio vs. burst time history for GB821104 (from Norris et al 1986)

THE TGNS PARADIGM

Fig.6 illustrates the standard arguments in favor of a teragauss field neutron star origin of gamma ray bursts. Millisecond and submillisecond time variability constrains the size of the emission region to that of 10's to 100's of km. Cyclotron lines at 10's of keV show that at least the emission region of the soft gamma rays below 100 keV (not necessary the gamma rays above several MeV) must come from very strong field regions. If the annihilation line interpretation of the 400 keV features is indeed correct and the shift of line-center position is all due to gravitational redshift, then it is both a strong confirmation of the neutron star origin and a constraint on the neutron star structure. So far the only periods from GRBs that seem convincing are the 8 second period of GB790305 and the 2.2 second period of two recent bursts (Kouveliotou et al. 1988, Owens et al. 1990). But if future observations can firmly establish the existence of periods for more bursts, either from the gamma ray light curve or the longer x-ray tail, then it is another evidence in favor of strong fields.

Other arguments in favor of a strong magnetic field include the efficient acceleration of energetic (> 100 MeV) particles which is facilitated by the presence of ultrastrong fields, and the confinement of the radiating plasmas which would have expanded at the speed of light due to thermal and radiation pressure. None of the above arguments is watertight, of course, and there are always more counter examples than examples. But at this point a TGNS picture seems to be the most useful baseline framework from which we can build meaningful models and confront them with current and future observational data. However, it should be emphasized that even if the TGNS paradigm turns out to be correct for a large majority of classical GRBs, this does not exclude the existence of subclasses of GRBs (e.g. soft repeaters, short soft bursts etc) that have other origins.

The TGNS paradigm comes with many puzzles, both physical and astronomical. We list some of them here:

1. Why are periods so rare?
2. Why are there no attenuation of multi-MeV gamma rays from gamma-gamma and gamma-B pair production?
3. Why are there so few x-rays?
4. Why are the cyclotron lines so narrow if the underlying continuum is so hot?
5. If GRBs derive from old, isolated neutron stars past their pulsar phase, why are their fields so strong?
6. Why are cyclotron and annihilation features so common (if we believe KONUS statistics) in GRBs but so rare in radio pulsars?
7. Why is there no radio emission from GRBs?
8. If the blackbody interpretation of the x-ray tail is correct, then it is consistent with a typical source distance of $\gtrsim$ kpc unless only a tiny fraction of the stellar surface is emitting. How can this be reconciled with the isotropy, V/V_{max} and log N - log P results if the sources are Galactic neutron stars?

CONTINUUM DIAGNOSTICS

The interpretation of any spectral line observed in GRB spectra would depend on the model for the continuum. Hence we must first address the origin of the continuum spectrum, say from ~ keV out to $\gtrsim$ 100 MeV. At present there are at least two competing scenarios concerning the origin of the continuum emission: (a) surface or near-surface emission from beamed

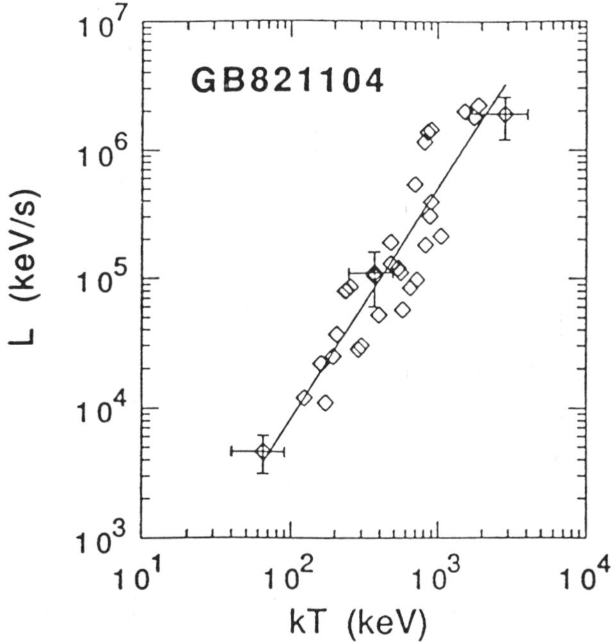

Fig. 5 Preliminary results of luminosity-temperature correlation based on 0.5 sec spectral fitting with bremsstrahlung model (Liang et al 1991, unpublished)

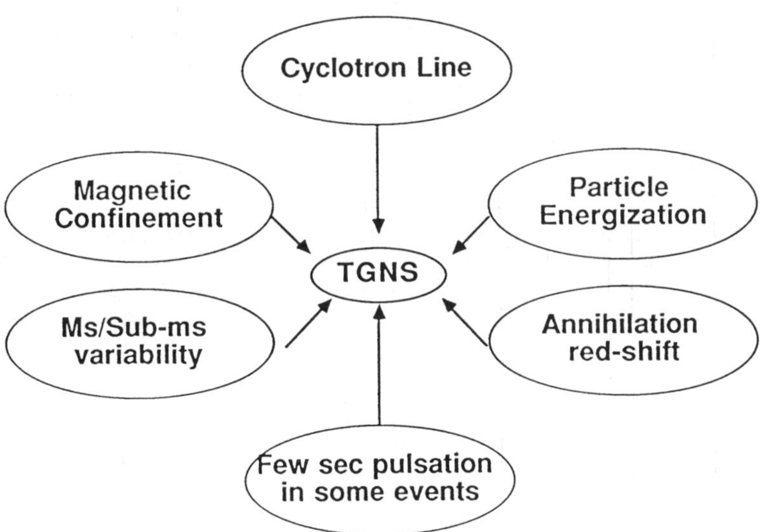

Fig. 6 Support for the TGNS paradigm

relativistic electrons (and pairs) for the entire spectrum, and (b) two component unbeamed emission with the sub-MeV soft component coming from near the surface and the super-MeV power-law tail coming from high in the magnetosphere away from the surface. Artist conceptions of these scenarios are depicted in Fig.7. The fundamental constraints that force us into one of these two scenarios are: a) lack of x-rays so that reprocessing from the stellar surface must be minimal; hence either the emitting particles are beamed away from the surface, or the primary emitting region must be far enough away from the surface so that only a small fraction of the energy release reaches the surface; b) lack of gamma-gamma or gamma-B e+e-pair production attenuation of the super-MeV gamma rays; this means that either the high energy gamma rays are emitted parallel to the field lines, or the field has to be weak enough ($\lesssim 10^{11}$G for isotropic photons). At this point both scenarios are still viable. But future high time resolution correlation study (say down to 10 μs) of the two spectral components should easily discriminate between them since in the former case both components are emitted simultaneously, at the same location and by the same bunch of particles, whereas in the latter case the soft component should lag the hard one by at least the light travel time. These and other consequences of the two scenarios and their implications for the spectral interpretations are summarized in Table 1.

The most advanced version of the former scenario is called the cyclotron or resonance up-scattering process (CUSP, Hameury et al. 1985, Dermer 1989, Ho and Epstein 1989, 1991, Vitello and Dermer 1991): thermal ($\lesssim$ keV) soft photons emitted by the stellar surface are up-scattered with the resonance cross section (~hundreds times Thomson) by outwarding streaming relativistic electrons or pairs (Lorentz factor >> 1) into forward-beamed gamma ray photons. To the extent that the pairs are injected with a power law, the output optically thin gamma spectrum will have a power low at high energies and break to a flat spectrum (photon index 0) if the photon target is thin and break to spectrum of index 1 if the photon target is thick. The break energy is given by ~ ν_B^2/ν_o in the former case where ν_B is cyclotron energy and by $\gamma_* \nu_B$ where γ_* is the low energy cutoff of the electron distribution in the latter case. Hence the most critical feature of this model is the existence of the break energy which is a measure of the soft photon temperature or the electron distribution cutoff. Note that if the spectral index below the break is flatter than 1 the photons must be thin target whereas if it is steeper than 1 it could be either thin or thick target because in that case the low energy component must be synthesized by superposition of spectra of different break energies. In most models the output spectrum will also show a soft x-ray bump corresponding to the unscattered input spectrum but in reality this may be washed out by optical depth effects. To summarize, in the context of the CUSP model the key spectral parameters are: break energy; spectral indices above and below the break and the location of the soft photon peak, if any. The emission parameters to be extracted from spectral fitting with this model are: electron injection index and low energy cutoff, soft photon temperature, B-field strength (from lines if detectable) and electron Thomson depth.

In the two component model the spectral break would be interpreted as the low energy cutoff of the energetic electrons emitting in the magnetosphere. The soft component below the break will be either reprocessed radiation or secondary emission with energy supplied directly to the surface (e.g by Alfven waves or particle flux if the primary energy source is external or by thermal diffusion if the primary energy source is internal). While most authors (e.g. Ruderman 1987, Liang 1987, Sturrock 1986, Harding and Sturrock 1989, Petrosian 1991) visualize the far-field high energy component to be some form of $E_{//}$-induced pair-cascade (from curvature, synchrotron and inverse Compton processes) due to the similarity of the index to those of pulars (~2-2.5), there is as yet no concensus as to the origin of the low-energy component. The only constraint is that it originates near the strong-field stellar surface in cases we see cyclotron absorption or redshifted annihilation features. It could be emitted by a combination of resonant scattering and synchrotron, or even bremsstrahlung and nonmagnetic Compton in cases where there is no evidence of strong field (e.g. B < 5×10^{10}G). The emitting particles could be

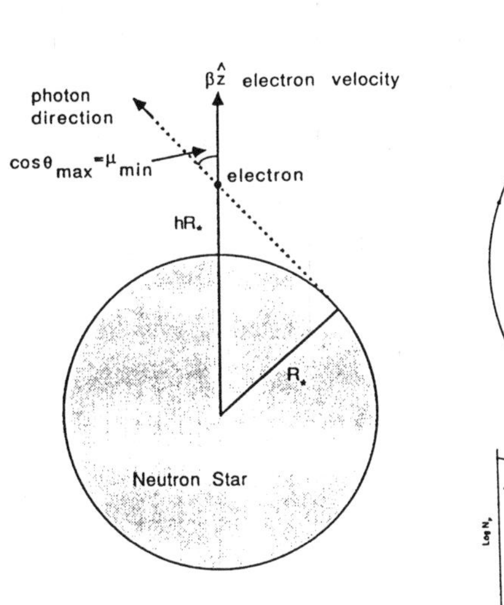

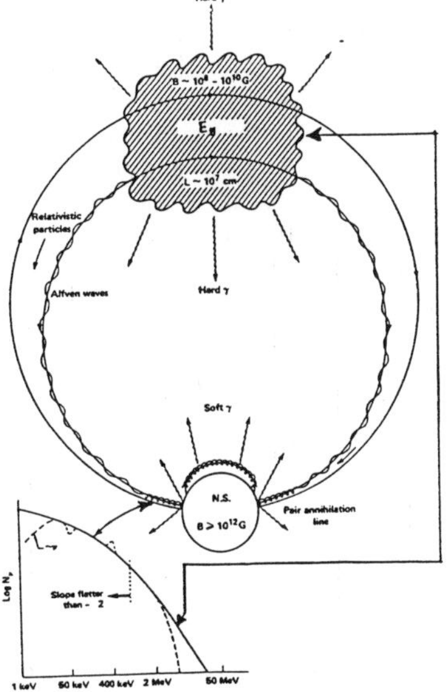

Fig. 7 Artist conception of two different scenarios of continuum spectra formation: (a) emissions by outwardly beamed electrons (pairs) near stellar surface (from Ho and Epstein 1989); (b) two component emission (from Liang 1989)

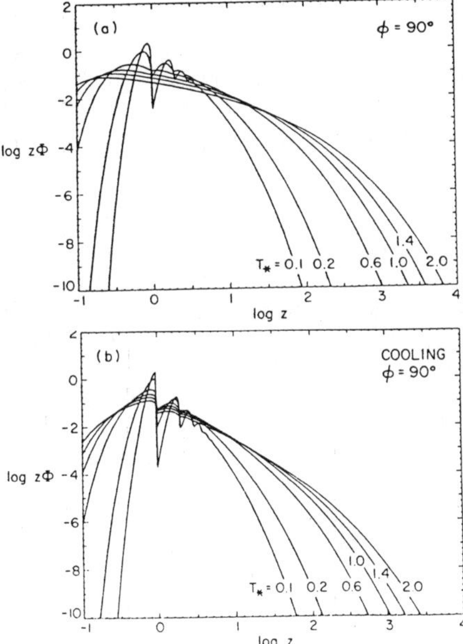

Fig. 8 Sample synchrotron emission spectra for thermal (a) and cooling (b) electron distributions (from Brainerd and Lamb 1987).

Maxwellian (due to collisionless relaxations) or nonMaxwellian (e.g. cooling) distributions. If the distribution of the low-energy electrons is indeed Maxwellian then the overall curvature of the low-energy component usually measured by bremsstrahlung fits will give a characterization of the mean electron energy. If the spectrum further turns over in the x-rays as hinted by some bursts (e.g. GB811016 Katoh et al. 1984 and GB880205) then the turnover energy also constrains the emission parameters. If the turnover is due to self-absorption or saturation of scattering we derive the emission measure or Thomson depth modulo temperature and field strength (if cyclotron line detected). If the turnover is due to first harmonic cutoff (and no line detected above turnover) we derive the field strength. If the turnover is due to plasma frequency cutoff (Razin-Tystovich effect) we derive the electron (pair) density. If the turnover is due to absorption by circumstellar or interstellar matter we derive the optical depth and composition of intervening matter, etc. Discrimination of these and other models can only be achieved by refined modeling and time-variability analyses.

In summary, continuum spectral measurements will give us a handle on a number of emission parameters. Interpretation of the high energy power-law component seems to be least sensitive to the choice of model or scenario, giving us unambiguous estimate of the electron index and low-energy cutoff. But parameters derived using the low energy component below a few hundred keV is sensitive to the specific emission model we choose. Discrimination among these models must await further study, including rapid time-variability correlation analyses.

LINE DIAGNOSTICS

Line features, when available, will give us critical informations which supplement the continuum parameters. However, their interpretation is again scenario dependent and the physical parameter derived in most cases is sensitive to the model adopted. In addition to cyclotron features at 10's of keV and annihilation features at ~ 400 keV, whose detections have some credibility, we will also give more speculative discussions of other possible lines and their use as diagnostic tools.

A. Cyclotron Features

Since their first discovery by KONUS (Mazets et al. 1980, Golenetskii et al. 1986) the reality or interpretation of these features have remained controversial until their recent confirmation by GINGA (Mumakami et al. 1988) with the discovery of multiple harmonics. The fundamental difficulty, of course, is observational uncertainty due to nonuniqueness or obligeness of the deconvolution procedure of count spectra to photon spectra. Even when one accepts the features as real there is still room for speculation as to their origin (e.g superposition of multicomponent continuua, Lasota and Belli 1983, K-edge absorption by heavy elements, Bisnovaty-Kogan 1987, synchotron emission instead of absorption, Brainerd and Lamb 1987, broad emission dip with line-centre reversal, Fenimore et al. 1988 etc.). But after many man-hours of detailed modeling, at this point by far the most popular and satisfactory interpretation is that these cyclotron features are produced by resonant scattering by cool electrons ($T_{//} \lesssim 5$ keV) in which photon-spawning from second harmonic decay partially fills in the first harmonic, producing the apparent strength of the second harmonic versus the first. Analyses invoking both harmonics (or maybe even the third) seem to give a scattering depth of 10^{21}-10^{22}cm^{-2} depending on the geometry, dynamics and view angle (Wang et al. 1989, Alexander and Meszaros 1989). The unambiguous parameter is the field strength, which is ~2 teragauss modulo any redshift correction. The narrowness of the line constrains both the absorbing medium temperature ($T_{//} \lesssim 5$ keV) and field gradient ($\Delta B/B \lesssim 30\%$). But at this point the absorption site, geometry and dynamics (wind or static medium) are all unknown.

Detection of absorption features means that at other view angles at other times, we should have a chance of seeing emission features. In addition to the narrow emission features predicted to come out from the cool scattering layer (Wang et al. 1989), broad emission features could also arise as part of the continuum when it is due to synchrotron emission if particles are energetized to high Landau levels (Fig.8, Brainerd and Lamb 1987). In addition to a measure of the field strength, such emission features would help to constrain the emission geometry and give consistency check of the continuum emission model. Table 2 summarizes the use of cyclotron features as diagnostic tools.

B. Annihilation Features

Besides the celebrated March 5 event (GB790305), there are only a few other widely recognized narrow annihilation feature in classical GRBs observed by several independent experiments, notably GB781119 (Fig. 9) (see Harding et al. 1986 for review). Besides the KONUS results which show annihilation features in a significant fraction of the bursts (Mazets et al. 1980) a good example of a broad feature at ~ 400 keV can also be found in the HEAO-A4 data (Fig. 10, Heuter 1987). Independent of the ultimate validity of such claims, it is meaningful to examine the implication of such line features and their potential as diagnostic tools.

The first important feature of the annihilation lines is that they appear to be redshifted. Liang (1986) analyzed the data of almost 40 events with apparent features and concluded that the redshift database is most consistent with a medium-soft neutron star equation of state. However this analysis is beset with caveats and uncertainties (Kluzinak 1989). Many factors can contribute to the shifting of the line centroid besides gravity. These include thermal, bulk Doppler (orbital or turbulent motion), Compton scattering, opacity effects, strong magnetic fields etc. Despite these uncertainties, future redshift measurements of GRB annihilation lines will remain the most powerful tool for the study of neutron star structure and relativistic effects.

Besides the line center position, two other observable parameters are the line width and line intensity. Like any other line diagnostics, these would provide constraints on the pair density, Thomson depth, field strength, temperature (of both the annihilating medium and any intervening scatterer), etc. For example, if the field is approaching the critical field of 4.4×10^{13}G, then the annihilation would be dominated by second harmonic (1-photon) emission at 1022 keV and the profile would be strongly asymmetric (Harding 1986). Ratio of the line to integrated power-law continuum intensity can potentially discriminate between the pair cascade and other models (Sturrock 1986). If the shift is due to Compton down-scattering, then the line center position is related to the width. Table 3 summarizes the implications of the 5ll keV line.

C. Other Lines

In addition to the annihilation feature, GB781119 (Fig.9) also shows evidence of a line at 740 keV and weaker features at higher energies. This could be interpreted as redshifted 870 keV Fe deexcitation line. Given the high temperature and high gamma intensity of the source it is quite plausible, or even likely that Fe nuclei, if abundant, would be excited, and the 870 keV and other Fe deexcitation lines (and those of other nuclei, including the 2.2 MeV line of D and 4.4 MeV line of C) would provide interesting diagnostics of the source. But we must first await the result of GRO and other future experiments.

So far no line feature has yet been discovered in the soft x-ray regime below 10 keV, primarily due to the lack of x-ray events and poor instrumental resolution. But if Fe is the abundant element in the burst environment, then at least during the precursor and x-ray tail when the temperature is cooler ($\lesssim$ 2 keV from GINGA), detection of Fe lines in the 6 - 7 keV range as well as spectral jump at the K-edge should be quite likely . Together these should provide a wealth of diagnostic tools for the x-ray emission region. Of course the x-ray emission region may be totally separate and different from the gamma ray emission region, but at least we will have some informations on the underlying neutron star environment, such as Fe abundance and

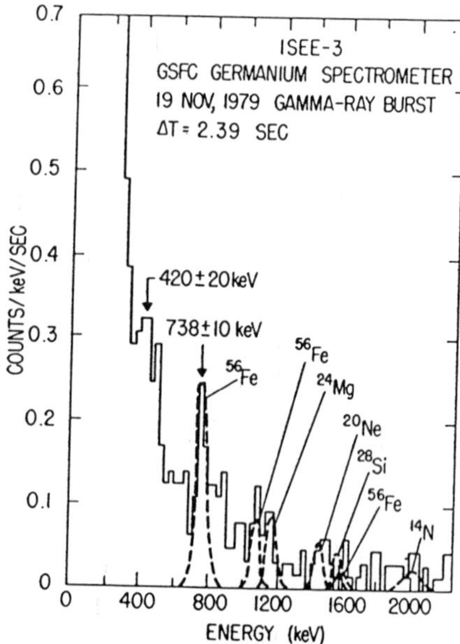

Fig. 9 Spectra of GB781119 showing emission features (from Teegarden and Cline 1980)

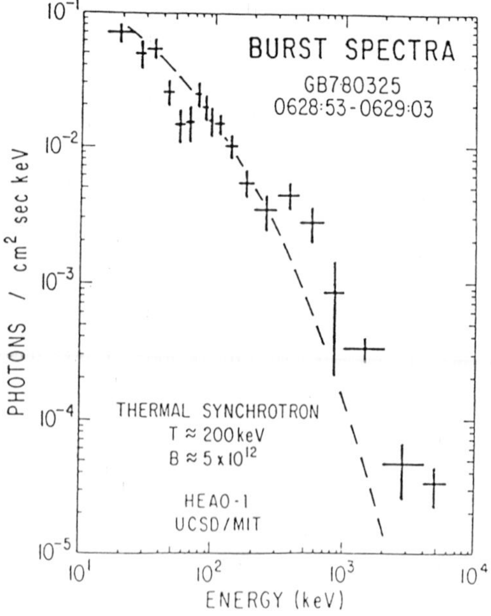

Fig. 10 HEAO-A4 spectra of GB780325 showing a braod bump at 400 keV (from Heuter 1987)

Table 1: Comparison of Two GRB Continuum Emission Scenarios

Single-Component Near-Surface Emission	*Two-Component Emission*
entire spectrum formed near surface	>MeV tail emitted at h>>10 km <100 KeV emitted near surface
particles and gamma rays outwardly beamed // local B	unconstrained
no time delay between hard and soft	soft lags hard by >h/c>>10 μs
requires soft (<KeV) x-ray source	not required
predicts outflowing wind	not predicted
prefers open flux tubes	prefers closed flux tubes
spectral hardness strong function of view angle	insensitive to view angle
break energy determined by field strength and soft photon temperature	break energy characterizes electron energy only
primary energy source likely interior	can be interior or exterior
emission region density likely high	density likely low
E// generation more difficult	E// generation easier

Table 2: Diagnostics with Cyclotron Lines

Measured	*Derived*
line center	B-field (module redshift, bulk Doppler)
line width	T of scatterer, ΔB/B
equivalent width (EW)	Thomson depth of scatterer
relative strength of harmonics	view angle, geometry

Table 3: Diagnostics with Annihilation Lines

Measured	*Derived*
line center	gravitational redshift (module bulk Doppler)
line width	T of pairs and scatterer
EW	pair density
profile asymmetry	compactness, B-field (if close to B_{crit})
relative strength of harmonics (511/1022)	B-field, view angle
ratio of line to continuum	injection Lorentz factor of pairs
if redshift~line width	hint of Compton downscattering
if blueshift~line width	hint of thermal broadening

Table 4: Diagnostics with Other Lines

Type of Line	*Derived Parameters*
nuclear deexcitation lines	abundance, T_{ion}, redshift
Fe x-ray lines	abundance, T_e, redshift, bulk Doppler, ionization states
plasma oscillation lines	electron density, T_e
ion cyclotron lines	B-field, T_{ion}, ion Thomson depth

redshift, winds or accretion flows. Just as Fe K-lines are expected to provide powerful diagnostic tools from x-ray binaries to clusters of galaxies (e.g Holt and McCray 1982), we anticipate Fe K-lines will also be a useful diagnostic for GRBs.

The acceleration of energetic particles responsible for the high energy continuum will likely lead to excitation of a variety of plasma instabilities either as a cause or consequence (e.g. Melrose 1980), most of which will end up dumping some of their energies into longitudinal electron plasma oscillations (Langmuir waves). A by-product of such excitations is the emission of plasma line radiation at ω_e (plasma frequency) and 2 ω_e (Liang 1985), as in the case of solar radio bursts. At present the plasma density of the emission region is unknown, but if it lies in the range 10^{18} - 10^{24} cm^{-3} deduced from various considerations (see e.g. Liang 1985), then the plasma frequency (= 0.4 eV $(n/10^{20})^{1/2}$) would lie in the IR to XUV range. Future optical-UV monitoring of GRB (e.g. with HETE) might have a chance of detecting this.

Another line that may lie in the UV range is that of ion cyclotron emission or absorption (Kirk and Stoneham 1982) in a 10^{12} G field. Since the opacity is much lower than that of Thomson it might be difficult to detect. Table 4 summarizes the results for these other lines.

CONCLUSIONS AND SUMMARY

Line and Continuum spectroscopy of GRBs is expected to provide powerful diagnostic tools of the emission and absorption regions. Within the context of a specific model or scenario, line and continuum data should be able to provide us with rich informations about the emission site, geometry, and physical parameters such as density, temperature or electron distribution, optical depth, magnetic field strength, bulk motion etc. Time evolution of the spectra will give further informations about the dynamics and energetics of the source. With the observations of GRANAT, GRO, ULYSSES, HETE and other missions we have high hopes that spectroscopic diagnostic of gamma ray bursts will become a reality. This will allow us to not only clarify the astronomical and astrophysical issues of GRBs, but also use GRBs as laboratories for the study of strong-field, relativistic plasma physics found nowhere else in the universe.

ACKNOWLEDGEMENTS

This work was performed under the auspices of the U.S. Department of Energy by the Lawrence Livermore National Laboratory under contract number W-7405-ENG-48.

The author thanks C.D. Dermer and D. Hartmann for helpful conversations and the conference organizers for partial support.

REFERENCES

Alexander, S. G. and Meszares, P., 1989, Ap. J. Lett. **344**, L1.

Atteia, J. L. et al. 1987, Ap. J. Supp. **64**, 305.

Barat, C. et al. 1983, in AIP Conf. No. 101, ed. M. Burns, A. Harding and R. Ramaty (AIP, NY).

Bisnovaty-Kogan, G. S., 1987, in Proc. IAO Symp. 125, ed. D. Helfand (Reidel, Holland).

Brianerd, J. and Lamb, D. Q. 1987, Ap. J. **313**, 231.

Dermer, C. D. 1989, Ap. J. **360**, 197.

Fenimore, E. et al. 1988, Ap. J. Lett. **335**, L71.

Golenetskii, S. et al. 1984, Nature **307**, 41.

Golenetskii, S. et al. 1986, Ap. Sp. Sci. **124**, 243.

Hameury, J. M., 1985, Ap. J. **239**, 56.

Harding, A. K. 1986, Ap. J. **300**, 167.

Harding, A. K. et al. 1986, Ch. 2 in AIP Conf. Proc. No. 141, ed. E. Liang and V. Petrosian, p. 75 (AIP, NY).

Harding, A. K. 1991, Physics Reports to appear.

Harding, A. K. and Sturrock, P. 1989, Ap. J. to appear.

Higdon, J. and Lingenfelter, R., 1990, Ann. Rev. Ast. Ap. **28**, 401.

Ho, C. and Epstein, R., 1989, Ap. J. **343**, 277.

Ho, C. and Epstein, R. 1991, in Proc. of Taos Workshop on Gamma Ray Bursts, ed. Ho. C. and Epstein R. (AIP, NY).

Holt, S. and McCray, R. 1982, Ann. Rev. Ast. Ap. **20**, 323.

Hueter, G. J., 1987, Ph.D. Thesis, U.C. San Diego, pp. 304.

Hurley, K. and Lamb D. Q. 1991, Physics Reports to appear.

Katoh, et al. 1984, in AIP Conf. Proc. No. 115, p. 390, ed. S. Woosley (AIP, NY).

Kirk, J. and Stoneham, R. J., 1982, MNRAS. **201**, 1183

Kluzniak, W. 1989, Ap. J. **336**, 367.

Kouveliotou, C. et al. 1988, Ap. J. Lett. **330**, L101.

Lamb, D. Q. et al. 1990, Ap. J. **363**, 670.

Lasota, J. and Belli, 1983, Nature **304**, 139.

Liang, E. P. 1985, Nature **313**, 202.

Liang, E. P. 1986, Ap. J. **304**, 682.

Liang, E. P. 1987, Comm. Ap. **12**, 35.

Liang, E. P., 1989, GRO Science Workshop, Proc. p. 4-397, ed. W. N. Johnson (NASA).

Liang, E. P. et al. 1991, unpublished.

Liang, E. P. and Petrosian, V., 1986, editors, AIP Conf. Proc. No. 141 (AIP, NY).

Mazets, E. P. et al. 1980, Ap. Sp. Sci. **80**, 3.

Melrose, D. 1980, Plasma Astrophys. (Gordon & Breach, NY).

Mitrofanov, I. G. 1991 in Proc. Taos Workshop on Gamma Ray Bursts, ed. C.Ho and R. Epstein (AIP, NY).

Murakami, T. et al. 1988, Nature **335**, 234.

Murakami, T. 1991, in Proc. of Taos Workshop on Gamma Ray Bursts, ed. C. Ho and R. Epstein (AIP, NY).

Nolan, P. et al. 1984, Nature **311**, 360.

Norris, J. et al. 1986, Ap. J. **301**, 213.

Owens, A. et al. 1990, Ap. J. **352**, 741.

Petrosian, V. 1991 in Proc. Taos Workshop on Gamma Ray Bursts, ed. C. Ho and R. Epstein (AIP, NY).

Ruderman, M. 1987 in Cargese NATO ASI Proc., ed. F. Pacini (Reidel, Holland).

Schaefer, B. 1981, Nature **294**, 722.

Shore, G. et al. 1986, Adv. Sp. Res. **6** (4), 15.

Sturrock, P. 1986, Nature, **321**, 47.

Teegarden, B. J. and Cline, T. L., 1980, Ap. J. Lett. **236**, L67.

Vitello, P. and Dermer, C. D. 1991, Ap. J. submitted.

Wang, J. C. L. et al. 1989, Phys. Rev. Lett. **63**, 1550.

CYCLOTRON LINES IN GAMMA-RAY BURSTS AND MAGNETIC FIELD DECAY

Dieter H. Hartmann

Lawrence Livermore National Laboratory
Department of Physics, Livermore, CA 94550

and

Department of Physics and Astronomy
Clemson University, Clemson, SC 29634

ABSTRACT

The presence of cyclotron lines in a significant fraction of γ-ray burst spectra suggests a source association with strongly magnetized neutron stars. The burster distance scale is not known but constrained by their angular- and brightness distribution. Using pulsar data we calculate the corresponding age distribution of Galactic neutron stars and apply an exponential field decay model to test whether the observed incidence rate of cyclotron lines is consistent with decay time scales derived for radio pulsars. We find that the statistical properties of γ-ray bursts are inconsistent with the idea of events originating at arbitrary times on neutron stars whose fields decay on time scales shorter than $\sim 10^9$ yrs. Possible interpretations of this inconsistency with radio pulsar field decay models are discussed.

INTRODUCTION

Neither the nature of γ-ray burst sources nor their distances are known. No counterpart has been unambiguously identified in any wavelength band. The event distribution on the sky is isotropic so that proposed sites may range from local sources in our Galaxy to events associated with the large scale structure of the universe. However, there is circumstantial evidence suggesting that γ-ray bursts are associated with neutron stars. Spectral absorption features observed in a significant fraction of all bursts[3,10,11,29,33] imply magnetic fields in excess of 10^{12} gauss, if one interpretes the lines as cyclotron resonances. Whether these teragauss neutron stars (TGNS) are located in our Galaxy might be demonstrated by future space missions such as Gamma Ray Observatory (GRO) or the High Energy Transient Experiment (HETE). For a complete overview of all aspects of γ-ray bursts a number of recent review articles[13,14,21,23,24,27] is available in the literature.

If the TGNS paradigm proves correct, γ-ray burst observations open an exciting new window for studying the evolution of neutron stars on time scales perhaps as long as $\sim 10^{10}$ yrs. With more than 100 non-recurring bursts per year, statistical analysis of these events could be superior to what can be learned from observations of pulsars, X-ray binaries. Analysis of redshifted pair annihilation lines can be used to constrain the nuclear matter equation of state[26,50], although the redshift interpretation must consider detailed line formation physics.[25] X-ray observations bear on the thermal evolution of neutron stars, constraining cooling calculations for models with or without internal heat sources or external heating due to accretion.[6,7] These observations could also provide an ideal tool to study the global properties of the ISM.[15] Here we emphasize the use of cyclotron line measurements to study magnetic field evolution. Of course we do not know the ages of neutron stars undergoing γ-ray bursts, thus we proceed statistically.

DISTANCE CONSTRAINTS

To derive limits on the sampling depth of current γ-ray burst detectors we calculate the spatial distribution of Galactic neutron stars and test it against the observed properties of the γ-ray burst population. In particular, the angular distribution on the sky and the radial distribution constrain the distance scale to a relatively small fraction of the Galaxy. The lack of multi-pole moments of their angular distribution,[16] the absence of angular clustering,[15,19] and their almost uniform radial distribution[20,22,31] place stringent constraints on the spatial distribution of burst sources. The procedure has been outlined by Hartmann, *et al.*[17,18] (HEW) and by Paczynski.[35] However, there remain many caveats in this analysis: a) Gamma-ray bursts might not descend from pulsars, b) Population II or III neutron stars are not included, c) The time span during which a neutron star is a likely burst source has been assumed to be the stars lifetime. Models in which bursts occur only in a restricted age range[47] or where the burst recurrence time depends on an accretion rate that in turn depends on the velocity and position of the neutron star in the Galaxy[4] could give quite different results. To test the hypothesis that γ-ray burst sources are distributed uniformly in space, Schmidt, *al.*[41] pointed out the advantages of the so-called V/V_{max} test. Subsequently, this test has been applied to data from the KONUS experiment[22] aboard Venera 11/12, the SMM burst database[31] the PVO data set[20], and the french-soviet experiments SIGNE[2] (on Venera 13-14) and LILAS/APEX (on PHOBOS). The measured V/V_{max} distributions for all data sets mentioned above suggest that the source distribution favors nearby sources. This is consistent with the predictions of Galactic neutron star simulations[17,18,35] implying a small radial decrease of the neutron star density. The observed angular distribution is consistent with the simulated Galactic neutron star population if current detectors do not sample bursts to distances much in excess of $\sim$ 2 kpc. For larger distances observable anisotropies and clustering would occur. At distances larger than $\sim$ 10 kpc, the "edge" of the neutron star distribution would also have been detected by the $V/V_{\rm max}$ test.

NEUTRON STAR FIELD DECAY

Direct measurements of neutron star field strengths have been obtained for a few X-ray pulsars[8,34,49] that exhibit cyclotron spectral features in their spectra. Typical values derived from these observations are $\sim 4\times10^{12}$ gauss. The prototype "cyclotron object" of this kind[43,46,48] is Her X-1. The largest set of indirect field measurements stems from radio observations of pulsars. The large number of known pulsars allows a detailed statistical search for possible field decay.[12,28,42,44] The observations indicate typical values of 10^{12} gauss, with a spread by a factor $\sim$ 10. We take these values as representative of neutron star "initial" conditions. Born within $\sim$ 100 pc of the Galactic disk, pulsars migrate away from the plane with space velocities of order 100 km s^{-1}. A comparison of their dynamical ages (deduced from their heights above the plane) with their characteristic ages (determined from $P/\dot{P}$) suggests that the spin down torque of pulsars decays with age. Whether this decay is due to field decay, alignement of magnetic axis and rotation axis, or a combination of both, is not clear. For pure field decay scenarios exponential decay laws have been suggested with time scales of $\sim 10^7$ yrs. Additional support for field decay stems from a comparison of temporal and spectral properties of massive and low-mass X-ray binaries.[45,47] The physical details of field decay are poorly understood and there is no reason to exclusively consider exponential decay. In fact, Sang and Chanmugam[38,40] have shown that Ohmic decay of crustal fields, leading to non-exponential decay, is consistent with current pulsar statistics. Field evolution history may be further complicated by field topology changes, mass accretion onto the neutron star, and rotational as well as thermal effects. We assume exponential decay only for simplicity.

Figure 1 shows the neutron star age distribution as a function of heliocentric distance. The volume averaged mean age exceeds the global mean age at $\sim$ 3 kpc. This is the result of the

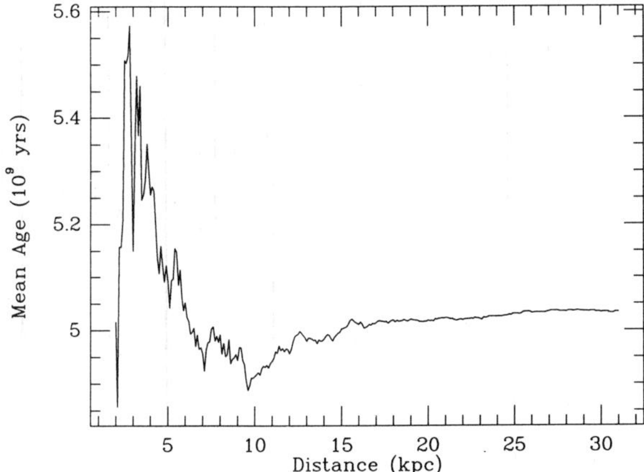

Fig. 1—Volume integrated mean age of neutron stars as a function of heliocentric distances. Locally, the mean is close to the global mean (~5×10^9 yrs) obtained when averaging over the whole Galaxy. The increase at ~ 3 kpc reflects the peak of the birth rate at galactocentric radius ~ 5 kpc.

radial birth rate distribution peaking at ~ 5 kpc galactocentric distance. Thus many young neutron stars born in that ring are excluded from the sampled volume. Because of the mild variation in mean age with sampling depth, field decay constraints are not very sensitive to the sampling depth derived from angular and brightness statistics dicussed in the previous section. This age distribution is then used to test the TGNS paradigm of γ-ray bursts for consistency and to constrain decay parameters. We assume that the burst probability is independent of age.

Recent cyclotron line observations in γ-ray bursts obtained with the GINGA satellite[10,33,51] suggest that at least 10% of all bursts show evidence for magnetic fields of ~10^{12} gauss. This fraction is comparable to that derived from KONUS.[11,29] If one considers the reported fields as an average surface value the SMM observations of high energy emission extending to ~ 100 MeV would be in conflict with this interpretation because of optical depth constraints due to one photon pair creation.[30] However, detailed ratiation transport calculations including relativistic effects along a curved photon path indicate that the SMM data are in fact consistent with a surface field strength distribution extending from 4×10^{11} gauss to ~ 5×10^{12} gauss, as expected for decaying fields.[32] More recently, independent support for fields of this magnitude was obtained with the Franco-Soviet LILAS experiment aboard the PHOBO 2 spacecraft.[3] Three absorption features at energies 10.4, 18.3, and 25.8 keV were reported for the event GB880806. From the relativistic expression of resonance energies[1]

$$E_n = [-m_e c^2 + \left(m_e^2 c^4 + 2\, m_e c^2\, E_c\, n\, \sin^2\theta\right)^{1/2}]/\sin^2\theta \ , \tag{1}$$

where n is the Landau principal number, E_c = 11.6 B_{12} keV is the cyclotron fundamental energy, we can simultaneously derive the field strength and angle θ between magnetic field and line of sight (Fig. 2). Ignoring possible red- or blue shifts, the field strength is B ~ 10^{12} gauss and the angle is $\theta \sim 7.5^o$. Detailed fits to the observed count spectrum should be carried out to determine the optical depth of the resonantly scattering region from line strengths and shapes.

For the intrinsic field at birth we assume a Gaussian distribution with mean B_0 and a relative dispersion resembling that of radio pulsars ($\sigma(\log B) = 1$). The observed fraction of bursts that exhibit cyclotron lines thus constrains the typical birth field B_0 and the exponential decay time scale, τ. Fig. 3 shows the predicted fraction of bursts with observable fields ($B/B_{obs} \geq 1$) as a function of the exponential decay time, τ. The curves are for different ratios B_{obs}/B_o. Using the GINGA limit f = 0.1 we conclude that decay time scales shorter than 10^9 yrs are not consistent with current observations if the birth fields are of order 10^{12} gauss. Suggested time

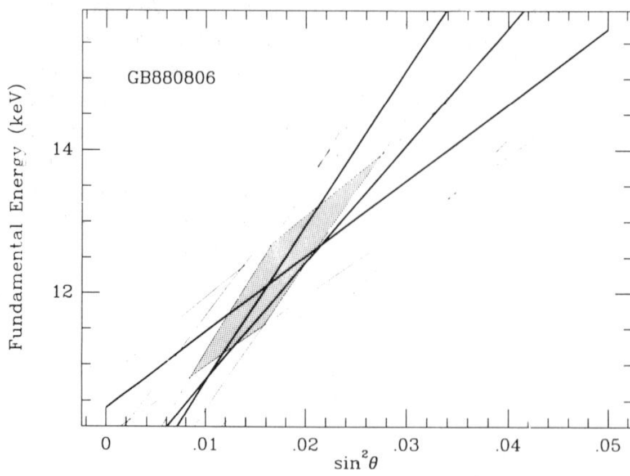

Fig. 2—Resonance energies measured for GB880806 determine the fundamental energy (E = 11.6 B_{12} keV) and viewing angle, θ. The three solid lines correspond to the three reported line energies. Thin lines mark the acceptable range due to uncertainties in the line center measurements. The intersect region (shaded) gives the allowed phase space in B-θ.

scales as short as $\tau = 10^7$ years are excluded unless mean birth fields exceed the critical value $B_{cr} = 4.414 \times 10^{13}$ gauss ($\hbar\omega_c = m_e c^2$), which appears unlikely.

This result can be interpreted in a number of different ways. Perhaps fields of neutron stars associated with γ-ray bursts decay on time scales that are long compared to those associated with pulsars. Thus, bursters may not evolve from standard radio pulsars but from a distinct pulsar sub-class.[36] If fields decay much more slowly after a few "e-folding" time scales the fraction of stars with large field strength could be increased. Nothing is known about the exact time dependence of field decay, the exponential form is frequently used because of its simplicity. The most restrictive assumption underlying all arguments presented here is that bursts can occur at any time during the stellar lifetime. Obviously, this simplification could be incorrect. To retain a larger fraction of high field neutron stars in spite of rapid field decay might be accomplished if bursts are most common during the teen years of neutron stars. However, if bursters are associated with a young population of neutron stars, the distance limits based on the observed angular distribution become even more severe. Of course, there is the possibility that nature is fooling us in believing that the observed spectral features are due to cyclotron resonances. The detection of double-features[10,33] has reduced the likelihood for that solution and the observed triple-feature in GB880806 appears to rule it out completely[3]. Alternatively, one could speculate that the observed fields do not reflect the true surface field but only indicate particular sites in the magnetosphere. The observed narrow range of field strength perhaps supports this point of view. However, this would make the discrepancy even worse because the measured energies would then only provide lower limits on the surface fields. Also, there remains the possibility that the observed fields do not reflect a global dipole component relevant to pulsar activity, but rather are a manifestation of localized strong field regions near the surface, analogous to solar sunspots. In these regions the field strengths could exceed the mean surface field by many orders of magnitude There is a variety of proposals for generation of such spots, ranging from the thermo-electric effect[5] to neutron star plate tectonics.[37] In this case the overall dipole field could decay rapidly but teragauss local surface fields could be generated until old age. However, if we are indeed observing the spectral fingerprints of localized spots, distance limits derived from optical depth constraints[9] might become smaller than the minimum distance required to sustain the observed burst rate without obvious recurrences (~ 200 pc: HEW). Detailed simulations must be carried out to address some of these possibilities.

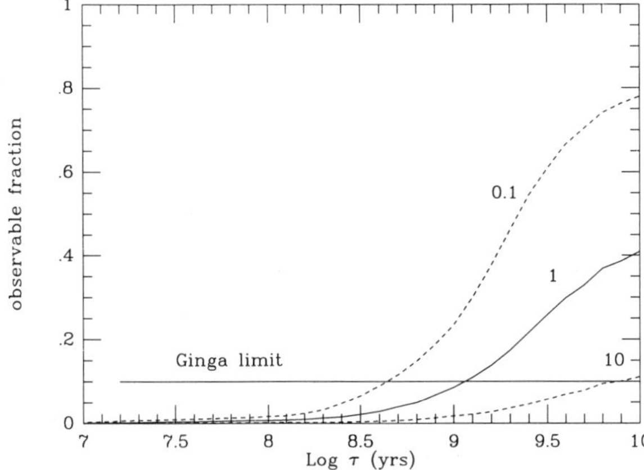

Fig. 3—Fraction of neutron stars with observable cyclotron lines as a function of the decay time scale. Different curves correspond to different ratios B_{obs}/B_0. Ginga observations suggest f ≥ 0.1, implying a lower limit on τ of $\sim 10^9$ yrs for a standard birth scenario.

SUMMARY

We have discussed statistical constraints on the γ-ray burst spacial distribution using their observed angular and brightness distribution. Stellar orbit simulations were used to derive distance limits for a class of models in which bursts occur randomly on Galactic neutron stars. To be consistent with the observed properties the sampling depth of current detectors is constrained to $\sim$ 2 and 10 kpc from angular and brightness data, respectively. The age distribution for the neutron stars was calculated as a function of sampling depth. Assuming an exponential decay law, the observed magnetic fields in γ-ray bursts then constrain the parameter space for the birth field strength and decay time. We have shown that available data are inconsistent with the idea that γ-ray bursts are associated with neutron stars whose magnetic fields decay exponentially on time scales shorter than 10^9 yrs. This result suggests that either one or more assumptions of our study are incorrect or that field decay models deviate from those considered for pulsars. Alternatively, the observed fields might not reflect the true global fields or the cyclotron resonance interpretation of line features could be incorrect. Although our results are model-dependent, they demonstrate that spectroscopy of γ-ray bursts could significantly advance our understanding of field evolution in neutron stars.

Acknowledgement: It is our pleasure to thank the symposium organizers for their support and hospitality. We thank Edison Liang for many valuable contributions. This work was performed under the auspices of the U.S Department of Energy by the Lawrence Livermore National Laboratory under contract No. W-7405-ENG-48. This research was also supported by NASA grant GRO-90-10.

REFERENCES

1. Alexander, S. G., and Meszaros, P. 1989, ApJL, 344, L1
2. Atteia, J.-L., *et al.* 1990, *Nature*, submitted.
3. Barat, C., *et al.* 1990, Taos Workshop on *Gamma-Ray Bursts*, submitted
4. Blaes, O., Blandford, R., Madau, P., and Koonin, S. 1990, ApJ, 363, 612
5. Blandford, R., Applegate, J. H., and Hernquist, L. 1983, MNRAS, 204, 1025
6. Boer, M., *et al.* 1988, A&A, 202, 117
7. Boer, M., *et al.* 1990, A&A, in press
8. Clark, G. W., *et al.* 1990, *Ap. J.*, **353**, 274.
9. Epstein, R. I. 1985, *Ap. J.*, **297**, 555.

10. Fenimore, E. E., *et al.* 1988, *Ap. J. (Letters)*, **335**, L71.
11. Golenetskii, S. V., *et al.* 1986, *Ap. Space Sci.*, **124**, 243.
12. Gunn, J. E., and Ostriker, J. P. 1970, *Ap. J.*, **160**, 979.
13. Harding, A. K. 1991, *Phys. Rep.*, in press.
14. Hartmann, D. H., and Woosley, S. E. 1988, in *Multiwavelength Astrophysics*, ed. F. Córdova, (Cambridge: Cambridge Univ. Press.), p. 189.
15. Hartmann, D. H. 1989, in *Evolution of the ISM*, ed. L. Blitz, PASP Proc., in press.
15. Hartmann, D. H., and Blumenthal, G. R. 1989, *Ap. J.*, 342, 521
16. Hartmann, D. H., and Epstein, R. I. 1989, *Ap. J.*, 346, 960
17. Hartmann, D. H., Epstein, R. I., and Woosley, S. E. 1989, *Nucl. Phys.*, **10B**, 27.
18. Hartmann, D. H., Epstein, R. I., and Woosley, S. E. 1990, ApJ, 348, 625: HEW.
19. Hartmann, D. H., Linder, E. V., and Blumenthal, G. R. 1991, ApJ, in press.
20. Hartmann, D. H., *et al.* 1990, Taos workshop on *Gamma-Ray Bursts*, submitted
21. Higdon, J., and Lingenfelter, R. 1991, ARA&A, 28, 401
22. Higdon, J., and Schmidt, M. 1990, ApJ, 355, 13
23. Hurley, K. 1989, *Ann. N.Y. Acad. Sci.*, 571, 442
24. Hurley, K., and Lamb, D. Q. 1991, *Phys. Rep.*, in preparation
25. Kluzniak, W. 1989, ApJ, 336, 367
26. Liang, E. P. 1986, ApJ, 304, 682
27. Liang, E. P., and Petrosian, V. 1986, *Gamma-Ray Bursts*, AIP Proceedings No. 141.
28. Lyne, A. G., Manchester, R. N., and Taylor, J. H. 1985, *MNRAS*, **213**, 613.
29. Mazets, E. P., *et al.* 1981, *Nature*, **290**, 378.
30. Matz, S. M., *et al.* 1985, ApJL, 288, L37
31. Matz, S. M., *et al.* 1990, Taos workshop on *Gamma-Ray Bursts*, submitted
32. Meszaros, P., Bagoly, Z., and Riffert, H. 1989, ApJL, 337, L23
33. Murakami, T., *et al.* 1988, *Nature*, 335, 234
34. Nagase, F. 1989, *Publ. Astr. Soc. Jap.*, 41, 1
35. Paczynski, B. 1990, ApJ, 348, 485
36. Ruderman, M., and Cheng, K. S. 1988, ApJ, 335, 306
37. Ruderman, M. 1990, preprint
38. Sang, Y., and Chanmugam, G. 1987, ApJL, 323, L61
40. Sang, Y., and Chanmugam, G. 1990, preprint
41. Schmidt, M., Higdon, J., and Hueter, G. 1988, ApJL, 329, L85
42. Shrinivasan, G. 1989, A&A Rev., 1, 209
43. Soong, Y., *et al.* 1990, ApJ, 348, 641
44. Stollman, G. M. 1987, A&A, 178, 143
45. Taam, R. E., and Van den Heuvel, E. P. J. 1986, ApJ, 305, 235
46. Trümper, J., *et al.* 1978, ApJL, 219, L105
47. van Paradijs, J. 1989, MNRAS, 238, 45p
48. Voges, W., *et al.* 1982, ApJ, 263, 803
49. White, N. E., Swank, J. H., and Holt, S. S. 1983, AJ, 270, 711.
50. Wiringa, R. B., Fiks, V., and Fabrocini, A. 1988, *Phys. Rev. C*, **38**, 1010.
51. Yoshida, A., *et al.* 1990, Taos workshop on *Gamma-Ray Bursts*, submitted

Gamma-Ray Lines from Accreting Neutron Stars

Lars Bildsten
Center for Radiophysics and Space Research
Cornell University, Ithaca, NY 14853 USA

ABSTRACT

The current generation of γ-ray telescopes (SIGMA, GRO) provides a new opportunity for observing red-shifted γ-ray lines from the atmospheres of accreting neutron stars. A successful observation would provide important information about how the accretion stream settles onto the neutron star and might allow limits to be placed on the nuclear equation of state. Thus, we have undertaken a theoretical re-analysis of different γ-ray emission mechanisms. This paper describes our results on the 4.438 MeV γ-ray line emission from ^{12}C and ^{16}O and outlines the ongoing calculation of the 2.2 MeV D-recombination line flux expected from the spallation of incident Helium. We show that a neutron star accreting material of solar abundances will produce a 4.438 MeV γ-ray line flux that is below the current observational limits.

INTRODUCTION

It has been realized for some time that nuclear $\gamma-$ray lines could be produced in accretion onto neutron stars, and that observations of these lines could provide important probes of the emission region[1-5] and possibly of neutron star structure and gravitation physics as well.[6,7] In order to determine the fluxes, the atmospheric composition must be known, and in the past was arbitrarily chosen. The choices ranged from a pure CNO atmosphere[2], to abundances similar to that in the incident beam[3,5], and therefore resulted in different values of the line flux. Rather than impose a composition on the neutron star atmosphere, we have computed it self-consistently as a function of depth in the atmosphere and accretion rate.[8] This atmospheric structure allows a determination of the γ-ray line fluxes from both radiative decay of excited nuclei and D-recombination of spallation produced neutrons (i.e. $n + p \rightarrow D + \gamma$).

GAMMA RAYS FROM ^{12}C AND ^{16}O

Material accreting onto a neutron star arrives at the atmosphere with characteristic energies of 100-300 MeV per nucleon. When the material is stopping through Coulomb collisions, the total atmospheric column density traversed by an incident particle of mass Am_p, charge Ze and initial energy E_i is

$$y_s \simeq 14 \frac{\text{gr}}{\text{cm}^2} \left(\frac{E_i}{200\,\text{MeV/nucleon}}\right)^2 \left(\frac{10}{\ln \Lambda}\right) \frac{A}{Z^2}, \qquad (1)$$

where $\ln \Lambda$ is the Coulomb logarithm. One key aspect to determining the abundance of heavy elements in the upper atmosphere is the A/Z^2 scaling of this Coulomb stopping depth. Although the Coulomb stopping depths for protons and ^{4}He are identical, heavier elements such as CNO do not penetrate as deeply. As a result, the upper layers of the atmosphere may be dominated by heavy elements, which can be excited or destroyed by collisions with the still-fast protons. It is the interplay between the rate at which the CNO elements drift downward through the atmosphere and the rate at which they are destroyed and excited by the beam that determines the abundances and resulting gamma-ray line emission.

Figure 1 is a sketch of the upper layer of an accreting neutron star atmosphere. The horizontal lines designate the stopping points for different species of incident particles. Below the proton stopping point, there are no fast particles, and the thermal plasma flows downward at an advection speed much less than thermal. The mechanism for ^{12}C excitation is shown on the left, where an incident ^{12}C thermalizes at the higher altitude, drifts downward and is suddenly struck by an incident proton, producing a 4.428 MeV γ-ray. In the upper reaches of the atmosphere, where the incident heavy elements thermalize, the incident protons and ^4He still have nearly the infall energy. At these energies the most prevalent interaction is spallation, rather than γ-ray production. The destruction of the elements is a very important effect that substantially reduces the γ-ray emission. The residence time of a nucleus in this hazardous region of the atmosphere determines the abundances, survival probabilities and γ-ray emission.

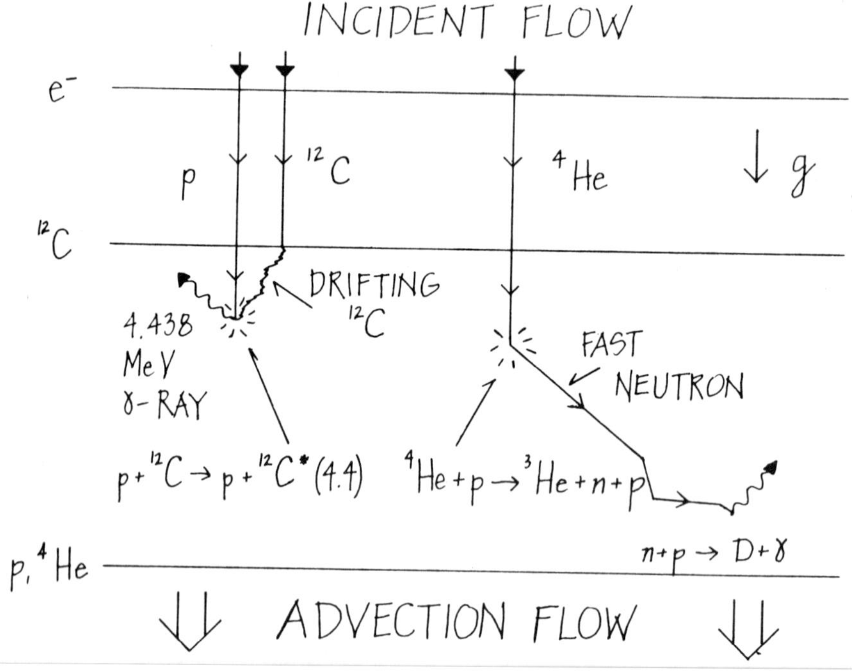

Figure 1: A sketch of the two primary γ-ray emission mechanisms. The horizontal lines designate the stopping depths for electrons, Carbon, Helium and protons.

The deposited ions are collisionally coupled to the atmosphere through Coulomb collisions with the atmospheric protons. This provides the drag force that determines the equilibrium drift velocity from a knowledge of the macroscopic force on the particle (i.e. gravity and the electrostatic field). We have determined the residence time through a solution of the drift/diffusion equation for each ion component. These calculations, coupled with the nuclear effects induced by the proton beam, determine three quantities: 1) the abundance of ^{12}C throughout the atmosphere, 2) the probability that a given ^{12}C nucleus reaches great depths in the atmosphere without being destroyed, and 3) the emission of 4.438 MeV γ-ray photons. Two parameters determine these results, the initial energy E_i, and an accretion rate dependent parameter,

$$\beta \approx 36 \left(\frac{R}{10\ \mathrm{km}}\right)^{7/4} \left(\frac{1.4M_\odot}{M}\right)^{11/8} \left(\frac{T_{\mathrm{eff}}}{T}\right)^{3/2} \left(\frac{\dot{M}}{\dot{M}_{\mathrm{Edd}}}\right)^{5/8} . \qquad (2)$$

where T_{eff} is the effective temperature set by the accretion rate, $\dot{M}$; M and R are the neutron star mass and radius, and $\dot{M}_{\mathrm{Edd}}$ is the Eddington accretion rate. Figure 2 plots the number of 4.438 MeV γ-rays emitted per accreted ^{12}C (solid) and ^{16}O (dashed) that reach the outer edge of the atmosphere unscattered (i.e. includes the flux dilution in the line due to Compton scattering). A 4.438 MeV photon is also emitted during the spallation of ^{16}O via $^{16}O(p,p\alpha)^{12}C(4.438\ \mathrm{MeV})$.

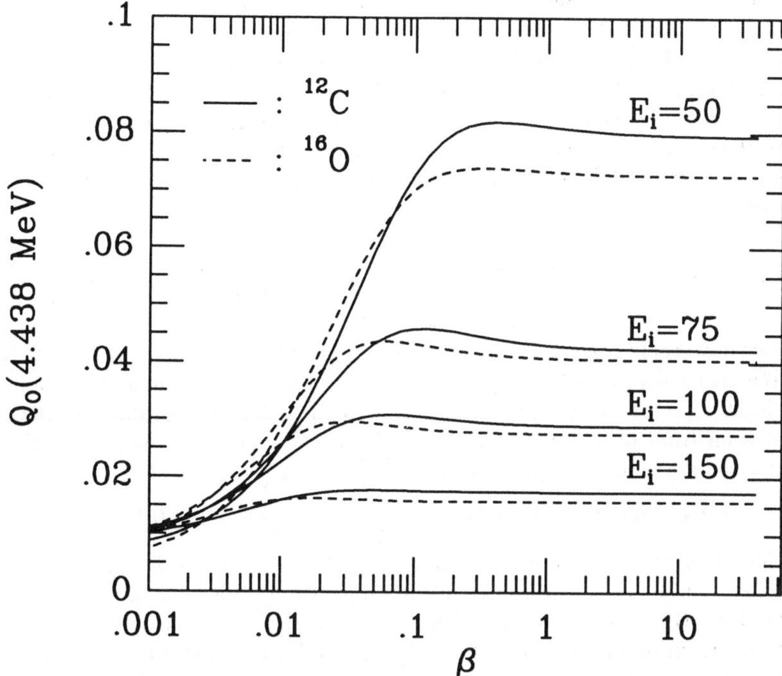

Figure 2: The number, Q_o, of observable 4.438 MeV γ-ray photons produced per accreted ^{12}C (solid) and ^{16}O (dashed) as a function of β. Higher accretion rates are to the right. The curves are labeled by the incident energy per nucleon in MeV/nucleon.

Most importantly, the yield, Q is typically less than 10%. This results from both the spallation effects and the drift time. At these energies, spallation is the predominant nuclear interaction, making it impossible for the nucleus to survive intact after a few collisions. This forces the yield, Q, to always be less than unity, a point made earlier (in slightly different terms) by Aharonian and Sunyaev.[9] The spallation effects are so strong that nearly all of the accreted CNO elements are destroyed. This has important consequences for the theoretical modeling of X-ray bursts, where the CNO elements determine the steady burning rate of Hydrogen in the deep atmosphere.[8]

OBSERVATIONAL PROSPECTS

The number of γ-rays emitted per accreted ^{12}C or ^{16}O is nearly independent of the accretion rate in the limit of high accretion rates (i.e. $\beta \gg 1$). The γ-ray line flux can then be directly determined from the observed X-ray flux. This eliminates the need to know the distance to the source, and likewise, the actual value of the accretion rate. The γ-ray line number flux is

$$F_\gamma = \frac{Q(E_i)f_i}{E_i}F_x, \tag{3}$$

where E_i is the infall energy, f_i is the elemental abundance relative to incident protons, and F_x is the X-ray energy flux. The thermal ^{12}C produces reasonably narrow lines of width $\Delta E/E \simeq 5\%$. The resulting 4.438 MeV line flux from the brightest accreting neutron star, Sco X-1, assuming solar abundances and an incident energy of 100 MeV, is $3.3 \times 10^{-8} \gamma\,\mathrm{cm}^{-2}\,\mathrm{s}^{-1}$. This line flux is much lower than the observational limit for GRO, $F_\gamma > 1 \times 10^{-5}\ \gamma\,\mathrm{cm}^{-2}\,\mathrm{s}^{-1}$. *For the collisional stopping scenario, an enhancement of nearly 100 is needed in the* ^{12}C *and* ^{16}O *abundance to produce an observable 4.4 MeV line.* In general, an X-ray source with flux F_x, accreting an element with an abundance f_i relative to protons provides an observable line flux to GRO when

$$f_i Q(\dot{M}, E_i) > 6.3 \times 10^{-3} \left(\frac{F_x(\mathrm{Sco\ X}-1)}{F_x}\right)\left(\frac{E_i}{100\mathrm{MeV}}\right)\left(\frac{F_\gamma}{10^{-5}\gamma\,\mathrm{cm}^{-2}\mathrm{s}^{-1}}\right), \tag{4}$$

where we have scaled the X-ray flux to the brightest object in the sky, Sco X-1. Since it is difficult to have $Q > 0.1$ for any nuclear line, the heavy element abundance must be at least a factor of 100 above solar to provide a flux observable to GRO. Since the emission from other lines originating from heavy elements will suffer this same difficulty, we are at present not investigating these processes.

NEUTRON PRODUCTION AND THE 2.2 MeV LINE

The upper atmosphere is very under-abundant in ^{4}He nuclei since only a small fraction of the incident ^{4}He (determined by p-^{4}He nuclear elastic scattering) is deposited above the proton stopping point. The majority of the incident ^{4}He thermalize at the proton stopping point (since $A/Z^2 = 1$) and are pulled into the deep layers of the atmosphere by the bulk advection. (Very few of the ^{4}He will diffuse upwards.) Neutrons are liberated from ^{4}He nuclei as the incident ^{4}He slows and thermalizes in the atmosphere. The total inelastic cross-section is roughly 100 mb. Comparing this cross-section with the Coulomb stopping depth determines the probability for a nuclear reaction to occur while the nucleus is slowing and thermalizing in the atmosphere. The product is given by

$$\sigma_{dest} y_s \simeq 0.8 \left(\frac{E_i}{200\,\mathrm{MeV/nucleon}}\right)^2 \left(\frac{10}{\ln \Lambda}\right). \tag{5}$$

Since this number is of order unity, a sizeable fraction of the incoming ^{4}He suffer spallation reactions prior to stopping.

The right hand-side of Figure 1 shows the results of this spallation. The most dominant inelastic reaction is p+^{4}He →p+n+^{3}He, producing a fast neutron and a residual ^{3}He. The ^{3}He thermalizes in the upper atmosphere and is then subjected to further destruction by the beam. The fast neutron slows via nuclear elastic scattering with thermal protons, and once thermalized will either recombine with a proton, or charge exchange on ^{3}He (^{3}He(n,p)^{3}H). The branching between these two options depends directly on the local ^{3}He abundance, which must be self-consistently determined. The initial input to this problem is the neutron and ^{3}He energy-angle spectra from the spallation reaction, which have recently been determined.[10] These effects and the full neutron transport are now being studied.[11]

STRANGE ABUNDANCE OBJECTS

The lower limit on the abundance in equation (4) is rather strong in the non-magnetic cases we are currently considering, and suggests that the most likely sources are those with abundances much greater than solar. There are three sources for which theoretical arguments imply that the mass-providing companion must be made of very hydrogen-poor material, and therefore rich in nuclei. An analysis of the LMXRB's 4U1916–05 and 4U 1626–67, concluded that the very short orbital periods (less than an hour) and large mass transfer rates require the mass-providing stars to be severely depleted in hydrogen.[12] Likewise, stable mass transfer in the 685 second orbital period binary, 4U1820–30 requires a helium white dwarf companion of 0.055 $M_{\odot}$.[13] *If the accreting material is predominantly Helium, or heavier elements, then these objects might be copious emitters of gamma-rays.* The X-ray burst source, 4U1820–30 is the brightest amongst these sources, and we strongly recommend that it be observed. The other two are less bright, but should still be considered, particularly 4U 1626-67, which is an X-ray pulsar.

Since the pure Helium (or pure ^{12}C) accretion is a much different problem than we have considered, we cannot provide flux estimates. Simple dimensional arguments show that it is an interesting observational prospect. The X-ray flux from 4U 1820-30 is in the range 4×10^{-9}–2×10^{-8}erg s^{-1} cm^{-2}, where the upper limit is nearly the Eddington limit.[13] From equation (4), the required efficiency to produce a 2.2 MeV line observable to GRO is

$$f_i Q(\dot{M},E_i) > 0.08\left(\frac{2\times10^{-8}\text{erg s}^{-1}\,\text{cm}^{-2}}{F_x}\right)\left(\frac{E_i}{100\text{MeV}}\right)\left(\frac{F_\gamma}{10^{-5}\gamma\,\text{cm}^{-2}\text{s}^{-1}}\right), \quad (6)$$

where we have scaled to the maximum flux. *To provide a 2.2 MeV line flux observable to GRO, 1 out of 5 accreted neutrons must provide an unscattered γ-ray. Compton-scattered γ-rays could contribute to an observable continuum.* We cannot say whether this is a reasonable expectation, though we know that more than 20% of the incident Helium is broken up during accretion into a pure proton atmosphere. In the case of a pure Helium beam, the atmospheric abundances depend on exactly how much destruction is occurring, and the amount of destruction depends on the atmospheric abundances.

CONCLUSIONS

Assuming solar abundances, our predictions for the 4.438 MeV γ-ray line fluxes from accreting neutron stars are below the current observational limits. Thus, we strongly encourage the observers to look at "strange" abundance objects in addition to the brightest X-ray sources. Other stopping scenarios must also be investigated. In particular, if the accretion flow settles onto the star after passage through a collisionless shock,[14] the line emission mechanism will be very different and possibly more efficient. We have yet to consider magnetic field effects, which are crucial to the understanding of line emission in the X-ray pulsars. In particular, the high magnetic field allows pair production, which can serve to "funnel" nearly all γ-rays of $E_\gamma > 2m_e c^2$ into a 511 keV annihilation line. Both of these scenarios might avoid the difficulties with normal abundance sources and provide an observable feature.

This research was supported through NASA grant NAGW-666, and NSF grants AST 87-14475 and AST 89-13112. I also thank the Fannie and John Hertz Foundation for Fellowship support.

REFERENCES

1. V. F. Shvartsman, *Astrophysics*, **6**, 56 (1972).
2. R. Ramaty, G. Borner, and J. M. Cohen, *Ap. J.*, **181**, 891 (1973).
3. C. Reina, A. Treves, and M. Tarenghi, *Astr. and Astrop.*, **32**, 317 (1974).
4. J. C. Higdon, and R. E. Lingenfelter, *Ap. J. (Letters)*, **215**, L53 (1977).
5. K. Brecher, and A. Burrows, *Ap. J.*, **240**, 642 (1980).
6. K. Brecher, *Ap. J. (Letters)*, **215**, L17 (1977).
7. K. Brecher, *Ap. J. (Letters)*, **219**, L117 (1977).
8. L. Bildsten, I. Wasserman, and E. E. Salpeter, *Ap. J.*, to be submitted (1991).
9. F. A. Aharonian, and R. A. Sunyaev, *Mon. Not. Roy. Astr. Soc.*, **210**, 257 (1984).
10. L. Bildsten, I. Wasserman, and E. E. Salpeter, *Nuc. Phys.*, **A516**, 77 (1990).
11. L. Bildsten, I. Wasserman, and E. E. Salpeter, *work in progress.*
12. L. A. Nelson, S. A. Rappaport, and P. C. Joss,*Ap. J.*, **304**, 231 (1986).
13. L. Stella, W. Priedhorsky, and N. E. White, *Ap. J. (Letters)*, **312**, L17 (1987).
14. S. L. Shapiro, and E. E. Salpeter, *Ap. J.*, **198**, 671 (1975).

GAMMA-RAY EMISSION FROM BLACK HOLES

James C. Ling
Jet Propulsion Laboratory, California Institute of Technology, Pasadena, CA. 91109 &
NASA Headquarters, Washington D.C. 20546

ABSTRACT

Strong continuum gamma-ray emission at ~1 MeV possibly correlated with a narrow annihilation line at 511 keV has been observed from both Cygnus X-1 and the Galactic Center. Such correlated emission has been interpreted as a unique gamma-ray signature for theoretically predicted relativistic, positron-electron pair-dominated plasma in regions surrounding the black holes. In this paper, I review primarily the Cygnus X-1 results, which have provided important new insights about the source. Cygnus X-1 may be considered a canonical reference stellar black hole whose spectral and temporal characteristics can be used for comparison with those of other black-hole candidates including the Galactic Center and AGN.

INTRODUCTION

Since the last review paper on the subject of "Gamma-Ray Emission of Cygnus X-1" [1], voluminous new results on this topic have become known[2-13], significantly enhancing our knowledge of the source. Specifically, there is now some evidence for a narrow annihilation feature[5] correlated with the MeV emission observed by HEAO 3 in 1979[14]. The MeV emission provides strong evidence for the theoretically predicted relativistic pair dominated plasma in black-hole systems. The narrow annihilation feature is produced by positrons which either escape the system or are formed by γ-γ interactions outside the system and then slow down and annihilate in the surrounding cooler medium. Evidence of such correlated line and continuum emission has also been reported from the Galactic Center[15-16]. Such a spectral signature provides unique information about the physical conditions in the innermost region near the event horizon which cannot be obtained at other wavelengths. The fact that this may have been seen in the brightest two black-hole candidates in our galaxy suggests the unique role that gamma-ray astronomy can contribute to the study of black-hole physics.

Recent observations of the Galactic Center region by both balloons and satellite gamma-ray experiments have yielded several interesting new results on the compact source. These include (1) several reports of variable 511 keV annihilation emission from the Galactic Center[17-20] with time scales ranging from days to months, and (2) the indication of 1E1740.7-2942[19-20] as the point source responsible for the variable positron-electron annihilation radiation. In addition, the x-ray luminosity of 1E1740.7-2942 was shown to be comparable to that of Cygnus X-1[21], suggesting that the source could also be a stellar-mass black hole.

Since many of the new Galactic Center results appear in these Proceedings, this paper focuses primarily on Cygnus X-1 with the objective that, as we increase our knowledge about the Galactic Center compact source (e.g., 1E1740.7-2942) and other black-hole objects, Cygnus X-1 can be used as a reference source for comparison.

BROAD-BAND γ-RAY EMISSION

During its ~170-day coverage of Cygnus X-1, HEAO 3 observed intense MeV emission (Figure 1b) during a two-week period (γ_1) in the fall of 1979 when the hard

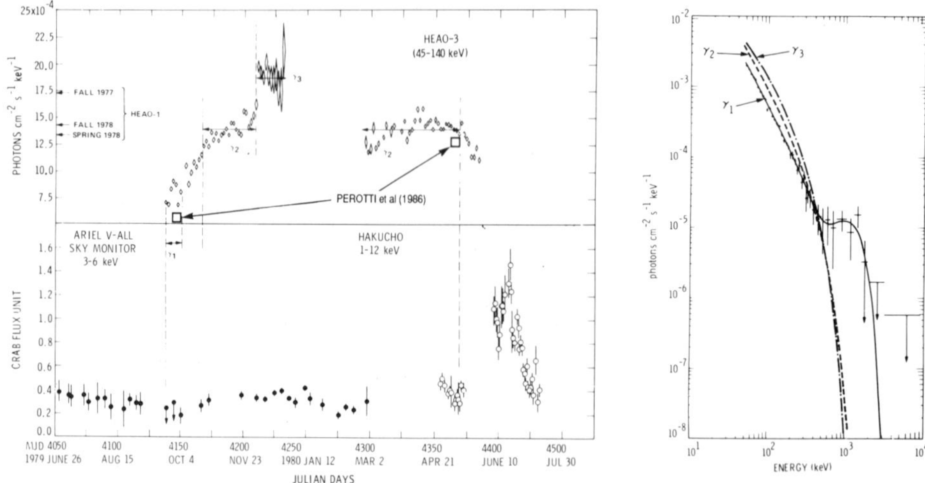

Figure 1: a) HEAO 3 observed MeV emission in 1979 when the hard x-ray flux was unusally weak (γ_1). Hard x-ray flux variations were confirmed by MISO balloon observations (open squares) on 1979 October 1 and 1980 May 18. b) The MeV flux vanished as the hard x-ray flux increased to the γ_2 and γ_3 levels[14].

x-ray flux was unusally weak (Figure 1a). As the source evolved to higher hard x-ray flux levels ("γ_2" and "γ_3") during the following eight months, the MeV emission disappeared while the spectrum softened, pivoting at E~0.4 MeV[14]. There was no significant change (<30%) of the luminosity in the 0.05-10 MeV range during this period. The variation of the hard x-ray fluxes between 1979 and 1980, and the anticorrelation between the gamma-ray and the hard x-ray fluxes (Figure 2) was independently confirmed by MISO balloon observations (open squares in Figure 1a) on 1979 October 1 (γ_1) and 1980 May 18 (γ_2)[2,22]. Because of the short observing time and therefore limited statistics, the two spectra measured by MISO can generally be fitted with a simple power-law. The integral flux in the 0.4-1.5 MeV band measured in 1979, however, is consistent with that obtained by HEAO 3. After examining the world's γ-ray database on Cygnus X-1, Bassani et al.[2] suggested that some weak evidence for anticorrelation between the MeV and the hard x-ray fluxes was also present in these data.

Another balloon observation of the MeV gamma rays from Cygnus X-1 was reported by the University of New Hampshire (UNH) investigators[3]. Their 1989 October 1-2 balloon spectrum is generally consistent with other measurements at the γ_2 level in the energy region below 1 MeV. However, above 1 MeV, the UNH result showed a hard spectral component extending to 9 MeV. The average flux in the 2-9.3 MeV band is (7.4±2.5) x 10^{-7} photons cm^{-2} s^{-1} keV^{-1}. Figure 3 compares the UNH results with those of other groups. Their 2-6 MeV fluxes are lower than those measured by Baker et al.[23], but consistent with the 2σ upper limits measured by HEAO 1[24-25] (not included in the figure) and HEAO 3 in the γ_2 period, and they are higher than the upper limits measured by the UC Riverside group[26]. Taken collectively, these data suggest that the MeV emission is highly variable.

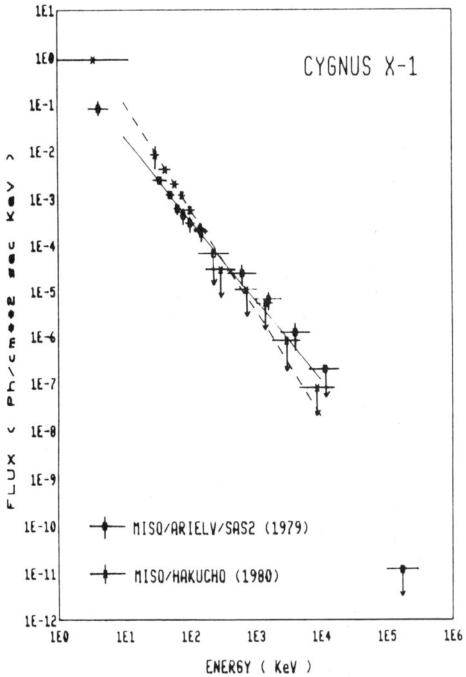

Figure 2: Anticorrelation between x-ray and γ-ray fluxes observed by HEAO 3 in 1979-1980 was confirmed independently by the MISO observations[2].

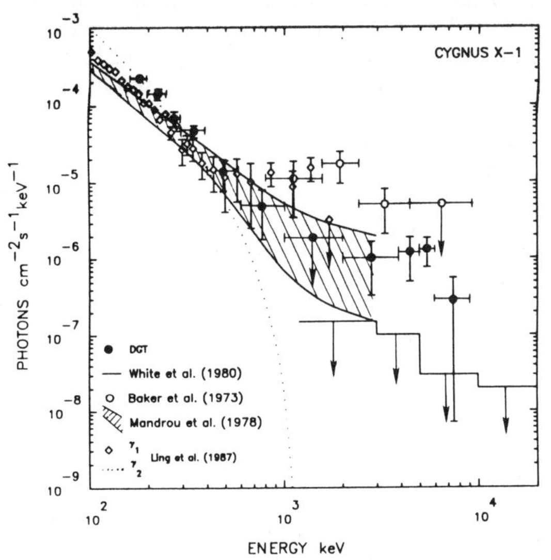

Figure 3: McConnell et al.,[3] of UNH observed a hard spectral component extending to 9 MeV (solid dots).

A GAMMA-RAY MODEL

The MeV gamma rays could result from either thermal or nonthermal processes, producing electron-positron pair-plasmas in the inner region of the accretion disk surrounding the black hole. To date, only one comprehensive model[27] has been proposed for interpreting the HEAO 3 results. Liang and Dermer[27] argued that gamma rays observed during the γ_1 period are produced in an inner ~500 km size spherical region where the temperature is about 4 x 10^9 K and the pair density $n_+ = 6 \times 10^{16}$ cm^{-3}, while the x rays are produced in the outer ion-dominated thin disk where the temperature is less than 10^9 K. During the "γ_1" period, there may have been a change in properties of the accreted matter, e.g., a large decrease of the magnetic field in the accretion stream[9], which would cause a depletion of soft synchrotron photons required to Compton cool the hot electrons. The inner region then became extremely hot and emitted by thermal bremsstrahlung primarily gamma rays and only weakly in x rays. This hot "bubble" therefore displaced a region which normally emits x rays, thus reducing the total x-ray luminosity of the source, and causing an intrinsic energy shift from x rays to gamma rays as the source evolved from the normal γ_2-like condition to γ_1. Liang[8] suggested that the two regions must be decoupled and separated by an intermediate region or the spectrum would

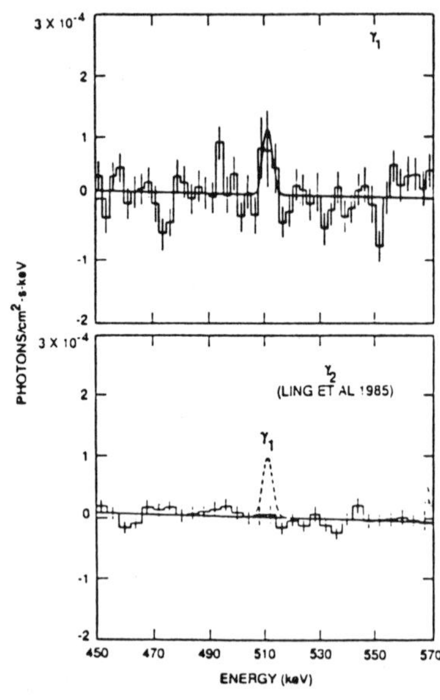

Figure 4: A possible narrow annihilation feature correlated with the MeV emission was observed in the γ_1 but not the γ_2 spectrum of Cygnus X-1[5].

be significantly distorted from that observed.

NARROW ANNIHILATION FEATURE CORRELATED WITH MEV EMISSION

Dermer and Liang[28] estimated that nonthermal pairs may be produced by photon-photon interactions of the gamma rays with hard x-rays from the accretion disk at a rate of $\leq 10^{41}$ $e^+ s^{-1}$. If some of the pairs escape and stop in any cold surrounding material, a narrow 511 keV annihilation feature with line flux of $\leq 3 \times 10^{-4}$ photons cm^{-2} s^{-1} may be observed in association with the MeV flux. Ling and Wheaton[5] reexamined the γ_1 spectrum of Cygnus X-1 and found some evidence, at 1.9σ statistical significance, for such a feature (Figure 4). The observed line flux of $(4.4 \pm 2.4) \times 10^{-4}$ photons cm^{-2} s^{-1} is consistent with the $\leq 3 \times 10^{-4}$ photons cm^{-2} s^{-1} computed by Dermer and Liang[28]. The feature was clearly not present in the γ_2 spectrum (Figure 4, lower), when no MeV emission was observed[29], to a 3σ upper limit of 3×10^{-4} photons cm^{-2} s^{-1}.

A similar correlation between the 511 keV line and continuum MeV fluxes was also observed from the Galactic Center[15,16]. Riegler et al.[16] observed a decrease in MeV flux between 1979 and 1980 which generally parallels the apparent decrease in 511 line flux reported over the same period. Lingenfelter and Ramaty[4] attempt to decompose the gamma-ray emission from the Galactic Center region into two components: a steady diffuse 511 keV line with an associated orthopositronium continuum, and a variable 511 keV line associated with high energy MeV emission from a few hundred solar mass black hole. They determined the flux due to the compact component by subtracting the estimated diffuse flux from the HEAO 3 data measured in the fall of 1979[16]. A comparison of the Cygnus X-1 (γ_1) and the Galactic Center compact source spectra, both observed during periods of intense MeV emission, is shown in Figure 5. The remarkable similarity suggests that similar processes occur in both systems.

Variable 511 line radiation from the Galactic Center has again been reported from two balloon experiments: GRIS[17] and HEXAGONE[18] and from the French/Soviet SIGMA experiment[19,20]. Preliminary SIGMA results[19,20] identify 1E1740.7-2942 as a likely source of 511 keV radiation from the Galactic Center region. The hard x-ray spectrum of 1E1740.7-2942 measured by the Caltech group[21] shows a strong resemblance with that of Cygnus X-1, suggesting that the source could be a stellar-mass black hole like Cygnus X-1.

5.6-DAY ORBITAL PERIOD MODULATION

Because of its low significance, the Cygnus X-1 511 keV line flux observed during the γ_1 period[5] can only be treated as interesting and suggestive and certainly requires future confirmation with more sensitive instruments. If this line is real, the annihilation site needs to be investigated. Whether the annihilation occurs in the

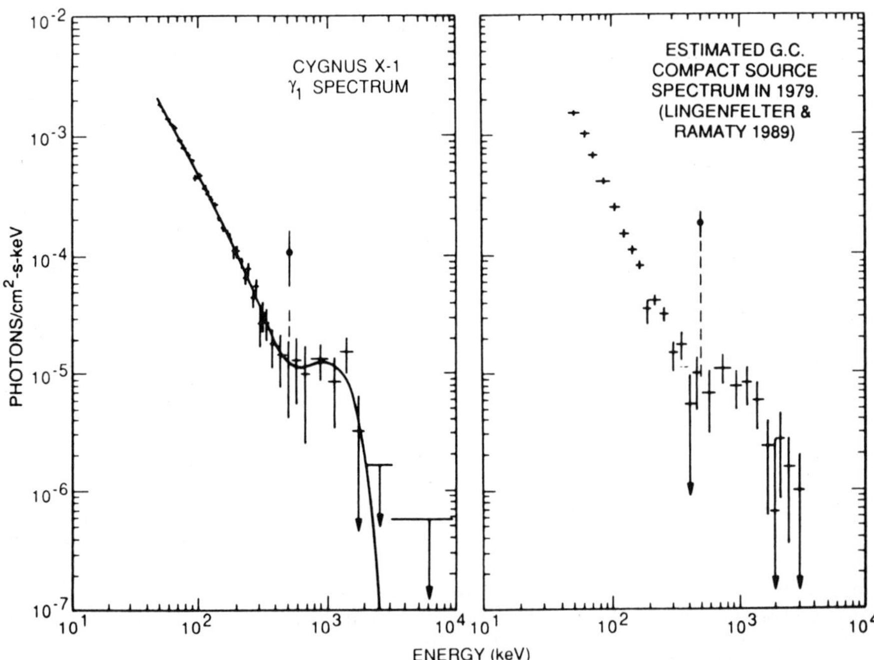

Figure 5: Cygnus X-1[5] and Galactic Center (as estimated by Lingenfelter and Ramaty[4]) compact source spectra show strong similarity, suggesting that similar processes are at work in both systems.

photosphere of the primary star, or in regions between the compact object and the star, e.g., the accretion stream or wind, we expect some modulation of the line flux with the 5.6-day orbital period. HEAO 3 observed the 5.6-day binary orbital period of Cygnus X-1 in the 50-140 keV range during the γ_2 period[7]. The data include about 21 binary cycles within the 36.6 binary cycles spanned by the observations covering the period between 1979 October 27 and 1980 May 5. Figure 6 shows the 5.6-day light curve. The solid line is a fit to a sinusoidal function plus a constant, $F(\phi) = A + B \sin [2\pi(\phi - \phi_o)]$. The χ^2 of the fit is 7.5 for 5 degrees of freedom. The formal significance, based on Poisson statistics, is ~7σ. The peak-to-peak amplitude variation is about 6%. There is possibly a phase shift of $\Delta\phi = 0.1$ of the minimum and maximum from the superior conjunction ($\phi = 0$) and inferior conjunction ($\phi = 0.5$), respectively. This requires further assessment and confirmation. The 5.6-day modulation has also been observed previously by the Ariel 5 All Sky Monitor[30] and by the HEAO 1 Hard X-Ray and Low Energy Gamma-Ray Experiment[24]. Ariel 5, a pinhole camera with a 3-6 keV proportional counter, detected a secular increase in the amplitude of the 5.6-day modulation preceding the low-to-high state transition of 1975. HEAO 1, a scanning 15-180 keV scintillation spectrometer collimated by crossed slats, observed Cygnus X-1 during three separate periods of ~30 days each, in the fall of 1977, the spring of 1978, and the fall of 1978. Only data from the second half of the fall of 1978 clearly show the 5.6-day effect, with a peak-to-peak amplitude of about 10%. Both Ariel 5 and HEAO 1 found the minimum at superior conjunction. The approximately zero phase of flux minimum observed by HEAO 1 and Ariel 5 may be consistent, within

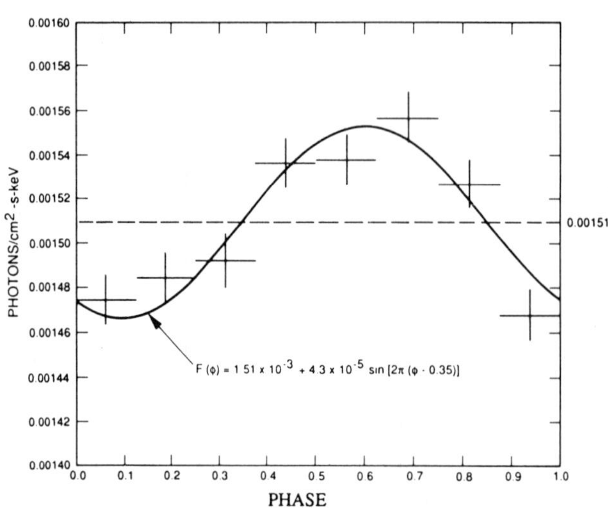

Figure 6: The 50-140 keV γ_2 flux folded modulo the 5.6-day binary period shows a 6% peak-to-peak modulation at 7σ significance[7]. The solid line is the best-fit sinusoid.

errors, with the shift of $\Delta\phi=0.1$ observed by HEAO 3.

Ling and Wheaton[6] further examined the 5.6-day orbital modulation of the 511 keV line flux in the γ_1 period. Figure 7a shows the 5.6-day light curve for the flux in the 507.5-515.0 keV band. The solid line is a fit to a constant term. The χ^2 of the fit is 7 for 7 degrees freedom. So based on these analyses, there is no significant evidence for modulation. However, it is interesting to note that the average flux appears to be larger on one side of the orbit than the other. Figure 7b shows a comparison of the corresponding spectra. The solid lines show fits to a linear continuum plus a gaussian line centered at 511 keV (4 parameters). The best-fit line fluxes are $(7.7 \pm 3.2) \times 10^{-4}$ and $(0.7 \pm 3.4) \times 10^{-4}$ photons cm^{-2} s^{-1}, respectively. The low significance of 1.5σ for the difference of the two flux-values again requires confirmation for the modulation in future observations.

SEARCH FOR γ_1-LIKE OCCURRENCES IN CYGNUS X-1

The Hard X-Ray Burst Spectrometer (HXRBS) onboard the Solar Maximum Mission (SMM) made 24 hard x-ray measurements of Cygnus X-1 from 1986 December 22 through 1987 April 3, and six additional observations from 1988 January 28 through March 8[11]. These results show that the average 45-140 keV flux in 1986-1987 was between the γ_2 and γ_3 level observed by HEAO 3[14]. This level is similar to those observed in 1977 by OSO 8[31] and HEAO 1[24]. In 1988, however, the average flux decreased to the normal γ_2 level. There was an outburst on 1987 February 24 when the 45-140 keV flux increased to a level above the γ_3 level. There was no evidence that the source was in the γ_1 level during these periods.

The High Energy X-Ray Experiment (HEXE) on the Soviet Mir-Kvant module made several hard x-ray observations (20-200 keV) of Cygnus X-1 from 1987 July through 1988 September[10]. During the period between 1987 July and 1988 April, the flux remained fairly constant near the γ_2 level, consistent with the level observed by SMM/HXRBS in 1988[11]. However, between 1988 April 28 and May 8, a few days after the last SMM/HXRBS observation in 1988, the flux seemed to triple to a level significantly higher than γ_3, followed by a sharp decrease. The former was the highest flux level observed to date. These results, combined with those from HEAO 3, indicate the wide dynamic range of the hard x-ray flux variability of the source. The low flux

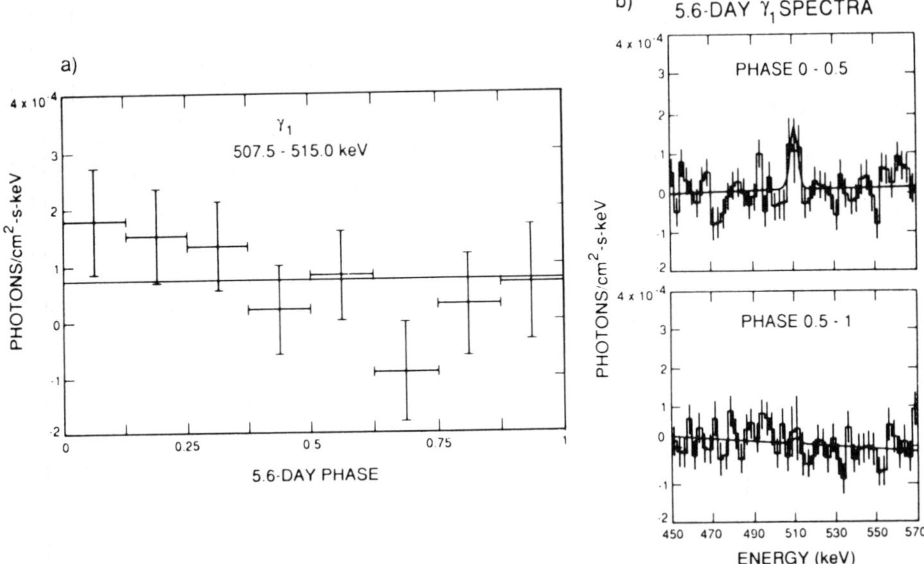

Figure 7: (a) Results of folding the γ_1 511 keV flux modulo the 5.6-day binary period[6] (b) the 511 keV flux observed in the γ_1 spectrum seems to be associated more with one side of the orbit than the other.

level observed after May may be questionable due to the failure of the TTM startracker. Whether the source was in the γ_1 level during this period has yet to be determined.

Finally, there were two balloon measurements reported by the European MIFRASO[12] and the POKER[13] investigators. The MIFRASO observation was made on 1987 July 15 and showed that Cygnus X-1 was in the "super-low" (γ_1) state. The POKER observation, made on 1985 August 5, showed that the 45-140 keV flux was between the γ_1 and γ_2 levels. The observation was made approximately two months earlier than another balloon observation reported by a Chinese team[32] (see Ling[1] 22-year light curve) which showed that the source was in the γ_1 level also. These results, if real, are certainly interesting because (1) the two observing periods in 1985 and 1987, respectively, for the γ_1 occurrences are roughly separated by two years, similar to the period of separation between the two possible repeating patterns seen in 1977 and 1979[1], and (2) the 1985 POKER and Chinese measurements may reflect the downward leg of a transition of the hard x-ray flux from the γ_2 to the γ_1 level with a time scale of the order of months.

SUMMARY

This paper provides a summary of the most recent work on gamma-ray emission from Cygnus X-1 which I hope will serve as a useful reference source for comparison in the future with data collected on other black-hole objects. The MeV emission and its possible correlated narrow annihilation feature observed in the spectrum of both Cygnus X-1 and the Galactic Center, if confirmed, could be a unique gamma-ray signature for identifying black holes. There are many questions related to such possible correlation seen in Cygnus X-1. For example, (1) What is the mechanism responsible for the formation of the γ-ray emitting environment? (2) Is there any regularity to the pattern, e.g. periodic behavior, associated with its long-term temporal characteristic? Is there a repetition of the pattern with a two-year interval? (3) What do we know about the annihilation site? Is there a 5.6-day modulation of the 511 keV line-flux? (4) Is there a positronium component associated with the annihilation line? I encourage future high energy astrophysics research to consider follow-up investigations of Cygnus X-1 and other black holes (e.g., Galactic Center and AGN) the top priority objective in this decade.

I want to thank C. Dermer, W. Mahoney and Wm. Wheaton for their valuable comments on this manuscript. The research described in this paper was carried out by the Jet Propulsion Laboratory, California Institute of Technology, under contract with the National Aeronautics and Space Administration.

REFERENCES

1. Ling, Nucl. Spec. of Astroph. Sources, ed. N.Gehrels and G.Share, AIP #170, 315, (1988).
2. Bassani et al., Ap. J., 343, 313, (1989).
3. McConnell et al., Ap. J., 343, 317, (1989)
4. Lingenfelter and Ramaty, Ap. J., 343, 686, (1989).
5. Ling and Wheaton, Ap. J. (Lett.) 343, L57, (1989).
6. Ling and Wheaton, 21st ICRC Proceedings, OG 3.3-3, 1, 189, (1990).
7. Ling et al., 21st. ICRC Proceedings, OG 3.3-5, 1, 197, (1990)
8. Liang, Astron. Ap., 227, 447, (1990).
9. Dermer, 14th Texas Sym. on Relativistic Astrophysics, ed. E. Fenyves, 513, (1989).
10. Dobereiner et al., ESA SP-296, 387, (1989)
11. Schwartz et al., to be published in Ap. J. (1991)
12. Ubertini et al., Ap. J., 366, 544, (1991a)
13. Ubertini et al., to be published in Ap. J. (1991b).
14. Ling et al., Ap. J.(Lett.), 321, L117, (1987).
15. Riegler et al., Ap. J. (Lett.), 248, L13, (1981).
16. Riegler et al., Ap. J. (Lett.), 294, L13, (1985).
17. Gehrels, these Proceedings, (1991).
18. Matteson, these Proceedings, (1991).
19. Paul, these Proceedings, (1991).
20. Sunyaev, these Proceedings, (1991).
21. Cook et al., 21st ICRC Proceedings, OG 3.3-12, 1, 216, (1990)
22. Perotti et al., Ap. J., 300, 297, (1986).
23. Baker et al, Nature, 245, 18, (1973)
24. Nolan, UCSD Ph. D. thesis, (1982)
25. Nolan and Matteson, Ap. J., 265, 389, (1983)
26. White et al., Nature, 284, 608, (1980).
27. Liang and Dermer, Ap. J. (Lett.) 325, L39. (1988).
28. Dermer and Liang, Nucl. Spec. of Astroph.Sources, ed. N.Gehrels and G.Share, AIP #170, (1988).
29. Ling et al., 19th ICRC Proceedings, La Jolla, (1985).
30. Holt et al, Ap. J. (Lett.), 203, L63, (1976).
31. Dolan et al., Ap. J., 230, 551, (1979).
32. Ma et al., preprint, (1986).

ELECTRON, PROTON BEAM INTERACTIONS WITH MATTER AND/OR RADIATION: APPLICATION TO CYG X-1, GEMINGA AND 3C 273

F. GIOVANNELLI

Istituto di Astrofisica Spaziale, CNR
C.P. 67, I 00044 Frascati, Italy

L. SABAU GRAZIATI

Laboratorio de Astrofísica Espacial y Física Fundamental
Instituto Nacional de Técnica Aeroespacial
E 28850 Torrejon de Ardoz, Madrid, Spain

W. BEDNAREK°^, S. KARAKULA°, W. TKACZYK°

° Institute of Physics, University of Lodz
Ul. Nowotki 149/153, PL 90-236 Lodz, Poland

^ International School for Advanced Studies
Strada Costiera 11, I 31014 Trieste, Italy

ABSTRACT

In this paper we present the use of the theoretical spectra derived in the most recent studies on the electron-proton beam interactions with the matter and/or radiation to fit the experimental data of Cyg X-1, Geminga, and 3C 273.

1. Origin of the High Energy Photons from Cyg X-1, Geminga and 3C 273

We want to show how the calculated photon energy spectra (Bednarek et al.(1); Bednarek and Calvani (2)) can be succesfully used in fitting the experimental data of three of the most interesting high energy γ-ray sources: Cyg X-1, 2CG 195+04 (Geminga) and 3C 273. A review on this topic can be found in the paper by Giovannelli et al.(3) and in the references therein.

1.1. CYG X-1

Bednarek et al.(4) suggest that the photon spectrum of Cyg X-1, in the MeV region, is non-thermal.The experimental high energy spectrum of Cyg X-1 above 1 MeV can be fitted very well with a comptonized black body spectrum ($T = 3{\times}10^{4}$ °K), coming from the supergiant companion HDE 226868, by relativistic electrons with Lorentz factor 600,emitted in our direction.

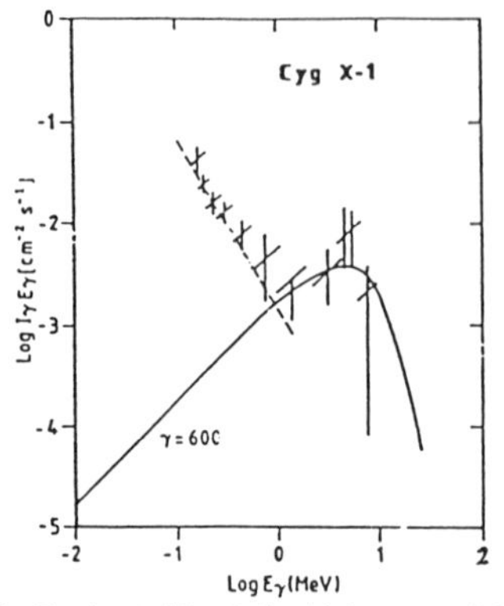

Fif. 1. The best fit of the high energy data of Cyg X-1.

Figure 1 shows the best fit to the experimental data of this calculated spectrum at energies above 1 MeV and with the derived spectrum from the single temperature inverse Compton model at energies below 1 MeV. From the observed intensity of the photons above 1 MeV, the lower limit on the emissivity of the relativistic electrons can be estimated. It is equal to 1.5×10^{38} electron/s, corresponding to an electron luminosity of 7.0×10^{34} erg/s.

1.2. 2CG 195+04 (GEMINGA)

Bednarek et al.(4) proposed that the high energy photons coming from Geminga originate in the interaction of beamed relativistic particles with the surrounding matter. Such a mechanism of high energy photon production can be realized in close binary systems, where beamed relativistic particles emitted by the collapsed object (e.g. a neutron star) interact with the matter coming from the optical companion.

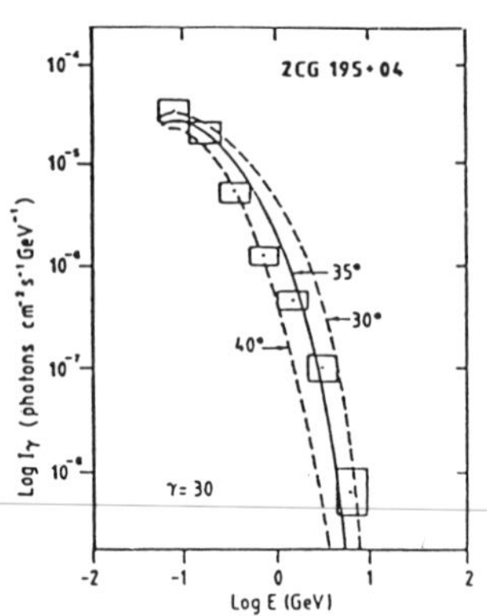

Fig.2. The best fit of the COS B data of Geminga.

Figure 2 shows the theoretical photon production spectra with the spectral measurements of Geminga from COS B satellite. One can see that the photon spectrum calculated for an angle $\alpha_1 = 35^\circ$ and the proton beam Lorentz factor $\gamma \geq 30$ fits very well the COS B data. In order to show how the fit is sensitive to the angle α_1, the theoretical curves for 30° and 40° are reported too in Fig.2, for the same value of Lorentz's factor. From the best fit, one obtains the emissivity of the proton beam equal to:

$$N_p = 7.5 \times 10^{-4} \times d^2/X_p \text{ particle/s}$$

where X_p is the column density (in g/cm^2) of the particles crossed by the proton-beam and d the distance to the source.This corresponds to a proton luminosity $L_p = m_p\gamma N_p$, where m_p is the proton rest energy.

1.3. 3C 273

Bednarek and Calvani (5) suggested the possibility that γ-rays from 3C 273 are produced in the jet. In particular they developed a non-isotropic model in which the γ-rays are produced by secondary electrons and positrons coming from $\pi^{\pm}$ and π° decay. They used their calculated spectrum of the bremsstrahlung radiation from secondary electrons and from π° decay (Bednarek et al.(1)) to fit the experimental data of 3C 273. They adopted $\gamma = 6$ and $\alpha_1 = 35^{\circ}$. The fit is shown in Figure 3. From this fit, it is possible to evaluate the total energy of the relativistic protons in the jet by using the relation:

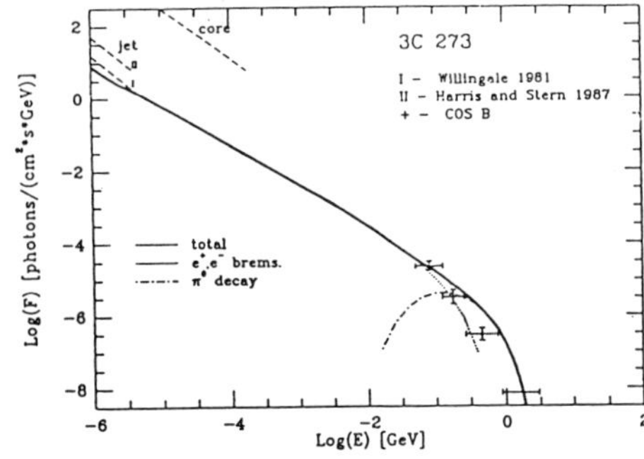

Fig. 3. The best fit of the experimental data of 3C 273

$$E \approx 10^{18}\gamma d^2 l/\lambda \quad \text{erg}$$

where d is the distance to the quasar, $l = d\sin\beta/\sin\alpha_1$ is the true length of the jet, β is the angular apparent length of the jet, λ is the column density in the jet (in g/cm^2).

For $\lambda \approx 70$ g/cm^2 (from the best fit), d = 860 Mpc and $\beta \approx 20''$, this energy is $E \approx 2 \times 10^{61}$ erg. The required power emitted in relativistic protons by the core of 3C 273 is $P = Ec/l \approx 1.5 \times 10^{48}$ erg/s, which corresponds to a mass loss in relativistic protons of ≈ 1.1 $M_{\odot}/yr$.

REFERENCES

1 Bednarek, W., Giovannelli, F., Karakula, S., and Tkaczyk, W. (1990a) Astron. Astrophys 236, 268.
2 Bednarek, W., Calvani, M., in preparation (1991).
3 Giovannelli, F., Sabau Graziati, L., Bednarek, W., Karakula,S., and Tkaczyk, W. (1991) in Cosmic Rays, Supernovae, and the Interstellar Medium, (eds. M.M. Shapiro, R. Silberberg, J.P. Wefel), Kluwer Academic Publisher, (in press).
4 Bednarek, W., Giovannelli, F., Karakula, S., and Tkaczyk, W. (1990b) Astron. Astrophys 236, 175.
5 Bednarek, W. and Calvani, M. (1991) in Frontier Objects in Astrophysics and Particle Physics II, (eds. F. Giovannelli and G. Mannocchi), Italian Physical Society, Bologna, Italy (in press)

SOLAR GAMMA-RAY LINES

GAMMA-RAY LINES FROM SOLAR FLARES

E. Rieger
Max-Planck-Institut für extraterrestrische Physik
8046 Garching, FRG

ABSTRACT

The observation of gamma-ray lines during solar flares is the most convincing evidence that nuclear reactions take place between high energy particles and the solar atmosphere. The application of data from the measurements of these lines is very wide spread. It ranges from information about the acceleration of charged particles to elemental composition determination of the various solar atmospheric layers. After a brief review of the production of the nuclear lines it is shown, how their measurements can be applied in different areas of the solar flare phenomenon.

INTRODUCTION

When measurements of high energy particles after solar flares were accurate enough to deduce spectral parameters, it was recognized that gamma-ray continuum and, most importantly, gamma-ray lines should be observed from balloon and satellite altitudes, if a sizeable fraction of these particles interacted in the solar atmosphere [1,2,3]. With the first unambignous discovery of nuclear lines from two large flares in Auguts 1972 [4,5], these expectations were verified. Especially a line at 2.223 MeV originating from the capture of free neutrons on hydrogen, determined to be the strongest line, was recorded with high statistical significance. Following this detection with the OSO7 gamma-ray spectrometer, the Sun was not continuously monitored, so that until 1979 only three more flares with gamma-ray line emission could be observed [6,7,8]. The number of gamma-ray line flares increased considerably, when after 1980 gamma-ray spectrometers on the dedicated solar missions SMM and Hinotori were brought into orbit. It was thus demonstrated that nuclear reactions of flare accelerated particles take place commonly during solar flares.

After a brief discussion of the production of gamma-ray lines we describe the implications of gamma-ray line measurements on different aspects of the solar flare phenomenon. The problem of how the charged particles are accelerated is not within the view of this paper. The reader is referred to [9,10,11,12,13,14].

PRODUCTION OF GAMMA-RAY LINES AND THE SOLAR FLARE GAMMA-RAY SPECTRUM

Gamma-ray lines result from the interaction of high energy protons or ions with nuclei of the solar atmosphere. Inelastic scattering of energetic particles and to a lesser degree spallation reactions of certain elements lead to excitation of nuclei to higher levels. Because the life times of the excited states are 10^{-12} sec or shorter, the lines are emitted without a measurable delay and are therefore called prompt lines. The cross-sections for the various nuclear de-excitation lines have been extensively discussed [15,16,17,18]. As the cross-section for excitation by protons peak around 10 MeV for the most abundant elements in the solar atmosphere, the de-excitation lines are produced by protons with relatively low energy of order 10-30 MeV. Shown in Figure 1 is the prompt nuclear de-excitation gamma-ray line spectrum [15] modified to the thick target case [15,19].

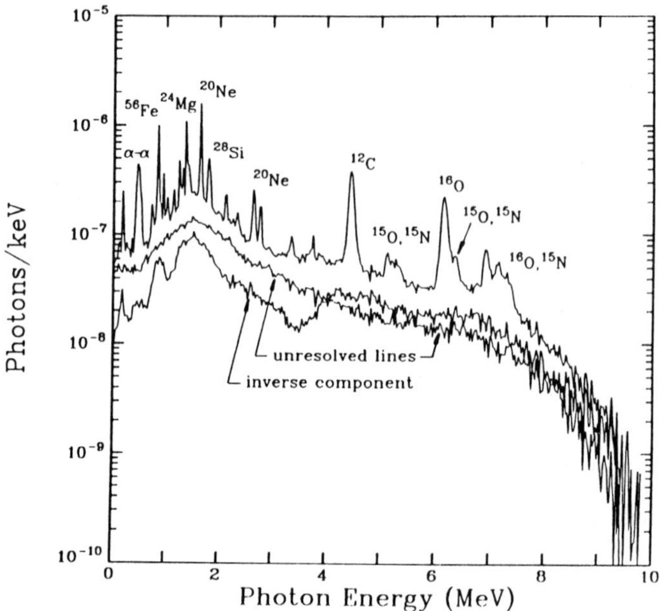

Fig. 1. Theoretical solar flare nuclear de-excitation gamma-ray spectrum (from [19]).

In a thick target model the energetic particles loose their whole energy in the ambient medium. The spectrum is shown to a resolution corresponding to a Ge-solid state detector. Not shown in this Figure are the delayed lines at 2.223 MeV from neutron capture on protons and at 0.511 MeV from positron annihilation. For flares not too close to the solar limb the 2.223 MEV line is by far the strongest among all the lines of the gamma-ray spectrum. The next strongest lines at 4.438 MeV and 6.129 MeV result from the de-excitation of ^{12}C and ^{16}O, respectively. The complex around 0.45 MeV (α - α) is a blend of the prompt lines of ^{7}Be at 0.429 MeV and ^{7}Li at 0.478 MeV. These elements are formed in flares by nonthermal fusion reactions of α-particles with ambient ^{4}He [20].

The lines mentioned are superimposed on a bremsstrahlungs spectrum from high energy electrons. It was pointed out [21,22] that at energies below 1 MeV electron bremsstrahlung and above 1 MeV radiation from nuclear lines should dominate the spectrum. These theoretical considerations have been confirmed by flare measurements in the recent past [9,13,23,24,25]. In order to illustrate this important fact, the energy loss spectrum recorded by the Gamma-Ray Spectrometer (GRS) on SMM of the end phase of the 6 March 1989 flare is shown in Figure 2.

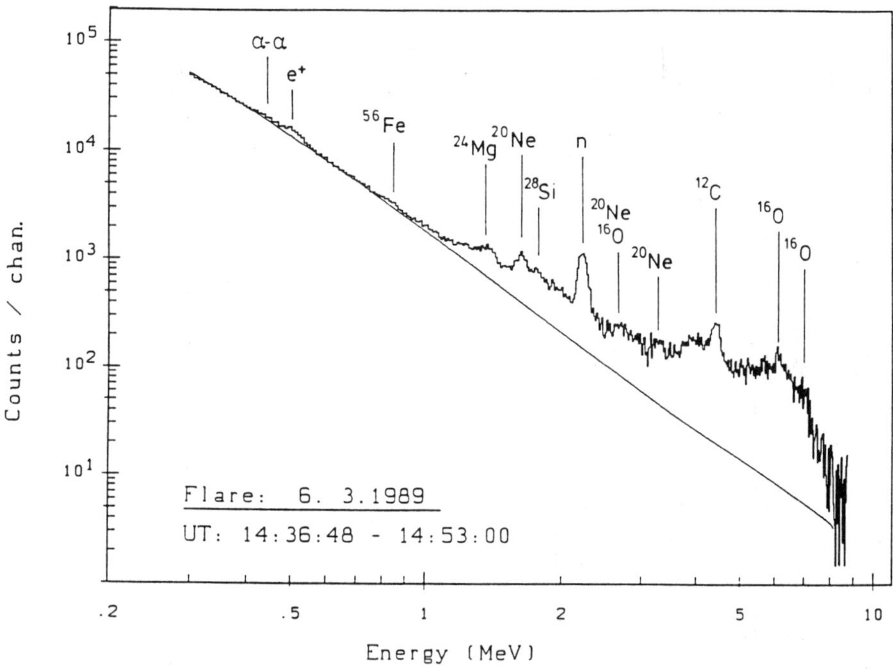

Fig. 2. Time integrated and background subtracted count spectrum of the 6 March 1989 flare (end phase). Note the drop of the signal above 7.2 MeV. Full line: best fit power law (< 1 MeV) in photon space transformed to count space (from [52]).

The contribution due to electron bremsstrahlung, resembling in photon space a power law in spectral shape, is indicated by a full line. The excess above this line is interpreted as the result of nucleonic interactions. This emission consists of three components [26,27]: (1) narrow lines from the interaction of accelerated protons and α - particles with heavy atmospheric nuclei, (2) broad lines from the interaction of energetic heavy nuclei with atmospheric hydrogen and helium, (3) a quasicontinuum from the blending of lines of high level transitions (see Figure 1). The broad line component is estimated to contribute about 25 % of the flux of the narrow line component [28], the quasicontinuum should be smaller than the broad line continuum above ≈ 4 MeV [15,19]. Despite the much lower spectral resolution of the GRS (it uses NaI (Tl) as detector material) than of a Ge-detector several lines are clearly visible. It is of importance to note that above 7.2 MeV the emission shows a cutoff. This is due to the fact, that above this limit intense nuclear lines are absent [16]. The excess above a power law extrapolation is especially prominent in the energy interval from 4-7 MeV, which contains the strong nuclear lines of carbon and oxygen. It is therfore often taken as a measure of the nucleonic interactions [23,29]. It must be pointed out, however, that the extrapolation of the power law, determined at energies below 1 MeV, to higher energies is the simplest approach to separate the nucleonic from the electronic component. If the bremsstrahlung should flatten above 1 MeV, the ionic contribution would be systematically overestimated.

The production of neutrons by the interaction of accelerated ions with the atmosphere has been studied extensively [26]. The most prolific neutron producing reaction is that of protons on ^{4}He in a wide spectral range of the energetic particles with a threshold of about 30 MeV. These neutrons can have different fates:

- they escape from the Sun (≈ 60 - 70 %) [10] and those reaching the earth before decaying can be detected with suitable detectors in space [30,31], and on the ground [32,33,34,35,36,37]. Neutrons leaving the Sun which do not survive can be identified through their energetic decay protons, which are then trapped by the interplanetary magnetic field [38,39,40,41].

- neutrons which remain at the Sun can be captured by nuclei before their decay. ^{3}He which is present in the solar atmosphere in low concentration is an effective neutron sink. The reaction $^{3}He(n,p)\,^{3}H$ proceeds without the emission of radiation, whereas the reaction $^{1}H(n,\gamma)^{2}H$ produces the 2.223 MeV line. Because the capture takes place in the photosphere this line is strongly attenuated for flares at the solar limb [42,43]. For flares off the limb it is by far the strongest line (see Figure 2). Oppositely to the nuclear de-excitation lines, which are emitted promptly, the 2.223 MeV neutron capture line is emitted with a delay, which originates from the finite capture time of the neutrons on protons (the probability for elastic scattering is much larger than the capture probability). This delay is evident in Figure 3, which shows the dynamic spectrum of the 24 April 1984 flare in the energy range from 0.3 to 9 MeV. The decay of the line can be clearly seen for about 20 min. The capture time of the neutrons on ^{3}He and hydrogen depend on the density of the respective elements. A study of the time history of the line can therefore yield information on the abundance of ^{3}He and the depth of the atmosphere where these reactions take place. The analysis of the decay of the line of the 3 June, 1982 flare shows that the capture occurs mainly in the photosphere at hydrogen densities of $\approx 1.3 \cdot 10^{17}$ cm^{-3} [44,45]. The $^{3}He/H$ ratio derived from the 2.223 MeV time history of this flare is $(2.3 \pm 1.2) \approx 10^{-5}$ [46].

Interactions of high energy ions with the elements of the solar atmposhere can create nuclei which decay by emitting positrons (β^{+}). The most prolific β^{+}-emitters are ^{11}C, ^{14}O and ^{15}O [26,47]. Another production mode of positrons is the decay of positively charged pions created by the interaction of very high energy ions with matter. After slowing down, the positrons can annihilate with ambient electrons to produce two 0.511 MeV photons per annihilation, or they form a positronium atom [10,48]. This atom is similar to the hydrogen atom except that the proton is replaced by a positron. 25 % of the positronium consists in the singulett state which produces two 0.511 MeV photons per annihilation. The other 75 % are in the triplett state which annihilate into three photons with energies less than 0.511 MeV. This decay into three photons forms a continuum shortward of the 0.511 MeV line. The energies of positrons from β^{+}- and pion-decay are ≈ 1 MeV and 10-100 MeV, respectively. The slowing down and annihilation time are inversely proportional to the ambient density. It is short compared to the decay times of the radioactive nuclei, so that the delay seen in Figure 3 reflects the half lifes of the β^{+}-emitters.

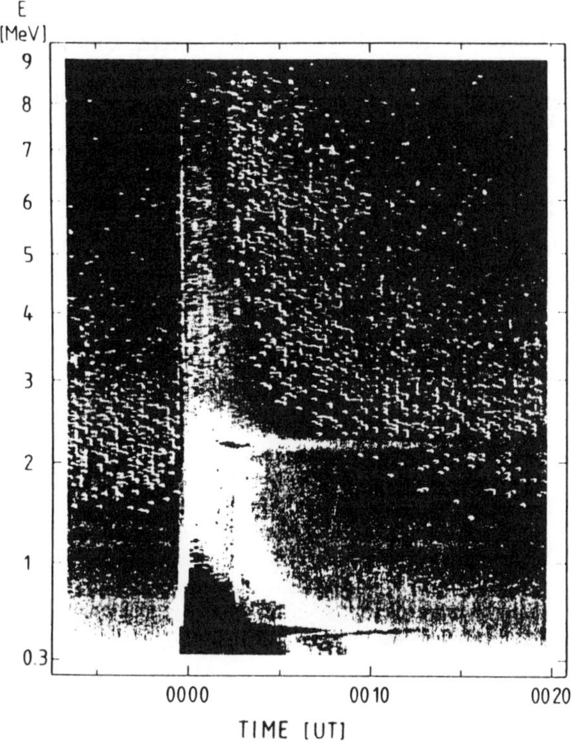

Fig. 3. Spectrum of the 24 April 1984 flare, recorded by the GRS. Note the "afterglow" of the delayed lines at 0.511 and 2.223 MeV [13].

It was shown [29,47] that the decay of the 0.511 MeV line of the 21 June 1980 and 3 June 1982 flares [49] is consistent with these considerations. Another interest of observation is the width of the 0.511 MeV line which is determined by the temperature of the environment where the positrons annihilate. This effect was studied by Bussard et al. [50] who determined a width of 3.5 keV and 11 keV at 10^5 and 10^6 K, respectively. It is clear that these small line widths cannot be resolved by the SMM- and Hinotori gamma-ray spectrometers.

TIME HISTORY OF LINE EMISSION AS A CONSTRAINT FOR ION ACCELERATION AND THE INTERACTION REGION

The study of the flare time history in different energy bands provides a possibility to investigate the acceleration of charged particles. It must be born in mind, however, that the emission versus time is a complicated superposition of acceleration-, transport-, and energy loss processes and that the last two have a tendency to smooth out quick temporal changes. To approach the problem of particle acceleration the times of peak emission in different energy intervals are compared. It is customary to take the emission at medium energy X-rays (30-100 keV) originating from bremsstrahlung of subrelativistic electrons as a reference time for the impulsive phase [51]. Since the launch of the SMM- and Hinotori spacecrafts the nuclear energy range (1-7 MeV) is accessible with high time resolution, which opens up the possibility to study the acceleration of ions. Unfortunately, the strongest line at 2.223 MeV cannot be used for this purpose because of its delayed character. The next strongest lines are the prompt de-excitation lines of ^{12}C (4.439 MeV) and ^{16}O (6.129 MeV).

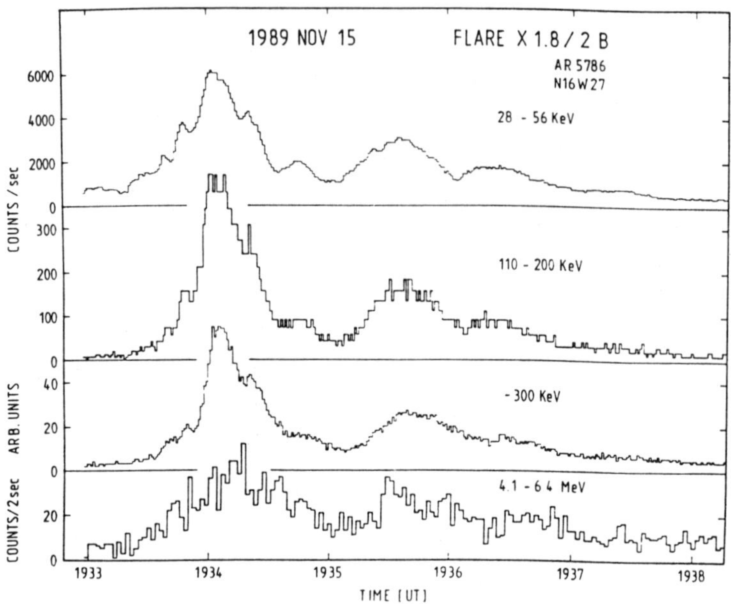

Fig. 4. Time history of the 15 November 1989 flare in different energy bands.

As an example, the time history of the last big event recorded by the GRS is shown in Figure 4 at medium and high energy X-rays and at 4.1 - 6.4 MeV (containing the ^{12}C and ^{16}O lines). At all energies two major bursts are seen. The burst at 1934 UT shows a peak delay of about 15 sec in the nuclear energy band with respect to the X-rays, whereas for the second burst the delay seems to be reversed. It is also of importance to note that the raise to the first peak occurs simultaneously in all energy bands implying that the acceleration of ions is a rather prompt process. Observational facts like these are important ingredients in the discussion of particle acceleration, their storage in flux tubes, transport and energy loss. These topics are not treated in this review. The reader is referred to other papers [11,13,14,53,54,55,56,57,58,59].

Contrary to X-ray energies imaging in the gamma-ray regime above 1 MeV, where most nuclear lines can be observed, is not possible. Information about the location of the interaction region can only be obtained by indirect methods, especially by the determination of the ambient density, where the lines are produced. As the atmospheric density depends strongly on altitude, this method is a sensitive tracer to locate the interaction region. A lower limit of $\approx 10^{12}$ cm^{-3} is obtained by the measurement of the slowing down and annihilation time of the positrons [60], determined from the observed time variation of the 0.511 MeV line of the 21 June 1980 flare [49]. Similar constraints are set by the near simultaneous peaking (≈ 2 sec) at X-ray and gamma-ray energies observed in the 7 June 1980, 21 June 1980, and 12 May 1983 events [13,61], if one takes into account that the ≈ 10 MeV protons loose their energy within ≈ 2 sec [62]. In a recent investigation [56] the maximum of the nuclear line generation is placed to densities of ≈ 10^{15} cm^{-3} by modelling the decay phases of certain very impulsive flares. This puts the interaction region to the lower chromosphere. Upper limits of ≈ $5 \cdot 10^{15}$ cm^{-3} for the density (the upper photosphere) can be set by the observation that from a limb event (27 April, 1981) appreciable attenuation of the 0.847 MeV ^{56}Fe line compared to lines at higher energies was not recorded [63]. It must be noted, however, that this event belonged to the more gradual type of flares [25] which could have their interaction sites at lower ambient densities. An upper limit of ≈ $5 \cdot 10^{15}$ cm^{-3} is also set by the observation of neutrons escaping from the Sun after the impulsive limb flare of 21 June, 1980 [30,64].

NUCLEAR GAMMA-RAY LINE RATIOS AS A DIAGNOSTIC OF SPECTRA OF ACCELERATED PARTICLES

If the cross sections of two lines have widely different thresholds in energy, then their flux or fluence ratios can give information about the spectrum of the high energy particles. As already mentioned, the cross sections for the de-excitation lines of ^{12}C (4.439 MeV) and ^{16}O (6.129 MeV) have a similar energy dependence and peak around 10 MeV if excited by protons. The flux ratio is therefore only weakly dependent on the energy spectrum of the particles [10]. On the other hand, the most prolific neutron production reaction of protons on ^{4}He has a threshold of ≈ 30 MeV.

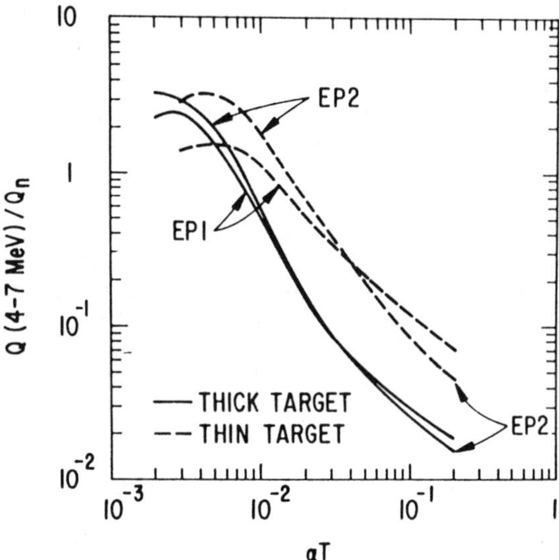

Fig. 5. Ratio of the photon yield in the energy range 4-7 MeV to the neutron yield assuming that the energetic particles have a Bessel function spectral shape. EP1 and EP2 designate different compositions of the energetic particle spectrum (from [10]).

The ratio of the neutron yield and the carbon and oxygen lines will therefore depend on the energetic particle spectrum [10]. This is shown in Figure 5 where a Bessel function is assumed for the spectrum of the energetic particles [65]. To obtain the 2.223 MeV line fluence the neutron yield has to be multiplied by the neutron-to-2.223 MeV photon conversion factor which depends upon the spectrum and the angular distribution of the ions and the position of the flare at the Sun [29,43]. The ratio of the 2.223 MeV and 4-7 MeV fluence is a measure of the steepness of the particle spectrum predominantly in the energy range 10-100 MeV. Up to now the fluence ratios for 12 gamma-ray line flares have been published (see [43]). Assuming isotropic distribution of the particles the parameter αT for a Bessel function spectral shape is between 0.018 and 0.034 [13,43].

Although, the energy dependence of the cross sections of the most important de-excitation lines are remarkably similar, the ^{12}C-nucleus itself presents an exception with the ground state transitions from the isospin level zero (4.438 MeV) and one (15.11 MeV). The thresholds of the cross sections are widely spaced at 5 MeV and 16 MeV, respectively. This case has been investigated by Crannell et al. [16]. It was assumed that the ratio for production of the two lines from oxygen spallation is the same as that for production from carbon. This simplification bears the additional advantage that uncertainties of the abundances of ^{12}C and ^{16}O do not influence the flux ratio of both lines. A later examination [66] using new gamma-ray line production cross sections, however, showed that this approximation is too crued. The spallation reactions of ^{16}O contribute relatively more to the excited level at 4.438 MeV than to the 15.11 MeV level. This is caused by energetic α-particles whose interactions with ^{16}O contribute only to the 4.438 MeV line. Saltzberg et al. [67] have estimated the thick target line fluence ratio as a function of the spectral parameter of the energetic particles by including the interaction of both protons and α-particles. It is shown that the 15.11 MeV to 4.438 MeV line fluence ratio is between $\approx$ 1 % and 2 % for a soft and hard particle spectrum, respectively. Thus, the 15.11 MeV line of carbon should be strong enough to be detected by large area spectrometers comparable to those on the Gamma-Ray Observatory, and the ^{12}C line pair would serve as an additional diagnostic of accelerated particle spectra.

NUCLEAR LINE FLUENCES AS A MEASURE OF THE NUMBER OF INTERACTING PARTICLES

In the previous chapter it was shown that the ratios of the fluences of certain nuclear lines can be used to get information on the spectrum of the accelerated particles. From the absolute values of the nuclear line fluences the number of the interacting particles can be determined. Murphy and Ramaty [29] have calculated the yields of the ^{12}C and ^{16}O de-excitation lines (contained in the 4-7 MeV band), neutrons, pions and positrons as a function of the primary proton spectrum, assuming that the interacting energetic protons have an isotropic distribution and that they loose all their energy in the ambient medium (thick target case). The yields are normalized to one proton with an energy > 30 MeV. The results are shown in Figure 6. The calculations have been performed for a hard (power law) and a soft (Bessel function) particle spectrum. It is evident that the 4-7 MeV yield has a tendency to decrease as the spectrum gets harder, whereas the neutron yield increases. This diverse behaviour determines the slope of the graph of Figure 5. To calculate the number of accelerated particles from the line fluences, the spectral parameter must be known. As explained in the previous chapter, this can be obtained by determining the fluence ratio of the 2.223 MeV neutron capture line and the Carbon and Oxygen de-excitation lines (contained in the 4-7 MeV band). If, on the other hand, the neutron capture line cannot be observed, because the flare is too close to the limb, the 4-7 MeV fluence alone is also a useful measure for the number of the interacting particles, because the 4-7 MeV yield is relatively independent on the spectral parameter. For most flares the number of interacting particles is calculated by using the 4-7 MeV fluence [11]. There are flares, for which also the number of the escaping particles have been measured. In most cases the number of the interacting protons is much higher than the number of the escaping ones [11].

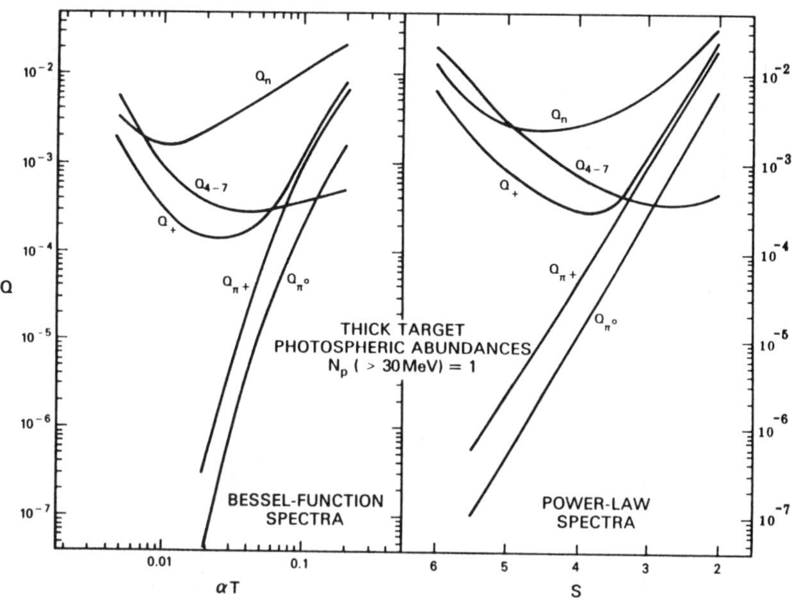

Fig. 6. Neutron, 4-7 MeV nuclear gamma-ray, positron π^+ and π^0 production versus the spectral parameter of a Bessel function and power law for the energetic protons (from [29]).

NUCLEAR LINE SPECTROSCOPY

As the Sun is one of the primary sources for information about elemental abundances of the cosmos, it was one of the principal goals of the Solar Maximum Mission to extend this information into the gamma-ray regime with the Gamma-Ray Spectrometer (GRS). The paper by Ramatey et al. [15] laid the theoretical foundations for the computation of elemental abundances from the intensities of gamma-ray lines. As already mentioned, the intensities of the de-excitation lines are determined by inelastic scattering of the energetic particles and by spallation reactions of heavier elements. Because the excitation cross sections of the most important elements have similar energy dependences, the line ratios due to scattering reactions are only insignificantly influenced by the energetic particle spectrum. In this case the relative line intensities depend directly on the composition of the ambient medium. The fractional contribution of spallation reactions to a given line, however, varies from element to element, being zero for Fe and more than 30 % for the 4.438 MeV line of ^{12}C [17]. Moreover, since the cross sections for spallation reactions typically have higher thresholds, the spallation fraction generally increases as the particle spectrum hardens. Thus, the line intensity ratio gets energy dependent to a certain extent.

A flare which was investigated in great detail is the line rich event of 27 April 1981. The count spectrum obtained by the GRS on SMM, integrated from 0804-0836 UT and subracted for background is shown in Figure 7 [25].

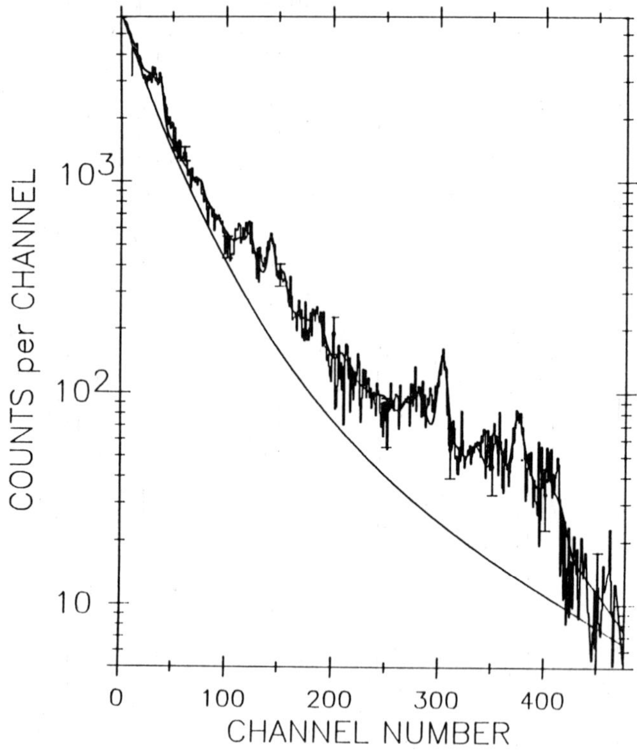

Fig. 7. Time integrated and background subtracted count spectrum of the 27 April 1981 flare, fitted by a nuclear line model. The neutron capture line at 2.223 MeV (channel 183) is attenuated because the flare occurred at the solar limb (from [25]).

Because the flare occurred at the solar limb, the 2.223 MeV neutron capture line is strongly attenuated in this event. The spectral parameter of the energetic particles therefore cannot reliably be determined. To calculate the abundance ratios a spectral parameter of $\alpha T = 0.025$ was assumed, which is an average value of parameters determined from not too limb near gamma-ray line events. Most prominent are the lines of ^{12}C (4.438 MeV) and ^{16}O (6.129 MeV). Other nuclear de-excitation lines, the 0.511 MeV annihilation line and possibly the α-α fusion line complex at $\approx$ 0.45 MeV can be figured out in the spectrum. The thick full line represents the best fit power law obtained between 0.3 and 1 MeV, which is due to electron bremsstrahlung. As already shown in Figure 2, line radiation dominates the emission above 1 MeV also in this event. To determine elemental abundances from the count spectrum shown in Figure 7, a solar flare photon spectrum (according to [15]) was constructed by calculating the de-excitation spectra resulting from thick target nuclear interactions of the 12 most abundant elements of the solar atmosphere [63,68]. To these, an electron bremsstrahlung spectrum taken as an unbroken power law with adjustable spectral index, a narrow 2.223 MeV neutron capture line and a positron annihilation spectrum, consisting of a line and a positron continuum, was included. This synthetical photon spectrum is then folded through the response of the GRS. The elemental abundances are obtained by fitting the computed to the measured count spectrum. To get the best fit to the observations, the power law index and 15 intensities are varied, until X^2 is minimal. The fit is shown in Figure 7 by a thin full line through the data. The elemental abundances determined in this way are then compared with the photospheric- (local galactic) and coronal abundances [69]. It is found that the abundances obtained from the gamma-ray measurements are in better agreement with coronal than with photospheric abundances. This is especially obvious for ^{12}C and ^{16}O which are by a factor 3 to 4 suppressed compared to their abundance in the photosphere. In this investigation the influence of unresolved lines and of the inverse component (heavy energetic nuclei on ambient H and He) on the shape of the solar flare gamma-ray spectum have been neglected. Only the influence of electron bremsstrahlung in form of an unbroken power law had been taken into account. It was therefore studied in a recent paper [19] to what extent the intensity of the narrow lines is influenced by the addition of the above mentioned components to the electron continuum. It is shown in this investigation that the intensities of strong lines can have model dependent differences of up to a factor of 2. Finally, the constraint that the accelerated particle composition be fixed, was relaxed [70]. It was found that the best fit to the data is obtained when both the ambient and the accelerated particle abundances are allowed to vary. In this case Magnesium, Scilicon, Neon and Iron referred to Carbon are enhanced compared to photospheric abundances.Compared to coronal abundances only the Neon to Carbon ratio is enhanced. If the ambient abundances were kept as photospheric or coronal, the derived beam composition turned out to be different to that found in large proton flares. It rather resembles that found in ^{3}He-rich events [71]. The analysis of the 27 April 1981 flare allowed for the first time to determine solar elemental abundances from gamma-ray measurements. More flares showing line emission with good statistical significance that occurred especially during the very active solar year 1989 are in the GRS data base. Analysis of these events will show if there are differences in the elemental abundances from flare to flare.

LINE-POSITIONS AND- SHAPES AS DIAGNOSTICS OF ACCELERATED PARTICLE ANISOTROPY

That flares, recorded by the GRS and showing emission above 10 MeV, are concentrated towards the solar limb, gave evidence for the first time that this high energetic radiation is emitted anisotropically [13,27,72,73]. As this radiation is produced predominantly by high energy electrons via bremsstrahlung, which is highly beamed in the direction of motion of the radiating particles, the observation implies that also the electrons move anisotropically. On the other hand, evidence for a corresponding anisotropy of motion of accelerated ions is less obvious. It was shown that the time history of the flux of neutrons emitted by the 21 June 1980 and 3 June 1982 flares, measured by the GRS [30,31] is consistent with calculations where the neutron producing particles had an isotropic or fan beam (parallel to the Suns' surface) distribution [74].

There are, however, also gamma-ray spectroscopic tests involving the position and the shapes of the lines which can give information about particle anisotropies. Forrest and Murphy [25] determined position and width of the three most intense lines (^{20}Ne, ^{12}C and

^{16}O) of the 27 April 1981 flare. It was found that the line centroids were within the accuracy of the measurements coincident with their nominal values. Ramaty and Crannell [75] calculated the Doppler shifts for the ^{16}O line at 6.129 MeV for an E^{-2} energetic particle spectrum and showed that it will produce Doppler shifts of 25 and 45 keV for proton or α-particle initiated reactions, respectively, when viewed along the particles' motion (assumed to be isotropic over 2π). Doppler shifts of this order would have been marginally visible with the resolution of the GRS. But, concerning the location of the flare at the limb, would have been unexpected.

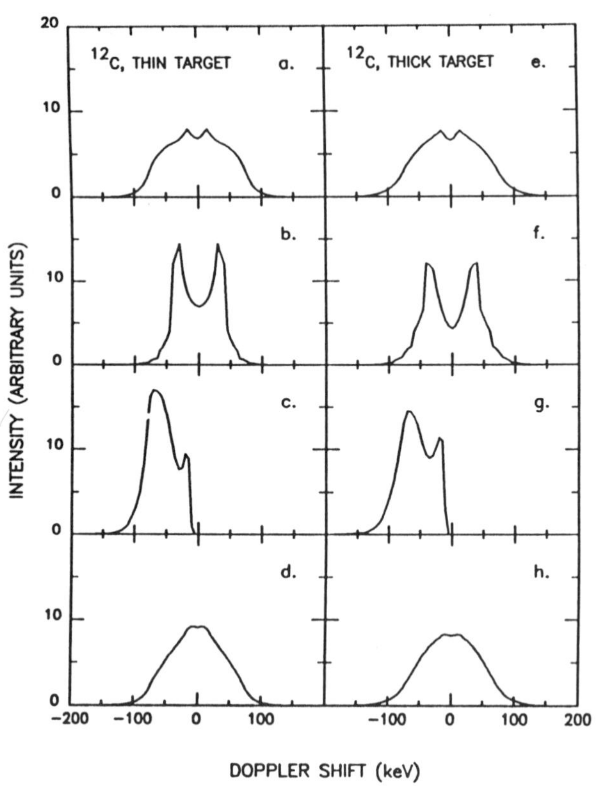

Fig. 8. Shape and position of the ^{12}C (4.438 MeV) de-excitation line for selected angular distributions of the high energy protons; (a,e) fan beam distribution of the protons for a limb flare; (b,f) downward directed proton beam for a limb flare and fan beam for a disk center flare; (c,g) downward directed proton beam for a disk center flare, and (d,h) isotropic distribution for any location on the solar disk (from [77]).

The 4.438 MeV carbon line is of special interest, because of its peculiar profile with respect to the direction of the energetic particles: when the gamma-ray emission is observed at 90° to a proton beam, the line profile has a minimum at its center and shows to symmetrical maxima, whose separation and height depend on the particle energy [17]. ^{12}C line shapes have been calculated for two extrem flare geometrics [76]: for flares occurring at the limb with a downward beamed distribution and for flares at the disk center with a perfect fan beam distribution of the energetic particles. Both distributions give the same line shapes. The dip in the center is somewhat reduced if spallation reactions are taken into account. Refined calculations of the line shapes and positions of ^{12}C (4.438 MeV) and ^{16}O (6.129 MeV) were carried out for thin and thick target cases [77]. Shown in Figure 8 are the results about the ^{12}C line shape under different geometries of the high energy particles, not including the contribution from spallation reactions. It is of special interest that for a flare at the disk center and assuming a downward particle beam (Figure 8c and 8g), the calculated redshift of the line would be of a magnitude, which should be measurable with the energy resolution of the GRS. There are some line flares in the GRS data base, especially the big near disk center X13/4B event of 19 October 1989 which should be investigated for Doppler shifts. On the other hand, the energy resolution of the GRS is not sufficient to show structures, like the dip in the center of the ^{12}C line. Large area Ge-detectors will be necessary to show such peculiarities and help distinguish between various particle geometries.

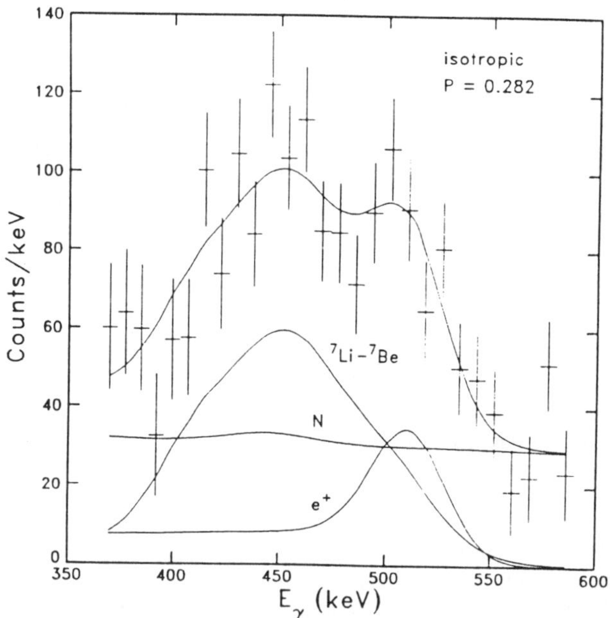

Fig. 9. Line spectrum of the 27 April 1981 between 350 - 600 keV fitted by a nuclear line model, assuming that the interacting particles have an isotropic distribution (from [78]).

Another spectral feature which is also sensitive to the angular distribution of the energetic particles is the line complex around 0.45 MeV. It consists of the de-exicitation lines of ^7Be and ^7Li at 429 keV and 478 keV, respectively. These elements which are very rare constituents of the solar atmosphere are produced by fusion reactions between accelerated α-particles and ambient ^4He. Line profiles have been calculated under the assumption of various particle geometries for disk and limb flares [76]. It was found that for disk flares a downward beamed distribution and a fan beam (parallel to the solar surface) both lines appear clearly separated, but redshifted for the downward beam. For an isotropic distribution of the energetic α-particles, however, both lines merge into each other showing a maximum between the nominal line centers. For a limb flare both lines appear separate only if the α-particles are beamed downward, whereas merging of the lines occurs for a fan beam or isotropic particle distsribution. In a subsequent paper [78] this investigation was continued by including a transport model [56] for the accelerated α-particles. Pitchangle scattering due to MHD turbulence in the corona was also taken into account. A comparison of these calculations with the spectrum of the 27 April 1981 flare shows that the ^{7}Li - ^{7}Be line feature really is present in this limb event (Figure 9). The data, however, are not capable to distringuish between limb flare spectra with and without pitchangle scattering. They are also consistent with an isotropic distribution of the α-particles, but a downward beamed distribution (which is normal to the line of sight for a limb flare) creates a fit which is least acceptable.

A new method to diagnose the angular distribution of the ions using the Compton tail of the 2.223 MeV neutron capture line was proposed recently [79]. As the capture of neutrons on protons takes place deep in the photosphere, the 2.223 MeV photons have to traverse a layer of matter before they can be observed. On their way to the surface they are Compton scattered, thereby loosing energy. A Compton tail is therefore formed shortward of the narrow 2.223 MeV line. It will be more pronounced if the neutrons are captured deep in the atmosphere and this in turn depends on the energy and flight direction of the neutrons. It was shown that for realistic solar flare particle spectra the ratio of the Compton tail flux to the flux of the narrow line can serve as a diagnostic for spectral hardness and/or angular distribution of the accelerated particles. These calculations were applied to the 2.223 MeV line structure of the late stages of the intense 3 June 1982 flare [80,81]. It was found that the very strong Compton tail observed during the second extended phase of the flare requires an extremely hard particle spectrum. Concerning the angular distribution of the energetic particles a fan beam distribution in the lower chromosphere can be excluded.

CONCLUSION

The application of data from the measurement of gamma-ray lines in solar flares is indeed very widepread in solar physics. It ranges from information about the acceleration of charged particles to elemental composition determinations of the various solar atmospheric layers. This explains the early interest this subject has gained. Since the launch of the gamma-ray spectrometers on SMM and Hinotori, observational material was obtained which stimulated further theoretical investigations. Especially the GRS on SMM, due to its long life of nearly 10 years brought a whealth of new data. This detector spanned almost one cycle of solar activity. It commenced its measurements shortly ofter the maximum of cycle 21 and was still operational during the raising phase of cycle 22, when in 89 the Sun attained an activity level never seen before. It was especially during this late phase that a number of big flares showing line emission was recorded. Analysis of these and earlier line events provides therefore a possibility to study flare to flare differences of line intensity ratios and to investigate the influence of the flare location at the Sun on line positions. More observational material is now available to determine the ^{3}He abundance from the decay of the 2.223 MeV line. The recently proposed method to determine spectral parameters and/or the angular distribution of the primary particles from the ratio of the narrow line component and Compton tail flux of the neutron capture line will be applied to more events of the GRS data base.

The Gamma-Ray Observatory (GRO) and the Japanese satellite SOLAR-A, both planned to be launched in 1991, and equipped with NaI- and BGO-detectors, respectively, will add new and valuable data about solar flares. All these measurements, however, will not have the spectral resolution, to investigate such subtle features like the structure of the ^{7}Li-^{7}Be complex at 0.45 MeV, and of the 4.438 MeV ^{12}C line, which are promising diagnostic tools to determine the angular distribution of the energetic particles. It is clear that in order to fully utilize the capabilities of gamma-ray line observations and to build upon the discoveries of the SMM and Hinotori satellites the application of high spectral resolution Germanium detectors in long duration balloon flights or satellite missions will be necessary [82,83]. Efforts should be made to reach this goal at least for the next sunspot maximum.

ACKNOWLEDGEMENTS

This work was supported by the Bundesministerium für Forschung und Technologie under 010K017-ZA/WS/WRK0275:4 in Germany.

REFERENCES

1. E.L. Chupp, in The Physics of Solar Flares, W.N. Hess (ed.) NASA SP 50 (1964), p. 445.
2. J.E. Dolan and G.G. Fazio, Rev. Geophys. 3, 319 (1965).
3. R.E. Lingenfelter and R. Ramaty, in High Energy Nuclear Reactions in Astrophysics, B.S.P. Shen (ed.) (Benjamin, New York, 1967), p. 99.
4. E.L. Chupp, D.J. Forrest, P.R. Higbie, A.N. Suri, C. Tsai, and P.P. Dunphy, Nature 241, 333 (1973).
5. E.L. Chupp, D.J. Forrest, and A.N. Suri, in Solar Gamma-, X-, and EUV Radiation, S.R. Kane (ed), IAU Symp. No. 68, (D. Reidel Publ. Comp., Dordrecht, Holland, 1975), p. 341.
6. H.S. Hudson, T. Bai, D.E. Gruber, J.L. Matteson, P.L. Noland and L.E. Peterson, Astrophys. J. 236, L91 (1980).
7. G. Chambon, K. Hurley, M. Niel, R. Talon, G. Vedrenne, I.V. Estuline, and O.B. Likine, Solar Phys. 69, 147 (1981).
8. T. Prince, J.C. Ling, W.A. Mahoney, G.R. Riegler, and A.S. Jacobson, Astrophys. J. 255, L81 (1982).
9. E.L. Chupp, Ann. Rev. Astron. Astrophys. 22, 359 (1984).
10. R. Ramaty, in The Physics of the Sun, Vol II, Cap. 14, P.A. Sturrock, T.E. Holzer, D. Mihalas, and A.R. Ulrich (eds.) (D. Reidel Publ. Comp., Dordrecht, Holland, 1986), p. 291.
11. R. Ramaty and R.J. Murphy, Space Sci. Rev. 45, 213 (1987).
12. G.E. Kocharov, in Soviet Sci. Rev. Section E, Astrophys. Space Phys., Vol. 6, R.A. Syunyaev (ed.), (Harwood Acad. Publ. G.m.b.H., U.K., 1987), p. 155.
13. E. Rieger, Solar Phys. 121, 323 (1989).
14. E.L. Chupp, Science 250, 229 (1990).
15. R. Ramaty, B. Kozlovsky, and R.F. Lingenfelter, Astrophys. J., Suppl. 40, 487 (1979).
16. C.J. Crannell, H. Crannell, and R. Ramaty, Astrophys. J. 229, 762 (1979).
17. F.L. Lang, C.W. Werntz, C.J. Crannell, J.I. Trombka, and C.C. Chang, Phys. Rev. C35, 1214 (1987).

18. K.T. Lesko, E.B. Norman, R.-M. Larimer, S. Kuhn, D.M. Meekhof, S.G. Crane, and H.G. Bussell, Phys. Rev. C37, 1808 (1988).
19. R.J. Murphy, G.H. Share, J.R. Letaw, and D.J. Forrest, Astrophys. J. 358, 298 (1990).
20. B. Kozlovsky and R. Ramaty, Astrophys. J. 191, L43 (1974).
21. I.A. Ibragimov and G.E. Kocharov, Soviet Astron. Letters 3 (5), 221 (1977).
22. R. Ramaty, B. Kozlovsky, and A.N. Suri, Astrophys. J. 214, 617 (1977).
23. D.J. Forrest, in Electron and Positron Pairs in Astrophysics, M.L. Burns, A.K. Harding, and R. Ramaty (eds.) (AIP, New York, 1983), p. 3.
24. M. Yoshimori, H. Watanabe, and N. Nitta, J. Phys. Soc. Japan 54, 4462 (1985).
25. D.J. Forrest and R.J. Murphy, Solar Phys. 118, 123 (1988).
26. R. Ramaty, B. Kozlovsky, and R.F. Lingenfelter, Space Sci. Rev. 18, 341 (1975).
27. W.T. Vestrand, Solar Phys. 118, 95 (1988).
28. R.J. Murphy, Ph.D.-thesis, Univ. of Maryland (1985).
29. R.J. Murphy and R. Ramaty, Adv. Space Res. 4 (7), 127 (1984).
30. E.L. Chupp, D.J. Forrest, J.M. Ryan, J. Heslin, C. Reppin, K. Pinkau, G. Kanbach, E. Rieger, and G.H. Share, Astrophys. J. 263, L95 (1982).
31. E.L. Chupp, H. Debrunner, E. Flückiger, D.J. Forrest, F. Golliez, G. Kanbach, W.T. Vestrand, J. Cooper, and G.H. Share, Astrophys. J. 318, 913 (1987).
32. H. Debrunner, E. Flückiger, E.L. Chupp and D.J. Forrest, 18th International Cosmic Ray Conf. Papers 4, 75 (1983).
33. G.E. Kocharov, Invited Talks, 18th Eur. Cosmic Ray Symp. (Bologna, 1983), p. 51.
34. M.A. Shea, D.F. Smart, and E.O. Flückiger, 20th Int. Cosmic Ray Conf. Papers 3, 86 (1987).
35. K. Takahashi, M. Wanda, M. Yoshimori, M. Kusunose, and I. Kondo, 20th Int. Cosmic Ray Conf. Papers 3, 82 (1987).
36. A.T. Filippov, 21th Int. Cosmic Ray Conf. Papers 5, 133 (1990).
37. D.F. Smart, M.A. Shea, E.O. Flückiger, H. Debrunner, and J.E. Humble, Astrophys. J. Suppl. 73, 269 (1990).
38. P. Evenson, P. Meyer, and K.R. Pyle, Astrophys. J. 274, 875 (1983).
39. P. Evenson, R. Kröger, and P. Meyer, 19th Int. Cosmic Ray Conf. Papers 4, 130 (1985).
40. P. Evenson, R. Kröger, P. Meyer, and D. Reames, Astrophys. J. Suppl. 73, 273 (1990).
41. M.A. Shea, D.F. Smart, M.D. Wilson, and E.O. Flückiger, Geophys. Res. Letters, accepted for publication (1991).
42. H.T. Wang and R. Ramaty, Solar Phys. 36, 129 (1974).
43. X.-M. Hua and R.E. Lingenfelter, Solar Phys. 107, 351 (1987).
44. G. Kanbach, K. Pinkau, C. Reppin, E. Rieger, E.L. Chupp, D.J. Forrest, J.M. Ryan, G.H. Share, and R.L. Kinzer, 17th Int. Cosmic Ray Conf. Papers 10, 9 (1981).
45. T.A. Prince, D.J. Forrest, E.L. Chupp, G. Kanbach, and G.H. Share, 18th Int. Cosmic Ray Conf. Papers 4, 79 (1983).
46. X.-M. Hua and R.E. Lingenfelter, Solar Phys. 113, 229 (1987).
47. B. Kozlovsky, R.E. Lingenfelter, and R. Ramaty, Astrophys. J. 316, 801 (1987).
48. C.J. Crannell, G. Joyce, R. Ramaty, and C. Werntz, Astrophys. J. 210, 582 (1976).
49. G.H. Share, E.L. Chupp, D.J. Forrest, and E. Rieger, in Positron-Electron Pairs in Astrophysics, M.L. Burns, A.K. Harding, and R. Ramaty (eds.) (AIP, New York, 1983), p. 15.

50. R.W. Bussard, R. Ramaty, and R.J. Drachman, Astrophys. J. 228, 928 (1979).
51. T. Bai and R. Ramaty, Astrophys. J.227. 1072 (1979).
52. E. Rieger and H. Marschhäuser, in Developments in Observation and Theory of Solar Cycle 22, R.M. Winglee and A.L. Kiplinger (eds.), Proc. of MAX91 Workshop No. 3, in preparation.
53. L. Vlahos, et al., in Energetic Phenomena on the Sun, M. Kundu, B. Woodgate (eds.), NASA CP 2439 (1986), p. 2-1.
54. G.E. Kocharov, L.G. Kocharov, G.A. Kovaltsov, and N.Z. Mandzahvidze, Preprint 1258, Academy of Sciences of the USSR (A.F. Joffe Physico-Technical Institute, Leningrad, 1988).
55. R. Ramaty, J.A. Miller, X.-M. Hua, and R.E. Lingenfelter, in Nuclear Spectroscopy of Astrophysical Sources, N. Gehrels and G.H. Share (eds.), (AIP, New York, 1988), p. 217.
56. X.-M. Hua, R. Ramaty, and R.E. Lingenfelter, Astrophys. J. 341, 516 (1989).
57. R. Ramaty, J.A. Miller, X.-M. Hua, and R.E. Lingenfelter, Astrophys. J. Suppl. 73, 199 (1990).
58. M. Yoshimori, Astrophys. J. Suppl. 73, 227 (1990).
59. V.G. Guglenko, Yu.E. Efimov, G.E. Kocharov, G.A. Kovaltsov, N.Z. Mandszhavidze, M.M. Terekhov, and L.G. Kocharov, Astrophys. J. Suppl. 73, 209 (1990).
60. R. Ramaty and R.J. Murphy, in High Energy Transients in Astrophysics, S. Woosley (ed.), (AIP, New York, 1984), p. 628.
61. D.J. Forrest and E.L. Chupp, Nature 305, 291 (1983).
62. T. Bai and R. Ramaty, Solar Phys. 49, 343 (1976).
63. R.J. Murphy, D.J. Forrest, R. Ramaty, and B. Kozlovsky, 19th Int. Cosmic Ray Conf. Papers 4, 253 (1985).
64. R. Ramaty, R.J. Murphy, B. Kozlovsky, and R.E. Lingenfelter, Astrophys. J. 273, L41 (1983).
65. R. Ramaty, in Particle Acceleration Mechanisms in Astrophysics, C. Max and C. McKee (eds.), (AIP, New York, 1979), p. 135.
66. C.J. Crannell and F.L. Lang, in Nuclear Spectroscopy of Astrophysical Sources, N. Gehrels and G.H. Share (eds.), (AIP, New York, 1988), p. 240.
67. D.P. Saltzberg, J.D. Brown, A.E. Champagne, R.T. Konzes, C.J. Crannell, F.L. Lang, P.S. Shawhan, R. Starr, and C.W. Werntz, in Proc. Gamma-Ray Observatory Workshop, Neil Johnson (ed.), (Greenbelt, Maryland, 1989), 4-366-4-373.
68. R.J. Murphy, R. Ramaty, D.J. Forrest, and B. Kozlovsky, 19th Int. Cosmic Ray Conf. Papers 4, 249 (1985).
69. J.P. Meyer, Astrophys. J. Suppl. 57, 173 (1985).
70. R.J. Murphy, R. Ramaty, B. Kozlovsky, and D.V. Reames, in press Astrophys. J. (April 20, 1991).
71. D.V. Reames, H.V. Cane, and T.T. von Rosenvinge, Astrophys. J. 327, 259 (1990).
72. E. Rieger, C. Reppin, G. Kanbach, D.J. Forrest, E.L. Chupp, and G.H. Share, 18th Int. Cosmic Ray Conf. Papers 10, 338 (1983).
73. W.T. Vestrand, D.J. Forrest, E.L. Chupp, E. Rieger, and G.H. Share, Astrophys. J. 322, 1010 (1987).
74. X.-M. Hua and R.E. Lingenfelter, Astrophys. J. 323, 779 (1987).
75. R. Ramaty and C.J. Crannell, Astrophys. J. 203, 766 (1976).
76. R.J. Murphy, B. Kozlovsky, and R. Ramaty, Astrophys. J. 331, 1029 (1988).
77. C. Werntz, F.L. Lang, and Y.E. Kim, Astrophys. J. Suppl. 73, 349 (1990).

78. R.J. Murphy, X.-M. Hua, B. Kozlovsky, and R. Ramaty, Astrophys. J. 351, 299 (1990).
79. W.T. Vestrand, Astrophys. J. 352, 353 (1990).
80. D.J. Forrest, W.T. Vestrand, E.L. Chupp, E. Rieger, J.F. Cooper, and G.H. Share, 19th Int. Cosmic Ray Conf. Papers 4, 146 (1985).
81. W.T. Vestrand, D.J. Forrest, R.J. Murphy, E.L. Chupp, and G. Kanbach, 21th Int. Cosmic Ray Conf. Papers 5, 160 (1990).
82. N. Gehrels, C.J. Crannell, D.J. Forrest, R.P. Lin, L.E. Orwig, R. Starr, Solar Phys. 118, 233 (1988).
83. A. Owens, E.L. Chupp, and P.P. Dunphy, Experimental Astronomy 1, 1 (1990).

SOLAR ABUNDANCES FROM GAMMA-RAY SPECTROSCOPY: COMPARISONS WITH ENERGETIC PARTICLE, PHOTOSPHERIC, AND CORONAL ABUNDANCES

R. J. Murphy
Naval Research Laboratory, Washington, DC 20375

R. Ramaty
NASA/Goddard Space Flight Center, Greenbelt, MD 20771

B. Kozlovsky
Tel Aviv University, Tel Aviv, Israel

ABSTRACT

We have derived accelerated particle and ambient gas abundances using solar flare gamma-ray spectroscopy. We find that the derived accelerated particle composition is different from the composition observed in large proton flares; rather, it resembles that observed in ^{3}He-rich flares. Our analysis also suggests that the ambient composition differs from the composition of both the photosphere and the corona.

INTRODUCTION

The compositions of the various regions of the solar atmosphere have been studied using atomic and molecular spectroscopy, and solar wind and solar energetic particle (SEP) observations (e.g., ref. 1). These studies have revealed considerable abundance variations. In particular, the SEP composition was found to be highly variable and two classes of SEP events have been identified[2]: (1)^{3}He-rich events which exhibit enhanced abundances of ^{3}He relative to ^{4}He and of elements heavier than O relative to C and (2) large proton flare (LPF) events with lower values of ^{3}He/^{4}He and lower heavy elements-to-C ratios.

Solar abundances have also been derived using nuclear gamma-ray spectroscopy[3,4]. The principal mechanisms for the production of gamma-ray lines are nuclear deexcitation, neutron capture and positron annihilation[5,6]. Nuclear deexcitation lines result from the bombardment of ambient gas with accelerated protons, α particles and ^{3}He nuclei (producing narrow lines), and from the inverse reactions where ambient hydrogen and helium are bombarded by accelerated C and heavier nuclei (producing broad lines). The strengths of the narrow lines depend primarily on the ambient gas abundances, while the strengths of the broad lines depend primarily on the accelerated particle abundances. The nuclear emission is superposed on a continuum due to electron bremsstrahlung. The bulk of the observed gamma-ray emission from solar flares is thought to be produced by particles accelerated and trapped in closed loops[7].

Using SMM data for the 27 April 1981 flare, it has been shown[8] that a theoretically-calculated[6] gamma-ray spectrum for which both the ambient gas and the accelerated particles have photospheric composition provides an unacceptable fit to the data. Using the same data, Murphy *et al.* (refs. 3 and 4) derived the best fitting ambient gas abundances. In their study the accelerated particle abundances were held fixed at LPF values and the ambient abundances were deduced. The results suggested that ambient Mg/C, Si/C and Fe/C are enhanced relative to the photosphere but are consistent with the coronal values; that Ne/C and Ne/O are larger than the coronal values; and that O/C is consistent with both the photospheric and coronal values.

In the present paper we relax the constraint that the accelerated particle composition is fixed and show that the SMM gamma-ray data are of sufficient resolution to allow the determination of the abundances of both these particles and the ambient gas. We compare the results to photospheric and coronal abundances, and SEP abundances in both ^{3}He-rich flares and large proton flares. The results presented here are fully discussed in ref. 9.

GAMMA-RAY SPECTRAL ANALYSIS

Measured nuclear cross sections and the relevant nuclear kinematics were incorporated[6,10] into a computer code which produces a complete deexcitation gamma ray spectrum for a given accelerated particle energy spectrum and given ambient and accelerated particle compositions. We construct a theoretical gamma-ray spectrum from a nuclear deexcitation spectrum (calculated by using an isotropic thick-target model), the very narrow lines at 2.223 MeV (due to neutron capture) and 0.511 MeV (due to positron annihilation), a bremsstrahlung continuum, and a continuum due to orthopositronium annihilation[7,11]. We assume that the bremsstrahlung continuum can be approximated by a power law and that the orthopositronium continuum corresponds to a positronium fraction of 2/3 (ref. 12). We also assume that all accelerated particle species have the same energy spectrum as a function of energy per nucleon given by the modified Bessel function characterized by the parameter αT (ref. 13).

Such a theoretical spectrum depends on 23 parameters: 16 abundances (ambient ^{4}He, N, O, Ne, Mg, Si, Fe relative to C; ambient ^{4}He relative to H, $[^4\text{He}/^1\text{H}]_{amb}$; accelerated N, O, Ne, Mg, Si, Fe relative to C; accelerated ^{3}He/^{4}He ; and accelerated α particles relative to protons, $[^4\text{He}/^1\text{H}]_{acc}$); 2 normalizations ($[n_C/n_H]\ N_p$ and N_C, where N_p and N_C are the total numbers of accelerated protons and C nuclei of energies greater than 30 MeV nucleon^{-1} incident on the thick target and n_C/n_H is the ambient C-to-H abundance ratio); the 2.223 and 0.511 MeV line fluxes; the amplitude and spectral index s of the power law; and the spectral parameter αT. We find that the results are essentially independent of $[^4\text{He}/^1\text{H}]_{amb}$ and so have fixed this parameter at the photospheric value of 0.1 (ref. 1). The calculated spectrum depends linearly on these parameters, except s, αT and $[^4\text{He}/^1\text{H}]_{acc}$. We consider 5 cases (Table I): In case 1, all 22 parameters are allowed to vary; in cases 2 and 3, the ambient abundances are fixed at photospheric[1] and coronal[14] values, respectively; in case 4, the accelerated particle abundances (except $[^3\text{He}/^4\text{He}]_{acc}$) are fixed at the LPF values[14] and $[^4\text{He}/^1\text{H}]_{acc}$ is fixed at 0.1; and in case 5 both the ambient and accelerated particle abundances (except $[^4\text{He}/\text{C}]_{amb}$, $[^4\text{He}/^1\text{H}]_{acc}$ and $[^3\text{He}/^4\text{He}]_{acc}$) are fixed at photospheric[1] and LPF[14] abundances, respectively.

TABLE I

Analysis Cases and Associated Qualities of Fit

Case	χ^2	ν	P_{χ^2}
1. Full Fit	459.2	427	0.136
2. Fixed Ambient Abundances[a]	484.0	434	0.049
3. Fixed Ambient Abundances[b]	487.6	434	0.038
4. Fixed Accelerated Particle Abund.[c]	533.3	434	8×10^{-4}
5. Fixed Ambient[a] and Accelerated[c] Particle Abundundances	990.9	439	$<10^{-6}$

[a]—Photospheric[1]
[b]—Coronal[14]
[c]—Large Proton Flare[14] (and $[^4He/^1H]_{acc}= 0.1$ for case 4)

We compare data[15] from the 27 April 1981 flare, obtained with the SMM gamma-ray spectrometer, with calculated spectra folded through the detector response function. We measure the quality of fit with a χ^2 statistic and the associated probability P_{χ^2}: the probability that random observations would produce a χ^2 as large or larger. It has been shown[10,16] that, for a given choice of s, αT and $[^4He/^1H]_{acc}$, χ^2 can be minimized by matrix inversion. The absolute minimal χ^2 can then be obtained by a systematic search in the 3-dimensional s-αT-$[^4He/^1H]_{acc}$ parameter space. The uncertainties of the gamma-ray derived results are estimated using $\Delta\chi^2 = 1$.

The minimal χ^2 values, the corresponding P_{χ^2}, and the numbers of degrees of freedom ν are given in Table I, except for case 5. In this case, the values correspond only to the minimal χ^2 achieved (at $[^4He/^1H]_{acc} = 0.1$ and $\alpha T = 0.010$) in the parameter space that we have searched ($0.1 <[^4He/^1H]_{acc}< 3.0$ and $0.010 <\alpha T< 0.050$); we do not expect that a search over a more extended space will significantly improve the quality of fit for this case.

The best fit (case 1) is achieved with $[^4He/^1H]_{acc} = 1$, a value larger than any observed in SEP events[17]. We note, however, that the fit is almost as good ($P_{\chi^2} = 0.131$) with a more reasonable value of 0.5. Hereafter, we refer to this latter calculation as case 1a. The qualities of fit for cases 2 and 3 are not as good as that of case 1, but the values of P_{χ^2} are not low enough to allow us to reject these cases. In both of these cases the best fitting values of $[^4He/^1H]_{acc}$ (0.1 for case 2 and 0.5 for case 3) are reasonable. Case 4, where the accelerated particle abundances are fixed at LPF values and $[^4He/^1H]_{acc}$ is fixed at 0.1, can probably be rejected. The value of 0.1 represents an upper limit to the range of $[^4He/^1H]_{acc}$ observed in LPF events and the fit worsens as $[^4He/^1H]_{acc}$ is reduced. Case 5, in which both the accelerated particle and ambient abundances are fixed, can also be rejected.

COMPARISONS OF THE DERIVED ABUNDANCES

The derived ambient abundances for case 1a are given in Table II, and compared with photospheric[1] and coronal[1,14] abundances. We see that Mg/C, Si/C and Fe/C are enhanced relative to the photosphere but not relative to the corona, while Ne/C is enhanced relative to both. We note that the photospheric Ne/C has not been directly measured; the value in Table II is actually the coronal value which is consistent with recent observations[18] of impulsive flare material shown to be similar to photospheric material. In addition, He/C is suppressed relative to both the photosphere and the corona. These differences should be contrasted with cases 2 and 3, for which the ambient abundances are photospheric or coronal. Since the qualities of fit for cases 2 and 3 are not much worse than that for case 1a, it is possible that the composition of the ambient medium is either photospheric or coronal.

TABLE II

Ambient Medium Abundances Compared with Photospheric[1] and Coronal[14] Abundances

	Gamma-Ray (Case 1a)	Photospheric Abundances	Coronal Abundances
^{4}He	108 ±15	$269\,^{+23}_{-21}$	$174\,^{+26}_{-23}$
C	1.00 ±0.19	$1.00\,^{+0.10}_{-0.09}$	$1.00^{+0.11}_{-0.10}$
N	0.23 ±0.48	$0.31\,^{+0.03}_{-0.03}$	$0.30^{+0.02}_{-0.02}$
O	2.37 ±0.27	$2.34\,^{+0.20}_{-0.18}$	$2.42^{+0.15}_{-0.14}$
Ne	0.91 ±0.15	$0.34\,^{+0.09}_{-0.07}$	$0.33^{+0.04}_{-0.03}$
Mg	0.35 ±0.13	$0.105^{+0.013}_{-0.011}$	$0.46^{+0.03}_{-0.03}$
Si	0.52 ±0.15	$0.098^{+0.012}_{-0.011}$	0.43
Fe	0.37 ±0.12	$0.130^{+0.009}_{-0.009}$	$0.54^{+0.07}_{-0.06}$

The derived accelerated particle abundances for cases 1a, 2 and 3 are given in Table III and compared with LPF[14,19] and ^{3}He-rich flare[17] abundances. We see that the gamma-ray derived abundances, in particular for cases 2 and 3 in which the ambient abundances are fixed and the uncertainties are smaller, are more consistent with the ^{3}He-rich SEP abundances than with LPF abundances.

The derived values of $[^4\mathrm{He}/^1\mathrm{H}]_{acc}$ for cases 1a, 2, and 3 are 0.5, 0.1 and 0.5, respectively, indicating that $[^4\mathrm{He}/^1\mathrm{H}]_{acc}$ is probably larger than 0.1. Since

$[^4He/^1H]_{acc}$ in LPF events is typically less than 0.1 while in ^{3}He-rich events $[^4He/^1H]_{acc}$ tends to be larger[17,20], this result is another indication that the composition of the gamma-ray producing accelerated particles is more similar to that observed in ^{3}He-rich events. Also, the derived values of $[^3He/^4He]_{acc}$ suggest an enrichment of ^{3}He relative to ^{4}He, although the uncertainty is quite large since ^{3}He interactions produce just one weak gamma-ray line, at 3.561 MeV from ^{6}Li (ref. 6).

TABLE III

Accelerated Particle Abundances Compared with Large Proton Flare[14,19] and ^{3}He-Rich Flare[17]

	case 1a*	case 2	case 3	Large Proton Flare (LPF)	^{3}He-Rich Flare
$^3He/^4He$	0.50±0.39	1.32 ±0.34	0.03 ±0.01	<0.05	>0.1
C	1.00 ±0.96	1.00±0.29	1.00 ±0.27	1.00 ±0.10	1.00
N	4.73 ±3.39	3.60 ±1.18	3.11 ±1.09	0.29 ±0.02	0.2–0.8
O	2.00 ±1.59	2.78 ±0.54	2.01 ±0.52	2.30 ±0.06	1–6
Ne	<1.47	0.71 ±0.39	1.28 ±0.35	0.33 ±0.03	0.3–5
Mg	4.94 ±1.26	3.73 ±0.39	2.05 ±0.37	0.45 ±0.02	0.5–4
Si	0.90 ±1.75	2.80 ±0.66	1.42 ±0.61	0.37	0.5–3
Fe	6.35 ±1.46	4.28 ±0.42	3.10 ±0.40	0.35 ±0.04	1–10

*—Case 1 with $[^4He/^1H]_{acc} = 0.5$

SUMMARY

Using gamma-ray observations of the 27 April 1981 flare, we have derived elemental abundances of both the ambient gas and the accelerated particles which interact with this gas to produce the gamma rays. The location of the gamma-ray producing site is not known. For impulsive flares, it is probably in the chromosphere, but for an extended flare (such as 27 April 1981), significant gamma-ray production could take place in the corona if the density exceeds 5×10^{11} cm^{-3}. The best fit (13% confidence) suggests that Mg/C, Si/C and Fe/C in the ambient gas are enhanced relative to the photosphere but not relative to the corona, and that the ambient Ne/C is enhanced relative to the corona. However, the cases in which the ambient abundances are fixed at photospheric or coronal values are also acceptable, but only at the 5% and 4% levels, respectively.

We found that the composition of the accelerated particles is probably different from that of LPF events. The cases in which the ambient abundances are photospheric or coronal yield accelerated particle abundances (including $[^4He/^1H]_{acc}$) which are more similar to ^{3}He-rich flare abundances. The accelerated particle abundances in the full fit case are not well determined but are probably not the LPF values. Since the accelerated particles which produce the gamma rays are most likely accelerated in closed flare loops[7], our results provide support to the suggestion[21,22] that the particles observed in ^{3}He-rich events are accelerated from hot flare plasma.

REFERENCES

1. E. Anders and N. Grevesse, *Geochim. Cosmochim. Acta*, **53**, 197 (1989).
2. D. V. Reames, *Ap.J. Suppl.*, **73**, 235 (1990).
3. R. J. Murphy, D. J. Forrest, R. Ramaty and B. Kozlovsky, *19th Internat. Cosmic Ray Conf. Papers*, **4**, 253 (1985).
4. R. J. Murphy, R. Ramaty, D. J. Forrest and B. Kozlovsky, *19th Internat. Cosmic Ray Conf. Papers*, **4**, 249 (1985).
5. R. Ramaty, B. Kozlovsky and R. E. Lingenfelter, *Sp.Sci.Rev.*, **18**, 341 (1975).
6. R. Ramaty, B. Kozlovsky and R. E. Lingenfelter, *Ap.J. Suppl.*, **40**, 487 (1979).
7. R. Ramaty and R. J. Murphy, *Sp.Sci.Rev.*, **45**, 213 (1987).
8. D. J. Forrest, in *Positron Electron Pairs in Astrophysics*, eds. M. L. Burns, A. K. Harding, and R. Ramaty (New York:AIP), p. 3 (1983).
9. R. J. Murphy, R. Ramaty, B. Kozlovsky and D. V. Reames, *Ap.J.* (in press, 1991).
10. R. J. Murphy, *Gamma Rays and Neutrons from Solar Flares*, Ph.D. Dissertation, University of Maryland (1985).
11. C. J. Crannell, G. Joyce, R. Ramaty and C. Werntz, *Ap.J.*, **210**, 52 (1976).
12. R. J. Murphy, X. M. Hua, B. Kozlovsky and R. Ramaty, *Ap.J.*, **351**, 299 (1990).
13. M. A. Forman, R. Ramaty and E. G. Zweibel, in *The Physics of the Sun*, eds. P. A. Sturrock *et al.*, Vol. II, p. 249 (1986).
14. H. H. Breneman and E. C. Stone, *Ap.J. Letters*, **299**, L57 (1985).
15. R. J. Murphy, G. H. Share, J. R. Letaw and D. J. Forrest, *Ap.J.*, **358**, 290 (1990).
16. J. I. Trombka and R. L. Schmadebeck, *A Numerical Least-Square Method for Resolving Complex Pulse Height Spectra*, NASA SP-3044 (1968).
17. D. V. Reames, H. V. Cane and T. T. von Rosenvinge, *Ap.J.*, **357**, 259 (1990).
18. U. Feldman and K. G. Widing, *Ap.J.*, (in press, 1990).
19. R. E. McGuire, T. T. von Rosenvinge and F. B. McDonald, *Ap.J.*, **301**, 938 (1986).
20. R. Ramaty *et al.*, in *Solar Flares*, ed. P. A. Sturrock, (Colorado Assoc. Univ.: Boulder), p. 117 (1980).
21. A. Luhn, B. Klecker, D. Hovestadt and E. Möbius, *Ap.J.*, **317**, 951 (1987).
22. R. P. Lin, *Rev. Geophys.*, **25**, 676 (1987).

SOLAR FLARE GAMMA-RAY LINE PROFILES

F. L. Lang*

Solar Physics Branch of the Laboratory for Astronomy and Solar Physics, Goddard Space Flight Center, Greenbelt, MD 20771, USA.

C. W. Werntz*

Physics Department, The Catholic University of America, Washington D.C. 20064, USA.

ABSTRACT

Solar gamma-ray lines are produced through collisions of pairs of positive ions whose center of mass energies are above the relevant thresholds for excitation of gamma-ray emitting states. Because of the low density of the solar plasma, prompt gamma rays are emitted by recoiling ions before significant energy loss has occurred. Thus, the lines are expected to be Doppler broadened to widths of the order of a hundred KeV. Gamma-ray lines resulting from proton and alpha-particle (^{4}He ion) beams on carbon and oxygen targets in the laboratory exhibit complex profiles which change rapidly[1] with the gamma-ray observation angle. Our earlier work has been limited to modeling laboratory and solar flare gamma-ray profiles produced by proton excitation[2], but laboratory profiles of the carbon 4.44-MeV line and oxygen 6.13-MeV line from alpha excitation have now been successfully modeled. Solar flare gamma-ray line profiles of the carbon and oxygen lines from heavy ions in the ambient solar medium interacting with representative high-energy proton and alpha particle populations will be presented.

INTRODUCTION

The class of reactions considered in this talk are direct excitation processes $N(x,\gamma_{N*})x'N$ with the projectile x representing either protons or alpha particles. The recoil distribution of the excited ions is determined by the center of mass inelastic differential cross section which, typically, resembles a diffraction pattern. For a given recoil direction the excited nuclear state is a coherent mixture of magnetic substates and the amplitudes of the substates determine the radiation pattern of the emitted gamma rays. The combined excited ion recoil distribution and the gamma-ray radiation pattern are defined by a doubly differential cross section $d^2\sigma/d\Omega_x d\Omega_\gamma(\Theta_p,\phi_p,\Theta_\gamma,\phi_\gamma)$ where the gamma-ray angles are in the frame of the recoiling ion. This cross section is determined[3] by the nuclear density matrix whose elements $\rho_{M,M'}(\Theta_p,\phi_p)$ are quadratic in the substate amplitudes. Nuclear deexcitation gamma-ray lines seen in solar flares are prompt gamma-ray lines because the excited states

*Work supported by NASA Grant NSG 5066.

have lifetimes of 10^{-11} – 10^{-14}s. Barring strong hyperfine interactions with unpaired 1s electrons or precession of the magnetic moment of the excited ion in a strong magnetic field, processes of importance only for the longer lived states, the coherent mixture of magnetic substates is unchanged from time of formation to time of decay.

In the frame of the excited ion an emitted gamma has an energy E_0. Because of the short lifetimes of the gamma-emitting states a recoiling excited ion does not slow appreciably in the solar plasma before emitting the gamma ray. Therefore, in the CM frame of the projectile-ion gamma rays can be observed over a range of energies E_γ, $(1 - \beta_N)E_0 \leq E_\gamma \leq (1 + \beta_N)E_0$, the exact energy depending upon the angle between the ion recoil direction ion and the gamma-ray emission direction. β_N is the original recoil velocity, in units of c, of the excited ion in the CM frame. If gamma rays alone are observed in the CM frame, not gamma's and scattered projectiles in coincidence, the gamma-ray line is broadened to a width of $2\beta_N E_0$ independent of the gamma-ray angle Θ_γ with respect to the incoming beam. The profile of the line is essentially the differential cross section $d^2\sigma/d\Omega_\gamma dE_\gamma(\Theta_\gamma,E_\gamma)$ which is obtained by integrating over all projectile scattering angles. In the frame in which the struck ion is initially at rest the gamma-ray line profiles are essentially the same, but the centroids of the lines are shifted by $\beta_{CM}E_0\cos\Theta_{CM}$ where β_{CM} is the velocity of the CM.

LABORATORY GAMMA-RAY LINE PROFILES FOLLOWING ALPHA EXCITATION

The feasability of calculating line profiles for gamma-rays from $N(p,\gamma_{N*})p'N$ and using these profiles to model solar-flare gamma-ray lines has been demonstrated[4]. Alpha particle projectiles are expected to play a significant role in the production of deexcitation gamma rays and there is good reason to believe that alpha particle excitation played a major role in the 27 April 1981 flare[5]. The most important $N(\alpha,\gamma_{N*})\alpha'N$ cross sections for carbon and oxygen have been measured[6] for alpha energies up to E_α = 26 MeV. We have been provided the multichannel analyser output from this experiment[7] which contains the line profiles and have calculated the line shapes by using the coupled channel code CHUCK 2[8] to calculate the excited nuclear density matrix. The real parts of the optical potentials were taken from calculations of rotational band energies in ^{16}O ($^{12}C + \alpha$)[9] and ^{20}Ne ($^{16}O + \alpha$)[10] while the strengths of the imaginary parts were adjusted to yield the correct total gamma-ray cross sections. Comparisons of typical experimental and theoretical profiles are shown in Figures 1. and 2. For most angles there is semi-quantitative agreement, in particular, the number of peaks in each line profile (about twice the number seen in lines from proton excitation)is reproduced by the calculations although the relative strengths are not always correct. However, optical potentials parametrized explicitly to fit inelastic differential cross sections should provide even better line profile fits.

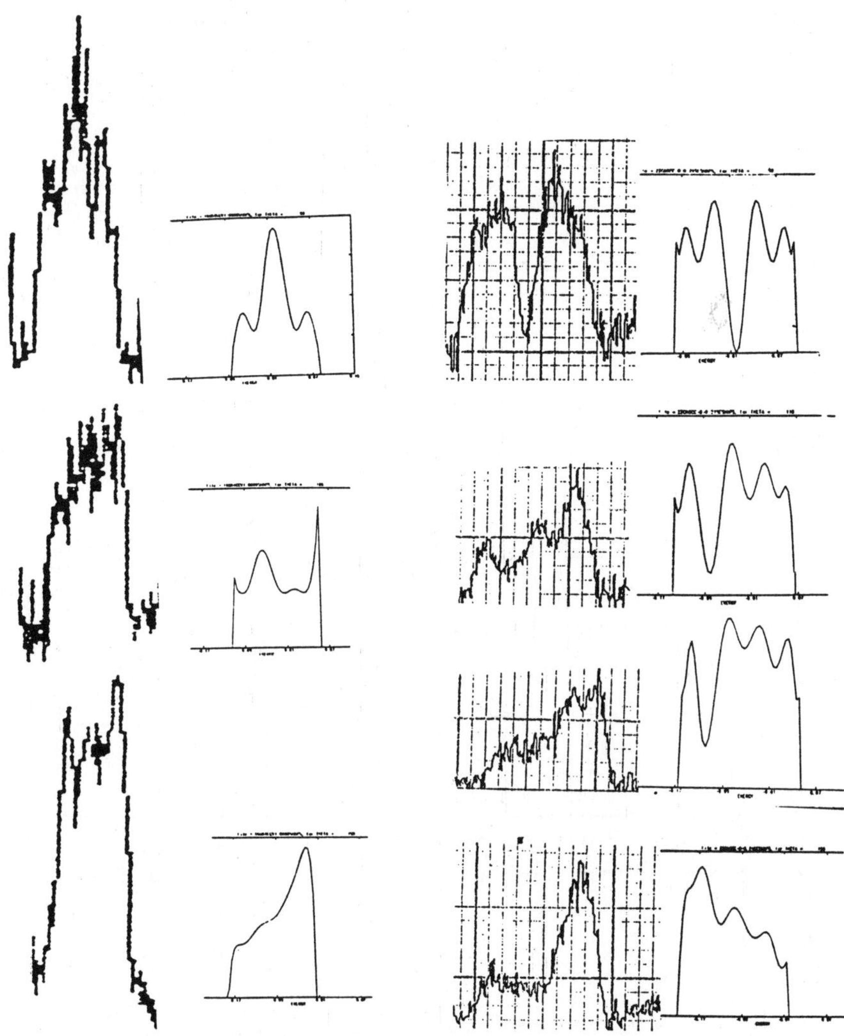

Fig. 1. Line profile of the $^{16}O^*$ (6.13 MEV) line produced by 16-MeV alpha particles on oxygen. The angles of the emitted gamma with respect to the beam are $90^0, 104^0$, and 150^0. The calculated intrinsic profile is compared to the first escape peak measured with 55-keV resolution (FWHM).

Fig. 2. Line profile of the $^{12}C^*$ (4.44 MeV) line produced by 20-MeV alpha particles on carbon. Caption the same as Fig. 1., except angles are 90^0, 110^0, 120^0, and 150^0, and the resolution is 27-keV.

SOLAR FLARE LINE PROFILES

Laboratory line profiles depend on the projectile energy and the angle between the observed gamma ray and the beam. To obtain a solar flare line profile the laboratory profiles must be weighted over a projectile energy spectrum and an angular distribution. Figs. 3 and 4 show our predictions of intrinsic line profiles for three extreme projectile distributions, pancake, representing particles mirroring at footpoints of coronal loops, downward beam perpendicular to the photosphere, and isotropic. Two different positions of the flare on the sun are considered, limb and disc center. The energy spectra are represented by second order Bessel functions[11] characterized by the parameter αT. The profiles in the middle column of graphs have been calculated with values of the

$\alpha/p = 0.$, $\alpha T = 0.030$ $\alpha/p = 0.5$, $\alpha T = 0.015$ $\alpha/p = 10^3$, $\alpha T = 0.015$

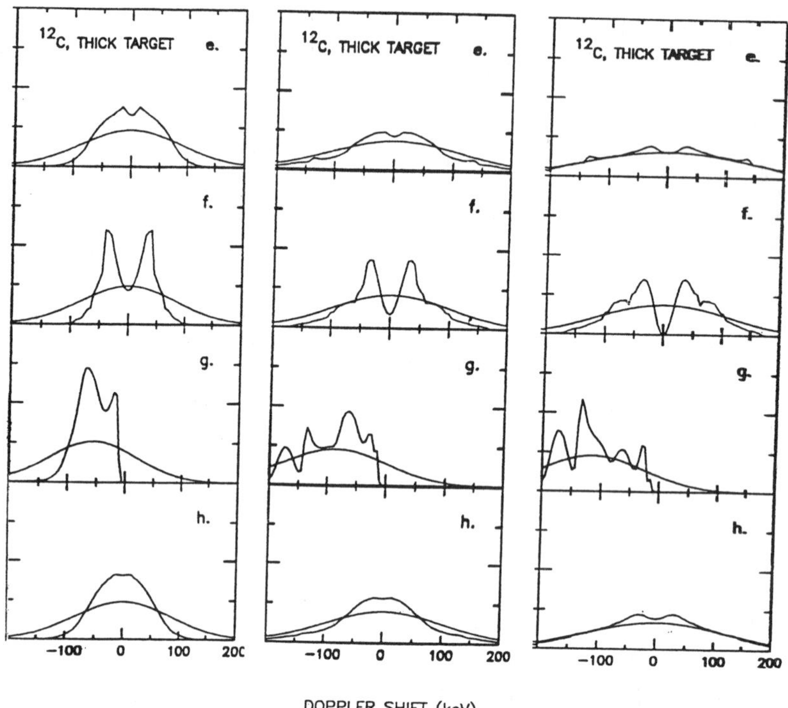

Fig. 3. Solar flare gamma-ray line profiles following excitation of $^{12}C^*$(4.44 MeV) by combined proton and alpha particle fluxes. The alpha-proton ratio and the spectral parameter αT (protons and alpha's have the same spectrum as a function of energy/nucleon) are displayed at the top of each column of graphs. The smooth lines show the intrinsic profiles folded with a Gaussian representing the resolution of the GRS instrument aboard SMM (see text). The directional distributions of the projectiles and the flare positions are: (e) pancake limb, (f) downward beam limb and/or pancake center, (g) downward beam center, (h) isotropic.

alpha particle-proton ratio and spectral hardness appropriate to the 27 April 1981 flare[5]. The smooth curves are obtained by folding the intrinsic shapes with Gaussian functions of 191-keV FWHM and 212-keV FWHM, appropriate to the resolution of the GRS aboard SMM for 4.44- and 6.13-MeV gamma rays, respectively[12].

$\alpha/p = 0.0$, $\alpha T = 0.030$ $\alpha/p = 0.5$, $\alpha T = 0.015$ $\alpha/p = 10^3$, $\alpha T = 0.015$

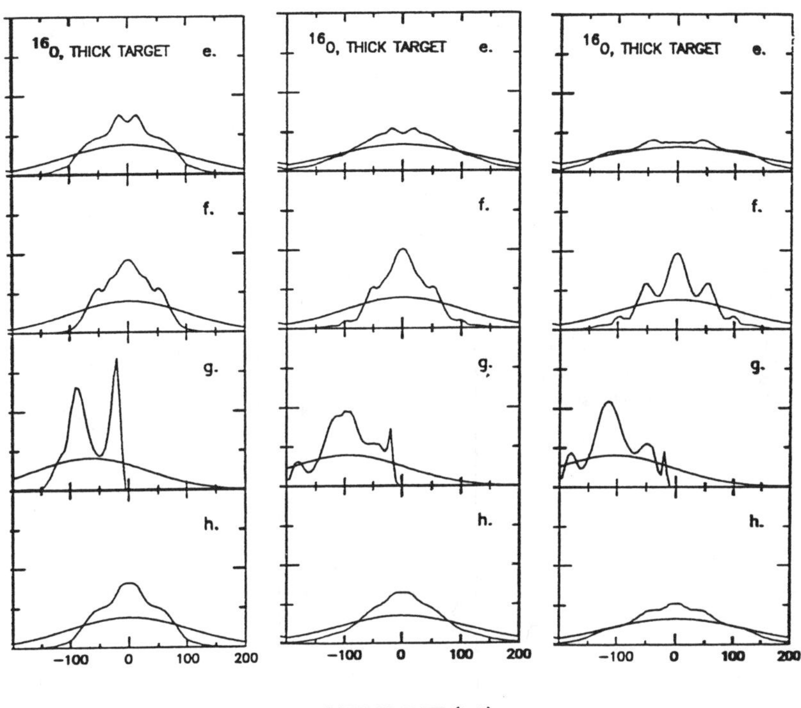

Fig. 4. Solar flare gamma-ray line profiles following excitation of $^{16}O*(6.13$ MeV) by combined proton and alpha fluxes. See caption of Fig. 3. for further description.

CONCLUSION

Gamma-ray line profiles in solar flares are expected to show complex structure which is directly related to heavy ion energy spectra and directionality. Variations in this structure are of the order of 10's of keV's in width and are observable only by high resolution germanium detectors. However, with good statistics, sodium iodide detectors should be capable of measuring intrinsic widths and centroid shifts which are most reliably calculated by our methods.

REFERENCES

1. J. J. Kolata, R. Auble, and A. Galonsky, Phys. Rev. 162, 957 (1967).
2. C. Werntz, F. L. Lang, and Y. E. Kim, Astrophys. J. (Suppl. Ser.) 73, 349 (1990).
3. G. R. Satchler, Direct nuclear Reactions (Clarendon, Oxford, 1983).
4. C. Werntz, F. L. Lang, and Y. E. Kim, Astrophys. J. (Suppl. Ser.) 73, 349 (1990).
5. R. J. Murphy, R. Ramaty, B. Kozlovsky, and D. V. Reames, Astrophys. J. , in press, 1991.
6. P. Dyer, D. Bodansky, D. D. Leach, E. B. Norman, and A. J. Seamster, Phys. Rev. C32, 1873 (1985).
7. Peggy Dyer, LASL, private communication.
8. P. D. Kunz, University of Colorado.
9. Richard A. Baldock, Brian Buck, and J. Alberto Rubio, Nucl. Phys. A426, 222 (1984).
10. B. Buck, C. B. Dover, and J. P. Vary, Phys. Rev. C11, 1803 (1975).
11. M. A. Forman, R. Ramaty, and E. G. Zweibel, The Physics of the Sun, (Reidel, Dordrecht, 1986) p. 249.
12. D. J. Forrest and R. J. Murphy, Solar Physics 118, 123 (1988).

POT-POURRI

On the $e^- - e^+$ Annihilation Line in 3C 273

J. Roland (1), R. Lehoucq (2) and G. Pelletier (3)

(1) IAP, Paris–France
(2) CEA, Saclay–France
(3) Observatoire de Grenoble–France

Summary

Assuming that the extended lobe of 3C 273 is a relativistic $e^- - e^+$ plasma with a low energy cut-off $E_o \simeq$ 1 MeV, we estimated the $e^- - e^+$ annihilation losses of the lobe. We found that it is $\dot{E}_{-+} \simeq 1.8 \times 10^{36}$ erg s^{-1}, which is well below the current detection limits. A careful determination of $\dot{E}_{-+}$ will need to take into account the thermal e^- inside the lobe and to use the two-fluid model to describe the formation of 3C 273.

However, this result suggests that the $e^- - e^+$ annihilation lines due to all active galaxies with redshifts $z \leq 4$ could produce a significant contribution to the X–ray background between 200 KeV and few MeV.

1 Introduction

Shklovsky[18] suggested that the asymmetrical structure of 3C 273 could be explained by a relativistic ejection of the radio plasma by the nucleus. For 20 years, VLBI observations of 3C 273 show superluminal expansions (see Zensus et al.[22] and references quoted). VLBI observations can be understood if the relativistic e^- and/or e^+, responsible for the synchrotron radiation are ejected with a bulk Lorentz factor $\bar{\gamma} \simeq 10$ at an angle of about 20 degrees with the line of sight[22]. Assuming an intrinsic asymmetry of the source, Shklovsky[19] argued that the extended lobe was likely an $e^- - e^+$ plasma. The extended lobe has been intensively observed with MERLIN[1,4,3,2] and the VLA[15]. The rotation measure RM of the extended lobe is[3] $|\mathrm{RM}| \simeq 1.8$ rad m^{-2} and it has been suggested by Davis et al.[2] that the jet powering the hot spot was relativistic, with a bulk Lorentz factor $\bar{\gamma} \simeq 2$. Consequently they deduced that 3C 273 was a one–sided radio source. These observations are consistent with the Shklvosky hypothesis.

However, Meisenheimer and Heavens[9], Meisenheimer et al.[10] and Fraix–Burnet and Pelletier[14] found that the formation of the hot spot in front of the extended lobe and its synchrotron spectrum can be explained by the interaction of a non-relativistic $e^- - p$ supersonic jet with the intergalactic medium. The velocities of the jet that they derived are respectively $v_j \geq 0.3$ c (Meisenheimer and Heavens[9]), $v_j \simeq 0.2$ c (Meisenheimer et al.[10]) and $0.2\ c \leq v_j \leq c/\sqrt{3}$ (Fraix–Burnet and Pelletier[14])

The results deduced from VLBI observations and those obtained from the hot spot formation models can be reconciliated using the two–fluid model. Indeed, superluminal sources associated with strong sources containing hot spots in their extended lobes can be explained by a two–fluid model[20,13]. The nucleus is supposed to eject two fluids, i.e.

i) a relativistic $e^- - e^+$ beam which bulk Lorentz factor is $\gamma_b \leq 10$ and which is responsible for the formation of superluminal radio sources,

ii) a non–relativistic $e^- - p$ jet which speed is $v_j \leq 0.45$ c and which is responsible for the formation of the hot spots and the extended lobes.

It is out of the scope of this note to investigate the description of 3C 273 using the two–fluid model and to discuss the origin of the asymmetry of the source on the kiloparsec scale. However, we will assume that the extended lobe is an $e^- - e^+$ plasma.

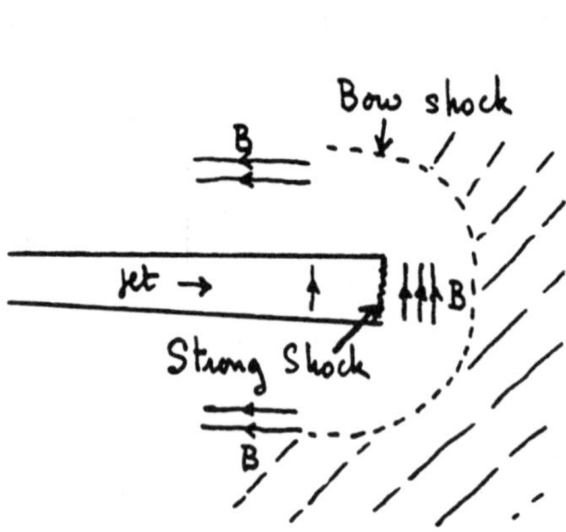

Fig. 1-a. The assumed magnetic field configuration in the jet, the hot spot and the cocoon.

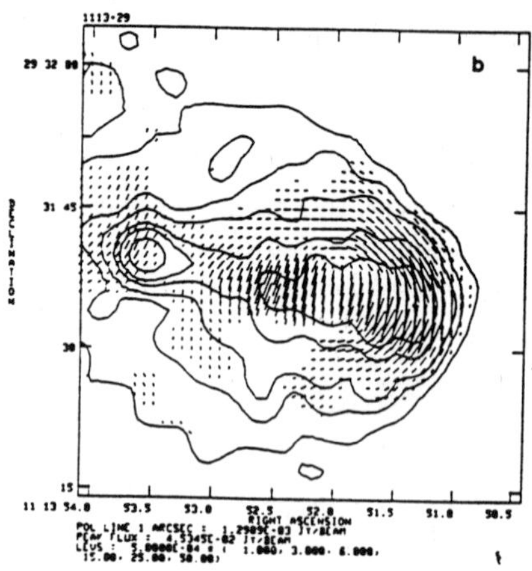

Fig. 1-b. The magnetic field configuration deduced from radio observations of B2 1113+29 by Morganti et al.[11].

Before estimating the $e^- - e^+$ annihilation losses from the lobe, let us make a general remark concerning the interpretation of the extended lobe. The formation of hot spots similar to those of Cygnus A can be understood if the magnetic field is quasi–perpendicular to the direction of the jet inside the hot spot and inside the jet[16,8]. A cocoon forms around the jet[12] and the magnetic field is parallel to the surface of the extended lobe inside the cocoon (Figure 1-a). This magnetic configuration is observed for radio sources with hot spots, see for instance the radio map of B2 1113+29 made by Morganti et al.[11] and presented in Figure 1-b.

This interpretation of the extended lobe of 3C 273 is suggested by the polarization observations[15,3] (Figure 2-a). It is also suggested by the steepening of the radio spectral index along the extended lobe from the hot spot to the nucleus[4]. This interpretation of the extended lobe of 3C 273 agrees with the general interpretation of other sources containing hot spots, but it is different from the adopted vue of 3C 273 for which the cocoon is assumed to be the jet (Figure 2-b).

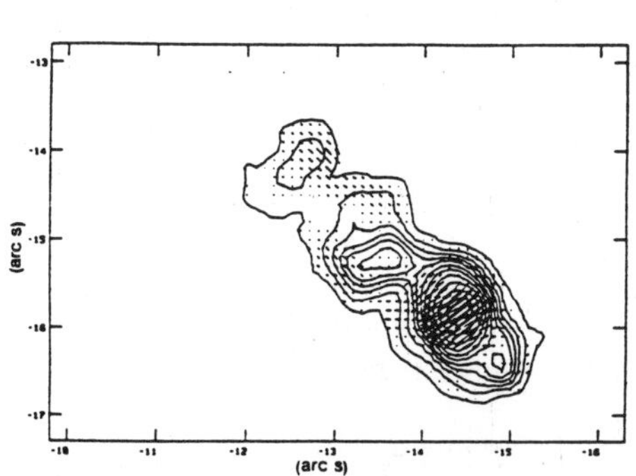

Fig. 2-a. The magnetic field configuration in the cocoon and in the hot spot of 3C 273 deduced from radio observations by Flatters and Conway[3].

Fig. 2-b. The extended radio lobe of 3C 273 from Davis et al.[2]. The cocoon is generally assumed to be the jet.

2 The e^- $-e^+$ Annihilation Line Due to the Extended Lobe

The redshift[17] of 3C 273 is $z \simeq 0.158$ and assuming for the Hubble constant $H_o \simeq 100$ km s^{-1} Mpc^{-1}, the distance of 3C 273 is $D \simeq 475$ Mpc and then $1'' \simeq 2.3$ kpc.

The length ℓ_{el}, of the extended lobe is $\ell_{el} \simeq 22''/\sin 20^o \simeq 148$ kpc.

Calling v_{ad}, the advance speed of the hot spot in the intergalactic medium, we have, independently of the speed of the jet, $0.01\ c \leq v_{ad} \leq 0.5\ c$. Thus, the age T of the source is 9.9×10^5 yr $\leq T \leq 4.9 \times 10^7$ yr.

The volume V_{el} of the extended lobe is $V_{el} \simeq \pi \times (1.4'')^2 \times 12''/\sin 20^o \simeq 7.1 \times 10^{67}$ cm^3.

Let us suppose that the extended lobe is a pure relativistic e^- $-e^+$ plasma and that it is characterized by a mean radio spectral index $\alpha \simeq 0.75$, a flux density $S \simeq 100$ Jy at 100 MHz and a low–energy cut–off $E_o \simeq 1$ MeV of the distributions of the relativistic e^- and e^+.

For the extended lobe, the parameter $\beta = p_{pa}/p_m$, where p_{pa} and p_m are the particle pressure and the magnetic pressure respectively, is of order of unity and we can deduce the mean magnetic field B_{el}, in the lobe from the synchrotron theory[5]. For $\beta \simeq 3$, it is $B_{el} \simeq 5.2 \times 10^{-5}$ G.

Now, if we suppose that there is an equilibrium between the pressure of the extended lobe and the pressure of the external medium, we have $p_{el} \simeq p_m + p_{pa} \simeq B_{el}^2/2\pi \simeq 2n_{ex}kT_x$. With $T_x \simeq 5 \times 10^7$ K, for the temperature of the intergalactic medium, we find $n_{ex} \simeq 3 \times 10^{-2}$ e^- cm^{-3}. Writing that the pressure inside the hot spot is about $10 \times p_{el}$, the advance speed of the hot spot is $v_{ad} \simeq 5B_{el}^2/\pi\rho_{ex} \simeq 0.01$ c and the age of 3C 273 is then $T \simeq 4.9 \times 10^7$ yr.

In the extended lobe, the total number N_T, of e^- or e^+ is $N_{T-} \simeq N_{T+} \simeq 7.3 \times 10^{63}$ and the densities of e^- or e^+ are $n_- \simeq n_+ \simeq 10^{-4}$ e^- cm^{-3}. From the age of 3C 273 we find that the mass ejected by the nucleus in the e^- $-e^+$ beam is $\dot{M}_b \simeq 1.3 \times 10^{-4}$ $M_\odot$ yr^{-1} and the kinetic power of the beam is $P_{kb} \simeq 8 \times 10^{43}$ erg s^{-1}.

Most of the e^- and e^+ have an energy $E_\circ \simeq 1$ MeV and the annihilation rate[21] is roughly $R_{-+} \simeq \pi\ n_+ n_- c\ r_e^2$ where $r_e = e^2/m_e c^2 \simeq 2.8 \times 10^{-13}$ cm. So, $R_{-+} \simeq 7.8 \times 10^{-23}$ cm^{-3} s^{-1}, and the total number of e^- $-e^+$ annihilations per second is $\dot{N}_{-+} \simeq 5.7 \times 10^{41}$ s^{-1}. The resulting e^- $-e^+$ annihilation line is a broad line because the e^- and e^+ are relativistic. The energy is radiated around 1 MeV and is $\dot{E}_{-+} \simeq 1.8 \times 10^{36}$ erg s^{-1}.

This value obtained for the extended lobe of 3C 273 is well below the current detection limits, but a careful determination of the annihilation rate requires to take into account the thermal e^-, which density n is $n \gg n_+ \simeq n_-$. This can be done using the two–fluid model. However, if the low energy cut–off is $E_\circ > 1$ MeV, we will have $B_{el} < 5.2 \times 10^{-5}$ G and $n_+ \simeq n_- < 10^{-4}$ e$^-$ cm^{-3}, but the value obtained before, i.e. $\dot{E}_{-+} \simeq 1.8 \times 10^{36}$ erg s^{-1} is probably a lower limit due to the annihilations with the thermal e^-.

3 Discussion and Conclusion

Assuming that the extended lobes of 3C 273 is a relativistic e^- $-e^+$ plasma and that the low energy cut-off of the distribution of the e^- and e^+ is $E_\circ \simeq 1$ MeV, we estimated the e^- $-e^+$ annihilation losses due to the lobe. We found it is $\dot{E}_{-+} \simeq 1.8 \times 10^{36}$ erg s^{-1} and the corresponding annihilation line is a wide line centered on $E_\circ$. However, a carreful description of 3C 273 needs to take into account:

i) the thermal $e^- - p$ component of the plasma, i.e. to use the two-fluid model to describe the formation of the hot spot and of the extended lobe. Consequently, the determination of the annihilation rate has to count the annihilations of the relativistic e^+ with the thermal e^-.

ii) a self-consistent determination of the low energy cut-off $E_\circ$.

If the value obtained for the extended lobe of 3C 273 is below the detection limits, the result suggests that the e^- $-e^+$ wide annihilation lines, due to all active galaxies with redshifts ≤ 4, could produce a significant contribution to the X–ray background between 200 KeV and few MeV, particularly the contribution of the compact sources associated with active nuclei, as it has been previously suggested by Kazanas and Shafer[6].

References

1 - Conway, R.G., Davis, R.J., Foley, A.R., Ray, T.P., 1981, Nature, **294**, 540
2 - Davis, R.J., Muxlow, T.W.B., Conway, R.G., 1985, Nature, **318**, 343
3 - Flatters, C., Conway, R.G., 1985, Nature, **314**, 425
4 - Foley, A.R., Davis, R.J., 1985, Monthly Notices Roy. Astron. Soc. **216**, 679
5 - Ginzburg, V.L. 1978, Physique Théorique et Astrophysique, Editions Mir, Moscow
6 - Kazanas, D., Shafer, R.A., 1983, AIP Conference Proceedings Number 101, Positron–Electron Pairs in Astrophysics, M.L. Burns, A.K. Harding and R. Ramaty Editors, A.I.P., New–York, p 343
8 - Lesch, H., Appl, S., Camendzind, M. 1989, Astron. Astrophys. **225**, 341
9 - Meisenheimer, K., Heavens, A.F., 1986, Nature, **323**, 419
10- Meisenheimer, K., Röser, H.J., Hiltner, P.R., Yates, M.G., Longair, M.S., Chini, R., Perley, R.A. 1989, Astron. Astrophys. **219**, 63
11- Morganti, R., Fanti, C., Fanti, R., Parma, P., de Ruiter, H.R., 1987, Astron. Astrophys. **183**, 203
12- Noorman, M.L., Smarr, L., Winkler, K.H.A., Smith, M.D., 1982, Astron. Astrophys. **113**, 285
13- Pelletier, G., Roland, J. 1989, Astron. Astrophys. **224**, 24
14- Fraix-Burnet, D., Pelletier, G. 1991, Astrophys. J. **367**, 86
15- Perley, R.A., 1984, in VLBI and Compact Radio Sources, R. Fanti, K. Kellerman and G. Setti Editors, Reidel, Dordrecht, p 153
16- Roland, J., Pelletier, G., Muxlow, T.W.B. 1988, Astron. Astrophys. **207**, 16
17- Schmidt, M., 1963, Nature **197**, 1040
18- Shklovsky, I.S., 1965, Sov. Astron. **9**, 22
19- Shklovsky, I.S., 1982, in Extragalactic Radio Sources, D.S. Heeschen and C.M. Wade Editors, Reidel, Dordrecht, p 475
20- Sol, H., Pelletier, G., Asseo, E. 1989, Monthly Notices, Roy. Astron. Soc. **237**, 411
21- Svensson, R., 1982, Astrophys. J. **258**, 321
22- Zensus, A.J., Unwin, S.C., Cohen, M.H., Biretta, J.A., 1990, Astron. J. **100**, 1777

High Energy γ-Ray Lines : a Probe for a Cold Dark Matter Halo *

Pierre Salati [a] † and Pascal Chardonnet [b].

a) Theory Division, CERN, CH-1211, Geneva 23, Switzerland
and
b) LAPP, Chemin de Bellevue, BP110, 74941 Annecy-le-Vieux Cedex, France

Abstract

We study the possibility of detecting halo cold dark matter through the annihilation process $\chi\chi \to \gamma\gamma$. This process produces monoenergetic γ-rays, and may be a clear signature of particle dark matter. If there is a closure density of photino – so far a favoured candidate – we show that it will be very difficult to observe this annihilation line from a satellite or space station borne experiment. On the contrary, triplet neutrinos which have recently been suggested as a plausible solution to the dark matter conundrum, whilst elusive at accelerators, should produce a clear γ-ray line signal, well above background.

1 - The CDM annihilation γ-ray line signal.

There is a considerable interest in the possible detection of dark matter [1] through astrophysical observations as well as terrestrial observations [2,3]. If dark matter is made of massive, weakly interacting particles, one of the best indirect signatures would be the detection of annihilation products of these particles [4]. It has been recently suggested [5,6,7] that if particles in the mass range 1 GeV to 100 GeV pervade the halo of our Galaxy and if these species self-annihilate, as most of the potential candidates do, then monoenergetic γ-rays should be observed when the cold dark matter particles annihilate directly into two photons. The energy of these γ's is very nearly equal to the mass of the (non-relativistic : $v/c \sim 10^{-3}$) parent particles. Very narrow γ lines in the range 1 GeV to 100 GeV would provide a clear signature for dark matter annihilation, if they stand out above the background.

We assume that the dark matter halo of our Galaxy is nearly spherical and nearly isothermal. Then, the dark matter density varies with the distance r to the galactic center as :

$$\rho(r) = \rho_{\odot} \frac{a^2 + r_{\odot}^2}{a^2 + r^2} \ , \tag{1}$$

where $\rho_{\odot}$ is the dark matter density in the solar neighbourhood and a is the core radius. We consider the annihilation process $\chi\chi \to \gamma\gamma$, the cross-section for which is $\sigma_{2\gamma}$. The mass of the dark matter species χ ranges from 1 GeV to 100 GeV according to specific

*Contribution to the International Symposium on Gamma-Ray Line Astrophysics, held at Saclay, Paris, France, on December 10-13 1990.

†On leave of absence from LAPP, BP110, 74941 Annecy-le-Vieux Cedex, France and from Université de Chambéry, 73000 Chambéry, France.

models, whereas their mean velocity in the halo is $v \sim 300\,\mathrm{km\,s^{-1}}$. The energy E_γ of the outgoing γ's is therefore practically equal to the particle mass m_χ. The number of photons received at a detector in the solar system, from pair annihilations taking place at a distance R and in the direction defined by galactic latitude b and longitude l is [8] :

$$F = \frac{\sigma_{2\gamma}\, v}{4\pi} \int_0^\infty \left[\frac{\rho(R,b,l)}{m_\chi} \right]^2 dR \;\; \mathrm{cm^{-2}\,s^{-1}\,sr^{-1}} \tag{2}$$

when summed over the distance R. This expression may be evaluated analytically and at high galactic latitudes ($b \sim 90^o$) – for which the galactic diffuse background is expected to be the faintest – the flux reduces to :

$$F \sim \frac{\sigma_{2\gamma} v}{16} \left(\frac{\rho_\odot}{m_\chi} \right)^2 a\,\sqrt{1 + A^2} \;\; , \tag{3}$$

where A denotes the ratio $r_\odot / a$. Taking $\rho_\odot = 0.4\,\mathrm{GeV\,cm^{-3}}$ and $a \sim r_\odot \sim 8$ kpc, and assuming a typical value of $\sigma_{2\gamma} v \sim 10^{-30}\,\mathrm{cm^3\,s^{-1}}$ for the γ-ray pair annihilation cross-section, we find that a detector of $1\,\mathrm{m^2\,sr}$ effective area, such as ASTROGAM or HR-GRAF [9], will collect N_{line} photons of energy $E_\gamma = m_\chi$ per year, where :

$$N_{line} \sim 110 \text{ photons} \left(\frac{1\,GeV}{E_\gamma} \right)^2 \; . \tag{4}$$

2 - The various γ backgrounds.

There is no measurement of γ-ray fluxes in the GeV-TeV energy range, so we must extrapolate from lower energy data. There are many different sources of background γ photons. The γ's from the decay of π^0's produced by collisions of cosmic ray protons with the interstellar medium are probably the dominant source of galactic background at energies larger than 1 GeV [10]. Bremsstrahlung γ's from cosmic ray electrons should be negligible above 1 GeV as well as the inverse Compton emission resulting from the interaction between ultra-relativistic electrons and the electromagnetic field of the galaxy. From the known distribution of cosmic ray protons, one expects a flux [11] :

$$\frac{dN_{galactic}}{dE_\gamma} \sim 8 \times 10^{-7}\,\text{photons}\,\mathrm{cm^{-2}\,s^{-1}\,sr^{-1}\,GeV^{-1}} \left(\frac{1\,GeV}{E_\gamma} \right)^{2.7 \pm 0.3} \tag{5}$$

at high galactic latitudes, and a correspondingly larger flux in the galactic plane.

The extra-galactic background is not well known, and different analyses do not agree on the extrapolation of COS-B and SAS-2 measurements to energies larger than 1 GeV [12]. Since it may drop sharply above 1 GeV, we do not consider it any further, but keep in mind the possibility that it could be of the same order of magnitude as the galactic background.

Dark matter provides its own background when it annihilates into fermion-antifermion pairs, which give γ-rays among their decay and annihilation end-products, but these γ's have an energy lower than m_χ (for trivial kinematical reasons), and do not contribute to the line background.

To summarize, we take Equ. 5 as representing the background, but we must keep in mind that it could be an order of magnitude smaller or larger. Since the aim is to detect a narrow line, much narrower than the energy resolution of the detector, the interesting quantity is the number of photons received in one energy bin during the time of observation. For a $1\,\mathrm{m}^2\,\mathrm{sr}$ detector with a 1% energy resolution (bin width $\Delta E_\gamma = 0.01\,E_\gamma$), one year of observation will give :

$$N_{background} \sim 2500\,\text{photons} \left(\frac{1\,GeV}{E_\gamma}\right)^{1.7\pm0.3} . \tag{6}$$

If one compares the expected signal N_{line} (Equ. 4) to the background $N_{background}$ (Equ. 6), the situation does not seem very promising. But this comparison is actually meaningless. We should *not* compare the expected signal to the background but to the *noise*, i.e. the error in measuring the background. The statistical uncertainty is the square root of the number of photons received :

$$Noise \sim 50\,\text{photons} \left(\frac{1\,GeV}{E_\gamma}\right)^{0.85\pm0.15} . \tag{7}$$

3 - The photino case.

The photino $\tilde{\gamma}$ is the neutral spin 1/2 supersymmetric partner of the photon, and one of the foremost candidates for dark matter because it naturally implies a near-critical density for the universe. Photino annihilation proceeds through the exchange of the so-called sfermions – the supersymmetric spin 0 partners of the conventional fermions such as the electron or the quarks – whose mass is generically denoted by $\tilde{M}$. The larger $\tilde{M}$, the lower the annihilation cross section and the larger the fossil density $\Omega_{\tilde{\gamma}}\,h^2$ [13]. The latter is actually related to the two-photon annihilation cross-section through :

$$\sigma_{2\gamma} v \sim 3.1 \times 10^{-31} \mathrm{cm}^3\,\mathrm{s}^{-1}\ (\Omega_{\tilde{\gamma}}\,h^2)^{-1} , \tag{8}$$

so that an ASTROGAM caliber γ-ray telescope, aimed toward the galactic pole, should collect each year :

$$N_{line} \sim 34\,\text{photons} \left(\frac{1\,GeV}{E_\gamma}\right)^2 (\Omega_{\tilde{\gamma}}\,h^2)^{-1} . \tag{9}$$

However, experiments at the Large Electron Positron storage ring, LEP, at CERN have recently set the constraint $\tilde{M} > 45$ GeV from their unavailing searches for supersymmetric fermions [14]. That lower bound on the generic sfermion mass $\tilde{M}$ translates into an upper bound on the annihilation cross-section $\sigma_{2\gamma} v$. For a 1 m^2 sr detector, the γ-ray line has signal to noise ratio constrained to be < 3. Even in the best case, *i.e.*, for $\Omega_{\tilde{\gamma}}\,h^2 = 1/40$, detection of photinos in the galactic halo would imply a significantly larger telescope area. In order to achieve such a detection at, say, the 5 σ level, a 3 m^2 sr γ-ray detector should operate during an entire year.

4 - Heavy neutrinos, LEP and the γ-ray line signal.

Experiments at LEP have also measured the total decay rate of the Z^0 boson [15]. From its invisible width, the conventional assumption that there are only three light neutrino families turns out to be distressingly valid. The LEP determination of the number of invisible species ruins an appealing explanation of the missing mass conundrum in terms of a fourth generation heavy neutrino. The latter should be heavier than ~ 45 GeV in order to escape detection at LEP. For such a large mass, neutrinos annihilate so efficiently in the early universe that their fossil density is too low to be cosmologically relevant [16].

There is however a possible way out. Note that the Dirac or Majorana neutrinos so far considered have always been assumed to be embedded inside a gauge *doublet* as the conventional electron neutrino ν_e and its associated electron are. Such an assumption may actually be relaxed. A heavy gauge *triplet* neutrino, N, has recently been suggested [17] as a plausible dark matter candidate. This species and its singly charged gauge partner E have gauge triplet structure : (E^+, N, E^-). Since the neutral particle N does not couple to the Z^0, it cannot be produced at LEP and is therefore invisible, at least in the first stage of the LEP experiments. N however couples to E through W boson exchange so that the delicate interplay between both species in the early stages of the Big-Bang leads to a closure fossil density of neutrinos, provided the neutrino mass m_N lies in the range 30–80 GeV. Note that if on the one hand N cannot be produced directly at accelerators, on the other hand its charged companion E is expected to be detected at LEP as soon as the beam energy is cranked up to ~ 100 GeV.

Another signature of the triplet neutrino model is precisely the fairly strong γ-ray line signal from galactic halo particles. Gauge couplings are significantly larger for triplet than for doublet species, so that the cross-section for the annihilation process $NN \to \gamma\gamma$ is quite large :

$$\sigma_{2\gamma} v \simeq \frac{96}{\pi^3} \alpha^2 {\rm G_F}^2 m_N^2 \ , \tag{10}$$

where $\alpha = 1/137$ is the electromagnetic fine structure constant, $\rm G_F$ is the Fermi constant and m_N is the neutrino mass. Note that m_N gracefully cancels out from expressions (3) and (10). The flux F_N of monochromatic photons with energy $E_\gamma = m_N$ due to neutrino N annihilations in the galactic halo does *not* depend therefore on the mass of this species : $F_N \simeq 10^{-10}$ photons $\rm cm^{-2}\,s^{-1}\,sr^{-1}$.

We infer a signal of order ~ 30 monochromatic photons (see Fig. 1) detected each year at a satellite or space station borne γ-ray telescope with a 1 $\rm m^2\,sr$ effective area, in the energy range between 30 and 80 GeV. At these energies, the corresponding background is fairly small. If the latter is well modelled by expression (5), a 5% energy resolution corresponds to, at most, ~ 40 background photons collected in the same energy bin as the annihilation line. Detection of the γ-ray line signature of galactic halo triplet neutrinos could therefore be achieved at least at the 5 σ level, a fairly comfortable situation. We therefore urge high energy astrophysicists to explore the γ-ray sky and to hunt for such a signal. Should a narrow γ-ray line be discovered, the nature of the galactic halo would be unravelled.

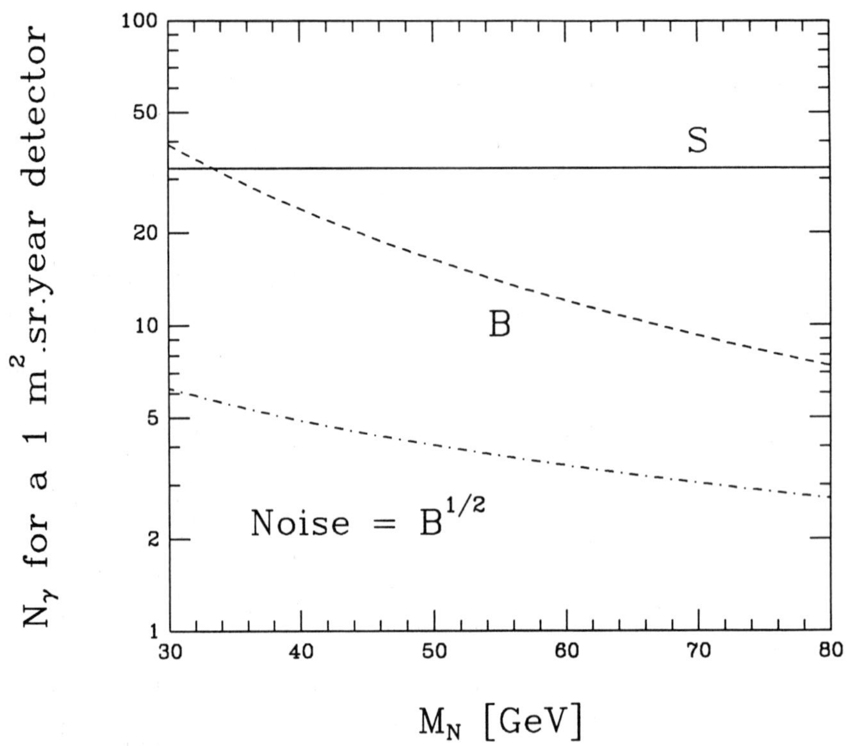

Figure 1

Expected signal (S), background (B) and noise for monoenergetic γ-rays due to pair annihilation of triplet neutrinos N in the halo of our galaxy. The full line displays the total number of annihilation γ's received for an entire year of exposure on a detector with a 1 m^2 sr detector surface and a 5% energy resolution. The short dashed line corresponds to the number of background photons collected in the same energy bin as the annihilation line, while the dotted - short dashed curve indicates the background noise. Note that if on the one hand the signal does not depend on the mass m_N of the heavy triplet neutrino, on the other hand, background and noise do decrease with line energy $E_\gamma = m_N$. The heavier the neutrino, the easier the detection. Even for $m_N =$ 30 GeV, detection will be achieved at the 5 σ level wilst for $m_N = 80$ GeV, the signal to noise ratio may reach up to $\sim$ 10.

We would like to express our gratitude towards C. Césarsky, P. Durouchoux and N. Prantzos for giving us the opportunity to contribute to the Proceedings of the International Symposium on γ-Ray line Astrophysics.

References

[1] J. Kormendy, and G.R. Knapp, *In* Dark matter in the Universe, IAU Symposium **117** (Reidel) (1987).

[2] J. Primack, D. Seckel, and B. Sadoulet, Ann. Rev. Nucl. Part. Phys. **38**, 751 (1988).

[3] P.F. Smith, and J.D. Lewin, Report Rutherford RAL-88-045 (1988).

[4] J. Silk, and M. Srednicki, Phys. Rev. Lett. **53**, 624 (1984).

[5] L. Bergström, and H. Snellman, Phys. Rev. **D37**, 3737 (1988).

[6] S. Rudaz, Phys. Rev. **D39**, 3549 (1989).

[7] A. Bouquet, P. Salati, and J. Silk, Phys. Rev. **D40**, 3168 (1989).

[8] M.S. Turner, Phys. Rev. **D34**, 1921 (1986).

[9] J.H. Adams *et al.*, Naval Research Laboratory proposal T-248-89 (1989); E.J. Fenyves, NASA Code E proposal UTD No. 890061 (1989).

[10] J.B.G.M. Bloemen, Astrophys. J. Lett. **317**, L15 (1987).

[11] F.W. Stecker, Astrophys. J. **223**, 1032 (1978).

[12] C.E. Fichtel and D.J. Thompson, Astron. Astrophys. **109**, 352 (1982).

[13] H. Goldberg, Phys. Rev. Lett. **50**, 1419 (1983).

[14] See for instance the DELPHI Collaboration, P. Abreu *et al.*, CERN-EP/90-79 preprint (1990).

[15] L3 Collaboration, B. Adeva *et al.*, Phys. Lett. **B231**, 509 (1989);
ALEPH Collaboration, D. Decamp *et al.*, Phys. Lett. **B231**, 519 (1989);
OPAL Collaboration, M.Z. Akrawy *et al.*, Phys. Lett. **B231**, 530 (1989);
DELPHI Collaboration, P. Aarnio *et al.*, Phys. Lett. **B231**, 539 (1989).

[16] K. Griest and J. Silk, Nature **343**, 26 (1990).

[17] P. Salati, preprint CERN-TH-5868/90 (1990); to be published in Phys. Lett. **B**.

SPECTROSCOPY WITH ENRICHED DETECTORS: DOUBLE BETA DECAY AND PERSPECTIVES IN ASTROPHYSICAL γ-RAY SPECTROSCOPY AND IN DARK MATTER DETECTION.

H.V. Klapdor-Kleingrothaus

Max-Planck-Institut für Kernphysik, Heidelberg, Germany

Abstract: A new era, of second generation experiments, using detectors made from enriched material is starting at present in double beta research. This allows to strongly improve the present limits on the neutrino mass. We review the present status, in particular of the HEIDELBERG-MOSCOW experiment, by which for the first time evidence has been given that the technology of enriched HP Ge detector production can be mastered nowadays. We further explore possible future application of this technology in astrophysics, in particular dark matter search and high-resolution c-ray spectroscopy in future ESA and NASA satellite missions. Use of enriched ^{73}Ge detectors for the former would allow to extend the present LEP limits for spin-interacting WIMPs considerably and use of enriched ^{70}Ge detectors in the satellite experiments would allow to drastically reduce the b-background in these missions.

1. Introduction

The recognition of the close connection between the laws of microphysics (nuclear and particle physics) and macrophysics (astrophysics and cosmology) is one of the most important discoveries of this century (see Fig. 1 and [1-4]).
The barriers between these 'classical' disciplines are becoming in some sense more and more meaningless. Nuclear physics, e.g.,may give - as non-accelerator particle physics - important contributions to beyond standard model physics

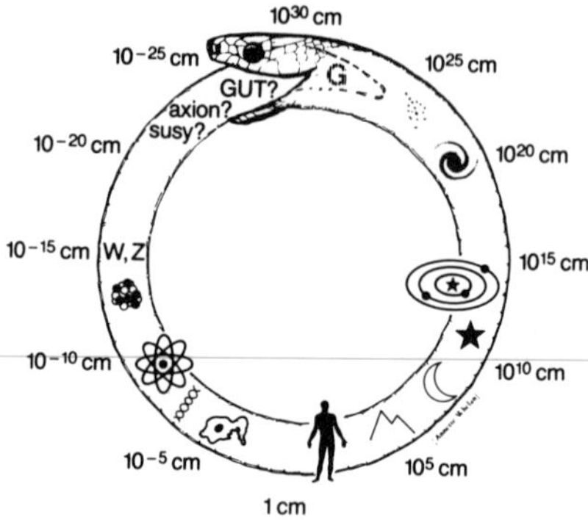

Fig. 1: Illustration of the interconnection between different physical disciplines (after Glashow, see [2]).

(double beta decay, proton decay,...). One of the best examples for this situation is given by the neutrino (Fig.2). The neutrino plays - by its nature (Majorana or Dirac particle) and its mass - a key role for the structure of modern particle physics theories (GUTs, SUSYs, SUGRAs,...), which is at least as important as the role it played at the time of Pauli and Fermi for the understanding of the weak

interaction. It is at the same time candidate for non-baryonic dark matter in the universe, and thus important for cosmology, it plays a decisive role for the energy transport in supernovae, and solar neutrino observation might help to answer the beyond standard model question of neutrino-oscillations. On the other hand the presently sharpest limits on the electron neutrino mass are coming from nuclear physics experiments: double beta decay, tritium decay and reactor neutrino oscillation experiments.
Single β-decay of nuclei far from stability is one of the most important input parameter for star explosion calculations and in particular for studies of the synthesis of heavy elements in the universe and of its age by cosmochronometers (see [2,5]), which again is connected by cosmology with the cosmological constant of general relativity and in this way with grand unified models. The nuclear beta strength distribution determines further the efficiency of solar neutrino detectors,

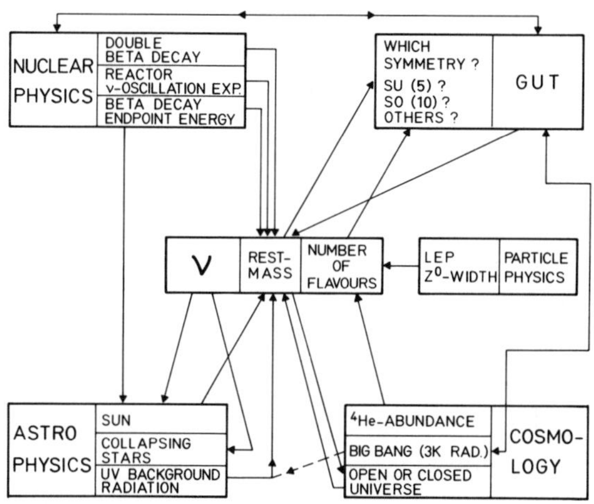

Fig. 2: The neutrino and its role in micro- and macrophysics.

such as the Gallium detector [6] - just to mention some of these interdisciplinary connections.
The use of enriched detectors in some experimental efforts in these fields allows a major step forward in sensitivity, and at the same time sometimes allows miniaturization of the experiments. Since the first application of enriched detectors is going on in double beta decay, we discuss in section 2 the status of this field. In sections 3,4 we discuss possible future applications in dark matter search and astrophysical high-resolution γ-ray spectroscopy in future ESA and NASA satellite experiments.

2. Double Beta Decay and Neutrino Mass

2.1. Neutrino mass and particle theories (GUTs, SUSYs)

The neutrino mass is one of the key quantities for the structure of grand unified theories (GUTs, SUSYs, SUGRAs) [1,2]. In GUTs, beyond minimal SU(5), e.g., SO(10), there exists a right-handed neutrino (the SU(5) singlet in Fig. 3), its Dirac mass being in lowest order of the order of the u quark mass. To obtain consistency with observation, the way of the see-saw mechanism [7] is to introduce an even larger right-handed Majorana mass term $m^M{}_R$ in the neutrino mass matrix which is, simplified to the one flavour case

$$(\bar{\nu_L}\overline{(\nu_R)^C})\begin{pmatrix} 0 & m_D \\ m_D & m_R^M \end{pmatrix}\begin{pmatrix} (\nu_L)^C \\ \nu_R \end{pmatrix} \qquad m_R^M \gg m_D \qquad m_D \simeq MeV$$

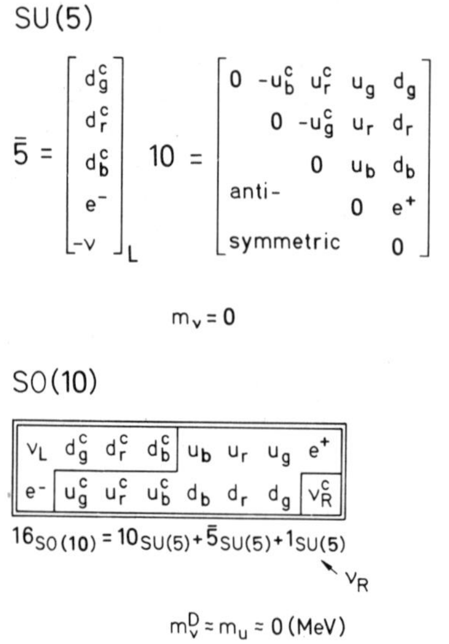

SU(5)

$$\bar{5} = \begin{bmatrix} d_g^c \\ d_r^c \\ d_b^c \\ e^- \\ -\nu \end{bmatrix}_L \quad 10 = \begin{bmatrix} 0 & -u_b^c & u_r^c & u_g & d_g \\ & 0 & -u_g^c & u_r & d_r \\ & & 0 & u_b & d_b \\ \text{anti-} & & & 0 & e^+ \\ \text{symmetric} & & & & 0 \end{bmatrix}$$

$m_\nu = 0$

SO(10)

ν_L	d_g^c	d_r^c	d_b^c	u_b	u_r	u_g	e^+
e^-	u_g^c	u_r^c	u_b^c	d_b	d_r	d_g	ν_R^c

$16_{SO(10)} = 10_{SU(5)} + \bar{5}_{SU(5)} + 1_{SU(5)}$ ↖ ν_R

$m_\nu^D \approx m_u \approx 0\,(\text{MeV})$

Fig. 3. Multiplets of the fermions of the first generation in the SU(5) and SO(10) model. The right-handed neutrino (left-handed antineutrino), not existing in SU(5), is a natural consequence in SO(10).

This leads to the phenomenologically required light (Majorana) neutrino

$$\nu_1 = \nu_L - \frac{m_D}{m_R^M}\nu_R \quad with \quad m_1 \approx \frac{(m^D)^2}{m_R^M} < (eV)$$

and to a heavy (Majorana) neutrino with $m_2{}^M \cong m_R{}^M$.

Introduction of $m_R{}^M$ is, on the other hand, equivalent to (B-L) nonconservation

and equivalent to the occurrence of neutrinoless double beta decay.

Since $m_R{}^M$ is produced in GUTs when breaking left-right symmetry, its value and

consequently the neutrino mass can range over a wide scale, depending on the energy scale at which left-right symmetry is broken [8]. This might happen in the range of the two

examples

$$SO(10) \xrightarrow{M>M_X} SU(5) \xrightarrow{M_X} SU(3)_c \otimes SU(2)_L \otimes U(1) \quad (a)$$

or

$$\mathrm{SO(10)} \rightarrow \mathrm{SU(4)_{EC}} \otimes \mathrm{SU(2)_L} \otimes \mathrm{SU(2)_R} \quad \text{(b)}$$

extension of strong interaction (extended color) with leptons as 4. color charge — right-handed bosons $W_R\pm$

between $\simeq 10^{15}$ GeV (a) or $\simeq 100$ GeV (Pati-Salam model, (b)). This is the energy range probed by $0\nu\beta\beta$ decay, which is reflected directly by the neutrino mass (see Table 1, from [8]).

Consequently the predictive power of GUTs for the neutrino mass is extremely weak: $m_\nu \simeq 10^{-11}$ - $\simeq$ 1eV, (see Table 1), which on the other hand reflects the

Table 1: Models of neutrino mass, along with their most natural scales for the light neutrino masses.

Model	m_{ν_e}	$\langle m_{\nu_e}\rangle$	m_{ν_μ}	m_{ν_τ}
Dirac	$1-10\ MeV$	0	$100\ MeV - 1\ GeV$	$1-100\ GeV$
pure Majorana (Higgs triplet)	arbitrary	m_{ν_e}	arbitrary	arbitrary
GUT seesaw ($M \sim 10^{14}\ GeV$)	$10^{-11}\ eV$	m_{ν_e}	$10^{-6}\ eV$	$10^{-3}\ eV$
intermediate seesaw ($M \sim 10^{9}\ GeV$)	$10^{-7}\ eV$	m_{ν_e}	$10^{-2}\ eV$	$10\ eV$
$SU_{2L} \times SU_{2R} \times U_1$ seesaw ($M \sim 1\ TeV$)	$10^{-1}\ eV$	m_{ν_e}	$10\ KeV$	$1\ MeV$
light seesaw ($M \ll 1\ GeV$)	$1-10\ MeV$	$\ll m_{\nu_e}$	-	-
charged Higgs	$< 1\ eV$	$\ll m_{\nu_e}$	-	-

importance of experimental information for selecting the "right" GUT model. In SUSY models such as a supersymmetric extension of the standard model

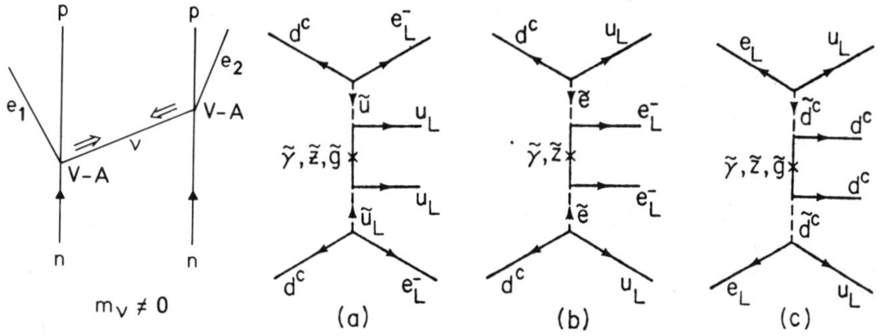

Fig. 4: Dominating transition modes for $0\nu\beta\beta$ decay in GUT models (left) and in supersymmedtric extension of the standard model. In the latter the neutrino exchange of the GUT models is replaced by gaugino (gluino, photino, zino) exchange.

neutrino exchange - dominating $0\nu\beta\beta$-decay in GUT models - would give only a small contribution to $0\nu\beta\beta$ decay, and the latter would be dominantly induced by gaugino (gluino, photino and zino) exchange [9] (see Fig. 4).
Thus, also in the case that the neutrino mass would be so small that it is not in the range which can be probed by $\beta\beta$ decay, such experiments would remain in any case a sensitive probe of physics beyond the standard model.

2.2. Double Beta Decay and Neutrino Mass

2.2.1. The required matrix elements

There are two main decay modes in $\beta\beta$-decay, two-neutrino (2ν) decay

$$ {}^A_Z X \rightarrow {}^A_{Z+2} X + 2e^- + 2\bar{\nu} $$

and neutrinoless (0ν) decay

$$ {}^A_Z X \rightarrow {}^A_{Z+2} X + 2e^- $$

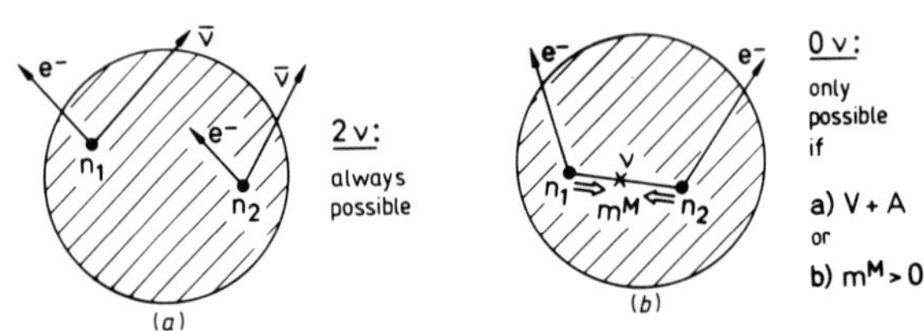

Fig. 5: 2ν and 0ν $\beta\beta$-decay, schematic.

The first process is a usual effect of second order in the classical weak interaction, the second, however, requires as a prerequisite a non-vanishing Majorana neutrino mass or a right-handed weak interaction (Fig. 5). In the framework of GUTs, or more general gauge theories, both of these conditions cannot be seen independently, but in any case a non-vanishing mass is required.
The half life for neutrinoless $\beta\beta$-decay $T_{1/2}$ is given by [10,11]

$$ \left[T^{0\nu}_{1/2}(0^+_i \rightarrow 0^+_f)\right]^{-1} = C_{mm}\frac{< m_\nu >^2}{m_e^2} + C_{\eta\eta} < \eta >^2 + C_{\lambda\lambda} < \lambda >^2 + C_{m\eta} < \eta > \frac{< m_\nu >}{m_e} $$

$$ + C_{m\lambda} < \lambda > \frac{< m_\nu >}{m_e} + C_{\eta\lambda} < \eta >< \lambda > $$

or, when neglecting the effect of right-handed weak currents, by

$$ \left[T^{0\nu}_{1/2}(0^+_i \rightarrow 0^+_f)\right]^{-1} = C_{mm}\frac{< m_\nu >^2}{m_e} = (M^{0\nu}_{GT} - M^{0\nu}_F)^2 G_1 \frac{< m_\nu >^2}{m_e} $$

where G1 denotes the phase space integral.

Thus a reliable mass can be extracted from a measured rate only if the nuclear matrix element $M^{0\nu} = (M^{0\nu}_{GT} - M^{0\nu}_{F})$ can be reliably calculated.
In the calculation of $2\nu\beta\beta$ decay one finds a strong sensitivity of the calculated matrix elements on details of the wave functions of initial and final states - and a cancellation reminiscent of a broken symmetry requiring $M^{2\nu}=0$. In 0ν decay this sensitivity fortunately is not present, the essential reason being the excitation of states of various multipolarities in the intermediate nucleus in this case. For details we refer to [12-15].

We have calculated for 2ν and 0ν decay with a realistic interaction (G matrix from Paris potential) the $M^{0\nu}$ for all potential $\beta\beta$-emitters with $A > 70$ [12,16] and particularly for the seven nuclei of main experimental interest, and extracted limits for a (Majorana) neutrino mass from experimental half-life limits. The results are shown in Table 2 (for references concerning the experimental limits see [11]). Where we can compare to the Tübingen calculation [17], also using a realistic interaction (G matrix from Bonn potential) we find good

Table 2. Present status of (effective) neutrino masses deduced from double beta decay experiments, for the nuclear matrix element of various nuclear structure calculations. Contributions of right-handed currents are neglected. Experimental $0^+ \rightarrow 0^+$ 0ν $\beta\beta$ decay half-life limits are given in the last line.

Refs.	^{48}Ca	^{76}Ge	^{82}Se	^{128}Te	^{130}Te	^{100}Mo	^{136}Xe	^{150}Nd
HS84	<40	<1.8	< 7.3	<0.90	<10	--	--	--
GK86								
no pp force		<0.7	< 2.9	<0.43	< 5.4	< 8.7	<13.3	<3.2
TF87		<2.0	< 7.4	<1.4	<19	--	--	--
MBK89		<2.1	< 7.4	<1.2	<18	<54	<36	<3.8
EVZ88+		<7-17	<32-64	<2.2-3	<29-39	<66	<61-80	--
$T^{0\nu}_{1/2}[y]$	$>2x10^{21}$	$>5x10^{23}$	$>1.1x10^{22}$	$>5x10^{24}$	$>1.5x10^{21}$	$>4.4x10^{20}$	$>1.7x10^{21}$	$>2.3x10^{21}$

+Not realistic (zero range) interaction

agreement (see Table 2). The 0ν results remain also stable against different choices of the renormalized effective interactions, e.g. Bonn-A and Reid potentials instead of the Paris potential [18]. A discrepancy occurs only when a not realistic interaction is used (as in [19]).

Concluding, 0ν $\beta\beta$ decay remains the most powerful tool for probing the mass of the electron neutrino.

2.2.2 Future experimental possibilities. The Heidelberg-Moscow experiment using enriched ^{76}Ge

Double beta decay yields at present the sharpest limit on the Majorana mass of the neutrino (see Table 2). Among the direct (non-geochemical) experiments the most stringent limit is given by 0ν $\beta\beta$ decay of ^{76}Ge: $m_\nu < 1.4$ eV, corresponding to the claimed present half-life limit of $T^{0\nu}_{1/2} > 1.2 * 10^{24}$ years [20,21].
In next generation $\beta\beta$ decay experiments ^{76}Ge will play also the most important role. Use of enriched ^{76}Ge (86%) [22] instead of the presently used detectors from natural Ge (containing 7.8% of ^{76}Ge) could explore the half-life of 0ν $\beta\beta$ decay up to $\sim 10^{25}$ years and correspondingly the neutrino mass down to $\sim 10^{-1}$ eV,

probing a class of left-right symmetric GUT models, with a right-handed Majorana mass term of about 1 TeV, based on $SU(2)_L * SU(2)_R * U(1)$ (see Table 1).

The Heidelberg-Moscow $\beta\beta$ experiment [22] makes use of 16.9 kg of ^{76}Ge metal enriched to 86%, corresponding to 14.5 kg of the isotope ^{76}Ge. The full amount of ^{76}Ge has been transferred from Moscow to Heidelberg. Up to now one enriched detector of ~1 kg and 1 crystal of 3.4 kg (the largest ever produced p-type Ge crystal) have been produced. The detector is running in the GRAN SASSO Underground Laboratory in Italy since end of July 1990. Figures 6,7 show the Heidelberg-Moscow $\beta\beta$-Laboratory built generously by the INFN in the GRAN SASSO, and the first detector in its extreme low-level shielding from electrolytic copper and special lead. The results of the first 108 days of measuring are $T_{1/2} > 5.66 * 10^{23}$ (90% confidence limit) or $> 1.19 * 10^{24}$ years (68% confidence limit). The corresponding limits for the neutrino mass are $m_\nu < 2.0$ and < 1.4 eV. The background level which characterizes the quality of the setup is 0.45 ± 0.14 events/kg year keV (Fig. 8) in an 80 keV interval around the hypothetical 2041 keV $0\nu\beta\beta$-line (10 events after 108 days). The prospects of the experiment, compared to standard experiments using natural Ge, are illustrated in Fig. 9.

Fig. 6. The $\beta\beta$-laboratory of the HEIDELBERG-MOSCOW experiment in the GRAN SASSO near Rome.

The new experiment indeed will represent a major step forward to higher sensitivity and lower neutrino mass. The time scale for the full experiment is approximately 5 years.

In the context of this conference the important message is that the results of the HEIDELBERG-MOSCOW experiment show for the first time that the technology of production of enriched HP Ge-detectors can be handled. In the next section we explore possible consequences for dark matter search and galactic γ-spectroscopy satellite missions (see also [3]).

Fig. 7. The worldwide first enriched HP-^{76}Ge detector (with Si cup) in its extreme low-level shielding.

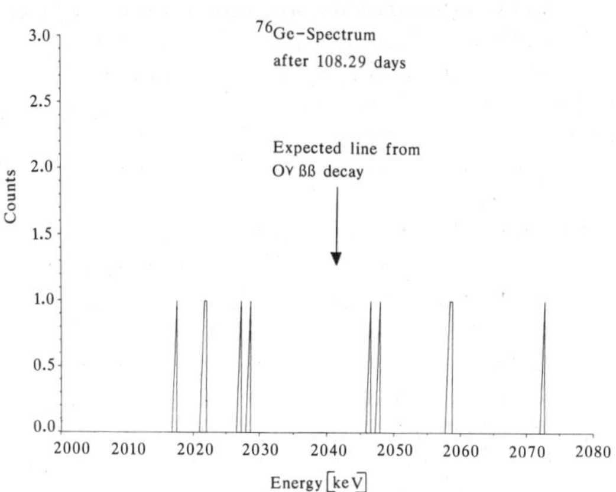

Fig. 8. The measured spectrum with the first HP-^{76}Ge detector after 108 days of measurement in the range of the hypothetical 0ν $\beta\beta$ line at 2041 keV.

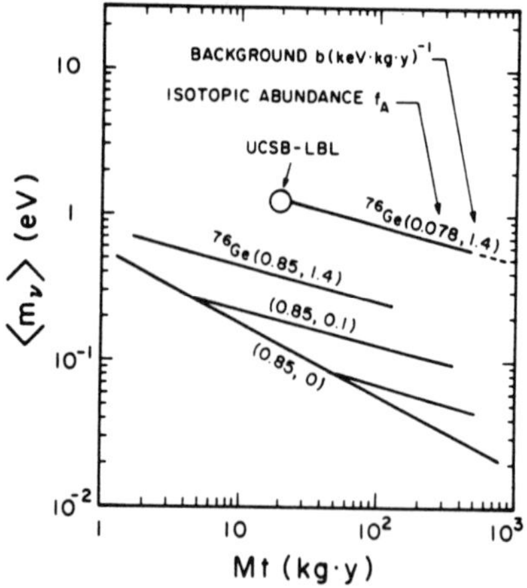

Fig. 9. The ranges of neutrino mass which can be probed with Ge-detectors of different enrichment in 76Ge, as function of the product detector mass times measuring time (kg y) (from Moe, 1990).

3. Possible applications of enriched detector technology

3.1. Dark matter search

An early inflationary phase of the evolution of the universe favours a present mass density of the universe which should agree with the critical density (if a vanishing cosmological constant is assumed, which is however not cogent [2,23]). The mass density of baryonic matter which can be deduced from early nucleosynthesis [2,4] is however much smaller (~10% of the critical value). This is seen by many astrophysicists and cosmologists as an indication of the existence of nonbaryonic dark matter. There are many candidates for dark matter, including neutrinos, monopoles, or certain particles from the SUSY models (e.g. LSP, the lightest supersymmetry theory predicted particle). Dark matter in the form of the above WIMPs (weakly interacting massive particles) may, under certain circumstances, explain the recently discovered large scale structure of the universe (see [2,4]).

One method to search for spin-independently interacting dark matter is using normal Ge detectors working in underground experiments. Search for this type of WIMPs by such $\beta\beta$-type experiments [24,20] excluded several types of WIMPs and a wide range of masses between 12 GeV and 3 TeV (and excluded spin-independent cosmions as a solution of the solar neutrino problem). They extended the limits obtained by LEP by an order of magnitude (Fig. 10). However, a large class of particle candidates proposed for the dark matter solution are only interacting via a spin-dependent interaction. To these belong Majorana fermions and the LSP. All presently existing limits concerning spin-dependent dark matter candidates come from ^{73}Ge (I=9/2) in normal Ge-detectors (natural abundance 7.8%). The Gotthard group [24] and the UCSB group [25] used about 70g and 140g of ^{73}Ge for their published results. So use of enriched ^{73}Ge detectors of about 10 kg would increase the present limits by about two orders of magnitude. Complementary motivation for search of dark matter with ^{73}Ge comes from LEP at CERN. From the measurements of the invisible Z^0 partial width, it can be concluded (in addition to three families of 'light' neutrinos) that there is no additional Majorana fermion lighter than 14 GeV. To say it in an other way: There is no additional particle with mass smaller 45 GeV, having at least 10% of the neutrino -Z0 coupling strength. Thus, if there exists supersymmetry, the LSP should be heavier than 20 GeV and couple weakly to ordinary matter. An enriched ^{73}Ge detector would extend (Fig.11) the range of sensitivity from masses more than 12 GeV, overlapping with the LEP results to masses of several TeV [27]. There exists also the possibility of spin-dependent cosmions, not only solving the dark matter problem, but also the solar neutrino problem. Some models introduce a separate gauge boson Z' which couples only weakly to the Z0 and are in agreement with all experimental data. They can only be ruled out by a ^{73}Ge experiment.

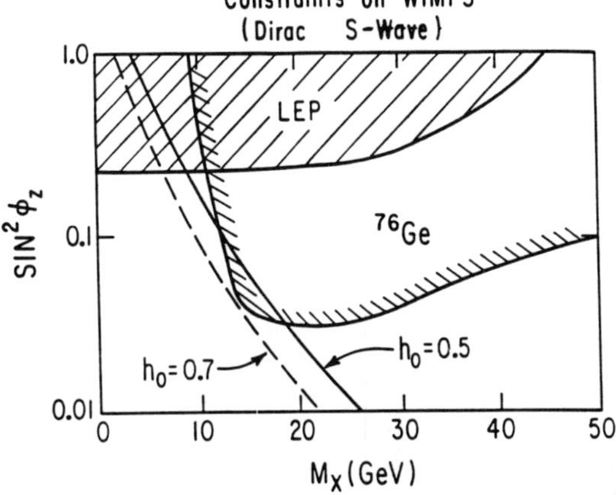

Fig. 10: Constraints on WIMPs of mass M_x versions $\sin^2 \phi_z$, the relative coupling to the Z°. The constraints are shown assuming Dirac particles (s-wave interactions). The diagonal lines show the combinations of M_x and $\sin^2 \phi_z$, that yield $\Omega=1$. The cross-hatched region is what is ruled out by the current LEP results. The ^{76}Ge region is that ruled out by the Caldwell et al. $\beta\beta$-style experiments (from [26]).

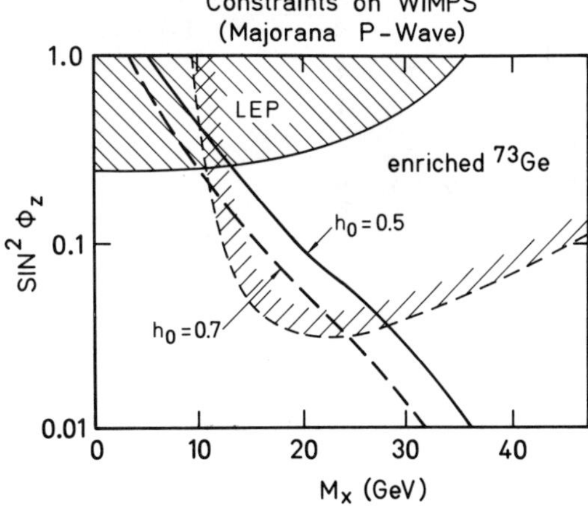

Fig. 11: Same as Fig. 10, but for Majorana particles (s-wave interactions). Note that $\Omega=1$ with h_0 =0.5 is possible only if M_x > 15 GeV and $\sin^2 \phi_z$ < 0.3. The use of about 2 kg of ^{73}Ge would extend the LEP limits to the marked borders (from [27]).

3.2. High Resolution γ-spectroscopy experiments using satellites
There are plans for new high resolution γ-spectroscopy experiments in the energy range 0.01 - 10 MeV using satellites. Such projects are the NASA project NAE (Nuclear Astrophysics Explorer) and the ESA project INTEGRAL (International Gamma Ray Astrophysics Laboratory) planned for the end of this decade. They plan to launch satellites with a spectrometer (Fig. 12) consisting of 9 large low-level Ge-detectors to investigate exciting questions like the variable positron annihilation source in the galactic center, nucleosynthesis products in the Galaxy and many other problems [28,29]. The high-resolution γ-spectrometer HEAO-3 has already made important discoveries such as the diffuse galactic 1.809 MeV line emission from the nucleosynthesis product 26Al [30] and the variable, compact 511 keV positron annihilation source in or near the galactic center [31]. Later balloon experiments with high-resolution Ge detectors yielded in addition important measurements of the ^{56}Co-γ-lines from the supernova SN 1987 A [32,33]. NAE and INTEGRAL are planned to extend these first observations. The sensitivity of the spectrometers is planned to be up to two orders of magnitude larger than those of the earlier missions (GRO = Gamma Ray Observatory, HEAO-3). This shall be realized by use of extremely low-background Ge-detectors. A decisive contribution to this aim could be use of detectors enriched in ^{70}Ge in the NAE/INTEGRAL missions. This would yield a considerable reduction of the background [34] compared to the use of normal detectors (Fig. 13), essentially since deenrichment in ^{74}Ge would reduce the β-background from the reaction ^{74}Ge (n,γ) ^{75}Ge (β^-) dominating in orbit. A corresponding agreement to build such detectors from 86% enriched ^{70}Ge and to explore their effectiveness in balloon experiments in 1991 or 1992 has recently been made between the NASA/ESA groups, and the HEIDELBERG-MOSCOW double beta decay cooperation.

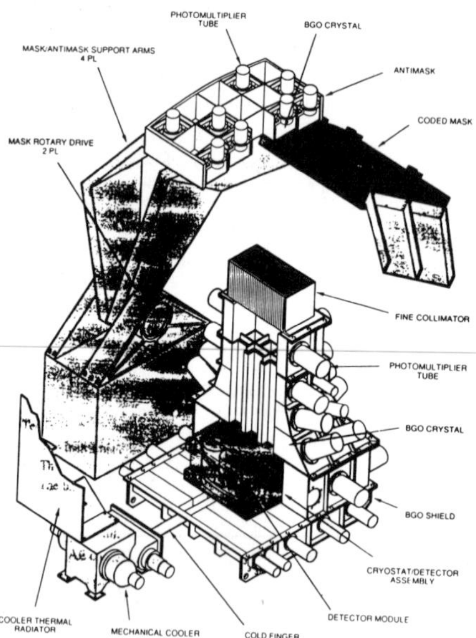

Fig. 12: View of the Nuclear Astrophysics Explorer (NAE) instrument, containing 9 large Ge detectors for high-sensitivity, high-resolution γ-ray spectroscopy. Enriched ^{70}Ge detectors would allow to drastically reduce the β-background (from [28]).

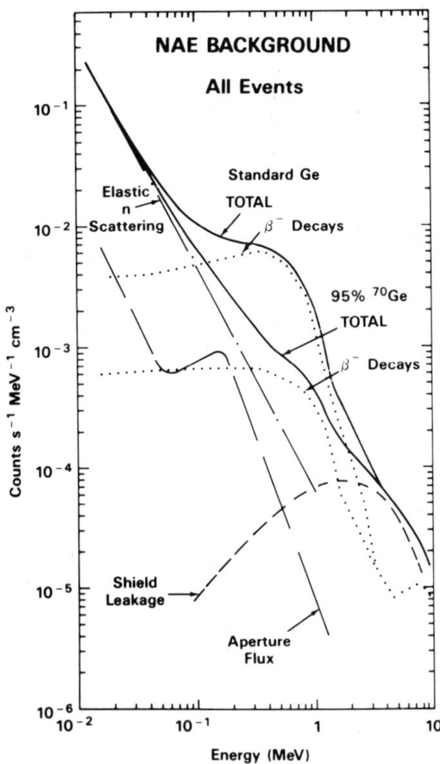

Fig. 13: The calculated NAE background for standard and enriched ^{70}Ge detectors (from [34]).

REFERENCES

[1] H.V. Klapdor (ed.), Neutrinos, Springer, Heidelberg, New York, 1988
[2] K. Grotz, H.V. Klapdor, The Weak Interaction in Nuclear, Particle and Astrophysics, Adam Hilger, Bristol, New York, 1990
[3] H.V. Klapdor-Kleingrothaus, Summary Talk given at 14th Europhysics Conf. on Nucl. Phys., 22- 26 Oct. 1990, Bratislava
[4] E.W. Kolb, M.S. Turner, The Early Universe, Addison-Wesley, 1990
[5] H.V. Klapdor, Progr. Part. Nucl. Phys. 10 (1983) 452, and 17 (1986) 419
H.V. Klapdor-Kleingrothaus, Proc. 7th Int. Symp. Capt. c-Ray Spectroscopy, Asilomar, California, 14-19 Oct. 1990
[6] K. Grotz, H.V. Klapdor, J. Metzinger, Phys. Rev. C33 (1986) 1263
[7] M. Gell-Mann, P. Ramond, S. Slansky, in: "Supergravity," ed. P. van Nieuwenhuizen, D.Z. Freedman, North-Holland, Amsterdam, 1979.
[8] P. Langacker, in: "Neutrinos," Springer, Heidelberg, New York, 1988, ed. H.V. Klapdor, p. 71.

[9] R.N. Mohapatra, Phys.Rev. D34 (1986), 3457
[10] M. Doi, T. Kotani, E. Takasugi, Progr. Theor. Phys. Suppl. 83 (1985) 1.
[11] K. Muto, H.V. Klapdor, in: "Neutrinos," Springer, Heidelberg, New York, 1988, ed. H.V. Klapdor, p. 183.
[12] K. Muto, E. Bender, H.V. Klapdor, Z. Phys. A334 (1989) 177 and 187.
[13] K. Grotz, H.V. Klapdor, Nucl. Phys. A460 (1986) 395. H.V. Klapdor, K. Grotz, Phys. Lett. 142B (1984) 32. K. Grotz, H.V. Klapdor, Phys. Lett. 153B (1985) 1, and 157B (1985) 242.
[14] H.V. Klapdor-Kleingrothaus, Invit. Talk at 14th Europhys. Conf. on Nucl. Phys., 22-26 Oct. 1990, Bratislava
[15] H.V. Klapdor-Kleingrothaus, Proc. Int. School-Seminar on Heavy Ion Physics, Dubna, 3-12 Oct. 1990, p 440-461. Invit. Talk at Int Conf. Nucl. Spectr. and Shape of the Atomic Nucleus, Leningrad, April 10-13, 1990.
[16] A. Staudt, K. Muto, H.V. Klapdor-Kleingrothaus, Europhys.Lett. 13 (1990) 31
[17] T. Tomoda, A. Faessler, Phys. Lett. 199B (1987) 475.
[18] A. Staudt, T.T.S. Kuo, H.V. Klapdor-Kleingrothaus, Phys. Lett. B.242, (1990) 17.
[19] J. Engel, P. Vogel, M.R. Zirnbauer, Phys. Rev. C37 (1988) 731.
[20] D.O. Caldwell et al.,Proc. 14th Europhys. Conf. on Nucl. Phys., 22-26 Oct. 1990, Bratislava
[21] I.V. Kirpichnikov, Proc. 14th Europhys. Conf. on Nucl. Phys., 22-26 Oct. 1990, Bratislava
[22] Heidelberg-Moscow-Cooperation: H.V. Klapdor, A. Piepke, G. Heusser, A. Buchner, A. Müller, U. Schmidt-Rohr, H. Strecker (MPI); S.T. Belyaev, A. Balish, A. Gurov, A. Demehin, I. Kondratenko, V.I. Lebedev (Kurchatov Institute), Proc. Internat. Sympos. on Weak and Electromagn. Interactions in Nuclei, Montreal, 1989 (WEIN '89).
[23] H.V. Klapdor, K. Grotz, Astrophys. J. 301 (1986) L39
[24] F. Boehm, et.al., Proc. Theor. and Phenomenological Aspects of Underground Physics, L'Aquila, 1989, Ed. Frontières, Gif sur Yvette
[25] D. Caldwell, Proc. La Thuile Workshop, March 1990
[26] D.N. Schramm, Proc. La Thuile Workshop, March 1990, J. Ellis, et al., Phys. Lett B245 (1990) 251
[27] K. Zuber, H.V. Klapdor-Kleingrothaus, Internal Report, Sept. 1990
[28] Nuclear Astrophysics Explorer Final Concept Study, Phase A Study Team, July 1989, Contract NAS5-30338
[29] INTEGRAL Proposal, 1990
[30] W. A. Mahoney et al.,Astrophys. J. 286 (1984) 578
[31] G.R. Riegler et al., Astrophys. J. 248 (1981) L13
[32] W. G. Sandie et al., Astrophys. J. 334 (1988) L91
[33] B.J. Teegarden et al., Nature 339 (1989) 122
[34] N. Gehrels, Nucl. Instr. Meth. A292 (1990) 505

TRANSFER OF γ RAYS: DETERMINISTIC SOLUTIONS BY MEANS OF THE DISCRETE-ORDINATE-MATRIX-EXPONENTIAL METHOD

Rainer Wehrse & Michael Hof
Institut f. Theoret. Astrophysik, Im Neuenheimer Feld 561, D 6900 Heidelberg, Germany

ABSTRACT

We consider a medium of non-negligible optical depth in which high energy line photons are Compton scattered. For the calculation of the angle and energy dependent specific intensities we first cast the correlations between the angles and energies of the incoming and the outgoing photons (as determined by the Compton conditions) into the form of a redistribution function, which is weighted by the Klein-Nishina cross-section in order to account for the energy dependence. We then discretize the angle $\times$ energy space so that the radiative transfer equation transforms from an integro-differential equation into a system of ordinary differential equations subject to boundary conditions on both sides. This system is subsequently solved in a numerically stable way without any further approximation by means of the discrete-ordinate-matrix-exponential ("DOME" [1]) method so that all emergent intensities and the energy converted into heat are obtained.

Since the redistribution function is very complex and involves singular hyperplanes, our algorithm requires a relatively large memory space whereas the CPU times are rather modest, e.g. for 16 angles and 25 energies in both the input and the output channel, 30 Mby of memory and 30 sec of CPU time are needed on an IBM 3090 VF independent of the actual optical depth.

First results for a plane-parallel medium, which consists of cold electrons and which is irradiated from one side by γ line photons, are presented and the accuracies of the derived radiation fields are discussed.

INTRODUCTION

Compton scattering plays a very important role in the radiative transfer of high energy photons. For a scattering medium of light elements ($Z \leq 24$) Compton collisions are the predominant interaction (compared to photoelectric absorption and pair-production processes) in the photon energy range between 0.1 and 10. MeV.[2] However, in the modeling of the corresponding radiation fields difficulties arise from the fact that for the scattering processes strong correlations exist between the directions and energies of the incoming and outgoing γ-photons.

Recently, Monte Carlo methods are often used to treat the γ-transport, in particular such calculations have been performed for the γ-ray spectra of supernovae (see e.g. Ref. 3). These methods have the advantage of being relatively simple to incorporate in a computer code. However, the slowly decreasing statistical errors ($\propto N^{-1/2}$) set certain limits on the accuracy of the results. This is the main problem for optically thick media where multiple scattering processes strongly increase the time for following the path of each photon. Another method often used applies the Fokker-Planck approximation for the treatment of Compton scattering by a diffusion equation (see e.g. Ref. 4). However, the approximations are only valid for photon energies that are small compared with the rest mass of the electron.

We describe here a new approach which originates from the radiative transfer in the optical wavelength range and in which a redistribution function for Compton scattering is used to obtain a deterministic solution of the transfer equation.

SOLUTION OF THE TRANSFER EQUATION

We write the radiative transfer equation for the specific intensity I at a Thomson optical depth τ, frequency ν and direction $\mu = \cos\vartheta$ in a medium of plane-parallel geometry

$$\frac{\partial I(\nu,\mu,\tau)}{\partial\tau} = \int_0^\infty d\nu' \int_{-1}^{1} d\mu' \left\{\left[-\frac{\nu/\nu'}{\mu\,\sigma_T} R(\mu,\nu,\mu',\nu') + \delta_{\nu\nu'}\,\delta_{\mu\mu'}\frac{\sigma^*_{KN}(\nu)+\varepsilon(\nu)}{\mu}\right] \times I(\nu',\mu',\tau)\right\} - \frac{\varepsilon(\nu)}{\mu}B(\nu,T) \tag{1}$$

where the following notation is used: σ^*_{KN}= total Klein-Nishina cross section in units of the Thomson cross section σ_T; $\varepsilon(\nu)$= absorption coefficient correspondingly normalized; $B(\nu,T)$= Planck function for thermal sources. The second term in the square brackets denotes the outscattering and absorption terms, while the first term contains the inscattering contributions expressed by the redistribution function

$$R(\mu,\nu,\mu',\nu') := \frac{r_0^2\frac{1}{\nu'\epsilon'}\left[\frac{\epsilon}{\epsilon'}+\frac{\epsilon'}{\epsilon}+2\left(\frac{1}{\epsilon'}-\frac{1}{\epsilon}\right)-\left(\frac{1}{\epsilon'}-\frac{1}{\epsilon}\right)^2\right]}{\sqrt{2\left(\frac{1}{\epsilon'}-\frac{1}{\epsilon}\right)(\mu\mu'-1)-(\mu-\mu')^2-\left(\frac{1}{\epsilon'}-\frac{1}{\epsilon}\right)^2}}\,. \tag{2}$$

R contains the differential Klein-Nishina cross section (with $\epsilon = h\nu/m_e c^2$ and r_0 = classical electron radius) and is normalized such that the photon number is conserved.

The transfer equation (1) is an integro-differential equation that by the discretization of the angle $\times$ energy space is transformed into a system of ordinary differential equations

$$\frac{\partial I_k}{\partial\tau} = \sum_{k'} G_{kk'}I_{k'} + Q_k\,. \tag{3}$$

The vector (I_k) contains the specific intensities at the various angle and energy grid points and (Q_k) contains the thermal sources. The redistribution terms are included in the matrix $(G_{kk'})$ which we consider to be independent of the optical depth. Without any further approximation this system is solved in a numerically benign way by means of the DOME method[1].

EXAMPLES OF CALCULATED SPECTRA AND DISCUSSION

In this paper we present first calculations of Comptonized spectra in which we have made the following simplifying assumptions: a homogeneous plane-parallel layer of a given optical depth which contains a cold electron gas is considered; pair-production and photoelectric absorption processes are neglected, only a very small constant artificial absorption coefficient is assumed on numerical reasons; the configuration is irradiated from one side by an isotropic γ-line source.

As examples we show in figures 1 and 2 the Comptonized spectra of 847 keV line radiation (this line appears in context with radioactive decay of ^{56}Co in supernovae)

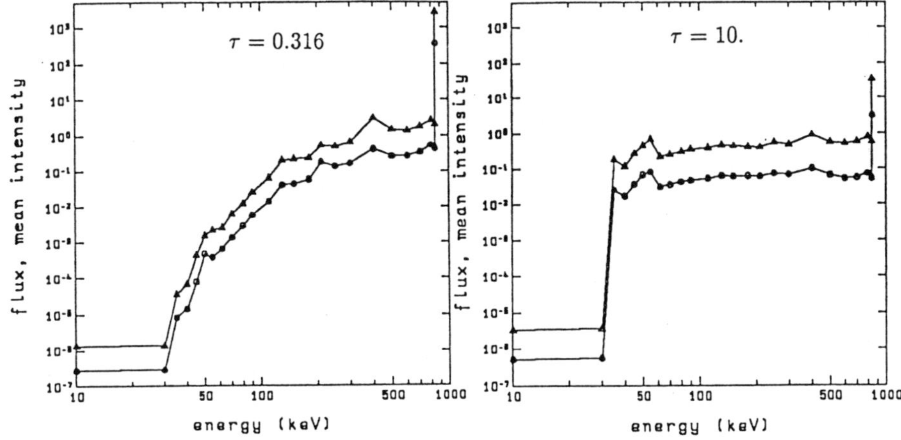

Figure 1 and 2: Comptonized spectra of line radiation in arbitrary units (triangles: flux; circles: mean intensity), see text.

which penetrates a layer of optical depth $\tau = 0.316$ and $\tau = 10.$ respectively.[5] On the right hand side the unscattered line is well recognized and to the left the scattered radiation is seen. For the calculations a discretization of 16 Gaussian angles and 25 energy points was chosen giving a matrix dimension of 400. The step-like feature in fig. 2 arises from the fact that the density of the energy grid points is not high enough to obtain a coupling between neighbouring points in the lower energy range. This implies that it is an important point in the calculations to use an adequate discretization scheme for the considered energy region. Limitations for the use of this method are given by the memory space and the CPU time ($\propto$ (dimension of matrix)^3) which are available on the particular machine. The example was calculated with about 30 Mby memory space in approximately 30 sec on an IBM 3090 VF computer.

Generalizations of the method can easily be made without affecting the performance of the method, e.g. inhomogeneous media are treated by dividing them into several homogeneous layers and absorption processes can be considered realisticly by using an appropriate $\varepsilon(\nu)$ in equation (1). Note also that as a by-product of the calculations one can get heating rates for all layers by taking the difference of incoming and outgoing total fluxes.

REFERENCES

1. M. Schmidt, R. Wehrse, *Numerical Radiative Transfer* (ed. W. Kalkofen, Cambridge Univ. Press, 1987) p.341
2. R.D. Evans, *Handbuch der Physik*, Vol. **XXXIV** (Springer, Berlin, 1955) p.218
3. K. Ambwani, P. Sutherland, *Astrophys. J.* **325**, 820 (1988)
4. A.I. Shestakov, D.S. Kershaw, M.K. Prasad, *J. Quant. Spectrosc. Rad. Transfer* **40**, 577 (1988)
5. M. Hof, diploma thesis (University of Heidelberg, 1990)

INSTRUMENTAL

The INTEGRAL Mission

Christoph Winkler*
ESA/ESTEC, Astrophysics Division, Noordwijk, The Netherlands

*on behalf of the INTEGRAL Science Team:
T.J.-L.Courvoisier, A.J.Dean, Ph.Durouchoux, N.Gehrels, J.Grindlay, J.L.Matteson, W.A.Mahoney, B.McBreen, J.O'Brien, O.Pace, T.A.Prince, V.Schönfelder, G.Share, G.K.Skinner, B.J.Teegarden, G.Vedrenne, G.Villa, S.Volonté

Abstract

The **Int***ernational* **G***amma-***R***ay* **A***strophysics* **L***aboratory* **(INTEGRAL)** *is a candidate for the next ESA medium-size scientific mission to be launched in 2000.* **INTEGRAL** *addresses the fine spectroscopy and accurate positioning of celestial gamma-ray sources in the energy range 15 keV to 10 MeV. A joint ESA-NASA assessment study has been completed recently. The scope of this study covered the mission's scientific prospects; the overall payload concept; payload accommodation and system analysis and an investigation of possible mission scenarios within the framework of an international collaboration.*

1. Introduction

The **INTEGRAL** mission, **INTE**rnational **G**amma-**R**ay **A**strophysics **L**aboratory, was proposed in response to the ESA M2 call for proposals, and is dedicated to the fine spectroscopy and imaging of celestial gamma-ray sources in the energy range 15 keV to 10 MeV. The instruments on INTEGRAL will achieve a gamma-ray line sensitivity of $3\text{x}10^{-6}$ ph $\text{cm}^{-2}\text{s}^{-1}$, a continuum sensitivity of $3\text{x}10^{-8}$ ph $\text{cm}^{-2}\text{s}^{-1}\text{keV}^{-1}$ at 1 MeV, and imaging with an angular resolution of 17′. This represents an order of magnitude improvement over the Gamma Ray Observatory (GRO) in line sensitivity, energy resolution and angular resolution. Comparison with SIGMA also shows a major advance: the continuum sensitivity improvement is at least one order of magnitude between 100 keV and 1 MeV; and the narrow line sensitivity is increased by nearly two orders of magnitude. Through the earlier GRASP and NAE studies, both the European and US communities have demonstrated the need for an advanced gamma-ray astronomy facility as a follow-on to the two major gamma-ray missions of the nineties, GRANAT, which includes the SIGMA instrument, and the Gamma Ray Observatory.

The scope of the ESA/NASA assessment study, carried out in 1990, covered the mission's scientific prospects and potential returns following SIGMA and GRO, an assessment of the overall payload concept, and a detailed investigation of possible mission concepts within the framework of a possible ESA-NASA agreement.

2. Overview Science Objectives

The scientific objectives of the mission will build upon the comprehensive observing programs already suggested in Europe and the US for GRASP and NAE, with the additions and improvements requested by the community at large, on such occasions as the 1988 Meudon GRASP

Workshop. INTEGRAL will address the fine spectroscopy with imaging and accurate positioning of celestial gamma-ray emission and include a hard X-ray monitor such that the gamma-ray observations can be placed in the context of simultaneous measurements at lower energies. In the 15 keV - 10 MeV region line-forming processes such as nuclear excitation, radioactivity, annihilation, cyclotron emission and absorption become important, and when used as astrophysical tools, are almost certain to lead to fundamental new discoveries. Unique astrophysical information is contained in the spectral shift, line width, and line profiles. Detailed studies of these processes require the resolving power (E/ΔE=500) of a germanium spectrometer such as that employed on INTEGRAL. Lower resolution spectrometers (e.g.SIGMA, OSSE, COMPTEL) will not have sufficient resolution to study the parameters of these lines. The last high resolution space instrument on HEAO-3 in 1979-80, was 100 times less sensitive than INTEGRAL. Solid observational and theoretical grounds already exist for predicting detectable emission from such varied celestial objects as the Galactic Centre, the interstellar medium, compact galactic objects, novae and supernovae (e.g. SN 1987A) and a variety of active galactic nuclei (AGN).

The gamma-ray emission from the Galactic Plane will be mapped on a wide range of angular scales from a few arc minutes to degrees in both discrete nucleosynthesis lines, such as ^{26}Al and the positron annihilation line, as well as the wideband continuum. Mapping diffuse galactic emission at the degree level will permit maps of the Galaxy similar in quality to the now-famous COS-B one to be drawn in the light of nucleosynthesis lines. At the same time, source positioning at the arc minute level within a wide field of view, of both continuum and discrete line emissions, is required to allow an extensive range of astrophysical investigations to be performed on a wide variety of sources, both targetted and serendipitous, with a good chance of identification at other wavelengths.

The study of active galaxies in both fine and broadband spectroscopy will yield unprecedented knowledge of the particle interactions which take place in the region where the central engine's energy meets the galaxy's matter. Because of the greatly improved sensitivity of INTEGRAL, sub-degree resolution imaging is absolutely necessary to avoid source confusion from the large population of AGN and to associate gamma-ray sources unambiguously with their optical, infrared and radio counterparts.
During the mission lifetime a significant number of transient events are expected to occur. Some of these will be gamma-ray bursts within the field of view, whilst supernovae events out to and including the Virgo cluster will be studied by periodic re-orientation of the telescope.

3. Model Payload

INTEGRAL consists of two main instruments: a germanium *SPECTROMETER* and a caesium iodide *IMAGER* (Figs. 1, 2; Table 1). These instruments are supplemented by two monitors: a *X-RAY MONITOR* and an *OPTICAL TRANSIENT CAMERA*. This configuration constitutes the first high resolution spectral imager to operate in the gamma-ray region, and will have the following features (Table 1):

- A wide operational bandwidth (4 keV to 10 MeV) which links X- and gamma-ray astronomy
- High resolution spectroscopy from 15 keV to 10 MeV (E/ΔE = 500 at 1 MeV).
- Accurate source positioning ($\sim 1'$) within a field of view of 50 square degrees.
- High sensitivity for both extended and point sources (typically 3x10^{-8} ph cm^{-2}s^{-1}keV^{-1} at 1 MeV)
- Optical, X- and gamma-ray studies of gamma-ray bursters and other transients.

The INTEGRAL sensitivities are displayed in Figures 3 and 4. The side by side payload arrangement, as described here, reflects the complementary nature of the spectroscopy and imaging mission objectives. The layout is based on the concept of a separate science experiment module, comprising an optical bench plus instruments, which is designed to be integrated and tested separate from the spacecraft and to have as simple as possible an interface with it.

4. System Aspects

The conceptual approach of a separate experiment science module applies to either the LEO (Low Earth Orbit) or HEO (High Eccentric Orbit) mission scenarios outlined in Table 2, and is similar to that currently planned for the XMM project; i.e. ESA responsibility for the experiment module definition and testing.

As may be seen from Table 2, two options have been studied for the INTEGRAL mission scenario. Both are capable of delivering the full scientific requirements :
HEO: This concept employs the XMM Common Bus (Fig.5) in combination with a Titan III or equivalent launcher. The orbit has been chosen to be a 24 hour, high perigee (>4000km), in order to provide long (>15 hour) uninterrupted observation periods at nearly constant background and away from trapped radiation. ESA would be responsible for the bus, the experiment module (see above), and mission control, with one European ground station baselined. A second station could be provided by NASA. A back up scenario envisages an Atlas IIAS as the launch vehicle, a configuration which restricts the payload mass to $\sim$ 1400 kg, and results in a sensitivity loss of typically a factor of 2 or more compared to the full payload.
LEO: This option employs the NASA Expendable Explorer Spacecraft (EES) or equivalent bus and an Ariane 4 to launch the payload into a near equatorial orbit which provides a low background environment, essentially free from passages through the South Atlantic Anomaly (SAA). A dedicated control moment gyro arrangement will be employed to maximise the observing efficiency to typically 75%. Mission control would most likely be the responsibility of NASA via TDRSS. However, an ESA ground station could be used. ESA would have responsibility for the experiment module.

The mission scenarios have been devised for an equitable division of responsibilities between ESA and NASA, and within the overall individual M2 and Explorer constraints. It is clear that a range of practical combinations of responsibilities exist from which a solution, attractive to both agencies, could emerge. It is anticipated that one agency, ESA or NASA, would provide the spacecraft whilst the other would provide the launch. It is clear that a joint mission allows greatly enhanced scientific returns.

5. Science Operations

INTEGRAL will accommodate the needs of the large number of guest observers who will use the majority of the telescope time. A Guest Observer program will follow an approach similar, for example, to that put forward in the Science Management Plan for ESA's X-ray astronomy cornerstone, XMM. For maximisation of scientific return, the concept of dual data centres, one in Europe and one in the US, is envisaged. Such centres could be managed by the scientific collaboration and serve the entire guest observer community.

6. Outlook

The INTEGRAL - together with five other candidate missions - will be presented to the ESA advisory bodies and the scientific community at large in March 1991. A maximum of four projects will be selected for a 15 months Phase A study which will start at the end of 1991, leading to the final selection of the next ESA M2 mission in 1993 and a scheduled launch in 2000.

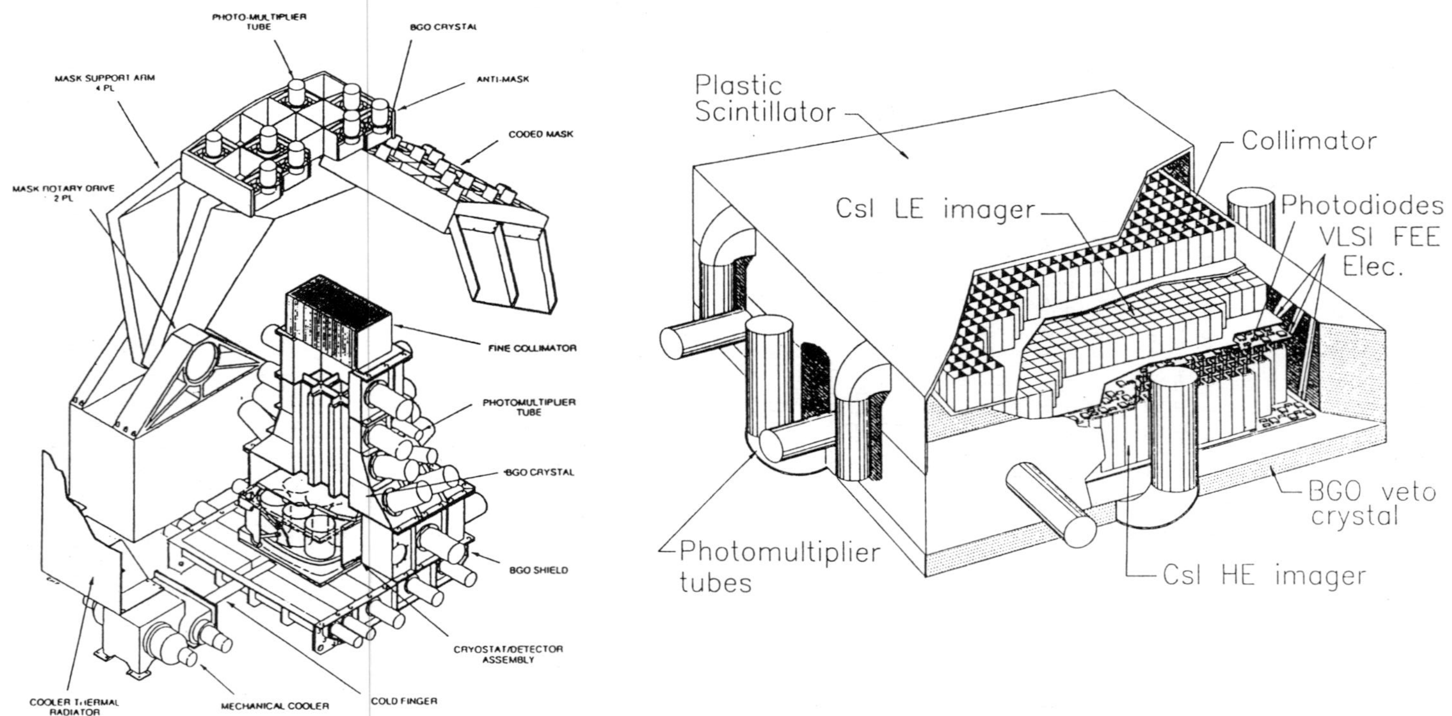

Figure 1: *Cut-away view of the high-resolution SPECTROMETER.*
The BGO shield, fine collimator and mask/anti-mask support arms have been cut away to show the cryostat/detector assembly. This contains 9 large Ge detectors cooled by a split Stirling Cycle mechanical refrigerator. Low background is obtained by surrounding the detectors with a 10 cm thick BGO shield. Apertures in the shield define a 10° field of view. Motion of the mask/antimask modulates the incoming beam to map its intensity and suppress background systematics. The fine collimator can be moved into the aperture to reduce the field of view to 1° for low energy observations.

Figure 2: *Cut-away view of the IMAGER.*
The (3 cm thick) BGO shield and plastic veto scintillator have been cut away so as to show, schematically, the arrangement of both the low-energy (2.5 cm thick) and high-energy (7.5 cm) CsI(Tl)/photodiode arrays, with their associated front end electronics, as well as the low energy collimator. The coded aperture mask assembly is not shown in this diagram.

Performance Parameter	Gamma-Ray Spectrometer	Gamma-Ray Imager
Energy Range	15 keV → 10 MeV	50 keV → 10 MeV
Detection Area	327 cm^2	2500 cm^2
Spectral Resolution ($E/\Delta E$) at 1 MeV	~ 500	~ 10
Field of View (FOV)	10° FWHM	8° (fully coded)
Angular Resolution FWHM	4°	17′
Point Source Location Capability (10σ)	30′	$\sim 1'$
Line Detection Sensitivity	(at 1 MeV, narrow line) (line width $\sim$ 2 keV)	(at 1 MeV, broad line) (line width $\sim$ 20 keV)
3σ in 10^5 s	10^{-5} ph cm^{-2} s^{-1}	3.8×10^{-5} ph cm^{-2} s^{-1}
3σ in 10^6 s	3×10^{-6} ph cm^{-2} s^{-1}	1.2×10^{-5} ph cm^{-2} s^{-1}
Diffuse Line Sensitivity 3σ in 10^6 s	(at 1 MeV, narrow line) 1.5×10^{-5} ph cm^{-2} s^{-1} rad^{-1}	
Continuum Sensitivity 3σ in 10^6 s	(at 1 MeV, $\delta E = 1$ MeV) 6×10^{-8} ph cm^{-2} s^{-1} keV^{-1}	(at 1 MeV, δE=1 MeV) 3×10^{-8} ph cm^{-2} s^{-1} keV^{-1}

Table 1: *Performance parameters for the SPECTROMETER and the IMAGER. The payload is supplemented by a X-RAY MONITOR (4 keV to 60 keV, 900 cm^2 detection area, 14° FOV, 5′ angular resolution) and an OPTICAL TRANSIENT CAMERA (450 nm to 750 nm, 385×288 pixel CCD, 10°× 12° FOV, 2.5′ angular resolution).*

	HEO	LEO
LAUNCHER	Titan III class (NASA)	Ariane 4 (ESA)
ORBIT	28°, 24^{hr}	$\sim 0°$, 90^{min}
	4000x70000 km	500 km
OBSERVATION EFFICIENCY	> 65%	75% (with CMG)
CONTINUOUS OBSERVATION	$\geq$ 15 h	$\leq$ 1 h
BUS	XMM (ESA)	EES (NASA)
EXPERIMENT MODULE	Optical Bench & Integration : ESA Instruments : European nations/NASA	
MISSION CONTROL	ESOC/NASA	NASA/ESOC
SCIENCE PAYLOAD (kg)	2141	2141
PROS	• Lower background > 1 MeV • Existing bus • Long continuous observations	• Lower background $\leq$ 1 MeV • Solar observations possible
CONS	• Solar obs require XMM-BUS modifications	• New bus required

Table 2: *INTEGRAL mission scenarios*

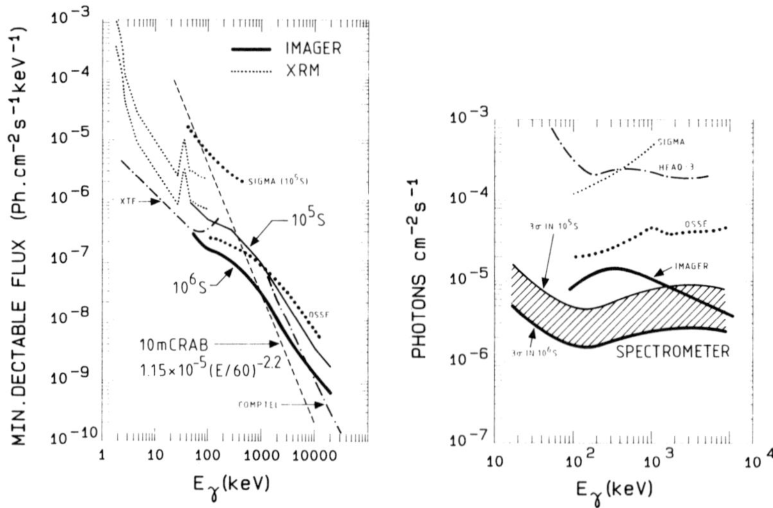

Figure 3: *INTEGRAL continuum sensitivity. For the purposes of this figure $\Delta E{=}E$ has been assumed (Mission option: LEO/Ariane).*

Figure 4: *INTEGRAL narrow-line sensitivity (Mission option: LEO/Ariane).*

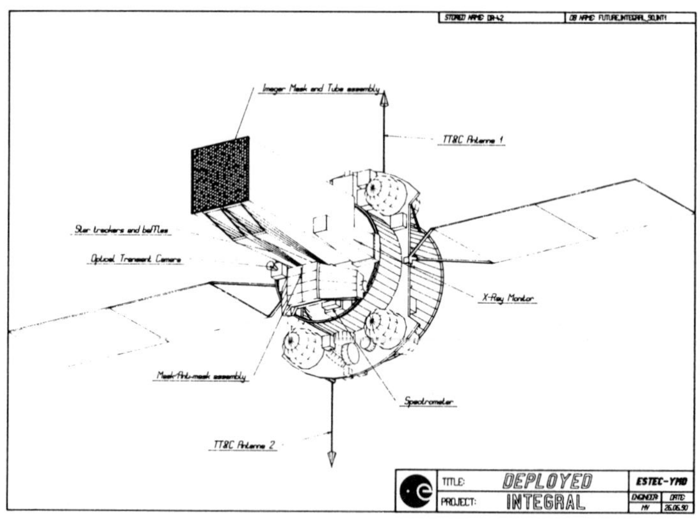

Figure 5: *INTEGRAL on XMM Common Bus in deployed configuration (HEO)*

THE IMAGING AND SPECTROSCOPY CAPABILITIES OF THE INTEGRAL IMAGER: MONTE-CARLO SIMULATION RESULTS

G. Malaguti[1,2], A.J. Bird[1], E. Caroli[2], A. J. Dean[1], G. Di Cocco[2], B.M. Swinyard[3] and G.E. Villa[4]

[1] Physics Department, Southampton University, U.K.
[2] Istituto TESRE/C.N.R., Bologna, Italy.
[3] Rutherford Appleton Laboratory, U.K.
[4] Istituto di Fisica Cosmica/C.N.R., Milano, Italy.

ABSTRACT

For the gamma-ray satellite INTEGRAL, a 3-D position-sensitive imaging detector has been proposed. The instrument is based on a two layer CsI(Tl) device coupled with a coded aperture mask placed at ~4 m from the detector. This particular configuration of discrete detection elements, together with the 3-D position sensitivity will ensure good imaging capabilities throughout a large energy range from 40-50 keV up to ~20-30 MeV. The imager will also have spectroscopy capability in the nuclear line region. In this work we present the first Monte-Carlo simulation results regarding the reconstruction of the shadowgram of the mask pattern. A simulated image of a 511 keV Galactic Centre point source is presented for both the low and the high state.

INTRODUCTION

Two of the more important aspects in γ-ray astronomy concern spectroscopy and imaging of the emitting sources. The former provides basic information on fundamental astrophysical problems such as nucleosynthesis phenomena and compact objects physics. Fine imaging, on the other hand, provides accurate positioning of the sources detected with in the instrument field of view and good angular resolution to map regions of diffuse emission and separate point source contributions.

The complementarity of these two aspects is particularly evident in the study of the 511 keV positron-electron annihilation line emission from the Galactic Centre. The gamma-ray observations suggest the existence of a point source of annihilation radiation in the region of the Galactic Centre[1,2]. The time variability of the emission from this source also suggests a point-like nature. Results coming from observation with the large field of view detector on the SMM satellite[3], indicate, on the other hand, the presence of a diffuse 511 keV component. Balloon observations with a high energy resolution Germanium spectrometer indicated a 511 keV line with a width of 3.6 keV in Autumn 1988 when the source was in the high state[1] and a narrower (<1.8 keV) line 6 months later when the point-like source was in the low state[4].

There is therefore a very clear need for a high level imaging mapping of the Galactic Centre region at γ-ray energies. The imager proposed for the joint ESA/NASA γ-ray mission INTEGRAL[5] will be able to locate the Galactic Centre 511 keV line to the arcmin level within a field of view of 6°, and, at the same time, provide a spectrum at a moderate resolution level (~10%). In this work we present the first Monte-Carlo results of the fine band imaging capability of the INTEGRAL imager: the simulated image of the Galactic Center 511 keV compact source is presented for two different values of the intensity.

THE SIMULATION OF THE IMAGER

The operational range of the imager proposed for INTEGRAL goes from 40-50 keV up to 20-30 MeV. The imager is constructed from a double-layer matrix of CsI(Tl) bars providing position sensitivity in 3 dimensions. The total sensitive area of the imager will be around 2500

cm^2. The bottom matrix will be 10 cm thick whereas the top layer will be only 3.5 cm thick in order to achieve a low energy threshold. At 511 keV the detection efficiency of the top layer is of the order of ~75%[6], so that, for the purpose of this work we neglected the bottom layer.

The simulation code is based on the *GEANT3* package[7] developed at CERN. The code takes into account all the principal interaction processes in the energy range between 10 keV and 10 TeV. All the secondary particles produced in the interactions are tracked throughout the simulated experimental setup. In order to decrease the necessary CPU time we have simulated a detector having a sensitive area that is a factor of two smaller than that of the imager proposed for INTEGRAL. Therefore the simulated geometric arrangement consists of an array of 34x38 CsI bars (1x1cm cross-section, 3.5cm thick), with a total area of 1292cm^2. The dimensions of the detector array have been chosen to maintain a 1:4 proportion between detector and mask pixel size in order not to lose signal in the reconstruction phase. The mask we used was a twin-prime mask of the type proposed by Fenimore and Cannon[8] having a basic pattern of 17x19 elements. In order to cover all the detector and provide the same angular resolution as INTEGRAL, the mask has been replicated 4 times and then placed at 2m from the detector.

SIMULATED OBSERVATION OF THE 511 keV GALACTIC CENTRE LINE

The simulated observations have been performed for the low and high state of the 511 keV Galactic Centre point-like source. The source was assumed to lie in the centre of field of view and the observations were assumed to last for 10^5 s. A uniform background was then added to the shadowgram obtained. In Tab. 1 we report the data used for the source flux in the low and high state and for the background flux. One of the critical parameter of an imaging instrument is the point-source location accuracy (PSLA). For a coded-mask telescope the PSLA depends[9] both on the signal to noise ratio (SNR) and on the spatial resolution of the position sensitive detector. For the INTEGRAL imager, since the mean path for a 511 keV photon in CsI(Tl) is very short, it is reasonable to assume a spatial resolution of detector equal to the size of the single discrete elements (1 cm). In Table 2 we report the results obtained for the signal to noise ratio (SNR) and for the maximum theoretically achievable point source location accuracy (PSLA) for the two possible states of the source. It is important to note that even when in the low state, the source will be detected at a satisfactory significance level and located at the arcmin level. This will permit to associate the emission to the hard X-ray source already detected in the Galactic Centre region[10].

Table 1: 511 keV Galactic Centre source and background fluxes used in the simulation

	Low state	High state
Source flux	2 10^{-4} ph cm^{-2} s^{-1}	1 10^{-3} ph cm^{-2} s^{-1}
Background	1 10^{-4} counts cm^{-2} s^{-1} keV^{-1}	

Table 2: SNR and theoretical PSLA obtained for the high and the low state

	High state	Low state
SNR (n_σ)	62	10
PSLA	~20"	~90"

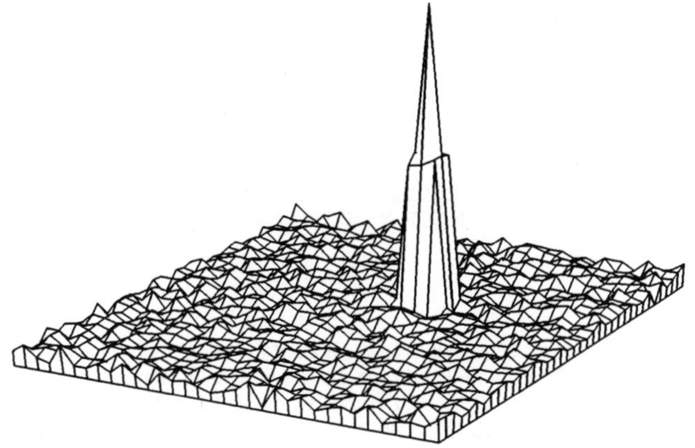

Fig.1: 511 keV Galactic Centre compact source. High state.

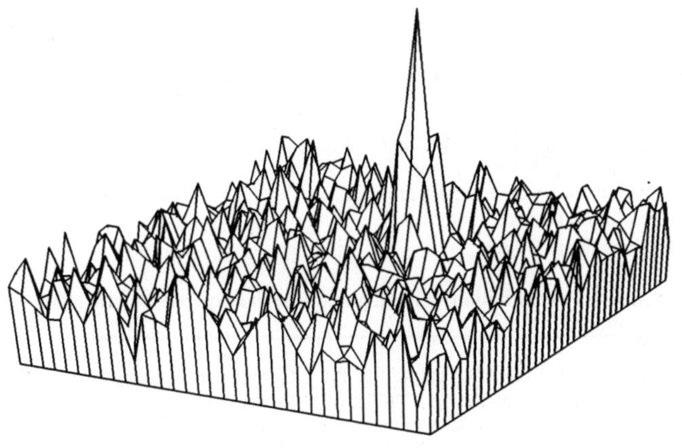

Fig.2: 511 keV Galactic Centre compact source. Low state.

In figure 1 and 2 we report the deconvolved images of the source in the case, respectively, of high and low state. In both the image of the source is clearly visible.

REFERENCES

1. M. Leventhal *et al.*, Nature 339, 36 (1989).
2. M. Riegler *et al.*, Astrophys. J. Lett. 294, L13 (1985).
3. G. Share *et al.*, Astrophys. J. 278 137 (1984).
4. J.L. Matteson, IAU Circular No. 4889, November 1980.
5. INTEGRAL Assessment Study Report, ESA, (1990), in press.
6. G. Malaguti *et al.*, Adv. Spa. Res., (1990), in press.
7. R. Brun *et al.*, GEANT3, CERN, (1987).
8. E.E. Fenimore and T.M. Cannon, Appl. Optics, 17, 3562, (1978).
9. E. Caroli *et al.*, Spa. Sci. Rev., 45 349 (1987).
10. G.K. Skinner *et al.* Nature, 330 544 (1987).

SIGMA : A SOFT GAMMA-RAY IMAGING TELESCOPE IN-FLIGHT PERFORMANCES

P. MANDROU, JP. CHABAUD, M. EHANNO, J. LANDE, M. NIEL, J.P. ROQUES, G. ROUAIX, P. SOULEILLE
CESR, 9 avenue du Colonel Roche, BP 4346 - 31029 TOULOUSE Cédex

J. PAUL, J. BALLET, M. CANTIN, B. CORDIER, A. GOLDWURM, Ph. LAURENT, F. LEBRUN, J.P. LERAY
SAp.CEA, CEN Saclay, 91191 GIF SUR YVETTE Cédex

ABSTRACT

The low energy gamma-ray telescope SIGMA, aboard the GRANAT observatory, is now entering in his second year of life. It has provided during the last year, high angular resolution images of the sky in the energy domain 35 keV-1.3 MeV. The data accumulated in different background conditions on series of astrophysical sites allow us to present the results on in-flight performances of this instrument.

INTRODUCTION

The Soviet GRANAT satellite, devoted to high-energy astronomy was launched on December, 1st, 1989 by a Soviet Proton rocket, from Baïkonour (USSR) to reach a highly excentric orbit, 2000 km perigee, 200 000 km apogee. The payload, made of seven different instruments, included the French coded-aperture telescope SIGMA.

- field of view : 4.3 deg x 4.7 deg totally coded
- partially coded : 12.9 deg x 14.1 deg
- point source location accuracy : 1.5 arc min
- intrinsic resolution power : 13 arcmin
- total detection area : 794 cm^2

For more details on the definition, general purpose of the experiment or detailed description see /1/2/3/ and references therein.

After the launch of the GRANAT satellite the different parameters of the telescope were adjusted, including the fine tuning of the 131 PMTs of both the detector and the anticoïncidence. After this period, the telescope was tested in the course of several Crab Nebula and Cyg X-1 observations, and its performances were determined. Several parameters are prominents for the detection sensitivity and imaging properties : the background, the energy resolution and the detector spatial resolution which is the key parameter of the contrast in the imaging procedure.

The mission of the SIGMA telescope is gamma-ray imaging of the sky between 35 keV and 1.3 MeV with an accuracy of about 1 arc min. The observation program includes galactic sites as well as extragalactic, but the most interesting purpose remain the mapping with a good precision of complex regions, taking advantage of high angular resolution.

During the last 1990 year, about 200 sessions of about 20 hours mean duration have been devoted to perform observation of different celestial objects.

DESCRIPTION OF THE EXPERIMENT - IN-FLIGHT PERFORMANCES

In order to fulfill the scientific objectives : high angular resolution and good sensitivity, a telescope has been studied and constructed which uses an association of a large area position sensitive NaI crystal detector 1.5 cm thick, developped on the principle of an Anger camera and a coded aperture mask whose pattern is 29 x 31 uniformly redundant array tungsten elements 1.5 cm thick located 2.5 m away from the detector (figure 1). A thick CsI anticoïncidence shield surrounds the detector to limit the field of view to about 1 ster and thus reduces the background to increase the sensitivity of the instrument.

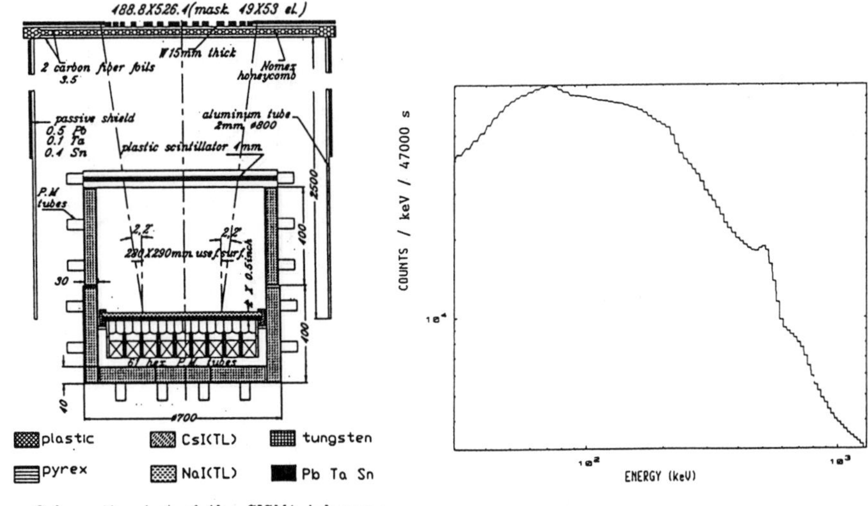

Schematic view of the SIGMA telescope
Figure 1

Figure 2 : In-flight background

The main characteristics can be summarized as :

The measured background : The background depends on the characteristics of the shield, the spacecraft orbit and the mass of the whole satellite. The telescope operates on a very excentric orbit whose period is 4.05 days. During one day, when the telescope is situated near the perigee, inside a sphere of 10 earth radii, the detectors are switched-off to avoid the effects of the radiation belts. During the three other days, the telescope can work in rather stable background conditions except during the first 10 hours where the counting rate decreases slowly in the low-energy channels corresponding to the de-activation of induced radioactive nuclei, produced when the satellite crosses the protons belts. The total counting rate integrated from 35 keV to 1.3 MeV is now of the order of 430 counts/sec (figure 2) but we note a long term decrease since the altitude of the perigee is constantly increasing due to the moon perturbation.

Sensitivity in the imaging mode : The performance of this telescope depends, as for any such instrument, on the detection efficiency, on the general background, but also on the precision of the photons localization on the detector plane which affect the contrast power of the images. Studying the results obtained on the Crab an Cyg X-1 observations, as well as the measured background along the orbits we have determined the parameters of the in-flight performances.

a) Energy resolution

The energy resolution was measured in the course of ground calibrations, using radioactive sources, when the telescope was mounted on the spacecraft, and ready to be launched. In-flight verifications are done on induced lines in the background spectrum. The result of this verification is presented on Figure 3 and compared to the ground measurements. We can conclude from this work that the detector energy resolution was not significantly modified in flight.

b) Spatial resolution

In order to determine the spatial resolution of the detector, we have used the observations both on the Crab Nebula and on Cyg X-1. In those two cases, the sources are sufficiently strong to project the mask on the detector with a high contrast. Sofisticated treatments using the bright to shadow transitions around the black or white mask pixels provided reliable results, presented in Figure 4, where the FWMH resolution in mm is given versus the photon energy.

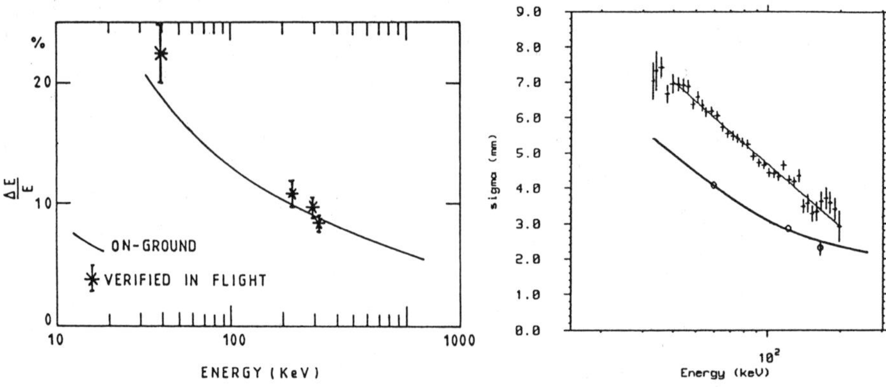

Figure 3 : Energy resolution

Figure 4 : Spatial resolution

We can observe that the spatial resolution is worse than ground calibration measurements. Extensive ground tests performed on the engeneering model have shown that a remanent light level in the crystal can explain perfectly the observed phenomenon. The introduction of light in the crystal decreases the spatial resolution more severely at low energy than at higher energy but does not affect the energy resolution.

We think now that the high number of protons events and charged particles in such a large crystal as observed in orbit (3500 counts/sec) can be at the origin of this effect. This phenomenon decreases by a factor of three our imaging sensitivity at 35 keV, but the telescope recovers its expected sensitivity beyond 150 keV. Taking into account all those parameters the 3 sigma sensitivity for a 10^5 sec exposure time is presented in Figure 5.

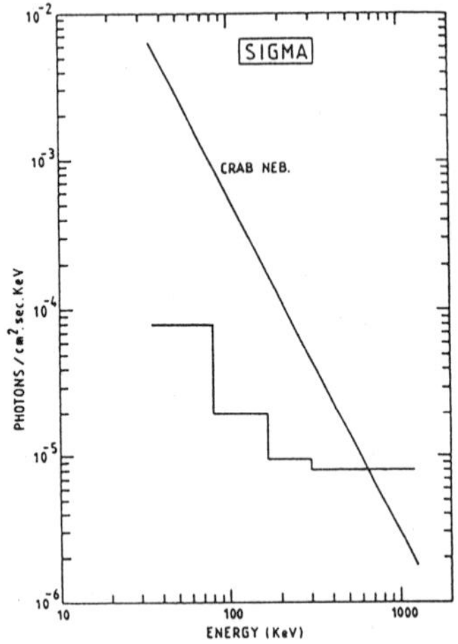

Figure 5 : 3 σ imagery sensitivity for 10^5 s

AKNOWLEDGEMENTS :

The authors thanks the SIGMA project group of the CNES-Toulouse, the staff of Lavotchkine Space Company, the Babakine Center, the Baïkonour Space Center, and the Evpatoria Ground Station for their unfailing support.

REFERENCES

1. P. Mandrou, Adv. Space Res., Vol 3, N°10-12, 525, 1984.

2. J.P. Roques, J. Paul, P. Mandrou, F. Lebrun, Adv. Space Res., Vol 10, N°2, 223, 1990.

3. J. Paul, P. Mandrou, J. Ballet, M. Cantin, J.P. Chabaud, B. Cordier, M. Ehanno, A. Goldwurm, A. Lambert, J. Lande, Ph. Laurent, F. Lebrun, J.P. Leray, B. Mena, M. Niel, J.P. Roques, G. Rouaix, L. Salotti, P. Souleille and G. Vedrenne, To be published in the proceedings of the XXVIII COSPAR, The Hague, 1990.

CALCULATION OF THE GAMMA-RAY LINE SENSITIVITY OF THE BATSE LARGE AREA DETECTORS ON GRO

G. N. Pendleton, W. S. Paciesas, University of Alabama in Huntsville

G. J. Fishman, R. B. Wilson, C. A. Meegan,
NASA/Marshall Space Flight Center

ABSTRACT

The gamma-ray line sensitivity of the BATSE detectors depends primarily upon the characteristics of the detectors' response and the nature of the backgrounds observed by the instruments. Using the example of the detectability of prompt annihilation radiation from nearby novae, the detector response characteristics and data analysis tools relevant to gamma-ray line and continuum sensitivity determination will be presented. Sensitivities to gamma-ray lines at various energies will be presented.

INTRODUCTION

The BATSE large area detectors (LAD's) are designed primarily for the measurement of the hard X-ray continuum of gamma-ray bursts. The LAD's are NaI(Tl) disks 25 cm in radius and 1.27 cm thick. The instrument described in detail elsewhere[1,2] is optimized for the 50 to 300 keV energy range and full sky coverage. However it produces data binned into channels several times narrower than the detector resolution making the search for line features possible. The gamma-ray line and continuum sensitivities presented here are calculated using simulation generated detector response matrices and an estimate of the background expected in orbit.

LAD DETECTOR RESPONSE MATRICES

The detector response matrices are generated using a modified version of EGS3[3] and a detailed geometry code written by the author specifically for the BATSE instrument. Their accuracy has been verified by extensive comparison with calibration data. The response is strong from 30 to 200 keV but drops off significantly at higher energies. At 500 keV a significant fraction of the photons that do interact in the crystal fail to deposit all their energy resulting in a significant Compton tail in the detector response at this energy.

THE BACKGROUND ESTIMATION

The background is estimated from balloon flight data[4] obtained using modified

LAD's combined with the detector's estimated response to the low energy diffuse cosmic background flux[5]. The detectors were facing upward so the background is considered to be that observed by a sky viewing detector. The low energy diffuse cosmic background is unable to penetrate the atmosphere to balloon altitudes. Therefore a component due to the cosmic background not observed by the detectors at balloon altitudes but detected in orbit was added to the balloon flight data for the background estimate.

CALCULATED SENSITIVITIES

Table 1 shows the line flux observable at a 3 sigma significance by an LAD for 100 sec. and 300 sec. exposures at various energies. These significances are for one LAD viewing the flux face on. A 300 sec. exposure could be obtained for a steady source using earth occultation. These sensitivities apply to a gamma-ray line that is not accompanied by continuum flux that is significant compared to the background flux.

Table 2 shows the LAD sensitivities to line features superimposed on a typical gamma-ray burst continuum for three different burst continuum fluences. The sensitivities in this table are given in seconds. These are the observation times required to obtain a 3 sigma significance measurement of a line feature whose equivalent width equals the FWHM of the detector response at that energy. The significance calculation is based on the assumption that a simple background subtraction is performed to extract the line feature using background intervals distributed around the line feature whose sum total width equals the FWHM of the detector at the given energy. These background intervals consist of the background continuum plus the burst continuum.

Table 3 shows the sensitivity in sigma above background of one LAD viewing the flux face on to the continuum produced by a typical nova[1,6] at 1 Kpc. 5 hours after ignition for three different integration times. The two shorter observations could be made using earth occultation. The longer observation would require the

TABLE 1

Line Energy (keV)	3 Sigma Line Strength (1sec)	3 Sigma Line Strength (300sec)
20	1.07	6.20E-2
30	2.19E-1	1.28E-2
60	9.49E-2	5.48E-3
100	6.59E-2	3.81E-3
200	6.45E-2	3.72E-3
500	1.32E-1	7.63E-3

Line Strengths in ph/cm^2/sec

use of a background estimation technique. Table 4 is the same as table 3 except that it applies to two LAD's viewing the flux at 35.26 degrees to the detector normal; this situation is not unlikely for the BATSE instrument.

The LAD's will provide measurements of the continuum flux from nearby novae but they cannot measure the 511 keV line flux as readily[1]. They will be able to measure line features in strong bursts. They will also be able to produce measurements hard X-ray continuum spectra of nearby novae that should provide constraints on nova model parameters.

TABLE 2

Sensitivites for Line Features in Gamma-Ray Bursts

Line Energy (keV)	Burst Strengths:	3 Sigma Observation Times(sec.) 10E-4 erg/cm^2	10E-5 erg/cm^2	10E-6 erg/cm^2
20		2.2E-3	3.8E-2	3.02
30		6.0E-4	7.2E-3	0.18
60		3.4E-4	3.6E-3	6.2E-2
100		3.7E-4	3.9E-3	5.8E-2
200		6.6E-4	6.9E-3	0.11
500		1.1E-2	8.3E-2	1.59

TABLE 3

Nova Continuum Sensitivities in Sigma above Background One Detector 0 Degrees

Energy Range	Exposure Duration 100 sec	300 sec	600 sec
20-30	0.28	0.48	3.32
30-50	1.35	2.35	3.32
50-100	5.40	9.36	13.24
100-200	5.01	8.67	12.27
200-500	3.76	6.52	9.22

TABLE 4

Nova Continuum Sensitivities in Sigma above Background Two Detectors 35.26 Degrees

Energy Range	Exposure Duration 100 sec	300 sec	600 sec
20-30	0.35	0.60	3.93
30-50	1.61	2.78	3.94
50-100	6.37	11.02	15.60
100-200	6.07	10.52	14.87
200-500	4.84	8.39	11.86

1. G.J. Fishman et al., poster paper these proceedings.
2. G.J. Fishman et al., GRO Sci. Wkshp., GSFC (1989).
3. R.L. Ford et al., SLAC-Report-210 (1978).
4. C.A. Meegan et al., Ap.J.,**291**, 479 (1985).
5. R.L. Kinzer et al., Ap.J.,**222**, 370 (1978).
6. M.D. Leising invited paper, these proceedings.

MART-LIME:
THE SPECTROSCOPIC HARD X-RAY IMAGER FOR SPECTRUM-X-GAMMA

P. Ubertini, A. Bazzano

Istituto di Astrofisica Spaziale (IAS), C.N.R.
C.P. 67 - 00044, Frascati, Italy

R.Sunyaev, N. Yamburienco

Space Research Institute (IKI), URSS Academy of Science,
Profsoyuznaya 84/32, 117810, Moscow, URSS

INTRODUCTION

The imaging and the spectroscopy of cosmic sources in the hard X-Ray/soft γ-Ray range is fundamental for high energy astrophysics. In fact most of the energy of several classes of galactic and extragalactic objects is emitted in this band and also the non thermal processes became very efficient. Large area, high resolution imaging telescopes are needed for hard X-Ray Astronomical missions to be launched in the 90's.

A high spatial resolution spectroscopic proportional counter has been jointly designed by Istituto di Astrofisica Spaziale (Italy) and Space Research Institute (Moscow) and proposed for the high energy experiment on board the Soviet Astronomical satellite SPECTRUM-X-Gamma shown in figure 1.

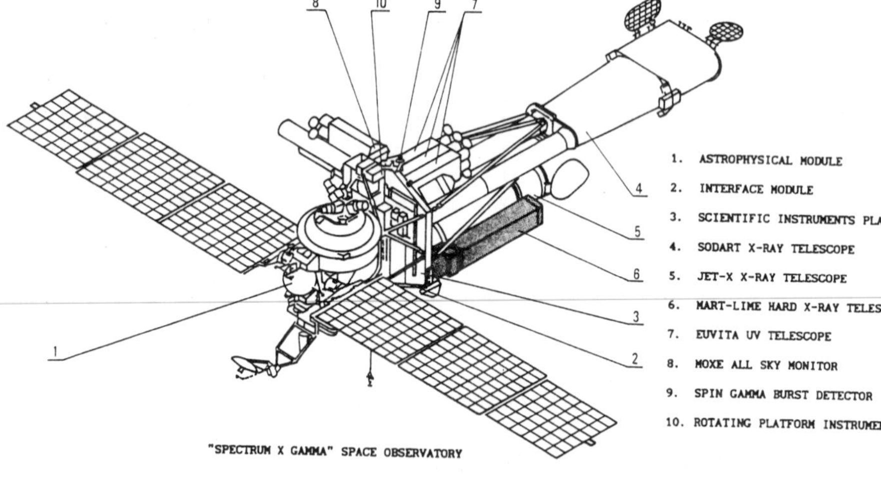

Figure 1. The Spectrum X Gamma Satellite

The instrument, named **MART-LIME** (**M**ask **A**stronomical **R**oentgen **T**elescope-**L**arge **I**maging **M**wpc **E**xperiment), shown in gray in figure 1, is operated with a fine coded mask placed at 3 meters from the detector. It will provide sub-arcminute spatially resolved images of the sky in the energy range 5 ÷ 150 keV, with good spectral resolution ($\lambda/\Delta\lambda>20$ for E>35 keV) and sub milliCrab sensitivity during a typical observation lasting from 10^4 to 10^5 seconds.

MAIN SCIENTIFIC OBJECTIVES

The main scientific objectives of this experiment are **Imaging and Spectroscopy** of Astrophysical Objects in the 5 ÷ 150 keV energy range.

Galactic Objects:

- Determination of the hard component in the spectra of galactic bulge X-Ray sources suggested as black hole candidate in crowded fields;
- High sensitivity pulse phase spectroscopy

Extragalactic objects:

- Thermal and non thermal processes studies of compact objects and AGN's probing the non thermal and high temperature thermal phenomena of these objects.
- Luminosity function for individual class of AGN; determination of variability and spectral parameters of different type of AGN (QSO, Seyfert, BL Lac etc) and their
- Contribution to the diffuse X-Ray background;

High sensitivity Hard X-Ray Sky Catalogue.

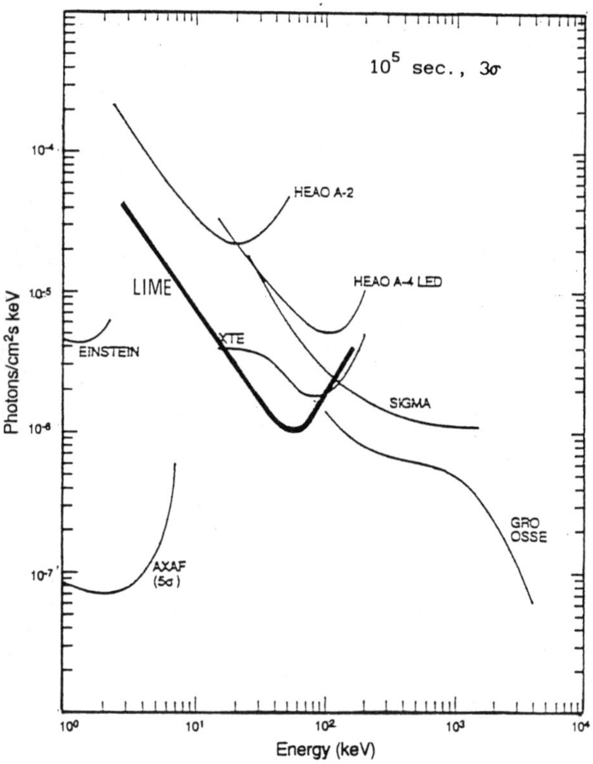

Figure 2

The predicted sensitivity curve of the experiment, shown in figure 2, is compared with other missions.

THE INSTRUMENT

The MART-LIME experiment is a third generation hard X-ray telescope for high energy astronomy observations. This instrument has been designed to obtain sky images in the hard X-ray range with high sensitivity, sub arcmin spatial resolution and high spectral resolution. As an imager will be able to produce wide field sky images (8.5 x 8.5 squared degree) with an angular resolution comparable to that one achieved by grazing incidence X-ray telescope (i.e. <1 arcmin), operative at lower energies.

From the spectroscopy point of view will be able to collect spectra with high resolving power of the order of $E/\Delta E \cong 20 \div 30$, therefore comparable in performances at the more sophisticated Germanium detectors ($\simeq$ 50 at 100 keV). Its large sensitive area and very good background rejection will allow a comparable sensitivity, with the same observing time. Its high continuum sensitivity make this instrument ideal, within the Spectrum X-Γ mission, to fill the "missing link" gap between the thermal processes, studied in detail with the two low energy telescopes SODART and JET-X, and the non thermal processes, not yet fully understood. Because of the mentioned characteristics, the MART-LIME experiment, is complementary at higher energies to the grazing incidence telescope based instruments, expected to fly during the next years (SAX, ASTRO D, AXAF etc). Figure 3 shows the Mass and Size Model of the MART-LIME Experiment.

Summary of the MART-LIME characteristics

Energy Range	5 ÷ 150 keV (10% eff.)
Energy Resolution	5% E>35 keV; $\Delta E/E=60/E^{1/2}$ E <35keV
Timing accuracy	1 ms (spectr./imag.) 100 μs (timing mode)
Field of view	8.5 x 8.5 degree
Sensitive area	2000 cm^2
Gas Mixture	Xe/Isobutilene/Ar
Filling Pressure	5 bar
Spatial Resolution	1 x 1 mm
Detector-mask dist.	2.7 m
Experiment weight	131 kg
Power	71 Watt
Telemetry rate	5 ÷ 20 kbit/s
On-board Memory	50 ÷ 200 Mbyte

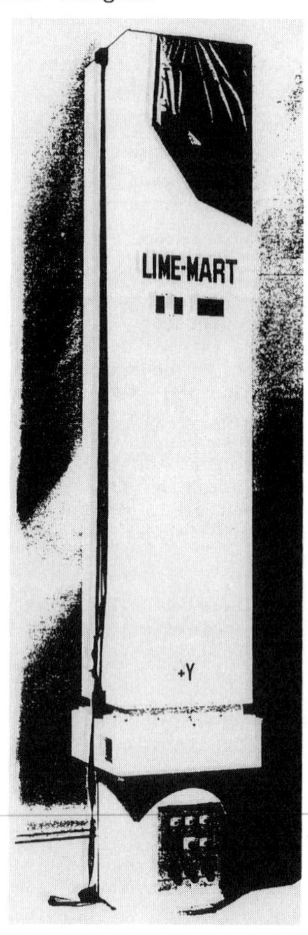

Figure 3.
The MART-LIME Mass Model

The authors acknowledge the technical and scientific staff of IAS and IKI for their support during several years of activity.

Advanced Ge Detectors for Gamma-Ray Astronomy

Larry S. Varnell
Jet Propulsion Laboratory
Pasadena, CA 91109 USA

ABSTRACT

Future observations of cosmic sources of gamma-ray lines will require instruments with sensitivities of the order of 10^{-6} photons/cm^2-s. To achieve this sensitivity, external background can be reduced by placing the Ge detector array inside a thick active shield, but the internal background produced by beta decay in the detector must be rejected by the Ge detector itself. In a collaboration between the High Energy Astrophysics Group at JPL and the Semiconductor Spectrometer Team at Lawrence Berkeley Laboratory (LBL), externally segmented coaxial detectors have been fabricated by dividing the outer electrode into five segments. The segmented detector is operated in a multi-segment mode to reject internal beta activity by requiring that events interact in two or more segments of the detector to be recorded. In this way, approximately 90% of the beta decays can be rejected, while the efficiency for detecting gamma rays remains high. To verify the operation of the first externally segmented detector, the full-energy-peak (FEP) efficiency was measured in this multi-segment mode as a function of the incident gamma-ray energy and of the discriminator threshold of the segments. To evaluate the beta rejection efficiency, the detector was irradiated with neutrons to produce ^{75}Ge ($t_{1/2}$ = 82.78 min) from neutron capture by ^{74}Ge(36.5% of natural Ge) in the detector itself. The beta activation spectra of the detector were then recorded by self counting in a series of measurements. Calculations with the Monte Carlo code ACCEPT were made of the gamma-ray FEP efficiency and of the beta rejection efficiency of the segmented detector. There was good agreement with the laboratory measurements, indicating that the code can be applied with confidence to model the response of more complicated detector configurations and environments.

I. INTRODUCTION

To achieve better sensitivity to weak cosmic sources, future Ge spectrometers will incorporate segmented detectors to reduce the radioactive background produced in space by energetic particle reactions with the Ge detectors. Although the externally segmented detector was introduced several years ago [1], there has been no balloon or space flight to demonstrate its effectiveness.

Alternative tests involve the use of particle beams from accelerators or neutrons from radioactive sources to irradiate the detector and produce internal radioactivity, simulating the energetic particle environment of space. Although these laboratory experiments cannot produce the complete range of the energetic particle environment experienced in balloon or space flight, they can measure the effectiveness of the segmented detector in rejecting internal background. The laboratory measurements can also be used to verify the accuracy of Monte Carlo calculations of the detector performance. In the first series of measurements, calibrated gamma-ray sources were used to measure the FEP efficiency of the segmented detector from 100 to 2600 keV in the ordinary mode, in which all events in the detector are accumulated,and simultaneously, in the multi-segment mode in which events are accumulated only if they deposit energy in two or more segments. Then an experiment was performed to measure the beta rejection efficiency by producing ^{75}Ge from neutron capture reactions with the detector material (36.5% ^{74}Ge). As in the efficiency measurements, two simultaneous spectra are taken of the activity in the detectors, an ordinary spectrum and a multi-segment spectrum. By comparing the spectra, the effectiveness of the detector in the multi-segment mode in rejecting beta events was measured. Then Monte Carlo calculations were made for comparison with the laboratory experiments. Good agreement between the results means that calculations can then be carried out with confidence for a wide range of detector configurations and induced activity.

II. MEASURED AND CALCULATED DETECTOR PERFORMANCE

The detector used for these measurements has been described previously [2]. It is a high-purity reverse electrode coaxial detector, 5.4 cm in diameter and 5.3 cm long, divided into five segments by masking the outer surface during boron ion implantation. This results in a 1 mm wide strip of high impedance Ge separating adjacent segments. The central hole is 1.12 cm in diameter and extends to within 0.7 cm of the closed end of the detector. The inner electrode is formed by Li diffusion so that the diameter of the hole plus the Li diffusion layer is 1.25 cm. The depletion voltage of the detector is 2300 V and it is operated at 3000 V. The efficiency of the detector is 26% of a standard 7.6 cm x 7.6 cm NaI detector. The total energy signal is taken from the inner electrode, while each segment has its own signal chain of preamplifier, amplifier and lower level discriminator taken from the external electrode. The discriminator threshold for each segment was set at 50 keV for these measurements. The five segment discriminator outputs are fed into a coincidence unit, which produces an output gate pulse when the number of simultaneous discriminator pulses equals or exceeds a selected minimum, such as two. The ordinary spectrum and the multi-segment spectrum are recorded simultaneously using the total energy signal from the inner electrode.

To obtain the relative FEP efficiency of the multi-segment to the ordinary mode, the spectra of gamma-ray sources with lines from 100 to 2600 keV were recorded. The peak area was measured for each line, and the relative FEP was obtained by dividing the area in the peak of the line in the multi-segment spectrum by that in the ordinary spectrum at each energy. Fig. 1 shows the results of the measurement. Also shown are the results calculated using the code ACCEPT [3], a three-dimensional coupled photon-electron Monte Carlo transport code. The errors in the measurements and the calculations are about 2%, so the agreement is quite good below 1500 keV. The effect of the setting of the LLDs is shown by plotting the result of a calculation with a setting of the LLDs of 10 keV. The effect of the LLD setting does not agree with the results shown in Fig. 4 of reference [2], where a higher efficiency was measured above 500 keV for a lower discriminator setting. A possible cause is crosstalk between the segment electronics. This may also be the cause for the lack of agreement in Fig. 1 between the measured and calculated efficiency above 1500 keV. A second segmented detector has been fabricated and mounted in a

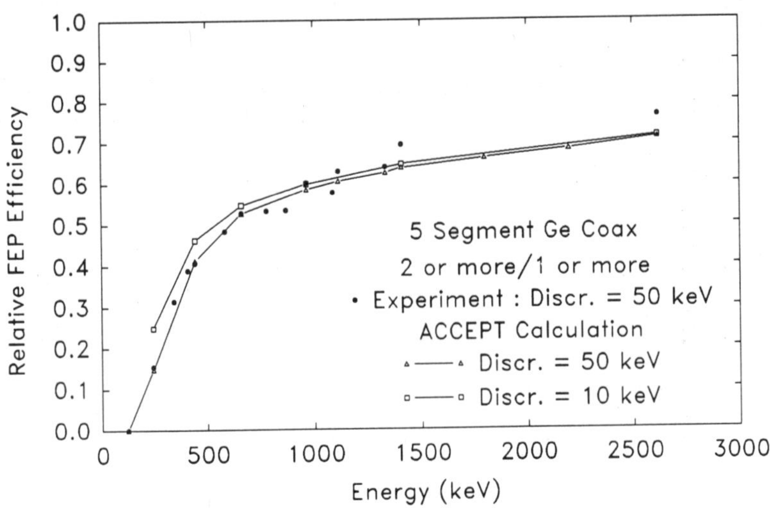

Fig. 1. Relative full-energy-peak (FEP) efficiency versus energy for the five segment coaxial detector. The relative efficiency is obtained by dividing the peak area of the multi-segment (two or more) spectrum by that of the ordinary (one or more) spectrum. The points calculated by the ACCEPT code are joined by line segments.

cryostat specially designed to minimize crosstalk, and measurements with this new system will begin in January 1991.

To demonstrate the effectiveness of the segmented detector in rejecting internal beta decay, the detector was irradiated with neutrons from a ^{252}Cf source and then placed in a shield of lead bricks to minimize the background. Multi-segment and ordinary spectra were recorded for intervals of 15 to 90 minutes for the first 6.3 hours, then for longer intervals to observe the longer lived activity. The shape of the beta spectrum from the decay of ^{75}Ge ($t_{1/2}$ = 82.78 min) can be clearly seen in the spectra, which have been reported previously [4]. In the beta decay of ^{75}Ge, 87.1% of the decays go to the ground state with an endpoint energy of 1188 keV. The calculated spectrum from this branch of the decay is shown in Fig. 2. From the calculations, 96% of these interactions are rejected in the multi-segment mode. In 11.5% of the decays, the transition is to the 264.6 keV level of ^{75}As, followed by a gamma ray of 264.6 keV. The calculations show that 50% of these decays will be rejected. The remaining 1.4% of beta transitions are followed by gamma emission and a similar fraction would be rejected. The sum of 96% of 87.1% and 50% of 12.9% gives 90% for the calculated rejection, slightly higher than the measured value of 85%.

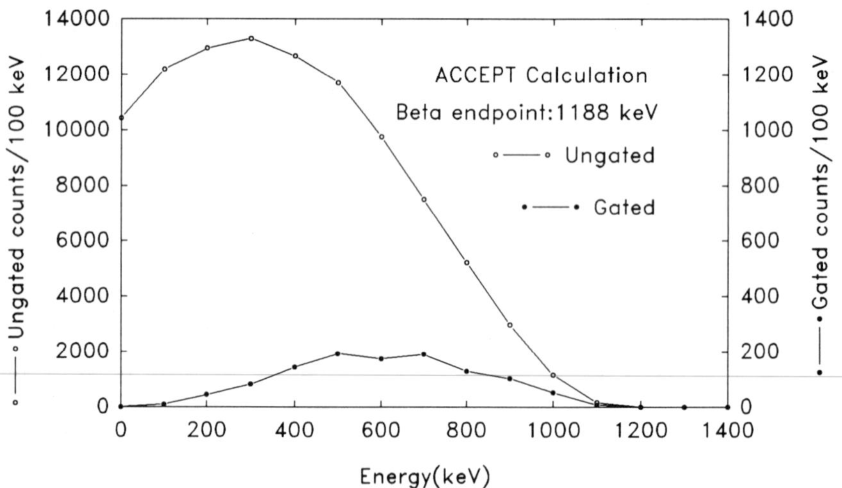

Fig. 2. Beta spectra calculated with ACCEPT for a five segment detector with radioactivity uniformly distributed, for an endpoint energy of 1188 keV with no following gamma ray, for the multi-segment (gated) and ordinary (ungated) modes. The ordinate of the gated counts is reduced by a factor of 10 compared to that of the ungated spectrum.

Taking this background reduction into account, the segmented detector in the multi-segment mode has a factor greater than two better sensitivity above 400 keV than an ordinary detector of the same size. At energies below 400 keV, the efficiency of this mode decreases significantly. In this energy range, the front segment only mode of the segmented detector is used to improve sensitivity. In this mode, events are counted if there is an interaction only in the front segment. For gamma rays of energy up to a few hundred keV, there is a high probability of complete absorption in the front segment (nearly 100% from 15 to 100 keV). The front segment only mode for low energy gamma rays gives a high efficiency with a background for a five segment detector only 1/5 that of an ordinary detector. Again, the sensitivity will be improved by a factor greater than two compared to an ordinary detector. Preparations are being made for a balloon flight in the Fall of 1991 with a segmented detector in a rugged mount and low-mass cryostat to measure the detector performance in a space environment.

The research described in this paper was carried out by the Jet Propulsion Laboratory, California Institute of Technology, under contract with the National Aeronautics and Space Administration.

REFERENCES

[1] L.S. Varnell, J.C. Ling, W.A. Mahoney, A.S. Jacobson, R.H. Pehl , F.S. Goulding, D.A. Landis, P.L. Luke and N.W. Madden, "A Position-Sensitive Germanium Detector for Gamma-Ray Astronomy," IEEE Trans. Nucl. Sci. **31**, 300 (1984).

[2] L.S. Varnell, J.C. Ling, W.A. Mahoney, R.H. Pehl, C.P. Cork, D.A. Landis, P.L. Luke, N.W. Madden and D.F. Malone, "Performance of a Five-Segment Coaxial N-Type Germanium Detector," in Nuclear Spectroscopy of Astrophysical Sources, ed. N. Gehrels and G. Share(New York:AIP),1988, pp. 490-497.

[3] J.A. Halbleib and T.A. Mehlhorn, "ITS: The Integrated TIGER Series of Coupled Electron/Photon Monte Carlo Transport Codes," Nucl. Sci. Eng.,**92**, No.2, 338(1986). ACCEPT is one of the ITS codes, available from the Radiation Shielding Information Center, ORNL, P.O. Box 2008, Oak Ridge, TN, 37831 USA. ITS Version 2.1 is CCC-467.

[4] L.S. Varnell, "Segmented Ge Detectors and Mechanical Coolers for Future Gamma-Ray Astronomy Instruments," in High-Energy Astrophysics in the 21st Century, ed. P.C. Joss(New York:AIP), 1990, pp. 252-255.

GERMANIUM DETECTOR VACUUM ENCAPSULATION

N.W. Madden, D.F. Malone, R.H. Pehl, C.P. Cork
P.N. Luke, D.A. Landis and M.J. Pollard
Lawrence Berkeley Laboratory
University of California
Berkeley, CA 94720

ABSTRACT

The encapsulation of germanium detectors has been a long sought after goal. We have begun to develop encapsulation technology that should significantly improve the viability of germanium gamma-ray detectors for a number of important applications. A specialized vacuum chamber has been constructed in which the detector and the encapsulating module are processed in high vacuum. Very high vacuum conductance is achieved within the valveless encapsulating module. The detector module is then sealed without breaking the chamber vacuum. The details of the vacuum chamber, valveless module, processing, and sealing method are presented in the paper.

INTRODUCTION

Germanium gamma-ray detectors are very environmentally sensitive devices. They require a very high quality vacuum to retain good diode characteristics. Encapsulating the detector in a small local vacuum chamber is a way to keep the detector in a stable "known" environment allowing it to be handled and thermally cycled. The encapsulating material must be very clean and not continue to outgas. Our encapsulation method uses a specialized vacuum chamber that allows the detector module to be vacuum pumped and sealed without opening the vacuum. The project that stimulated this development was the Transient Gamma-Ray Spectrometer (TGRS) that is scheduled to be flown on the WIND spacecraft[1] in 1992.

VACUUM CHAMBER

A drawing of the vacuum chamber is shown in Fig. 1. The vacuum chamber has a vacuum isolated hydraulic ram penetrating the base plate. This ram provides the force to compress and maintain the metal seal in the valveless module that encapsulates the detector. Special washers are provided to allow alignment of the detector module. Heaters are attached to the detector module mounting plate. There are quartz windows on the viewing ports to evaluate the advantages of using ultraviolet light to accelerate the desorption of water. Local ion gauging for total pressure measurement as well as a residual gas analyzer are provided through the side wall of the lower fixed portion of the vacuum chamber.

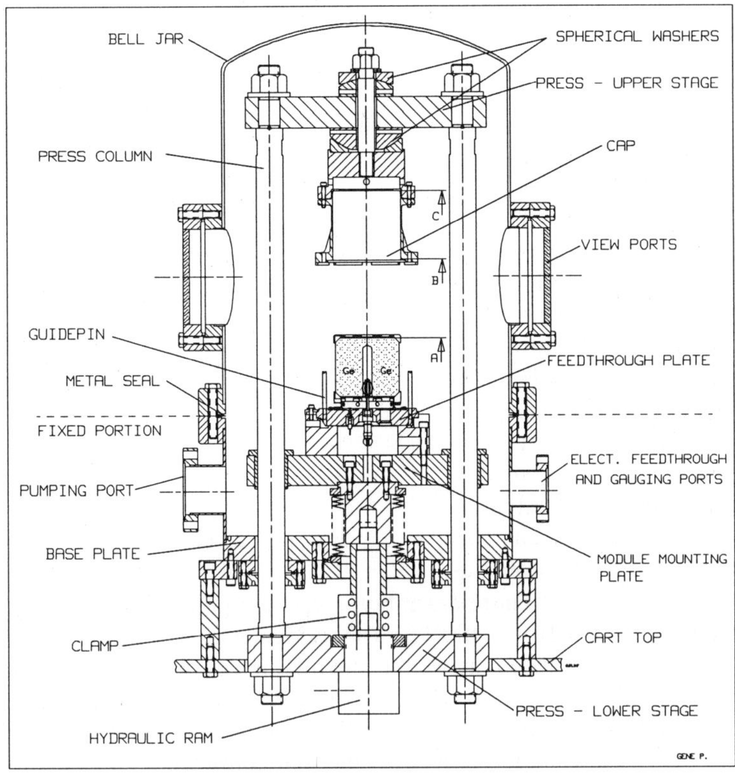

Fig. 1: Drawing of vacuum chamber

VALVELESS DETECTOR MODULE

A drawing of the valveless detector module is shown in Fig. 2. The module in our present test (Nov. 1990) is made of stainless steel. The TGRS flight module will be made of beryllium. The module is fabricated in two parts: one is the cap with the seal detail around the bottom and the other is the feedthrough plate with its seal detail and vacuum brased signal feedthroughs. There are no valves on the module. The detector has an outside bevel on each end to provide a way of clamping the detector in the module. A wave spring is used to provide pre-loading of the detector. The "delta seal"[2] is a special type that provides a metal seal but does not take a tremendous force to seal. The module contains 2 cm^3 of molecular sieve and H_2 gettering material to maintain high vacuum.

VACUUM PROCESSING AND SEALING METHOD

The two module parts are mounted in the vacuum chamber without the seal and detector in the position as shown in Fig. 1. Temporary guidepins are inserted in the feedthrough plate to assist radial alignment. The hydraulic ram moves the parts together and the sealing surfaces are aligned and clamped in place. The ram is then moved back to the starting position. The "delta seal", wave washer, detector, spacers, and conical rings are installed. The detector (Fig. 1, position A) is moved to position B by the ram. This position allows high conductance vacuum pumping. The vacuum chamber is then sealed and pumped. The outside of the vacuum chamber as well as the module mounting plate are heated to ~ 100°C. and the vacuum chamber is flushed with hot N_2 gas from liquid nitrogen boil off. The entire assembly is vacuum pumped until a very good vacuum is obtained (low 10^{-7} at 100°C). The residual gas analyzer is used to evaluate the dryness of the vacuum as well as the total pressure. The ram then compresses the wave spring and seal with a force of 6000 PSI. The clamp on the ram outside the chamber at the bottom is secured and the pressure on the ram is maintained. The vacuum chamber is opened, the guidepins are removed, and high strength screws are inserted and torqued down. The pressure on the ram is released and the module is removed.

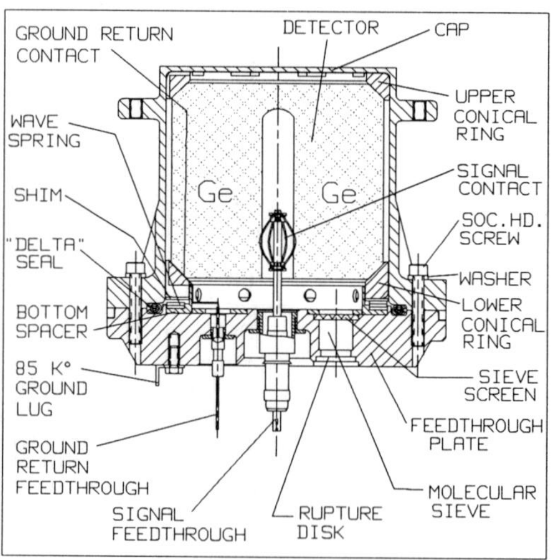

Fig. 2: Valveless module drawing.

ACKNOWLEDGMENT

This work was supported in part by NASA Contract No. W-14, 606, under Interagency Agreement with the Director's office of Energy Research, and by the Director, Office of Energy Research, Office of Health and Environmental Research, Physical and Technological Division, of the U.S. Department of Energy under Contract No. DE-AC03-76SF00098.

REFERENCES

1. A. Owens, et al, "The Transient Gamma-Ray Spectrometer," pres. at the 1990 IEEE Nucl. Sci. Symp., Arlington, VA; to be publ. IEEE Trans. Nucl. Sci. Proc. (1991).
2. "Aluminum Delta Seal," manufactured by Helicoflex Comp., Columbia, South Carolina.

ELECTRONIC CONSIDERATIONS FOR EXTERNALLY SEGMENTED GERMANIUM DETECTORS

N.W. Madden, D.A. Landis, F.S. Goulding, R.H. Pehl, C.P. Cork
P.N. Luke, D.F. Malone and M.J. Pollard
Lawrence Berkeley Laboratory
University of California
Berkeley, CA 94720

ABSTRACT

The dominant background source for germanium gamma ray detector spectrometers used for some astrophysics observations is internal β decay. Externally segmented germanium gamma ray coaxial detectors can identify β decay by localizing the event. Energetic gamma rays interact in the germanium detector by multiple Compton interactions while β decay is a local process. In order to recognize the difference between gamma rays and β decay events the external electrode (outside of detector) is electrically partitioned. The instrumentation of these external segments and the consequence with respect to the spectrometer energy signal is examined.

INTRODUCTION

The dominant source of instrument background between 200 keV and 2 MeV in well shielded and collimated germanium gamma ray spectrometers used for astrophysics observations is internal β decay created by cosmic rays undergoing nuclear reactions with germanium nuclei. One technique used to recognize this background source is shape recognition of the detector current pulse to determine if the event was a single point interaction or multiple interactions.[1,2] This technique is not discussed in this paper. A second technique is to partition the detector into many segments and determine if the event occurs in only one segment or in several.[3] This technique requires the connections to the segments to be on the outside of the coaxial detector (external segmentation) because of the room needed for the multiple connections. External segmentation technology is presently in the early stages of development even though significant effort has gone on in both the commercial and research sector. Fortunately, the fundamentals for the instrumentation of these detectors is better understood. It is the phenomenal resolution (.2% at 1 MeV) of these detectors that makes them the tool of choice for gamma ray astrophysics observations, but care must be taken not to degrade this resolution when segmentation is implemented.

THE SPECTROSCOPY CHANNEL

The normal detector-preamplifier configuration for high-resolution germanium spectrometers is shown in Fig. 1. Here, the center contact of the detector is DC coupled to the input field effect transistor (FET) of the charge sensitive preamplifier while the outside of the detector is biased at high voltage and bypassed to ground with a large value capacitor. The FET and feedback components are usually mounted very close to the detector to reduce the stray capacitance. The capacitance of the bypass capacitor must be made very large compared to the detector capacitance to reduce the amount of charge lost across it. This high voltage capacitor (2-5 kV) is physically large; it must be reasonably stable and exhibit low noise. It is usually mounted just outside the vacuum chamber. The detector-preamplifier configuration for outside segmentation is shown in Fig. 2. The center contact is at high voltage and is AC coupled to a spectroscopy quality preamplifier. The segments are DC coupled to their preamplifiers. This minimizes the number of coupling capacitors and detector load resistors. Large value capacitors would be necessary to minimize charge division problems. This is because in the non-segmented case a small amount of charge from the detector can afford to be lost as the same percentage is lost from each event, but when the detector has AC coupled segments the percent of charge lost will change if the capacitors have different values. This would broaden the resolution if the capacitor value is too small. There would also need to be many high voltage vacuum feedthroughs. The disadvantage with high voltage on the spectroscopy channel is that the microphonics could be much worse. The microphonics can be drastically reduced by installing an isostatic shield as shown in Fig. 2. The coupling capacitor in the energy channel need not be as large as it would be in the segment channels as there is only one and it is inside the feedback loop.

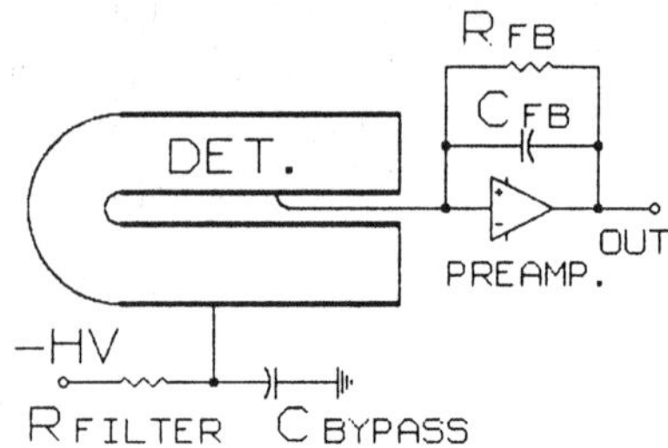

Fig. 1: Normal detector-preamplifier configuration for N-type coaxial Ge detector.

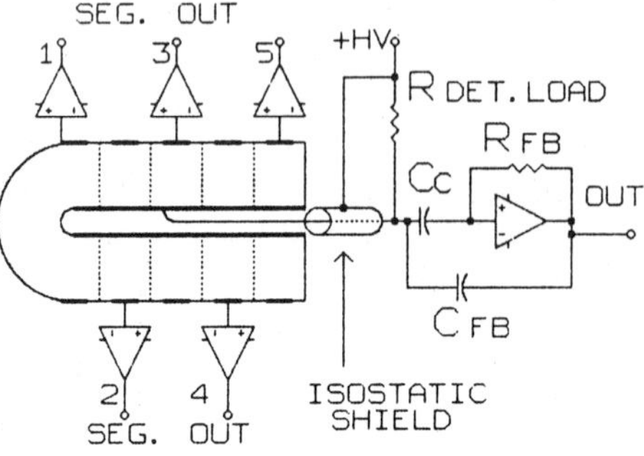

Fig. 2: Detector-preamplifier configuration for external segmentation.

SEGMENTED ELECTRODE PREAMPLIFICATION

The segment preamplifiers must present a low impedance to the current signal to prevent any voltage movement at the segment as it would cause crosstalk through the capacitance between segments. The energy channel preamplifier is not as critical because the crosstalk to the segments is the opposite polarity as the signal. The need for a low impedance connection to each of the several segments of the external electrode has already been established. Charge sensitive preamplifiers are often characterized as having an equivalent dynamic capacitance. This dynamic capacitance is roughly the open loop gain at a particular frequency times the feedback capacitance. Preamplifiers with an open loop gain of about 100 k and a feedback capacitance of a few pF have an adequate dynamic capacitance. The output of the charge sensitive preamplifiers are connected to simple shaping amplifiers that limit the bandwidth and then to discriminators. If the resistance between the segments of the detector is not high enough DC coupled charge sensitive preamplifiers may not work properly. A second form of preamplifier has been developed which is less sensitive to this problem. Technically, these preamplifiers are known as voltage sample-current sum feedback amplifiers. We refer to them as low Z preamplifiers. Unlike charge sensitive preamplifiers, low Z preamplifiers are current sensitive, and can have remarkably low input impedance to frequencies in excess of 30 MHz. However, these preamplifiers are more complex and consume more power. The output of these preamplifiers must be integrated and further bandwidth limited to obtain an adequate low energy threshold.

Until a significant number of externally segmented germanium gamma ray detectors are fabricated, the decision on the "correct" preamplifier for the segments remains open.

ACKNOWLEDGMENT

This work was supported in part by the JPL Director's Discretionary Fund via a NASA Interagency Agreement with the Director's office of Energy Research, and by the Director, Office of Energy Research, Office of Health and Environmental Research, Physical and Technological Division, of the U.S. Department of Energy under Contract No. DE-AC03-76SF00098.

REFERENCES

1. N. Gehrels, "Instrumental Background in Balloon-borne Gamma-ray Spectrometers and Techniques for its Reduction," Nucl. Instrum. and Meth., A239, 324-349 (1985).
2. P.T. Feffer, D.M. Smith, R.D. Campbell, J.H. Primbsch, R.P. Lin, P.N. Luke, N.W. Madden, R.H. Pehl and J.L. Matteson, "Pulse Shape Discrimination for Background Rejection in Germanium Gamma-Ray Detectors," SPIE, 1159 (1989).
3. L.S. Varnell, J.C. Ling, W.A. Mahoney, R.H. Pehl, C.P. Cork, D.A. Landis, P.N. Luke, N.W. Madden and D.F. Malone, "Performance of a Five-Segment Coaxial N-type Germanium Detector," AIP Conf. Proc., 170, 490-497, Washington, D.C. (1987).

RANDOM CHARGE PULSER: A DIAGNOSTIC TOOL

D.A. Landis, N.W. Madden and F.S. Goulding
Lawrence Berkeley Laboratory
University of California
Berkeley, CA 94720

ABSTRACT

Electronic instrumentation used in conjunction with germanium gamma-ray detectors for astrophysics observations must be tested under conditions similar to those encountered during actual operations. Simulation of cosmic-rays, gamma-ray bursts and solar flares is necessary to validate the performance of the entire spectroscopy chain. A random charge pulser used in conjunction with other diagnostic pulsers, and the techniques developed to verify the performance with and without a germanium gamma-ray detector are described.

INTRODUCTION

The performance of the electronic instrumentation used in conjunction with germanium gamma-ray detectors can be measured with electronic pulsers. The typical pulser produces a stable tail or square wave voltage to inject charge through a charge terminator (test capacitor) into the input of a charge sensitive preamplifier. The fast voltage rise of the stable pulser through the charge terminator produces an equivalent energy signal equal to $CV\epsilon/q$ where C is the capacitance of the charge terminator, V is the voltage rise of the stable pulser, ϵ is the average energy required to create a hole-electron pair in the detector, and q is the charge of the electron. A tail pulser produces a pulse having a fast rise and a long exponential decay. The advantage of a tail pulser is that only one polarity pulse is produced from a unipolar shaping amplifier, but an undesirable undershoot that cannot be canceled with the normal pole-zero adjustment is also produced. The undershoot is reduced as the exponential decay is made longer, but even a 1% undershoot of the shaping amplifier signal will degrade the energy resolution of random signals at a modest counting rate. The square wave pulser does not cause the undershoot problem, but it produces two pulses from the shaping amplifier, one from each edge of the square wave pulse of opposite polarity. The negative pulse (from the back edge of the pulse) is normally not a problem unless it overloads the amplifier or disturbs the base line restorer in the shaping amplifier.

RANDOM CHARGE PULSER

Another method of producing a charge pulse is to gate on for a short time a small current into the input of the preamplifier. The equivalent energy signal from

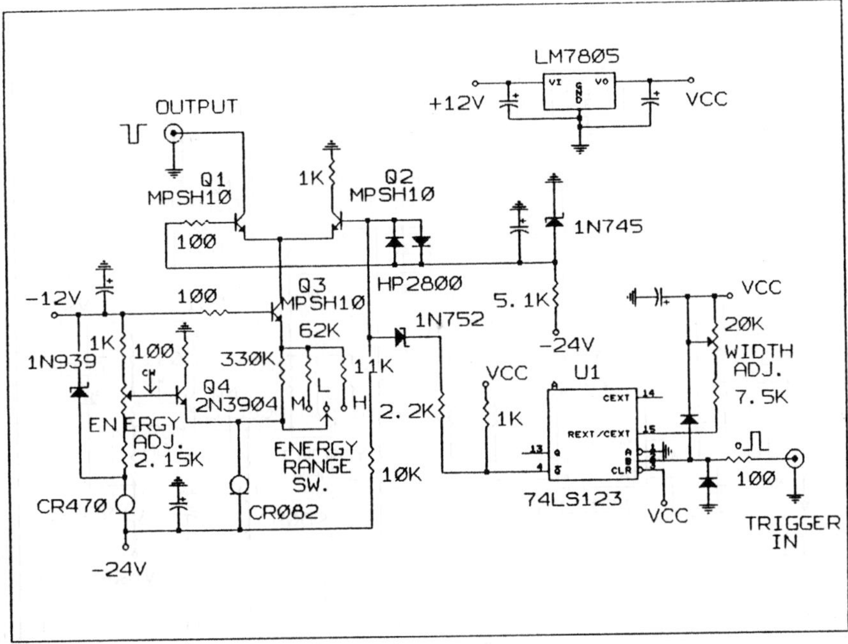

Fig. 1: Schematic of negative current charge pulser.

this type of pulser is equal to IT∈/q, where I is the small current and T is the time. For example, 1 μA for 100 ns is about 1.85 MeV for a germanium detector. This charge pulser, which produces a current pulse similar to the pulse from the detector, can be pulsed at a high random or periodic rate without producing any pole-zero problems. Figure 1 is a schematic diagram of the charge pulser. This pulser is for N type detectors (negative current). The current can be adjusted to produce an equivalent energy from a few MeV to many GeV, and because the current pulse width is only 100 ns, it can be pulsed at rates greater than 1 MHz.

This concept is not intended to provide a highly stable amplitude or to replace the stable amplitude test pulser used to calibrate spectrometers. Rather, it is designed to be an overload pulser, burst signal, or just random events. It can be mixed with a stable tail or square wave pulser fed through a charge terminator to measure peak shift and resolution change due to changing random rate conditions. The dead time of the system, including the ADC can also be measured under these conditions. This is accomplished by comparing the number of triggers of the square wave pulser with the number of counts in the spectrum measured by the ADC gated by the trigger pulse. We have used a similar pulser to evaluate high value feedback resistors. The pulser is connected to a charge sensitive preamplifier with external connections for the feedback resistor. The resistor to be tested is connected and the pulser is triggered with random pulses. The preamplifier is connected to a shaping amplifier with pole-zero adjustment. The output of the shaping amplifier is observed on a scope, and the pole-zero is adjusted for best results. The pole-zero adjustment will produce poor results if the feedback resistor value is frequency dependant. The transistors (MPSH10) used in the current switch of the charge pulser are very low capacitance and selected devices can be very low leakage (1 pA). Care must be taken in the construction to keep the wires short and have minimum capacitance on the emitters of Q1 and Q2. If the opposite polarity current switch is needed (P type detectors) the MPSH81 transistor can be used.

RANDOM SOURCE

The noise from a high gain fast amplifier is used as the source to generate random triggers. The noise signal from the amplifier is AC coupled to one input of a voltage comparator. The output of the comparator is the random trigger source and is also connected to a count rate meter circuit that produces an output DC voltage proportional to its input rate. This DC voltage is applied to the input of a difference amplifier along with the demand DC voltage of opposite polarity. The output of the difference amplifier goes to the other input of the voltage comparator. This negative feedback controls the trigger level of the comparator producing a random rate set by the demand voltage and measured by the count rate meter circuit. Care must be taken to filter the supply voltages to the fast amplifier to prevent feedback from the output.

ACKNOWLEDGMENT

This work was supported in part by the Director's Office of Energy Research, Office of High Energy and Nuclear Physics, Division of Nuclear Physics, by Nuclear Sciences of the Basic Energy Program of the U.S. Department of Energy, under Contract No. DE-AC03-76SF00098.

OPTIMISATION OF FIELD OF VIEW SIZE IN A SQUARE ELEMENT CODED APERTURE IMAGING SYSTEM INCORPORATING A CIRCULAR DETECTOR

K. Byard
Department of Physics, University of Southampton, Southampton. SO9 5NH U.K.

ABSTRACT

A new coded aperture detector design approach is discussed, whereby square element coded aperture unit patterns are mosaiced onto a circular detector with a view to maximising the area occupied by the unit pattern, and hence the size of the field of view. The results for these optimum patterns are presented for all URA and MURA unit patterns of order $v < 400$.

INTRODUCTION

Coded aperture imaging is now a well established method of producing images of self luminous X- and gamma-ray sources[1,2], and for astrophysical purposes the best aperture types to use are the uniformly redundant arrays (URA)[2], and the modified uniformly redundant arrays (MURA)[3]. In gamma-ray astronomical coded aperture systems, the size of the unambiguous field of view (FOV) is dictated by that of the unit pattern of the aperture and increases as the unit pattern area increases. To date most gamma-ray telescopes have used apertures based on square elements coupled with rectangular unit patterns. For example the gamma-ray telescope ZEBRA[4] has a 9×7 element rectangular unit pattern and a URA order of 63 elements. Although such a system is perfectly adequate for short duration balloon flights, a long term satellite mission must be as weight economical as possible, suggesting the use of a compact circular detector. However, if square detector pixels are a necessity of the telescope, possibly in the form of square cross section caesium iodide bars, then the resulting rectangular unit pattern is an inefficient one to use, leaving a large amount of wasted detector area when fitted into a circle and hence giving a limited FOV size. Even in the best theoretical case of the rectangular unit pattern being a square, the useful area is only 63.7% of the total circular detector area. The problem of maximising circular detector area usage to increase detection sensitivity has already been solved by the development of the technique of partial cycle averaging[5]. What follows here however, is the description of a new mosaicing method which, whilst still able to be used in conjunction with partial cycle averaging, provides the added bonus of a wider FOV by maximising the area of the detector occupied by the unit pattern.

BASIC STRATEGY

The basic idea behind maximising the occupied detector area lies in a unit pattern construction similar to that used to construct hexagonal uniformly redundant arrays, described by Finger and Prince[6]. Firstly choose a URA or MURA order v, determined by the detector size, and construct the one dimensional binary $(0, 1)$ URA or MURA array B of order v of the form:

$$B = \{a_0, a_1, a_2, \ldots, a_{v-1}\} \tag{1}$$

where a_i are the binary elements of the array. These arrays can be constructed in a number of ways[2,7,3]. Now, define $\vec{e_0}$ and $\vec{e_1}$ to be the lattice vectors in the x and y directions of a grid of square cells, and choose an integer r and label all the cells of the grid such that the cell centred at $i\vec{e_0} + j\vec{e_1}$ has a label

$$l \equiv i + rj (\mathrm{mod}\, v). \tag{2}$$

Next, mosaic the array B onto the lattice such that the cell labelled l contains the element a_l, and make all the cells with $a_l = 1$ open and those with $a_l = 0$ closed. Finally choose a unit pattern of any shape that contains every label modulo v once only. Note that the unit pattern does not necessarily have to be rectangular, as can be seen from the example of the URA of order $v = 59$ shown in Fig. 1, which has been constructed using the lattice multiplier $r = 8$. Despite its shape, this URA unit pattern still retains perfect imaging properties with zero sidelobes when balanced correlation[2] is used. The corresponding full aperture is easily constructed by simply repeating this unit pattern. Note from Fig. 1 that the basic geometrical property of the unit pattern is that it tessellates at a fixed orientation.

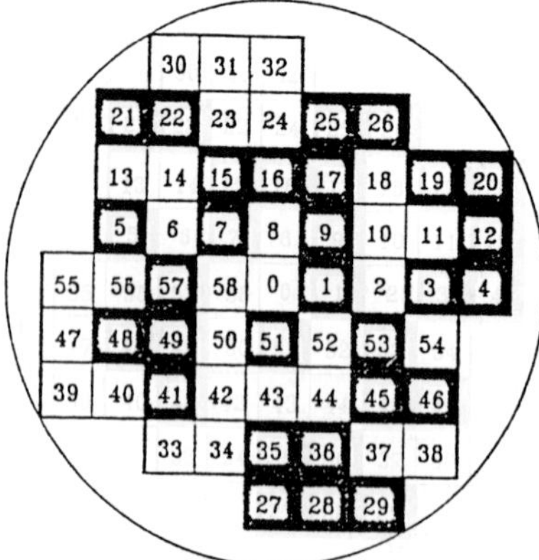

Fig. 1. Optimum unit pattern of a URA with parameters $v = 59$ and $r = 8$ fitted into a circular detector. The unit pattern occupies 70.9% of the area of the circle.

This property of perfect imaging with different shaped unit patterns has been used to find aperture configurations with a range of values of unit pattern order v, and multiplier r. These have been constructed with a view to finding the most compact configuration for each v, namely those which maximise the percentage area A of the circle occupied by the unit pattern, and hence give the largest FOV size.

RESULTS

Maximum percentage areas A were found for all possible URA and MURA unit patterns of order $v < 400$ as follows. For each value of v, every inequivalent value of lattice multiplier r (i.e. $1 \leq r \leq (v-1)/2$) was used to construct the corresponding square lattice. For each lattice the minimum radius of a circle which contains wholly at least one cell of each label l, $0 \leq l \leq v-1$, was calculated computationally by an iterative convergence technique. These radii were tabulated versus multiplier r, and then inspected to find the smallest overall radius for the order v. Knowing the value of overall minimum radius, the percentage area A could now be calculated by simple geometry.

When studying the results it should be borne in mind that the best possible case for a rectangular unit pattern is a square pattern, where the percentage area is $A = 63.7\%$. Also, because the unit pattern must tessellate, then the theoretical upper limit to the occupied area would occur for an infinite order unit pattern, with the elements arranged so that they collectively form a hexagonal shape. In this case the percentage area is $A = 82.7\%$.

As an example, the optimum unit pattern for the URA of order $v = 59$ is in fact that shown above in Fig. 1. The multiplier $r = 8$ used here is one of six which gives the same maximum percentage area of $A = 70.9\%$, the other five being 9,13,22,25 and 26. Note that this percentage area is an improvement on the 63.7% of a square unit pattern, and will also give a larger FOV than that produced by a square pattern.

Table I shows the maximum percentage area for each unit pattern of order $v < 100$, along with one multiplier for each v which produces an optimum unit pattern occupying this percentage area. The analysis has been extended further and maximum percentage areas have been calculated for all patterns with $v < 400$. The results are plotted on the graph in Fig. 2, with the URA results shown as circles and the MURA results as crosses. Also plotted on Fig. 2 are the values of A for a square unit pattern, and for the best possible "hexagon" configuration (dashed lines).

Table I Percentage area A occupied by optimum unit patterns for all $v < 100$. Also shown for each v is one lattice multiplier r which generates an optimum unit pattern.

v	A	r	v	A	r	v	A	r
5	63.7	2	31	53.3	7	63	75.7	14
7	49.5	2	35	76.8	10	67	65.6	7
11	53.9	3	37	63.7	6	71	69.5	8
13	63.7	5	41	70.5	6	73	71.5	11
15	56.2	3	43	66.8	12	79	68.9	9
17	63.7	4	47	66.5	7	83	72.4	11
19	71.2	4	53	75.0	8	89	66.7	9
23	58.6	4	59	70.9	8	97	72.6	10
29	63.7	5	61	73.3	8			

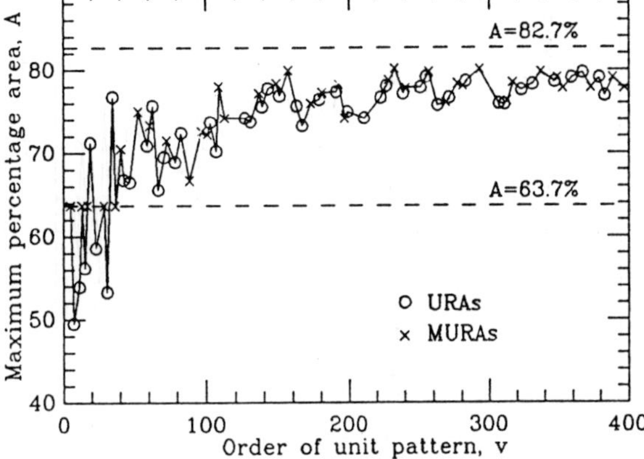

Fig. 2. Plot of maximum percentage area A of circles occupied by optimum URA and MURA unit patterns for all $v < 400$.

DISCUSSION AND CONCLUSIONS

After an initial period of large fluctuation at $v \lesssim 63$, the graph settles into an expected trend, with a general increase in A with increasing v, but always remaining below the theoretical maximum of 82.7%. The increase is due to the better "fitting" of the squares into the above mentioned hexagon shape at larger v. Because the size of the FOV increases with increasing percentage area, the results presented above therefore give configurations for which the FOV size is maximised for each value of unit pattern order v, in some cases giving up to 25% increase in FOV size over that of a square unit pattern.

As already pointed out, the problem of circular detector area usage has been solved by using partial cycle averaging[5]. However, partial cycle averaging can still be used here in conjunction with these optimum apertures to utilise more of the detector area, but with the added bonus that a wider FOV results, which always remains dependant upon the unit pattern area.

REFERENCES

1. R.H. Dicke, Astrophys. J. 153, L101 (1968).
2. E.E. Fenimore & T.M. Cannon, Appl. Opt. 17, 337 (1978).
3. S.R. Gottesman & E.E. Fenimore, Appl. Opt. 28, 4344 (1989).
4. G. Villa et al., IEEE Trans. Nucl. Sci. NS-34, 26 (1987).
5. G.K. Skinner et al., Astrophys. Spa. Sci. 136, 337 (1987).
6. M.H. Finger & T.A. Prince, Proc. 19th ICRC. 3, 343 (1985).
7. F.J. MacWilliams & N.J.A. Sloane, Proc. IEEE. 64, 1715 (1976).

Author Index

AIP Conference Proceedings

		L.C. Number	ISBN
No. 173	Cooperative Networks in Physics Education (Oaxtepec, Mexico, 1987)	88-72091	0-88318-373-0
No. 174	Radio Wave Scattering in the Interstellar Medium (San Diego, CA, 1988)	88-72092	0-88318-374-9
No. 175	Non-neutral Plasma Physics (Washington, DC, 1988)	88-72275	0-88318-375-7
No. 176	Intersections Between Particle and Nuclear Physics (Third International Conference) (Rockport, ME, 1988)	88-62535	0-88318-376-5
No. 177	Linear Accelerator and Beam Optics Codes (La Jolla, CA, 1988)	88-46074	0-88318-377-3
No. 178	Nuclear Arms Technologies in the 1990s (Washington, DC, 1988)	88-83262	0-88318-378-1
No. 179	The Michelson Era in American Science: 1870–1930 (Cleveland, OH, 1987)	88-83369	0-88318-379-X
No. 180	Frontiers in Science: International Symposium (Urbana, IL, 1987)	88-83526	0-88318-380-3
No. 181	Muon-Catalyzed Fusion (Sanibel Island, FL, 1988)	88-83636	0-88318-381-1
No. 182	High T_c Superconducting Thin Films, Devices, and Application (Atlanta, GA, 1988)	88-03947	0-88318-382-X
No. 183	Cosmic Abundances of Matter (Minneapolis, MN, 1988)	89-80147	0-88318-383-8
No. 184	Physics of Particle Accelerators (Ithaca, NY, 1988)	89-83575	0-88318-384-6
No. 185	Glueballs, Hybrids, and Exotic Hadrons (Upton, NY, 1988)	89-83513	0-88318-385-4
No. 186	High-Energy Radiation Background in Space (Sanibel Island, FL, 1987)	89-83833	0-88318-386-2
No. 187	High-Energy Spin Physics (Minneapolis, MN, 1988)	89-83948	0-88318-387-0
No. 188	International Symposium on Electron Beam Ion Sources and their Applications (Upton, NY, 1988)	89-84343	0-88318-388-9
No. 189	Relativistic, Quantum Electrodynamic, and Weak Interaction Effects in Atoms (Santa Barbara, CA, 1988)	89-84431	0-88318-389-7
No. 190	Radio-frequency Power in Plasmas (Irvine, CA, 1989)	89-45805	0-88318-397-8
No. 191	Advances in Laser Science–IV (Atlanta, GA, 1988)	89-85595	0-88318-391-9
No. 192	Vacuum Mechatronics (First International Workshop) (Santa Barbara, CA, 1989)	89-45905	0-88318-394-3

No. 193	Advanced Accelerator Concepts (Lake Arrowhead, CA, 1989)	89-45914	0-88318-393-5
No. 194	Quantum Fluids and Solids—1989 (Gainesville, FL, 1989)	89-81079	0-88318-395-1
No. 195	Dense *Z*-Pinches (Laguna Beach, CA, 1989)	89-46212	0-88318-396-X
No. 196	Heavy Quark Physics (Ithaca, NY, 1989)	89-81583	0-88318-644-6
No. 197	Drops and Bubbles (Monterey, CA, 1988)	89-46360	0-88318-392-7
No. 198	Astrophysics in Antarctica (Newark, DE, 1989)	89-46421	0-88318-398-6
No. 199	Surface Conditioning of Vacuum Systems (Los Angeles, CA, 1989)	89-82542	0-88318-756-6
No. 200	High T_c Superconducting Thin Films: Processing, Characterization, and Applications (Boston, MA, 1989)	90-80006	0-88318-759-0
No. 201	QED Stucture Functions (Ann Arbor, MI, 1989)	90-80229	0-88318-671-3
No. 202	NASA Workshop on Physics From a Lunar Base (Stanford, CA, 1989)	90-55073	0-88318-646-2
No. 203	Particle Astrophysics: The NASA Cosmic Ray Program for the 1990s and Beyond (Greenbelt, MD, 1989)	90-55077	0-88318-763-9
No. 204	Aspects of Electron–Molecule Scattering and Photoionization (New Haven, CT, 1989)	90-55175	0-88318-764-7
No. 205	The Physics of Electronic and Atomic Collisions (XVI International Conference) (New York, NY, 1989)	90-53183	0-88318-390-0
No. 206	Atomic Processes in Plasmas (Gaithersburg, MD, 1989)	90-55265	0-88318-769-8
No. 207	Astrophysics from the Moon (Annapolis, MD, 1990)	90-55582	0-88318-770-1
No. 208	Current Topics in Shock Waves (Bethlehem, PA, 1989)	90-55617	0-88318-776-0
No. 209	Computing for High Luminosity and High Intensity Facilities (Santa Fe, NM, 1990)	90-55634	0-88318-786-8
No. 210	Production and Neutralization of Negative Ions and Beams (Brookhaven, NY, 1990)	90-55316	0-88318-786-8
No. 211	High-Energy Astrophysics in the 21st Century (Taos, NM, 1989)	90-55644	0-88318-803-1
No. 212	Accelerator Instrumentation (Brookhaven, NY, 1989)	90-55838	0-88318-645-4
No. 213	Frontiers in Condensed Matter Theory (New York, NY, 1989)	90-6421	0-88318-771-X 0-88318-772-8 (pbk.)
No. 214	Beam Dynamics Issues of High-Luminosity Asymmetric Collider Rings (Berkeley, CA, 1990)	90-55857	0-88318-767-1
No. 215	X-Ray and Inner-Shell Processes (Knoxville, TN, 1990)	90-84700	0-88318-790-6

No. 216	Spectral Line Shapes, Vol. 6 (Austin, TX, 1990)	90-06278	0-88318-791-4
No. 217	Space Nuclear Power Systems (Albuquerque, NM, 1991)	90-56220	0-88318-838-4
No. 218	Positron Beams for Solids and Surfaces (London, Canada, 1990)	90-56407	0-88318-842-2
No. 219	Superconductivity and Its Applications (Buffalo, NY, 1990)	91-55020	0-88318-835-X
No. 220	High Energy Gamma-Ray Astronomy (Ann Arbor, MI, 1990)	91-70876	0-88318-812-0
No. 221	Particle Production Near Threshold (Nashville, IN, 1990)	91-55134	0-88318-829-5
No. 222	After the First Three Minutes (College Park, MD, 1990)	91-55214	0-88318-828-7
No. 223	Polarized Collider Workshop (University Park, PA, 1990)	91-71303	0-88318-826-0
No. 224	LAMPF Workshop on (π, K) Physics (Los Alamos, NM, 1990)	91-71304	0-88318-825-2
No. 225	Half Collision Resonance Phenomena in Molecules (Caracus, Venezuela, 1990)	91-55210	0-88318-840-6
No. 226	The Living Cell in Four Dimensions (Gif sur Yvette, France, 1990)	91-55209	0-88318-794-9
No. 227	Advanced Processing and Characterization Technologies (Clearwater, FL, 1991)	91-55194	0-88318-910-0
No. 228	Anomalous Nuclear Effects in Deuterium/Solid Systems (Provo, UT, 1990)	91-55245	0-88318-833-3
No. 229	Accelerator Instrumentation (Batavia, IL, 1990)	91-55347	0-88318-832-1
No. 230	Nonlinear Dynamics and Particle Acceleration (Tsukuba, Japan, 1990)	91-55348	0-88318-824-4
No. 231	Boron-Rich Solids (Albuquerque, NM, 1990)	91-53024	0-88318-793-4
No. 232	Gamma-Ray Line Astrophysics (Paris–Saclay, France, 1990)	91-55492	0-88318-875-9
No. 233	Atomic Physics 12 (Ann Arbor, MI, 1990)	91-55595	088318-811-2
No. 234	Amorphous Silicon Materials and Solar Cells (Denver, CO, 1991)	91-55575	088318-831-7
No. 235	Physics and Chemistry of MCT and Novel IR Detector Materials (San Francisco, CA, 1990)	91-55493	0-88318-931-3
No. 236	Vacuum Design of Synchrotron Light Sources (Argonne, IL, 1990)	91-55527	0-88318-873-2
No. 237	Kent M. Terwilliger Memorial Symposium (Ann Arbor, MI, 1989)	91-55576	0-88318-788-4